GW00838660

INTEGRATED FIBER-OPTIC RECEIVERS

THE KLUWER INTERNATIONAL SERIES IN ENGINEERING AND COMPUTER SCIENCE

ANALOG CIRCUITS AND SIGNAL PROCESSING

Consulting Editor
Mohammed Ismail
Ohio State University

Related Titles:

MODELING WITH AN ANALOG HARDWARE DESCRIPTION LANGUAGE, H. Alan Mantooth, Mike Fiegenbaum
ISBN: 0-7923-9516-6
LOW-VOLTAGE CMOS OPERATIONAL AMPLIFIERS: *Theory, Design and Implementation*, Satoshi Sakurai, Mohammed Ismail
ISBN: 0-7923-9507-7
ANALYSIS AND SYNTHESIS OF MOS TRANSLINEAR CIRCUITS, Remco J. Wiegerink
ISBN: 0-7923-9390-2
COMPUTER-AIDED DESIGN OF ANALOG CIRCUITS AND SYSTEMS, L. Richard Carley, Ronald S. Gyurcsik
ISBN: 0-7923-9351-1
HIGH-PERFORMANCE CMOS CONTINUOUS-TIME FILTERS, José Silva-Martínez, Michiel Steyaert, Willy Sansen
ISBN: 0-7923-9339-2
SYMBOLIC ANALYSIS OF ANALOG CIRCUITS: Techniques and Applications, Lawrence P. Huelsman, Georges G. E. Gielen
ISBN: 0-7923-9324-4
DESIGN OF LOW-VOLTAGE BIPOLAR OPERATIONAL AMPLIFIERS, M. Jeroen Fonderie, Johan H. Huijsing
ISBN: 0-7923-9317-1
STATISTICAL MODELING FOR COMPUTER-AIDED DESIGN OF MOS VLSI CIRCUITS, Christopher Michael, Mohammed Ismail
ISBN: 0-7923-9299-X
SELECTIVE LINEAR-PHASE SWITCHED-CAPACITOR AND DIGITAL FILTERS, Hussein Baher
ISBN: 0-7923-9298-1
ANALOG CMOS FILTERS FOR VERY HIGH FREQUENCIES, Bram Nauta
ISBN: 0-7923-9272-8
ANALOG VLSI NEURAL NETWORKS, Yoshiyasu Takefuji
ISBN: 0-7923-9273-6
ANALOG VLSI IMPLEMENTATION OF NEURAL NETWORKS, Carver A. Mead, Mohammed Ismail
ISBN: 0-7923-9049-7
AN INTRODUCTION TO ANALOG VLSI DESIGN AUTOMATION, Mohammed Ismail, José Franca
ISBN: 0-7923-9071-7
INTRODUCTION TO THE DESIGN OF TRANSCONDUCTOR-CAPACITOR FILTERS, Jaime Kardontchik
ISBN: 0-7923-9195-0
VLSI DESIGN OF NEURAL NETWORKS, Ulrich Ramacher, Ulrich Ruckert
ISBN: 0-7923-9127-6
LOW-NOISE WIDE-BAND AMPLIFIERS IN BIPOLAR AND CMOS TECHNOLOGIES, Z. Y. Chang, Willy Sansen
ISBN: 0-7923-9096-2
ANALOG INTEGRATED CIRCUITS FOR COMMUNICATIONS: Principles, Simulation and Design, Donald O. Pederson, Kartikeya Mayaram
ISBN: 0-7923-9089-X
SYMBOLIC ANALYSIS FOR AUTOMATED DESIGN OF ANALOG INTEGRATED CIRCUITS, Georges Gielen, Willy Sansen
ISBN: 0-7923-9161-6

INTEGRATED FIBER-OPTIC RECEIVERS

Aaron BUCHWALD
Hong Kong University of Science & Technology
Clear Water Bay, Kowloon, Hong Kong

Kenneth W. MARTIN
University of Toronto
Toronto, Ontario, Canada

KLUWER ACADEMIC PUBLISHERS
Boston/London/Dordrecht

Distributors for North America:
Kluwer Academic Publishers
101 Philip Drive
Assinippi Park
Norwell, Massachusetts 02061 USA

Distributors for all other countries:
Kluwer Academic Publishers Group
Distribution Centre
Post Office Box 322
3300 AH Dordrecht, THE NETHERLANDS

Library of Congress Cataloging-in-Publication Data

Buchwald, Aaron (Aaron W.)
Integrated fiber-optic receivers / Aaron Buchwald, Kenneth W. Martin.
p. cm. -- (The Kluwer international series in engineering and computer science ; 306. Analog circuits and signal processing)
Includes bibliographical references and index.
ISBN 0-7923-9549-2 (acid-free paper)
1. Integrated optics. 2. Integrated circuits--Design and construction. 3. Pulse amplitude modulation. I. Martin, Kenneth W. (Kenneth William), 1952- . II. Title. III. Series: Kluwer international series in engineering and computer science ; SECS 306. IV. Series: Kluwer international series in engineering and computer science. Analog circuits and signal processing.
TA1660.B82 1995
621.36'93--dc20 94-45250
CIP

Printed on acid-free paper.

Printed in the United States of America

To

Warren G. BUCHWALD

CONTENTS

PREFACE

This book covers many aspects of the design of integrated circuits for fiber-optic receivers and other high-speed serial data links. Fundamental concepts are explained at the system-level, circuit-level, and semiconductor-device-level. Several books have been published on the broad topic of fiber-optic communications, covering various aspects of optical systems, including, optical fiber technology, wave propagation in optical fibers, optical sources, optical detectors, optical receivers, coherent optical fiber communication, and applications of fiber-optics. Since these books cover a wide range of topics, the chapters on receiver design are necessarily abbreviated, and few books even mention the challenging problem of high-speed clock recovery. As it turns out, clock recovery is the most difficult task to perform in broadband receivers. In this book, which is devoted solely to discussing integrated optical receivers, techniques for extracting timing information from the random data stream will be described in considerable detail, as will all other aspects of receiver design. This book could be used as a text for graduate and upper undergraduate courses in both analog circuit design and communication systems. It is written in a tutorial form and should also prove useful to practicing engineers wishing to update their knowledge through self-study.

Intended Audience

Communications systems are becoming increasingly complicated and ever smaller. The personal communication revolution will see portable communication units fitting in shirt pockets. Advances in disk-drives for portable computers are resulting in higher bit-densities, requiring higher speed serial processing. As a result of this trend — of higher-speeds, coupled with smaller packages — more elements of the system are being implemented in integrated form, and smaller systems are becoming increasingly complex. This requires that the IC designer be sufficiently knowledgeable about systems theory at the global-level, and semiconductor physics at the micro-level, to provide a middle-ground for the development of monolithic systems. This *common-ground* is illustrated conceptually in Fig. 0.1.

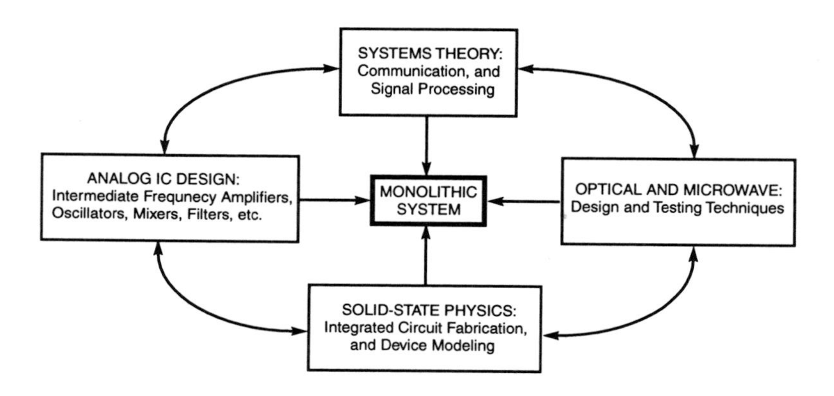

Figure 0.1 Illustration of analog circuit designers filling an important gap

Circuit designers are the intended audience of this book. These are the people who choose the circuit topology, transistor dimensions, current and voltage levels, and do the layout and testing of integrated circuit chips. It is hoped that this work will help to fill two serious gaps that the authors have perceived in the design of integrated systems. One is the gap between system designer and circuit designers. The second is the gap between designers of traditional analog circuits and microwave engineers. Traditionally, the design of communication systems begins with systems theorists who perform complex mathematical analysis and optimization on a global level. The system engineer produces a block diagram containing various circuit building blocks. Often microwave engineers design the front-end amplifier, mixer, and oscillator blocks, leaving the design of the intermediate frequency building blocks to a circuit designer experienced in standard analog techniques — a natural partition, since microwave and analog designers rarely speak the same language. Despite the various disciplines of engineering required for the design of a complete system, in the past, system-engineers needed only a limited knowledge of circuit design, and conversely, circuit designers needed only a limited knowledge of systems theory, for this division of tasks to fit seamlessly together. However, when the data-rate increases to a point where the limitations of the transistors are reached, this seam becomes ever wider. Various parasitics have a large effect on system performance and need to be taken into account in the system-design at the outset.

Design Philosophy

We contend that it is more appropriate for a skilled circuit designer to learn enough about system theory to make modifications in optimal architectures, that are realizable at high-speeds, than it would be for a systems-engineer to anticipate all potential problems in circuit design, and account for them a priori. The reasons for this statement are both philosophical and pragmatic. From a philosophical point of view, the design of a high-speed analog circuit is often as much a work of Art, as the result of a mathematical prescription. The Art comes in developing an intuition about what can be done in a given technology, making a leap of faith to a possible implementation, and then using analysis to fine tune the result. Often elegant analysis deriving an optimal structure come after the fact, and only serve to justify the validity of this intuitive leap. Optimizing a circuit on a systems level, without knowledge of the parasitic effects that can render the circuit useless, is usually a waste of time. From a much more practical standpoint, if a system is going to be designed on a single chip, it is chip-designers who are ultimately responsible for getting the system to work. The chip-designer, therefore, has no choice but to become, at least, a novice system architect.

To aid circuit designers in filling the gap between themselves and system engineers, Part I of this book explains the fundamentals of system theory required for the design of broadband receivers in a manner that makes sense to a circuit designer. To this end, emphasis is placed on intuition, and various illustrations are given to make results clearer. It is hoped that by presenting the fundamentals in an intuitive manner, a sufficient core knowledge of the subject can be digested to allow the reader to *leap* beyond the mathematics, and apply the intuition gained to improve future circuit designs. The mathematical development in Part I is rather lengthy, and the density of equations may scare away circuit designers, who typically like to see more hand-waving than at the launching of a cruise ship on its maiden voyage. Although the chapters are dense with equations, many of the intermediate steps in the derivation have been included. We believe this actually allows a longer book to be read faster, than if it were shorter. Also, fundamental results are enclosed in boxes to set them apart from steps in the derivation, and frequent rest-stops are encountered along the way to reflect on the results and give examples.

Outline of the Book

The book is organized into two parts. Part I covers the theory of communications systems as it applies to high-speed PAM (Pulse Amplitude Modulation) systems. The primary emphasis is on clock recovery circuits, and two chapters thoroughly cover this topic.

Theoretical concepts are generally grasped more easily by example. Therefore Part II is devoted to circuit design issues that illustrate example realizations of architectures described in Part I. Part II is not a comprehensive step-by-step guide for designing receiver ICs, but fundamental concepts are presented so that the reader can grasp the main ideas and begin to design circuits of his own.

Part I

The basic requirements of a fiber-optic receiver are briefly reviewed in **chapter 1**. This provides an overview of the problems that will be dealt with in considerably more detail in the remainder of the work.

Frequency domain analysis of random data, and data derived signals, is the topic of **chapter 2**. Although these results have appeared elsewhere, we found them difficult to understand and interpret from the point of view of a circuit designer. Therefore, We have presented results from first principles, in a tutorial form, with an emphasis on applications to receiver design. By the end of this chapter, the reader should have the analytical tools to answer important questions about receiver design trade-offs. More importantly, the reader should develop a *feel* for the characteristics of random data, and be able to predict the basic behavior of certain circuits by inspection.

In **chapter 3**, we address the problem of deriving an optimal receiver in the presence of both non-white noise, and phase-jitter. Several books on communication theory cover this topic adequately. Our focus will be to discuss the application of this theory to the design of high-speed IC receivers.

In **chapter 4**, the theory of clock recovery in a broadband system is presented. The recovery of a timing waveform from random data is the most difficult task that a broadband receiver must perform. The speed of clock recovery circuits often limits the maximum bit-rate of the receiver. Various clock recovery techniques are given, and the advantages and disadvantages of each method are discussed. In addition, clock recovery circuits based on maximum a posteriori (MAP) estimates in white Gaussian noise are considered, and the resulting architectures are compared to heuristic approaches.

In **chapter 5**, practical architectures for clock recovery at high-speeds are given. Some of these circuits are modifications of previously reported schemes, and others are novel. One novel technique in particular is outlined that has several desirable properties.

Part II

In **Part II** we present the transistor-level design, and measured results, of fundamental building blocks and test circuits. A brief review of high-speed IC processes, applicable to fiber-optic receiver design, is given in **chapter 6**. The theory and properties of HBTs (Heterojunction Bipolar Transistors) is presented. Typical models of GaAs and InP HBTs for SPICE simulations are given at the end of this chapter.

A detailed noise analysis of a transresistance preamplifier is given in **chapter 7**, showing the fundamental noise limitations of broadband receivers. Also, an InP preamplifier design is discussed and simulated results are given. The preamplfier circuit is integrated with a p-i-n photodiode for detection of light at a wavelength of approximately 1.3-μm. This wavelength is ideally suited to single-mode glass optical fibers, which display very low losses at wavelengths of 1.3-μm and 1.55-μm.

Test structures are essential for process evaluation and modeling. In **chapter 8**, we report on two voltage controlled oscillators (VCOs). The measured results of the oscillators were compared to SPICE simulations, and the model parameters of the HBTs were optimized to fit the observed data.

In **chapter 9**, the circuit design and measured results of a patented VCO and a 6-GHz phase-lock loop are presented. The VCO combines a ring oscillator with frequency doubling to produce quadrature outputs at twice the ring frequency, and a third output at four times the ring frequency. The PLL was designed using the VCO and demonstrates functionality of key circuit building blocks of a clock recovery circuit.

Finally, in **chapter 10**, the design of a complete clock recovery and data retiming circuit, based on the novel architecture of chapter 5, and utilizing circuits of chapters 7–9, is presented. Simulation results are given which show that the circuits are applicable to multi-gigabit-per-second communication systems.

It is our intention, that more than just reporting on the results of specific circuits, this book will serve as a tutorial on the the design of integrated high-speed broadband PAM data systems, such as, repeaters in long-haul, fiber-optic, trunk-lines, tranceivers for use in LANs and WANs, read-channels for high-density data-storage devices, and wireless communication hand-sets. We hope this work will provide a basis for improved designs of the future.

AARON BUCHWALD
KENNETH W. MARTIN

Hong Kong

ACKNOWLEDGMENTS

We would like to express our gratitude to the following circuit designers who reviewed the manuscript and provided background material, either directly or via their research papers: Hans Ransijn of AT&T Bell Labs, Rick Walker of Hewlett Packard, Thomas Lee of Stanford and Rambus, Mehran Begheri of Bellcore, Ansgar Pottbäcker of SICAN GmbH, Behzad Razavi of AT&T Bell Labs and Gabriel Li of National Semiconductor.

We also wish to thank our colleagues at our respective universities — The Hong Kong University of Science & Technology (HKUST), and The University of Toronto (UT). In particular several Professors at HKUST provided proofreading, consultation and encouragement. Thanks to Ross Murch, Man Wong, Tsz Mei Ko, and Mark Yau for proofreading, Jack Lau and Yitshak Zohar for support, and the rest of the E&EE faculty at HKUST. Thanks also to the E&EE students of HKUST, especially the group in the Analog Research Lab: Kenneth Choi, Darwin Cheung, Felix Cheung, Paul Chan, and Leo Leung. Special thanks to K.C. Smith and Laura Fujino who are associated with both HKUST and UT. K.C. is the Grand-Advisor of K.M. and the Great-Grand-Advisor of A.B. He has provided guidance on circuit design, book writing, and countless other topics.

Funding for the research that lead to the writing of this book was provided by TRW, Inc. We are grateful to Kevin Kobayashi, Liem Tran, Mike Kim and Gary Gorman for all their help. A very special thanks is due to Aaron Oki. His seemingly endless supply of energy and expertise were given gladly. Without his help this work would not have been possible.

Research for this book was performed at The University of California, Los Angeles (UCLA) in the Integrated Circuits and Systems Laboratory (ICSL), where K.M. was a Professor and A.B. was a Ph.D. student. We acknowledge the contributions from our former colleagues at UCLA: Profs. Asad Abidi, Henry Samueli, H.J. Orchard, Gabor Temes, Alan Willson, Jack Willis and all the students of ICSL.

The first draft of this book was the Ph.D. dissertation of A.B., which was supervised by K.M. at UCLA. In the following, A.B. would therefore like to express his gratitude to those who helped in the completion of his original dissertation. Because these comments are of a personal nature, with the readers indulgence, we will switch to first-person singular, where personal pronouns will refer to A.B.

First of all I'd like to thank the second author of this book, Ken Martin, for his advise and direction. He cared about the personal life of his students as well as their research, and I appreciated that. Much of the benefit derived from graduate school comes from interactions with fellow students. Scott Willingham and Mukund Padmanabhan played a significant role in my education and research; this project is no exception. Thanks also to fellow K.M. students Kevin Chan, and Tom Kwan for paving the way.

In the hectic final days of the dissertation, I appreciated the help of John Bain, Robin Joshi, Shrikanth Narayana, and Scott Willingham. Thanks also to Troy House for taking photos of the testing circuits. I owe a special debt to my brother Ted, who did most of the illustrations in chapters 6, 7, and 8, (all the good ones). He also helped me with proofreading and gave me the moral support to see this project through to the end.

I'm fortunate to have been able to spend time with my brothers in the past few years. Aside from Ted's direct involvement with this project, in the final month of my stay at UCLA, Ben took care of all the details of my move to Hong Kong, which allowed me to concentrate on finishing the dissertation. Jess was my roommate throughout my entire Ph.D. program; he tolerated my sometimes venomous disposition and gave me encouragement and support as only a brother could. My sister Lonnie also gave me valuable advise and support.

To Ma and Pa, who are so much a part of me, in many ways it seems that I never left home, because the care and support I received growing up is still felt daily — I don't know where I'd be without it. So to my parents, Warren G. and Peggy Jo, thanks for your love and guidance as you shaped me and set me off on my journey to become a real Paddle-to-the-Sea.

Finally, I offer my most sincere thanks to Daphne for her patience, love and understanding.

INTEGRATED FIBER-OPTIC RECEIVERS

Perhaps it will one day be said
that I have written something of substance,

something useful,
that I have entered the Mystery.

When cutting an axe handle with an axe,
surely the model is at hand.

Each writer finds a new entrance into the Mystery,
and it is difficult to explain.

Nonetheless, I have set down my thinking
as clearly as I am able.
— LU CHI, *Wen Fu*

PART I

SYSTEM CONSIDERATIONS

What we've got here — is a failure to communicate.
— DONN PEARCE, *Cool Hand Luke*

1

INTEGRATED FIBER-OPTIC RECEIVERS: AN OVERVIEW

1.1 INTRODUCTION

Multi-gigabit-per-second fiber-optic communications circuits were once considered to be the exclusive domain of high-cost telecom applications; only recently have such circuits found their way into a variety of datacom systems. In particular, a new class of networks is emerging, which uses SONET (Synchronous Optical Network) or SDH (Synchronous Digital Hierarchy) hardware and ATM (Asynchronous Transfer Mode) packet-switching for multimedia data communication. Plans to build avenues connecting this *information super-highway* to the public will create a large demand for fiber-optic communication systems. In addition, fiber-optics will be used widely in video distribution systems, and in wireless personal communication, where a large number of localized base-stations — separated by only a few hundreds of meters in densely populated areas — will likely communicate via high-speed optical networks.

With this increased demand for fiber-optic systems, focus has shifted from previous high-speed-at-any-cost approaches toward low-cost systems for high-volume production. Economic pressure is thereby exerted to move from discrete to fully-integrated receivers and transmitters. Previous receivers using manually tuned, expensive, discrete microwave components, for low-volume telecom circuits, are being replaced with low-cost integrated circuit transceivers. As a result, the task of receiver design now falls upon IC chip designers, who may not be as familiar as they would like to be with system-level issues and clock recovery difficulties. Conversely, system designers need an intimate knowledge of integrated circuit design to successfully implement a single-chip receiver. It is therefore evident that engineers of integrated communication systems require a good balance of systems- and circuit-knowledge. In this book we cover the relevant theory and discuss circuit design issues so as to equip IC designers with the necessary tools to realize next generation integrated fiber-optic receivers.

Scope of the Book

Several books on fiber-optic systems cover the subject thoroughly — from components and devices to applications. Four excellent books are those by Personick [1], Keiser [2], Green [3], and Senior [4]. In this book we will narrow our scope and be primarily interested in the design of high-speed integrated receivers for pulse amplitude modulated (PAM) transmission of digital data. We will only discuss direct-detection receivers; coherent systems will not be covered. By *high-speed*, we mean speeds close to the limitations of the transistors used. This implies data-rates of 1–2 Gb/s for fine-line CMOS, 2–10 Gb/s for advanced silicon bipolar, and 10 Gb/s and beyond for III–V FETs and heterostructure devices. By *integrated*, we mean a *high-degree* of integration, although we include multi-chip hybrids in this definition. This is in contrast to systems built primary with discrete microwave components, or with monolithic-microwave integrated circuits (MMICs), containing only a few active components per chip. Although MMIC techniques are not considered here, this does not exclude their usage in practical cost-effective receivers.

The circuits considered contain on the order of 100–1000 active devices, and the design methodologies use traditional analog techniques, relying on small intra-chip distances so that transmission-line effects can be ignored. Still, a multitude of problems arise at these very high speeds, making the design task difficult. The primary challenge of the design of high-speed integrated receivers, therefore, is to make the circuit insensitive to deleterious parasitic effects, which become increasingly troublesome at high-speeds. This is considered both from an overall system standpoint, by choosing an acceptable architecture, and from a physical standpoint in the IC layout. Most of the circuits presented in this book used III–V heterojunction bipolar transistor (HBT) structures (GaAs and InP). However, they are also directly applicable to Si-bipolar, and the design techniques and architectures presented can be realized using either CMOS or high performance FETs with appropriate circuit modifications.

Target Applications

Much of this book focuses on the design of circuits and development of architectures that will lead to the eventual implementation of a 10-Gb/s fiber-optic receiver for long-haul telecommunication trunking. Prototype circuits were designed to meet this objective. In what follows, the term *receiver* will refer to all the electronics, after, and including the photodetector. A block diagram of a typical fiber-optic receiver is shown in Fig. 1.1. Aside from the primary usage in telecom applications, the

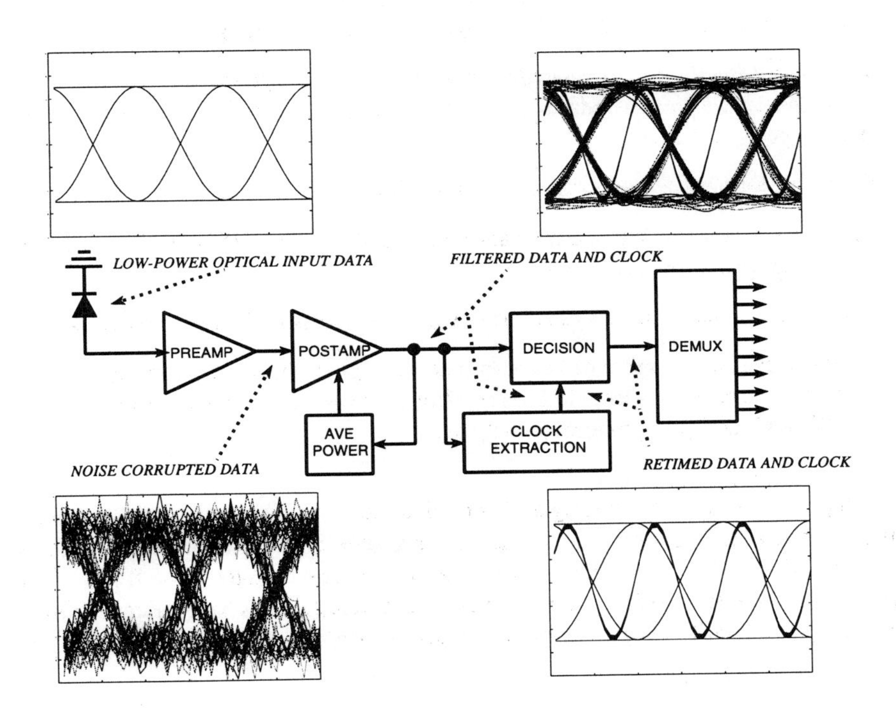

Figure 1.1 Block diagram of a fiber-optic receiver.

architectures and circuits presented here have a wider applicability to any high-speed PAM communication system; such applications include the following.

- LANs (Local Area Networks), providing broadband data communication links between computers over optical fibers such as FDDI (Fiber-Distributed Data Interface).
- WANs (Wide Area Networks) for multimedia applications (as mentioned above these can be based on SONET hardware with ATM switching).
- High-speed read/write channels for magnetic data-storage (as the bit-density of data storage devices is reduced, the serial data-rates are fast approaching the gigabit-per-second range).
- Date transfer between wireless-communication base-stations.
- High-speed serial data communication on metallic transmission media, such as coaxial cable and twisted pairs.
- Video-on-demand, Cable TV, and two-way video communications to the home.
- High-speed interconnections between integrated circuits, highly-parallel connections for neural networks, and conceivably, interconnections between electronic and biological signal processing systems.

An attempt was made in the writing of this book to keep the analysis, and design techniques as general as possible, making the results readily applicable to all applications requiring high-speed processing of serial data. In this first chapter we will present a brief overview of integrated fiber-optic receivers and note some of the challenges faced in the design of circuits for multi-gigabit-per-second systems.

1.2 ADVANTAGES OF FIBER-OPTICS

In recent years there has been a significant research effort in the area of high-speed electronics for communication. Higher speeds are required in order to take full advantage of the broadband capabilities of optical fibers. In particular integrated solutions are sought for practical systems to reduce cost and improve reliability. One of the target bit-rates for integrated fiber optic receivers is 10 Gb/s, which is consistent with the SONET hierarchical specification [5]; practical transmission systems at these extremely high data rates will open the way to unexplored territory in networking. Each of these systems will require high-speed, low-cost interface electronics.

Currently, the bandwidth of optical fiber (1400 GHz-km for 1.3 μm single-mode fibers) and low losses (0.15 dB/km) can not be fully exploited. A bottleneck in system throughput exists due to speed limitations of the electronics in the receiver and transmitter. This bottleneck can be circumvented by optically multiplexing several lower data-rate channels through a single fiber. Both a 9.6 Gb/s wavelength-division multiplexing (WDM) system [6], and a 20 Gb/s time-division multiplexing (TDM) system [7], have been demonstrated in laboratory experiments. These systems are capable of handling enormous data rates, because all of the high-speed processing, including amplification, can be done optically. These systems, however, are quite expensive and complicated.

1.3 STATUS OF INTEGRATED FIBER-OPTIC RECEIVERS

In the near term, optical communications systems must rely on electronic circuits for high-speed data processing. A low-cost solution to high-capacity fiber-optic transmission is to integrate high-speed electronic transmitters, and receivers onto a single chip, or a chip-set for use in a hybrid system. This requires circuits capable of processing multi-gigabit-per-second data. Several front-end circuits, such as: preamplifiers, postamplifiers, decision circuits, multiplexers and demultiplexers have been reported [8, 9, 10, 11, 12, 13, 14, 15], as shown graphically in Fig. 1.2. Although most of these circuits can process data at rates above 10 Gb/s, with others still capable of handling rates greater than 20 Gb/s [16, 17], little has been reported on fully-integrated clock extraction circuits above 2 or 3 Gb/s [18], with recent results of 8 Gb/s demonstrated in the laboratory [19].

In this book, new clock extraction architectures will be investigated, and transistor-level circuit solutions will be developed to enable the integration of a fiber-optic receiver operating in the multi-gigabit-per-second range. The IC technology used, and the maximum date rate will depend on the application. Bulk CMOS can be used for 622-Mb/s to 2.5-Gb/s systems (SONET levels 12–48). SONET and SDH levels are given in table 1.1. SONET is a hierarchical systems and development is underway for circuits operating at level OC–192 (STM–64) at a bit-rate of 9953.28-Mb/s ($\sim$10-Gb/s). These 10-Gb/s circuits could use silicon bipolar processes, GaAs FETs, BiCMOS, or SOI-CMOS (Silicon on Insulator)-CMOS. For even higher speeds, heterojunction devices such as HBTs (Heterojunction Bipolar Transistors) or HEMTs (High-Elector Mobility Transistors) could be used.

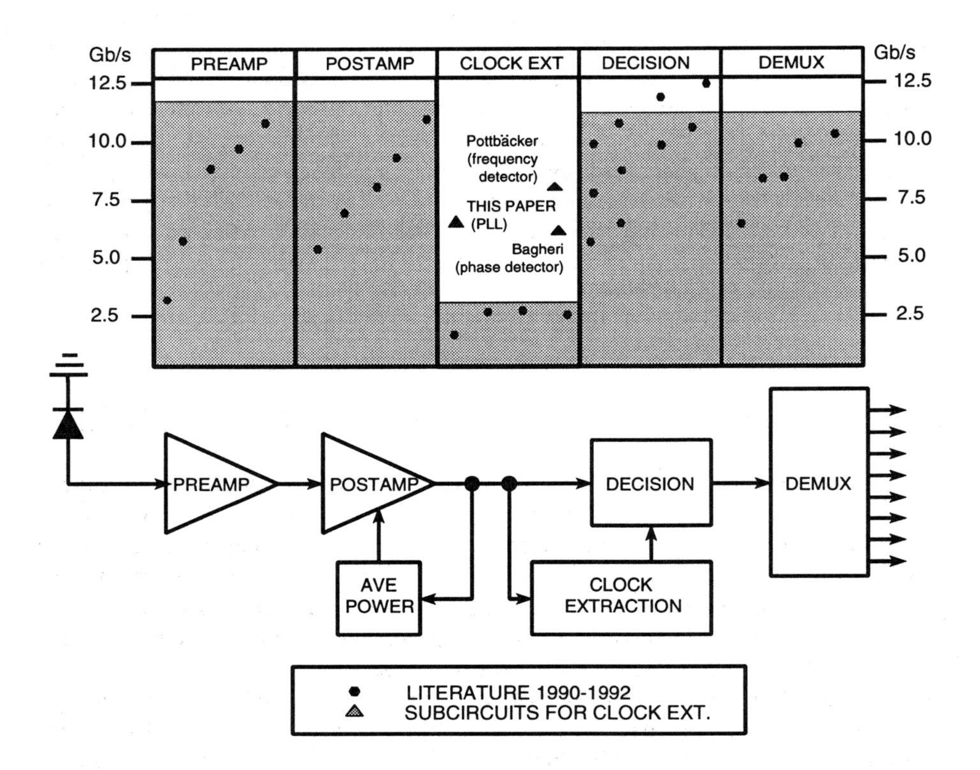

Figure 1.2 Status of fiber-optic receivers for nonreturn-to-zero digital data as of 1993.

Bit Rate (Mb/s)	SONET-Level	SDH-Level
51.84	OC–01	
155.52	OC–03	STM–1
622.08	OC–12	STM–4
1244.16	OC–24	STM–8
1866.24	OC–36	STM–12
2488.32	OC–48	STM–16

Table 1.1 Bit-rates and corresponding SONET (North America) and SDH (Europe) levels.

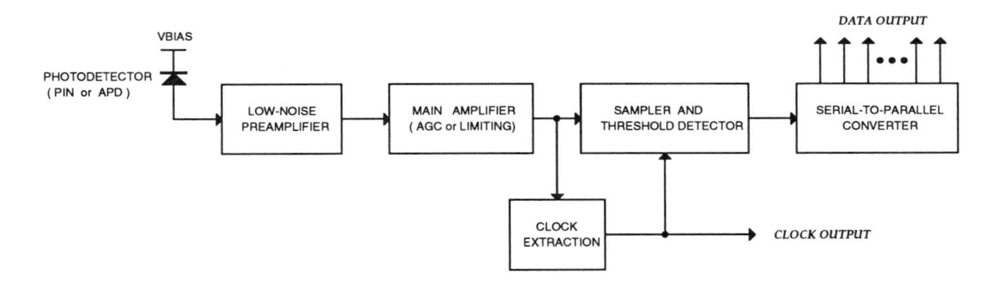

Figure 1.3 Block Diagram of a Fiber-Optic Receiver

1.3.1 High-Speed Integrated Circuit Processing Technologies

Most of the prototype circuits in this research were fabricated using TRW's AlGaAs/GaAs HBT process ($f_{max} \simeq$=40-GHz), which has consistently demonstrated a level of integration with over 1000 devices. Other smaller circuits were be realized in TRW's developmental indium-phosphide (InP) HBT process ($f_{max} \simeq$=80-GHz). Dissimilar materials are utilized in an HBT to form a heterojunction, such that the bandgap energy on the emitter side of the junction is larger than the base bandgap energy. This energy difference gives the process engineer an additional parameter for controlling device behavior. In particular, emitter-injection-efficiency is dominated by the bandgap energy difference, and is no longer controlled by the ratio of emitter-to-base doping levels. This allows doping levels to be optimized for high-speed performance, without being constrained by current-gain considerations. It is not uncommon for the base to have a higher doping concentration than the emitter, resulting in lower base resistances, and lower emitter junction capacitances, and thus higher speeds. Due to bandgap engineering, the HBT can have anywhere from a 20% to a 100% speed advantage over homojunction devices with similar dimensions. More will be said about HBTs in chapter 6.

1.4 OVERVIEW OF FIBER-OPTIC RECEIVER DESIGN

A simplified block diagram of a fiber-optic receiver is shown in Fig. 1.3. It consists of a high impedance detector at the front-end. This can be either a p-i-n diode, or an avalanche photodetector (APD). The low-level signal from the photodetector is amplified by a low-noise preamplifier, followed by a main amplifier with automatic gain control. A clock extraction and data regeneration circuit recovers the timing information from the random data, and samples the data stream at the appropriate instant. Finally, a serial to parallel converter demultiplexes the retimed serial data to a lower rate, where it can be processed by other circuitry. What follows is a brief description of each of these blocks, and the problems that must be solved to produce a successful receiver IC.

1.4.1 Photodetector

When light pulses, traveling down an optical fiber, reach their destination, they are focused onto a photodetector diode, which absorbs the light energy and generates electron-hole pairs. These electron-hole pairs are swept across the depletion region of the diode, resulting in a current that is proportional to the incident optical power. The absorption mechanisms of single-mode glass fibers are such that three separate wavelength windows exist, where the attenuation of light pulses in the fiber achieves a local minimum. These windows are at wavelengths of 0.82 μm, 1.3 μm, and 1.55 μm. For low impurity fibers, the dominant loss mechanism inside these windows is due to Rayleigh scattering. Since Rayleigh scattering is inversely proportional to the fourth power of the wavelength in a given material, the lowest loss is at the longest wavelengths, specifically 1.55 μm for glass fibers [1].

The wavelength of light absorbed by AlGaAs photodetectors is approximately 0.8 μm. This is well matched to the short wavelength low-loss window for glass fibers. However, the attenuation at this wavelength is about 10 dB higher than at 1.55 μm. Because the attenuation at 0.8 μm is relatively high, three separate implementations can be pursued with regard to the photodetector when using AlGaAs HBTs. The first is to integrate a p-i-n diode using AlGaAs with the receiver circuitry to obtain a lightwave communication system at a wavelength near 0.8 μm. This system will be capable of processing high data rates, but the scattering losses of the fiber will restrict the distance between repeaters to at most 10–20 kms, which is applicable to short-haul trunk-lines and local area networks. The second alternative is to use an external long-wavelength detector. Lower losses of the long wavelength transmission system will enable communication over a longer distance. However, the interconnect between the detector and preamplifier will increase parasitic capacitances and inductances, which can degrade both the noise performance, and the frequency response. As a third alternative, a photodetector and a low-noise preamplifier can be integrated in an InP based material system. InP has a bandgap energy that corresponds to a wavelength of about 1.3 μm. InP HBTs with extremely high f_ts (60–110 GHz) can be fabricated on the same chip with the photodetector. Although this technology is not very mature, ten transistor circuits can be fabricated with a reasonable yield. Using InP for the detector and the preamplifier will improve the noise performance, because the InP HBTs are faster than the GaAs HBTs. We will see in the next section that the noise of a preamplifier

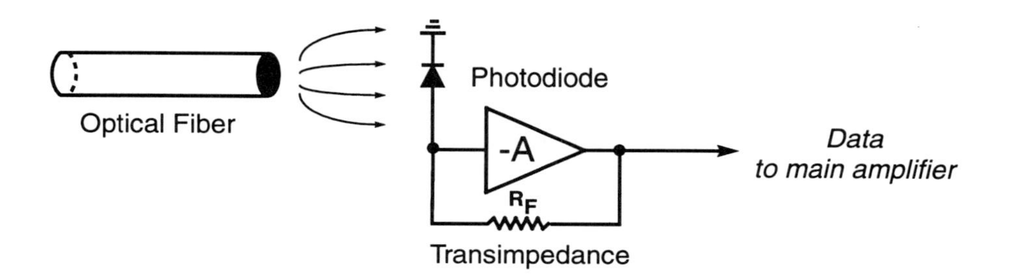

Figure 1.4 Block diagram of a transimpedance preamplifier.

at high-speeds is related to the maximum speed of the transistors. Integrating the photodetector with the amplifier eliminates interconnect problems, because interconnections are now made between the preamplifier output, and the postamplifier input, where impedance levels are much easier to control. Also, noise performance is not degraded at this point, because any added noise will be well below the noise floor.

1.4.2 Preamplifier

The low-level signal current from the photodetector must be amplified so that additional processing will not add significantly to the noise. A preamplifier is used to convert this current into a voltage for subsequent processing. The sensitivity of the receiver and the signal-to-noise ratio will be determined at this stage. Therefore, a very low-noise amplifier is required. A transimpedance amplifier, like the one shown in Fig. 1.4, has typically been used for this purpose, and its noise performance is well characterized [20, 21, 22, 23, 24, 25]. The input referred current-noise spectral-density for a preamp with a bipolar input device is given by

$$S_{n_B}(f) = \frac{4kT}{R_F} + \frac{2qI_C}{\beta} + 4kTr_b\,(2\pi f C_{ds})^2 + \left[2qI_C + \frac{4kT}{R_C}\right]\left(\frac{2\pi f C_{TB}}{g_m}\right)^2 \quad (1.1)$$

and for an FET input device the result is

$$S_{n_F}(f) = \frac{4kT}{R_F} + \left[4kT\Gamma g_m + \frac{4kT}{R_C}\right]\left(\frac{2\pi f C_{TF}}{g_m}\right)^2 \quad (1.2)$$

where R_F = feedback resistor

R_C = collector/drain resistor in first stage

r_b = base resistance

$$\begin{aligned} C_{ds} &= \text{detector plus stray capacitance} \\ C_{TB} &= C_{ds} + C_{\pi} + C_{\mu} \\ C_{TF} &= C_{ds} + C_{gs} + C_{gd} \\ \Gamma &= \text{FET excess noise factor.} \end{aligned}$$

The noise at lower frequencies can be shown to be dominated by thermal noise in the feedback resistor, and by the base-current shot-noise, for a bipolar front-end. Because an FET device lacks this base-current shot-noise term, it has generally been accepted that FET devices will exhibit superior noise performance. However, input noise levels comparable to, and even lower than FETs are obtainable using bipolar devices when the bandwidth is broadened [24]. This is possible because at higher frequencies, the collector current shot-noise becomes dominant, and the input-noise-current spectral-densities for a bipolar device reduce to

$$S_{n_B}(f) \approx 4kTr_b\left(2\pi f C_{ds}\right)^2 + \left[2qI_C + \frac{4kT}{R_C}\right]\left(\frac{2\pi f C_{TB}}{g_m}\right)^2, \tag{1.3}$$

and for an FET device,

$$S_{n_F}(f) \approx \left[4kT\Gamma g_m + \frac{4kT}{R_C}\right]\left(\frac{2\pi f C_{TF}}{g_m}\right)^2. \tag{1.4}$$

Since HBTs can be fabricated with very low base resistance, the first term in (1.3) can be made small. The remaining term is proportional to the square of a capacitance-transconductance ratio, or an effective time-constant. For a bipolar device with large bias current, this time constant asymptotically approaches τ_F, the forward transit time in the base, which can be quite small for high-speed HBTs (~1ps). Since HBTs have higher gain than FET devices, the same transconductance can be obtained at a much lower bias current. Therefore, at high data rates, where the collector-current shot-noise is dominant, an FET device will generally require significantly more bias current to reduce the term C_{TF}/g_m in order to achieve the same noise performance as a bipolar device at equal temperatures. However, since the noise power is proportional to temperature, the FET can have higher noise than an HBT of equal speed due to the increased power dissipation of the FET. Therefore, in a fully-integrated receiver, where power dissipation must be kept low, achieving low-noise with low bias currents is an extremely advantageous property. Aside from the noise penalty due to an increase in operating temperature, an FET device may never reach the same noise level of an HBT device with a low base-resistance, high β, and small τ_F, even when the bias current of the FET is raised beyond practical limits of a single-chip preamplifier (100–200 mA).

A schematic of an electro-optical InP integrated low-noise transimpedance preamplifier is shown in Fig. 1.5. This amplifier has a p-i-n photodetector integrated on the same

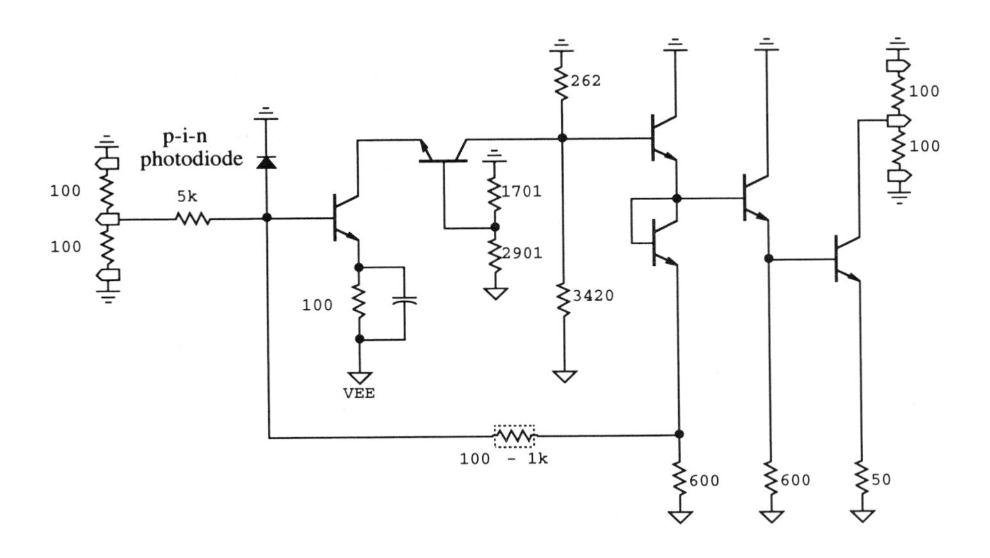

Figure 1.5 A low-noise InP transimpedance preamplifier.

chip. The design of this circuit, and a detailed noise analysis will be presented in chapter 7.

1.4.3 Main Amplifier

The main amplifier will act to buffer the circuit from process variations and changes in signal strength, and will also perform noise shaping. It must contain either a limiter, or an automatic-gain-control circuit to provide the proper signal level to the clock-extraction and data-recovery circuit, regardless of the output power of the preamp circuit. The single-ended signal from the preamplifier will be converted to a differential signal, and fully-differential circuits will be employed throughout the remainder of the receiver. The main amplifier circuit will make extensive use of adaptive biasing techniques to automatically adjust its dc levels to match the common-mode level of the clock extraction and data recovery circuit. The saturation characteristics of this stage will be considered carefully as they will affect the overall dynamic range of the receiver.

Specific challenges in this circuit are in providing dc level restoration. Since long sequences of data can be transmitted without transitions, the data can contain low-frequency information. Therefore, a dc restoration that subtracts the average-data

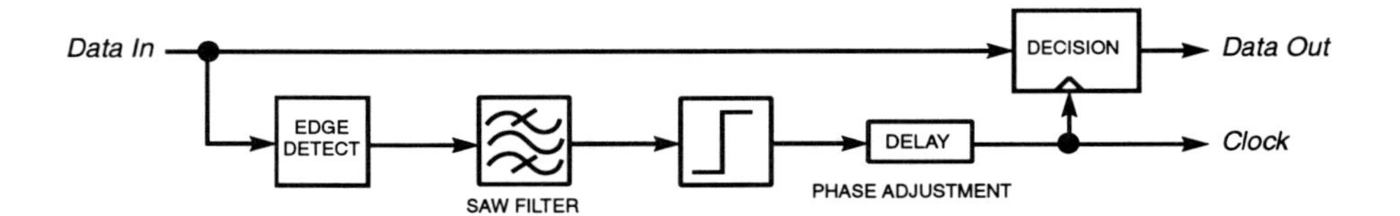

Figure 1.6 Block diagram of a clock recovery and data retiming circuit using a bandpass filter.

from the input data is forbidden. Another challenge is in conversion of the single-ended output from the preamplifier to a differential signal. At high-speeds, care must be taken to equalize the delays in the positive and negative paths.

1.4.4 Clock Extraction and Data Recovery

Clock extraction circuits for nonreturn-to-zero (NRZ) data can be grouped into two main categories: open loop filters, and closed loop synchronizes. Formally, filters have been used almost exclusively in high bit-rate receivers. With this open loop technique, the periodic timing information is extracted from the data by first using a nonlinear edge-enhancement circuit to generate a spectral line at the bit rate. The signal is then passed through a narrowband filter, centered at the bit-rate frequency, as shown in Fig. 1.6. The filter must be highly selective (high Q) in order to minimize the phase-jitter in the clock signal. Typically, surface-acoustic-wave (SAW) filters have been used for this purpose, however commercially available SAW filters are limited to a frequency of less than 3 GHz [26].

The open-loop technique is attractive because it doesn't suffer from instabilities and nonlinear problems, such as frequency acquisition and cycle-slipping. However, open-loop systems usually need to be manually adjusted to center the clock-edge in bit-interval. This one-time adjustment will not track phase offsets due to temperature variations and component aging. The filter is also external to the receiver electronics and bulky, leading to both packaging and interconnect problems.

In contrast to an open loop filter, a closed loop system is integrable, and can continually compensate for changes in the environment and the input bit-rate. This technique requires that a voltage-controlled oscillator (VCO) be tuned by a suitably filtered error signal, so as to align its transitions to the center of the bit interval. This is illustrated conceptually in Fig. 1.7. Although the loop has the desirable property of being self-adjusting, complications due to nonlinear frequency acquisition and tracking makes the circuit difficult to design.

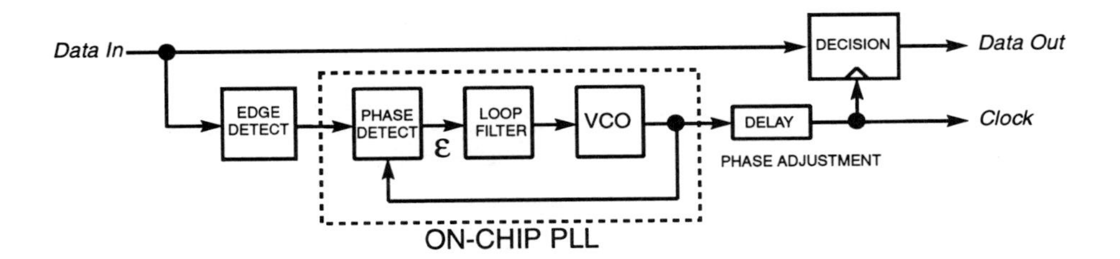

Figure 1.7 Block diagram of a clock recovery and data retiming circuit using a PLL.

Clock recovery circuits presently limit the obtainable data-rate of multigigabit-per-second integrated fiber-optic receivers. Currently, practical receivers that include methods for extracting the clock signal are limited to about 2.5 Gb/s, both for systems using a SAW filter for clock extraction [27, 28], and systems using a PLL [18], although recently reported experimental circuits are fast approaching the 10-Gb/s range [19].

Several groups are working to produce practical 10-Gb/s integrated fiber-optic receivers. Among them are: AT&T, Bellcore/Rockwell, NTT, NEC, Rühr Universität in Bochum Germany, and UCLA/TRW. Preamplifier and postamplifier ICs [29, 30], an amplifier and mixer [31], a demultiplexer and phase-aligner IC [32, 33], a phase/frequency-detector [34, 35], a PLL (phase-lock loop) [36, 37], and a clock-extraction and data-retimming circuit [19] are among the circuits presented recently. Thus far, all of the main functional blocks of a 10-Gb/s receiver have been demonstrated with one notable exception — the clock recovery circuit. This circuit is the most complicated, and the most difficult to design; it's not surprising that development of high-speed clock recovery has lagged behind development of the simpler amplifier and demultiplexer circuits.

One of the major thrusts of this book will be in developing the clock extraction and data recovery circuit. Several special challenges exist in designing a single chip system. In keeping with the goal of economy, the amount of external trimming should be minimized. For an integrated solution, a phase-locked loop will be used. Several advantages of integration will be exploited in this circuit. For example, simple oscillator circuits, such as multivibrators and ring oscillators, can be realized with sufficiently low phase-jitter, and PLLs can be used to further purify the spectrum and reduce low-frequency jitter and drift. Also, one can take advantage of the matching of devices to obtain continual phase alignment and frequency acquisition.

A conceptual diagram of a self-correcting clock-recovery and data-retiming circuit using this technique is shown in Fig. 1.8. The clock recovery loop measures the clock-phase and aligns it so as to minimize the bit-error-rate. Since we propose to design

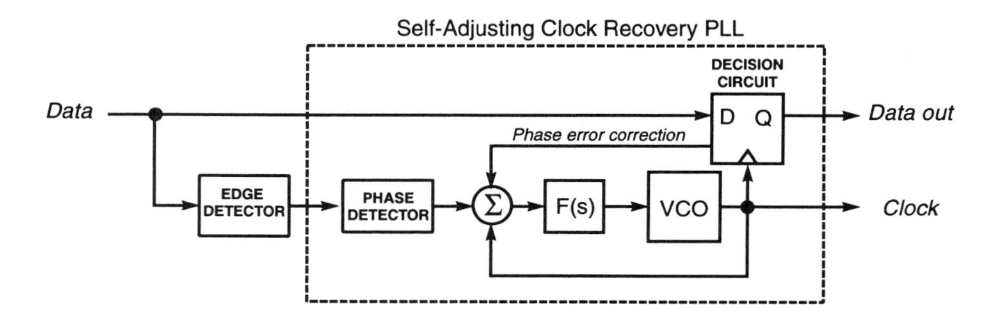

Figure 1.8 Block diagram of a self-adjusting clock recovery circuit.

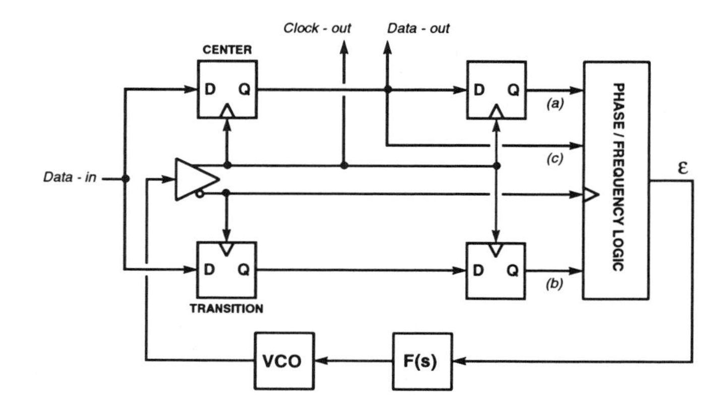

Figure 1.9 A self-correcting phase detector for a clock-recovery and data retiming circuit.

a fully-integrated receiver, no external delay lines can be used for tuning. Therefore, the optimal phase alignment of the clock recovery circuit must be done on chip; a self-correcting circuit additionally requires the decision circuit to be included in the feedback loop for final clock-phase adjustment. This is shown explicitly in Fig. 1.8 as the phase error correction signal.

Practical High-Speed Clock Recovery and Data Retiming Circuits

Clock recovery circuits are explained in considerable detail in chapters 4 and 5. Here we will briefly describe three self-adjusting circuits capable of high-speed operation. One method of recovering the clock was first described by Alexander [38]. A block diagram of this approach is shown in Fig. 1.9. The basic idea of this circuit is to use the decision flip-flop in conjunction with an identical reference flip-flop to obtain a

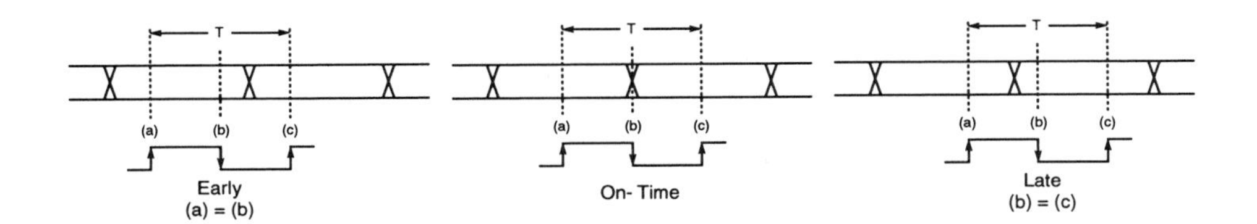

Figure 1.10 Illustration of timing of samples in Alexander's clock recovery and data retiming circuit.

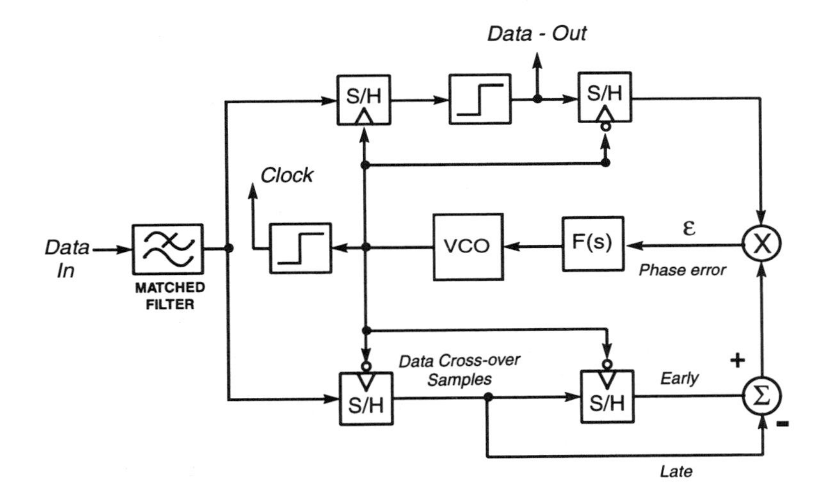

Figure 1.11 An early-late gate clock synchronizer for data retiming.

differential error signal. The sample (a) is the previous data symbol, and the sample (c) is the current data symbol. The reference sample (b) is taken at the data crossover. The timing of these three samples is illustrated in Fig. 1.10. The digital logic block looks at the three samples, and decides whether the clock was early, late, or indeterminate for each sampling interval. This decision is averaged, and used to control a VCO.

A second method is a variation on the early-late gate technique. This circuit, illustrated in Fig. 1.11, is similar to the one previously described, in that it uses identical decision circuits to arrive at a differential phase-error measure. In this circuit, data is detected using an early clock, a late clock, and an on-time clock. By subtracting the late from the early signal, and multiplying by the retimed data to remove random polarity variations, a phase-error signal is derived, which will go to zero when the early and late clocks are exactly centered about the optimal sampling point. The usual depiction

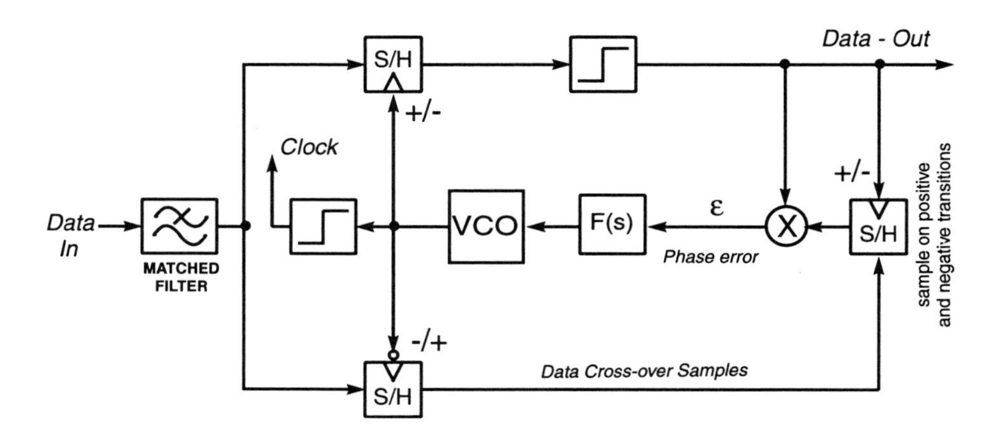

Figure 1.12 Block diagram of a data transition tracking loop for timing recovery and data regeneration.

of the early and late gates as dumped integrators has been replaced by a matched filter with sample-and-holds, which facilitates high-speed operation. An alternative implementation of this circuit could use two levels of bit-interleaving, so that dual track-and-holds can be multiplexed to perform the sample-and-hold function, and the VCO would run at half the data rate. Since the early- and late-gate correlators are matched to the decision circuit correlator, their delay times will track each other, and the circuit will be automatically and continually, phase aligned for optimal sampling.

A practical clock recovery circuit will require some type of frequency acquisition aid. A PLL-based clock recovery circuit is only capable of *pulling-in* a frequency error of the same order of the closed-loop bandwidth, which is typically a factor of 1000 less than the bit-rate. Therefore, without frequency acquisition aids, the VCO center frequency will have to be stable to within 0.1% over all processing and temperature variations, which is quite a stringent specification.

A third clock recovery circuit that was adopted for application to 10-Gb/s systems is known as a data transition tracking loop (DTTL) [39, 40]. A conceptual block diagram of DTTL circuit is shown in Fig. 1.12; this circuit is discussed in detail in chapter 5, and simulations results are given in chapter 10. A frequency discriminator was added to the DTTL to increase the *pull-in* range, and the circuit can be implemented using two levels of bit-interleaving. A block diagram of the interleaved DTTL with frequency detection is shown in Fig. 1.13. This circuit has several desirable properties as discussed in section 5.4; these advantages are briefly outlined in table 1.2.

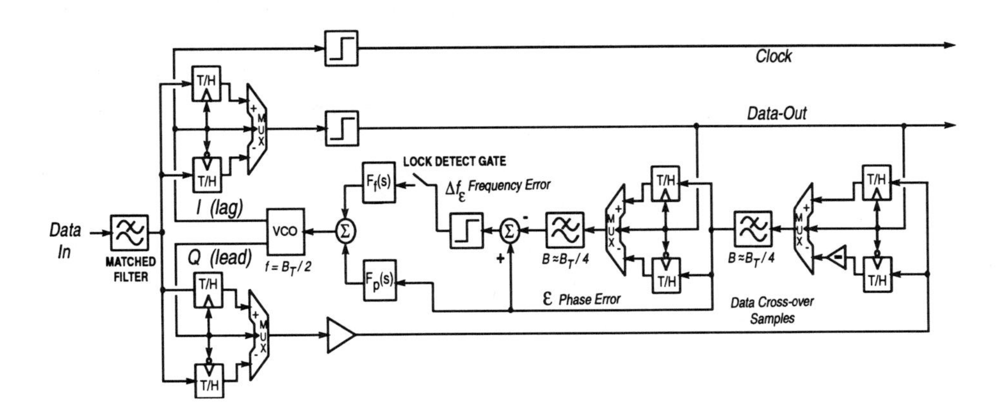

Figure 1.13 Block diagram of an interleaved DTTL with frequency detection.

Advantages of DTTL Clock Recovery Circuit

- Can function at very high-speeds
- Is inherently self-adjusting
- Using Sample-and-holds before decision circuits improves sensitivity
- Phase-detector function is monotonic over the bit interval $[-T/2, T/2]$, improving phase-tracking and frequency-acquisition
- The phase-error is independent of the transition density, eliminating pattern dependent jitter.
- Resampling the phase error only after a data transition eliminates ripple, and significantly reduces ripple-induced phase-jitter

---

Table 1.2 Advantages of data transition tracking loop for clock extraction and data retiming of random NRZ data.

In order to understand the design trade-offs employed in the optimization of circuit performance, a solid grasp of the fundamentals of communication theory, as it applies to high-speed, broadband digital receivers is required. This theory is outlined in the remainder of Part I, and special emphasis in placed on clock recovery in broadband systems. The circuit designs, and measured results of the fabricated test structures will be presented in Part II.

REFERENCES

[1] Stewart D. Personick. *Fiber Optics Technology and Applications*. Plenum Press, New York, 1985.

[2] Gerd Keiser. *Optical Fiber Communications*. McGraw-Hill, Inc., New York, second edition, 1991.

[3] Paul E. Green, Jr. *Fiber Optic Networks*. Prentice-Hall, Inc., Englewood Cliffs, New Jersey, 1993.

[4] John M. Senior. *Optical Fiber Communications Principles and Practice*. Prentice-Hall, Inc., New York, second edition, 1992.

[5] Bellcore, Morristown, NJ. *Synchronous Optical Network (SONET) Transport Systems: Common Generic Criteria*, TA-NWT-000253 6th edition, September 1990.

[6] H. Taga, Y. Yoshida, N. Edagawa, , S. Yamamoto, and H. Wakabayashi. 459km, 2.4Gbit/s four wavelength multiplexing optical fibre transmission experiment using six Er-doped fibre amplifiers. *Electron. Lett.*, 26(8):500–501, April 1990.

[7] G. E. Wickens, D. M. Spirit, and L. C. Blank. 20 Gbit/s 205km optical time division multiplexed transmission system. *Electron. Lett.*, 27(11):973–974, May 1991.

[8] Hans-Martin Rein. Silicon bipolar integrated circuits for multigigabit-per-second lightwave communications. *J. Lightwave Technol.*, LT-8(9):1371–1378, September 1990.

[9] Klaus Runge, Winston I. Way, Mehran Bagheri, James L. Gimlett, D. Clawin, Nim K. Cheung, Daniel J. Millicker, Detlef Daniel, and C. Snapp. Silicon bipolar integrated circuits for multi-Gb/s optical communication systems. *IEEE J. Select. Areas Commun.*, SAC-9(5):636–644, June 1991.

[10] Hiroshi Hamano, Takuji Yamamoto, Yoshinori Nishizawa, Akinori Tahara, Norihito Miyoshi, Kouichi Suzuki, and Akihito Nishimura. High-speed Si-bipolar IC design for multi-Gb/s optical receivers. *IEEE J. Select. Areas Commun.*, SAC-9(5):645–651, June 1991.

[11] Jens N. Albers and Hans-Ulrich Schreiber. A Si-bipolar technology for optical fiber transmission rates above 10 Gb/s. *IEEE J. Select. Areas Commun.*, SAC-9(5):652–655, June 1991.

[12] Kazuo Hagimoto, Yuuzou Miyagawa, Yutaka Miyamoto, Masanobu Ohhata, Tatsuhito Suzuki, and Hiroyuki Kikuchi. Over 10 Gb/s regenerators using monolithic IC's for lightwave communication systems. *IEEE J. Select. Areas Commun.*, SAC-9(5):673–682, June 1991.

[13] Kiyoshi Nakagawa and Katsushi Iwashita. High-speed optical transmission systems using advanced monolithic IC technologies. *IEEE J. Select. Areas Commun.*, SAC-9(5):683–688, June 1991.

[14] R. K. Montgomery et al. 10 Gbit/s high sensitivity low error rate decision circuit implemented with C-Doped AlGaAs/GaAs HBTs. *Electron. Lett.*, 27(11):976–978, May 1991.

[15] Junko Akagi, Yasuhiko Kuriyama, Kouhei Morizuka, Masayuki Asaka, Kunio Tsuda, Masao Obara, Hideaki Yamakawa, and Hiroyuki Ibe. AlGaAs/GaAs HBT receiver ICs for a 10 Gbps optical communication system. In *IEEE GaAs IC Symposium*, pages 45–48, New Orleans, Louisiana, October 1990.

[16] Hans-Martin Rein, J. Hauenschild, W. McFarland, and D. Pettengill. 23 Gbit/s Si bipolar decision circuit consisting of 24 Gbit/s MUX and DEMUX ICs. *Electron. Lett.*, 27(11):974–976, May 1991.

[17] J. Hauenschild, Hans-Martin Rein, W. McFarland, J. Doernberg, and D. Pettengill. Demonstration of retiming capability of silicon bipolar time-division multiplexor operating to 24 Gbit/s. *Electron. Lett.*, 27(11):978–979, May 1991.

[18] Hans Ransijn and Paul O'Connor. A 2.5 Gb/s GaAs clock and data regenerator IC. In *IEEE GaAs IC Symposium.*, pages 57–60, New Orleans, Louisiana, October 1990.

[19] Ansgar Pottbäcker and Ulrich Langmann. An 8 GHz silicon bipolar clock-recovery and data-regenerator IC. In *ISSCC Dig. Tech. Papers*, pages 116–117, San Francisco, California, February 1994.

[20] Bryon L. Kasper and Joe C. Campbell. Multigigabit-per-second avalanche photodiode lightwave receivers. *J. Lightwave Technol.*, LT-5(10):1351–1364, October 1987.

[21] Mike Brian and Tien-Pei Lee. Optical receivers for lightwave communication systems. *J. Lightwave Technol.*, LT-3(6):1281–1300, December 1985.

[22] Tran Van Muoi. Receiver design for high-speed optical-fiber systems. *J. Lightwave Technol.*, LT-2(3):243–267, June 1984.

[23] R. G. Smith and S. D. Personick. Receiver design for optical fiber communication systems. In Henry Kressel, editor, *Semiconductor Devices for Optical Communication*, chapter 4, pages 86–160. Springer-Verlag, New York, 1980.

[24] Siegfried G. Knorr, Osman Kaldirim, and C. Yeh. Low-noise fiber optics receiver with super-beta bipolar transistors. *Fiber and Integrated Optics*, 1(4):369–386, 1978.

[25] Stewart D. Personick. Receiver design for digital fiber optic communication systems, part I and II. *Bell Syst. Tech. J.*, 52(6):843–886, July 1973.

[26] Zhigong Wang, Ulrich Langmann, and Berthold Bosch. Mulit-Gb/s silicon bipolar clock recovery IC optical receivers. *IEEE J. Select. Areas Commun.*, SAC-9(5):656–663, June 1991.

[27] B. Wedding, D. Schlump, E. Schlag, W. Pöhlmann, and B. Franz. 2.24-Gbit/s 151-km optical transmission system using high-speed integrated silicon circuits. *IEEE J. Select. Areas Commun.*, SAC-8(2):227–234, February 1990.

[28] E. Schlag, B. Franz, and W. Pöhlmann. Integrierte Si-bipolar schaltungen für ein optisches übertragungssystem von 2.4 Gbit/s. In *Proc. ITG Fachtagung Mikroelektronik für die Informationstechnik*, pages 221–226, Stuttgart, Germany, October 1989.

[29] Masaaki Soda, Tetsuyuki Suzaki, Takenori Morikawa, Hiroshi Tezuka, Chihiro Ogawa, Sadao Fujita, Hisashi Takemura, and Tsutomu Tashiro. A Si bipolar chip set for 10 Gb/s optical receiver. In *IEEE ISSCC Dig. Tech. Papers*, pages 100–101, San Francisco, California, February 1992.

[30] Tetsuyuki Suzaki, Masaaki Soda, Takenori Morikawa, Hiroshi Tezuka, Chihiro Ogawa, Sadao Fujita, Hisashi Takemura, and Tsutomu Tashiro. Si bipolar chip set for 10-Gb/s optical receiver. *IEEE J. Solid-State Circuits*, 27(12):1781–1786, December 1992.

[31] Shuich Fujita, Yuhki Imai, Yasuro Yamane, and Hiroshi Fushimi. DC to 10-GHz mixer and amplifier GaAs IC's for coherent optical heterodyne receiver. *IEEE J. Solid-State Circuits*, 26(12):1847–1852, December 1991.

[32] Mehran Bagheri, Keh-Chung Wang, Mau-Chung F. Chang, Randy B. Nubling, Peter M. Asbeck, and Andy Chen. 11.6 GHz 1:4 demultiplexer with bit-rotation control and 6.1 GHz auto-latching phase-aligner ICs. In *ISSCC Dig. Tech. Papers*, pages 94–95, San Francisco, California, February 1992.

[33] Mehran Bagheri, Keh-Chung Wang, Mau-Chung F. Chang, Randy B. Nubling, Peter M. Asbeck, and Andy Chen. 11.6-GHz 1:4 regenerating demultiplexer with bit-rotation control and 6.1-GHz auto-latching phase-aligner IC's using AlGaAs/GaAs HBT technology. *IEEE J. Solid-State Circuits*, 27(12):1787–1793, December 1992.

[34] Ansgar Pottbäcker, Ulrich Langmann, and Hans-Ulrich Schreiber. A 8 Gb/s Si bipolar phase and frequency detector IC for clock extraction. In *ISSCC Dig. Tech. Papers*, pages 162–163, San Francisco, California, February 1992.

[35] Ansgar Pottbäcker, Ulrich Langmann, and Hans-Ulrich Schreiber. A Si bipolar phase and frequecny detector IC for clock extraction up to 8 Gb/s. *IEEE J. Solid-State Circuits*, 27(12):1747–1751, December 1992.

[36] Aaron W. Buchwald, Kenneth W. Martin, Aaron K. Oki, and Kevin W. Kobayashi. A 6GHz integrated phase-locked loop using AlGaAs/GaAs heterojunction bipolar transistors. In *ISSCC Dig. Tech. Papers*, pages 98–99, San Francisco, California, February 1992.

[37] Aaron W. Buchwald, Kenneth W. Martin, Aaron K. Oki, and Kevin W. Kobayashi. A 6GHz integrated phase-locked loop using AlGaAs/GaAs heterojunction bipolar transistors. *IEEE J. Solid-State Circuits*, 27(12):1752–1762, December 1992.

[38] J. D. H. Alexander. Clock recovery from random binary signals. *Electron. Lett.*, 11(22):541–542, October 1975.

[39] William C. Lindsey and Marvin K. Simon. *Telecommunication Systems Engineering*. Dover Publications Inc., New York, 1991. Dover edition first published in 1991 is an unabridged, unaltered republication of the work first published by Prentice-Hall, Inc., Englewood Cliffs, N.J., 1973 in its "Prentice-Hall Information and System Science Series.".

[40] T. O. Anderson, W. J. Hurd, and W. C. Lindsey. U.S. pat. no. 3,626,298; Transition Tracking Bit Synchronization System, December 1971.

2

MATHEMATICAL PRELIMINARIES: POWER SPECTRAL DENSITIES OF RANDOM DATA AND NOISE

In this chapter techniques for determining the power spectral density (PSD) of random data and random signals derived from the data will be presented. There exists a complete theory for determining the spectral content of random signals [1, 2, 3, 4]. However, the general theory involves a knowledge of probability distributions, and is restricted in application only to stationary, or wide-sense stationary random signals. The condition of stationarity is violated for random binary non-return-to-zero (NRZ) data, and the general theory cannot be directly applied to the problem at hand. However, an NRZ data stream in not totally random; such signals are termed cyclo-stationary because their statistics are cyclic. There exists a well defined structure in the data such that the absolute value of the signal in the bit period T is precisely known — only its polarity is random. Therefore it is reasonable to assume that the representation of this random data stream in the frequency domain can be obtained directly by applying the definitions of the Fourier series and Fourier transform, and problems with non-stationarity can be averted.

The spectral analysis of random data has been considered previously. Bennett's work on the statistics of regenerative digital transmission at Bell Labs was published in 1958 [5], and Titsworth and Welch of the Jet Propulsions Laboratory published their work on power spectra of random signals in 1961 [6]. These works are significant, but the average circuit designer will likely gain little insight from these formulations based on Markov chains and probability transitions matrices. Our goal in this chapter is not to repeat these works, but rather to illuminate their applicability to fiber-optic receiver design. To meet this goal, we will develop a frequency domain description of random data, and data-derived signals systematically, starting from first principles. The results will then be generalized, and related to the general theory of random signals. By presenting the power spectral densities in this manner, the interpretation of the results is straightforward. Moreover, intuition is enhanced that will enable us to find quick solutions to complex problems, especially when the data-derived signal results from a nonlinear operation on the data, as is required in clock recovery schemes.

2.1 ANALYTICAL EXPRESSIONS FOR NRZ BANDLIMITED DATA

A random data stream can be represented analytically as the sum of pulses shifted in time by a multiple of the bit-period T. If the data is binary and symmetric, then the pulse shape will be identical for each bit. Multiplication by a random variable $r_n(\cdot)$ determines the polarity, such that the data signal is given by[1]

$$d(t,\cdot) = \sum_{n=0}^{N-1} r_n(\cdot) p_T(t - nT). \tag{2.1}$$

If the data is NRZ, then the only pulse that can be used to represent the data in this manner is a rectangular pulse that is unity in the interval $[0, T]$ and zero elsewhere. Therefore,

$$p_T(t) = \text{rect}(t/T - 1/2), \tag{2.2}$$

where the rectangular function is defined by

$$\text{rect}(t/T) \triangleq \begin{cases} 1 & \text{for } |t| \leq T/2 \\ 0 & \text{for } |t| > T/2. \end{cases} \tag{2.3}$$

If the NRZ data has non-zero rise times, then memory must be introduced into the expression for $d(t,\cdot)$, because the shape of the function during a transition will depend on the previous data values. The data can then be represented analytically as the output of a linear filter with rectangular NRZ data as an input, such that the bandlimited data is represented by the convolution;

$$\boxed{d(t,\cdot) = h(t) * \left[\sum_{n=0}^{N-1} r_n(\cdot) p_T(t - nT)\right].} \tag{2.4}$$

[1] A random variable $r_n(\cdot)$ represents the entire ensemble of possible outcomes of random trials. If each random trial were given labels $[\xi_1, \xi_2, \xi_3, \ldots]$, then $r_n(\xi_1)$ is the value of the random variable resulting from the outcome of the first random trial. Likewise, a random process can be represented as $d(\cdot,\cdot)$. The interpretation of this notation is that $d(\cdot,\cdot)$ is an ensemble of all possible sample functions of the random process. At any given value of time, $d(t,\cdot)$ is a random variable. $d(\cdot,\xi_1)$ is the sample function, over all time, that corresponds to the outcome of the first random trial. Finally $d(t,\xi_1)$ is the value of the first sample function at time t. This notation may seem a bit cumbersome, but the authors have found it helpful in keeping track of which variables are random, and which are deterministic.

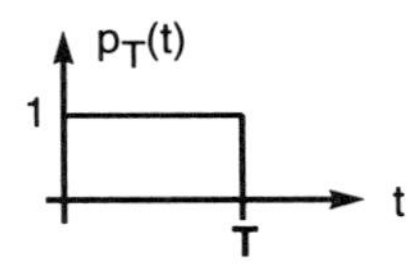

Figure 2.1 A rectangular data pulse

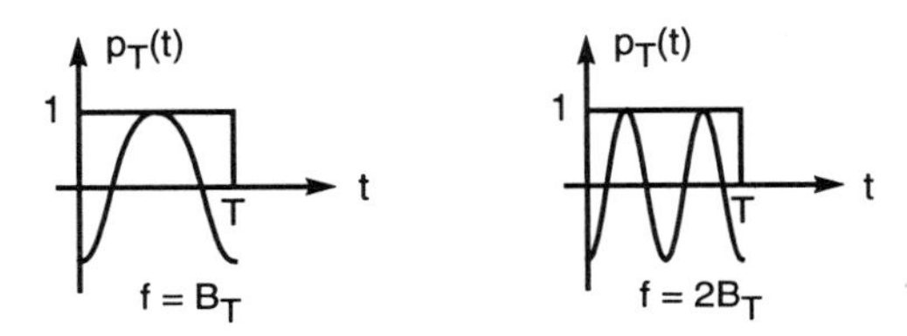

Figure 2.2 A rectangular data pulse superimposed with tones whose frequencies are a multiple of the bit-rate $B_T = 1/T$.

2.2 FOURIER SERIES FREQUENCY DOMAIN REPRESENTATIONS

Qualitative Analysis of Frequency Content Before getting bogged down in the details of determining the precise functional form of the PSD for NRZ data, we should spend a few moments to consider qualitatively what type of results to expect. For rectangular NRZ data the pulse shape is shown in Fig. 2.1. The frequency content of a signal is obtained by correlating the signal with tones of various frequencies. We can first consider any tone at a frequency that is a multiple of the bit-rate, as in Fig. 2.2. A tone at frequency B_T or any harmonic of this tone will complete an integer number of cycles within on bit-period. Since there is a positive portion to precisely cancel a negative portion of the signal in a time T, the correlation of these harmonic tones with the data pulse (the integral of the product of the tone with the data signal) is easily seen to be zero. Therefore, one would expect to find nulls in the PSD of the data steam at integer multiples of the bit-rate. Further, we can consider the contribution to the correlation integral when the frequency of the tone is increased. Fig 2.3 shows the data pulse superimposed with two tones of different frequencies. The symmetric portion of the integral is shown shaded; the residual unshaded portion is the contribution to the integral. As the frequency of the tone is increased, the portion of the signal that contributes to the integral is reduced in proportional to the reduction in the period. Therefore, we should also expect an envelope of the frequency spectrum proportional to the period of the tone, or $1/f$, where f is the frequency of the tone. Since the PSD

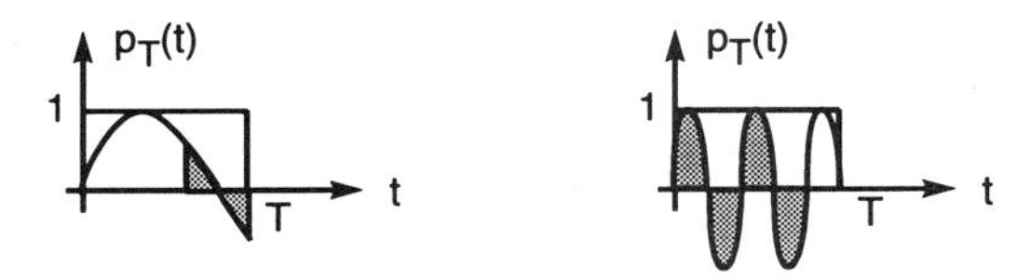

Figure 2.3 Rectangular data pulses superimposed with tones of different frequencies.

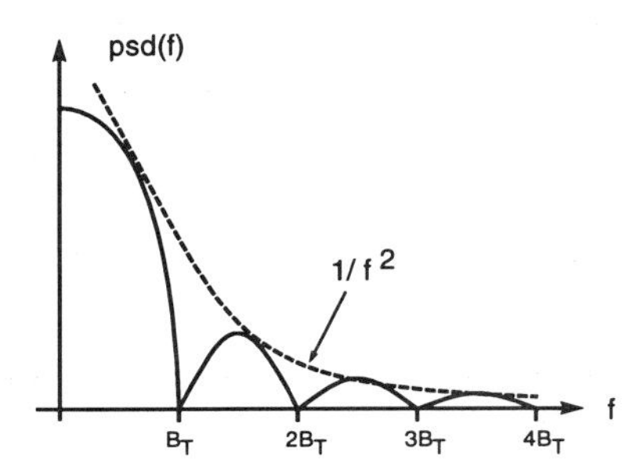

Figure 2.4 Approximate power spectral density of NRZ data based on qualitative arguments

is obtained by squaring the the frequency spectrum, the PSD will have an envelope proportional to $1/f^2$.

Now we can sketch the approximate shape of the frequency content of the data signal based on the previous qualitative observations which can be summarized as follows:

- The frequency spectrum has nulls at multiples of the bit-rate.
- The PSD has an envelope proportional to $1/f^2$.

This approximate PSD is plotted in Fig. 2.4. Based on previous experience, we might assume that the PSD has the form of a $(\sin(x)/x)^2$ function. The next few sections are devoted to deriving this functional form precisely, and interpreting exactly what it means to speak of a power-spectral-density when the time-signal is random data.

2.2.1 Fourier Series Representation of NRZ data

A rectangular NRZ random data stream $d(t, \cdot)$ of length N-bits has an analytical representation given in (2.1). A pseudo-random data sequence $d_N(t, \cdot)$ can be generated from $d(t,\cdot)$ by repeating the signal every N bits. Since $d_N(t,\cdot)$ is periodic with a period of NT, it can be represented by a Fourier series of the form

$$d_N(t,\cdot) = \frac{a_0(\cdot)}{2} + \sum_{m=1}^{\infty} a_m(\cdot)\cos\left(\frac{2\pi mt}{NT}\right) + \sum_{m=1}^{\infty} b_m(\cdot)\sin\left(\frac{2\pi mt}{NT}\right). \tag{2.5}$$

The coefficients of the Fourier series are random variables and can be extracted from the original signal. Since all harmonics of the fundamental frequency are mutually orthogonal when integrated over the period NT, random spectral coefficients are determined according to

$$a_m(\cdot) = \frac{2}{NT}\int_0^{NT} d_N(t,\cdot)\cos\left(\frac{2\pi mt}{NT}\right)dt \tag{2.6}$$

$$b_m(\cdot) = \frac{2}{NT}\int_0^{NT} d_N(t,\cdot)\sin\left(\frac{2\pi mt}{NT}\right)dt. \tag{2.7}$$

These coefficients of the Fourier series can be considered as "dot-products," or equivalently, projections of the data signal onto the orthogonal basis functions. Since the cosine and sine are quadrature signals, they are also mutually orthogonal, and both must be included in the Fourier series expansion, with the relative magnitudes of the coefficients $a_m(\cdot)$ and $b_m(\cdot)$ determining the phase. In the analysis that follows both $a_m(\cdot)$ and $b_m(\cdot)$ will be evaluated directly from the above definitions, and the interpretation of the result will be clear. Later the complex form of the Fourier series and negative frequencies will be introduced for analytical convenience.

The process of finding the power spectral density of the random data begins by evaluating $a_m(\cdot)$ directly. Applying the definition,

$$a_m(\cdot) = \frac{2}{NT}\int_0^{NT}\sum_{n=0}^{N-1} r_n(\cdot)p_T(t-nT)\cos\left(\frac{2\pi mt}{NT}\right)dt, \tag{2.8}$$

interchanging the order of integration and summation,

$$a_m(\cdot) = \frac{2}{NT}\sum_{n=0}^{N-1} r_n(\cdot)\int_0^{NT} p_T(t-nT)\cos\left(\frac{2\pi mt}{NT}\right)dt, \tag{2.9}$$

and recalling that the pulse $p_T(t)$ is rectangular, such that

$$p_T(t-nT) = \begin{cases} 1 & \text{for} \quad nT \le t \le (n+1)T \\ 0 & \text{elsewhere,} \end{cases} \tag{2.10}$$

$a_m(\cdot)$ can now be expressed as the sum of integrals

$$a_m(\cdot) = \frac{2}{NT}\sum_{n=0}^{N-1} r_n(\cdot)\int_{nT}^{(n+1)T} \cos\left(\frac{2\pi mt}{NT}\right)dt. \tag{2.11}$$

Evaluating the integrals

$$a_m(\cdot) = \frac{1}{\pi m} \sum_{n=0}^{N-1} r_n(\cdot) \left[\sin\left(\frac{2\pi m(n+1)}{N}\right) - \sin\left(\frac{2\pi mn}{N}\right) \right]. \quad (2.12)$$

To facilitate manipulation of the sinusoids we define

$$\theta_n \triangleq \frac{2\pi mn}{N} \quad \text{and,} \quad \phi \triangleq \frac{\pi m}{N}.$$

The result in (2.12) can now be simplified. Leaving in all of the intermediate steps,

$$a_m(\cdot) = \frac{1}{\pi m} \sum_{n=0}^{N-1} r_n(\cdot) \Im \left\{ e^{j\theta_n} e^{j2\phi} - e^{j\theta_n} \right\} \quad (2.13a)$$

$$a_m(\cdot) = \frac{1}{\pi m} \sum_{n=0}^{N-1} r_n(\cdot) \Im \left\{ e^{j\phi} e^{j\theta_n} \left[e^{j\phi} - e^{-j\phi} \right] \right\} \quad (2.13b)$$

$$a_m(\cdot) = \frac{1}{\pi m} \sum_{n=0}^{N-1} r_n(\cdot) \Im \left\{ e^{j\phi} e^{j\theta_n} \left[2j \sin(\phi) \right] \right\} \quad (2.13c)$$

$$a_m(\cdot) = \frac{2\sin(\phi)}{\pi m} \sum_{n=0}^{N-1} r_n(\cdot) \Im \left\{ j e^{j\phi} e^{j\theta_n} \right\} \quad (2.13d)$$

$$a_m(\cdot) = \frac{2\sin(\phi)}{\pi m} \sum_{n=0}^{N-1} r_n(\cdot) \cos(\theta_n + \phi). \quad (2.13e)$$

Using the definition of the sinc function

$$\text{sinc}(x) \triangleq \frac{\sin(\pi x)}{\pi x},$$

then the Fourier series coefficients $a_m(\cdot)$ are given by

$$a_m(\cdot) = \frac{2}{N} \text{sinc}(m/N) \sum_{n=0}^{N-1} r_n(\cdot) \cos\left(\frac{2\pi m}{N}(n + 1/2)\right). \quad (2.14)$$

The random coefficients $b_m(\cdot)$ can be found in a similar manner.

$$b_m(\cdot) = \frac{2}{NT} \int_0^{NT} \sum_{n=0}^{N-1} r_n(\cdot) p_T(t - nT) \sin\left(\frac{2\pi mt}{NT}\right) dt \quad (2.15)$$

This too can be expressed as the sum of integrals

$$b_m(\cdot) = \frac{2}{NT} \sum_{n=0}^{N-1} r_n(\cdot) \int_{nT}^{(n+1)T} \sin\left(\frac{2\pi mt}{NT}\right) dt. \tag{2.16}$$

The result of the integration is

$$b_m(\cdot) = -\frac{1}{\pi m} \sum_{n=0}^{N-1} r_n(\cdot) \left[\cos\left(\frac{2\pi m(n+1)}{N}\right) - \cos\left(\frac{2\pi mn}{N}\right)\right]. \tag{2.17}$$

From (2.13) it can be seen that $b_m(\cdot)$ can be expressed similarly,

$$b_m(\cdot) = -\frac{2\sin(\phi)}{\pi m} \sum_{n=0}^{N-1} r_n(\cdot) \Re\left\{je^{j\phi}e^{j\theta_n}\right\}, \tag{2.18}$$

and after simplifying

$$b_m(\cdot) = \frac{2}{N}\text{sinc}(m/N) \sum_{n=0}^{N-1} r_n(\cdot) \sin\left(\frac{2\pi m}{N}(n+1/2)\right). \tag{2.19}$$

The pseudo-random rectangular NRZ data stream has now been represented by a Fourier series expansion, where the coefficients given in (2.14) and (2.19) are random variables that depend on the actual data stream. It is desirable to find the average behavior of the data in the frequency domain so that the result would correspond to the output of a spectrum analyzer, averaging several sweeps. Each sweep measures the time-averaged power in a given bandwidth, and the final display is an average over several smaller segments of the complete data signal. To perform this operation analytically, we first need to find the time-averaged power of the random data in a given bandwidth. This power will be a random variable which also depends on the actual data sequence. By averaging over the ensemble of all possible data sequences the statistical average of the time-averaged power[2] can be determined.

For a deterministic signal of the form,

$$f(t) = a_m \cos(2\pi f_m t) + b_m \sin(2\pi f_m t), \tag{2.20}$$

[2] Thus far we have not defined the units of a_m and b_m. However, if we want to talk about power, then they clearly must have units proportional to $\sqrt{\text{Watts}}$. Normally we will consider the signal $f(t)$ to be a either a current or a voltage. Therefore, a resistance must be associated with the coefficients to obtain a power. If we associate a 1Ω resistor with each coefficient, then for a voltage signal the units are $\text{Volts}/\sqrt{1\Omega} = \sqrt{\text{Watts}}$, and for a current the units are $\text{Amps}\sqrt{1\Omega} = \sqrt{\text{Watts}}$. However, usually we will ignore the 1Ω normalization and still talk about the power when the units are actually Amps^2 or Volts^2 and not Watts. Although this is a misnomer, we will use the word "power," when it should be kept in mind that we actually mean the power dissipated in a 1Ω resistor.

the time-averaged power $\overline{P_m}$ is equal to

$$\overline{P_m} = \frac{a_m^2 + b_m^2}{2}. \tag{2.21}$$

A periodic deterministic signal can be represented by a Fourier series, such that

$$g(t) = \frac{a_0}{2} + \sum_{m=1}^{\infty} a_m \cos(2\pi f_m t) + \sum_{m=1}^{\infty} b_m \sin(2\pi f_m t). \tag{2.22}$$

Since the basis functions are mutually orthogonal, then the power in the mth harmonic is also given by (2.21), except at dc where the average power is

$$\overline{P_0} = \frac{a_0^2}{4}. \tag{2.23}$$

To facilitate power calculations it is convenient to express the Fourier series coefficients as the real and imaginary parts of a complex number. From (2.14) and (2.19) it can be seen that $a_m(\cdot)$ and $b_m(\cdot)$ can be expressed in the following form

$$a_m(\cdot) = \frac{2}{N}\text{sinc}(m/N)\Re\left\{\sum_{n=0}^{N-1} r_n(\cdot)e^{-j\phi}e^{-j\theta_n}\right\} \tag{2.24a}$$

$$b_m(\cdot) = -\frac{2}{N}\text{sinc}(m/N)\Im\left\{\sum_{n=0}^{N-1} r_n(\cdot)e^{-j\phi}e^{-j\theta_n}\right\}. \tag{2.24b}$$

Defining a complex Fourier coefficient as

$$c_m(\cdot) \triangleq \frac{a_m(\cdot) - jb_m(\cdot)}{2}, \tag{2.25}$$

then

$$c_m(\cdot) = \frac{1}{N}\text{sinc}(m/N)\left\{\sum_{n=0}^{N-1} r_n(\cdot)e^{-j\phi}e^{-j\theta_n}\right\}. \tag{2.26}$$

The squared magnitude of $c_m(\cdot)$ is proportional to the average power,

$$|c_m(\cdot)|^2 = c_m(\cdot)c_m^*(\cdot) = \frac{a_m(\cdot)^2 + b_m(\cdot)^2}{4} \tag{2.27}$$

so that

$$\overline{P_m(\cdot)} = \begin{cases} |c_m(\cdot)|^2 & \text{for } m = 0 \\ 2|c_m(\cdot)|^2 & \text{for } m \neq 0. \end{cases} \tag{2.28}$$

Now the magnitude of $c_m(\cdot)$ can be determined from (2.26).

$$|c_m(\cdot)|^2 = \left(\frac{1}{N}\text{sinc}(m/N)\right)^2 \sum_{n=0}^{N-1} r_n(\cdot)e^{-j\phi}e^{-j\theta_n} \sum_{n=0}^{N-1} r_n(\cdot)e^{j\phi}e^{j\theta_n}, \tag{2.29}$$

and writing this as a double summation we have

$$|c_m(\cdot)|^2 = \left(\frac{1}{N}\text{sinc}(m/N)\right)^2 \sum_{n=0}^{N-1}\sum_{k=0}^{N-1} r_n(\cdot)r_k(\cdot)e^{j(\theta_n-\theta_k)}. \tag{2.30}$$

Ensemble Expectations of the Average Power Continuing with the analysis we want to find the ensemble average of $\overline{P_m}(\cdot)$. To do this requires a knowledge of the statistics of the random variable $r_n(\cdot)$. Since $r_n(\cdot)$ represents the polarity of the pulse, it can only take on values of +1 or -1. It will be assumed, unless otherwise specified, that the data is equally likely to be positive as it is negative. Therefore,

$$r_n(\cdot) = \begin{cases} +1 & \text{with Probability 1/2} \\ -1 & \text{with Probability 1/2.} \end{cases} \tag{2.31}$$

As a result, the mean of the data is zero,

$$E[r_n(\cdot)] = 0. \tag{2.32}$$

It is further assumed that all data bits are uncorrelated, so that a knowledge of one, or more bits, gives no information about the value of any other bit. Therefore,

$$E[r_n(\cdot)r_k(\cdot)] = \begin{cases} 1 & \text{for } n = k \\ 0 & \text{for } n \neq k \end{cases} \tag{2.33}$$

Now the expected value of $|c_m(\cdot)|^2$ can be determined.

$$E[|c_m(\cdot)|^2] = \left(\frac{1}{N}\text{sinc}(m/N)\right)^2 \sum_{n=0}^{N-1}\sum_{k=0}^{N-1} E[r_n(\cdot)r_k(\cdot)]e^{j(\theta_n-\theta_k)} \tag{2.34}$$

The inner sum vanishes for all values of $k \neq n$, so that the double sum can be replaced by a single sum.

$$E[|c_m(\cdot)|^2] = \left(\frac{1}{N}\text{sinc}(m/N)\right)^2 \sum_{n=0}^{N-1} (1)e^{j0} \tag{2.35}$$

$$= \frac{1}{N}\text{sinc}^2(m/N) \tag{2.36}$$

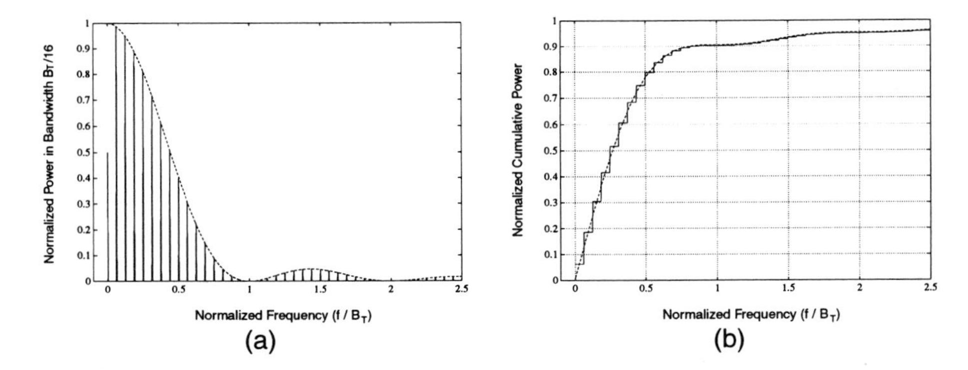

Figure 2.5 (a) Power of Fourier series constituent tones for $N = 16$, (b) Cumulative power.

Letting the mean of the of the time-averaged power be defined by

$$\overline{P_m} \triangleq E[\overline{P_m(\cdot)}],$$

then

$$\boxed{\overline{P_m} = \begin{cases} \dfrac{1}{N} & \text{for } m = 0 \\ \dfrac{2}{N}\text{sinc}^2(m/N) & \text{for } m \neq 0. \end{cases}} \tag{2.37}$$

$\overline{P_m}$ is plotted in Fig. 2.5a for the case of $N = 16$ bits. We recall that this is a plot of the expected value, over all possible periodic pseudo-random sequences with a period of 16 bits, of the time averaged power in each of the harmonics of the Fourier series representation of NRZ data. We can use Parseval's Theorem that equates the average power in the time and frequency domains to check this result.

$$P_{time} \triangleq E\left[\int_0^{NT} d_N^2(t,\cdot)dt\right] = 1 \tag{2.38}$$

$$P_{freq} \triangleq \frac{1}{N} + \frac{2}{N}\sum_{m=1}^{\infty} \text{sinc}^2(m/N) = 1, \tag{2.39}$$

and we see that the expected value of the time-averaged power is the same in both the time and frequency domains as anticipated. The cumulative power of the coefficients is plotted in Fig. 2.5b, where it can be seen that 80% of the expected signal power lies within a bandwidth of $B_T/2$. Since the frequency increment in the Fourier Series expansion is

$$\Delta f = B_T/N = 1/NT, \tag{2.40}$$

a spectral-density coefficient can be defined that gives the power in a given harmonic, divided by the frequency spacing;

$$\frac{\overline{P_m}}{\Delta f} = \begin{cases} T & \text{for } m = 0 \\ 2T\text{sinc}^2(m/N) & \text{for } m \neq 0. \end{cases} \tag{2.41}$$

Interpretation of Results

It is interesting to note that the power spectrum of the random NRZ data contains a component at dc, even though the data is balanced around zero, and is equally likely to be positive, as it is to be negative. The emergence of a dc component arises because $\overline{P_m}$ was derived using an ensemble expectation operator which averages the square of one particular sample function of the random process. Therefore, we need a moment to clarify what is actually meant by the presence of a "dc" term. The interpretation of the results is, for every possible combination of sequences N-bits long, there will not always be an equal number of positive and negative pulses ("ones" and "zeros") in the bit sequence. The average dc value of this sequence will be the difference between the number of "ones" and "zeros" $\Delta_{1/0}$ divided by the total bit length N.

Naturally, any deterministic, periodic sequence will have a well defined dc value. If the number of positive pulses is equal to the number of negative pulses, then $\Delta_{1/0} = 0$, and the dc value is also zero. However, if the sequence is very long, and is broken into several subintervals of length N_s, then each of these subintervals will have an average value that may not be zero. Therefore it can be considered to have a "dc" component, if the idea of "dc" is interpreted to mean a frequency that varies slower than can be observed in the given measurement interval. For example: if we have a 65.5 KHz clock and a pseudo-random sequence of length, $N = 2^{16} - 1$, then the sequence will repeat once every second. If the "dc" value is measured once every 0.1 s, squared, and averaged over a one-second interval, then a non-zero result can occur.

Finding the dc Power in the Time-Domain Instead of finding $\overline{P_0}$ as in (2.37) we could do it in the time domain. The average dc power for the sample function corresponding to the random event ξ_1 is then

$$\overline{P_0(\xi_1)} = \left(\frac{\Delta_{1/0}(\xi_1)}{N}\right)^2, \tag{2.42}$$

and the expected value of the dc power for all possible sample functions is

$$\overline{P_0} = E\left[\left(\frac{\Delta_{1/0}(\cdot)}{N}\right)^2\right]. \tag{2.43}$$

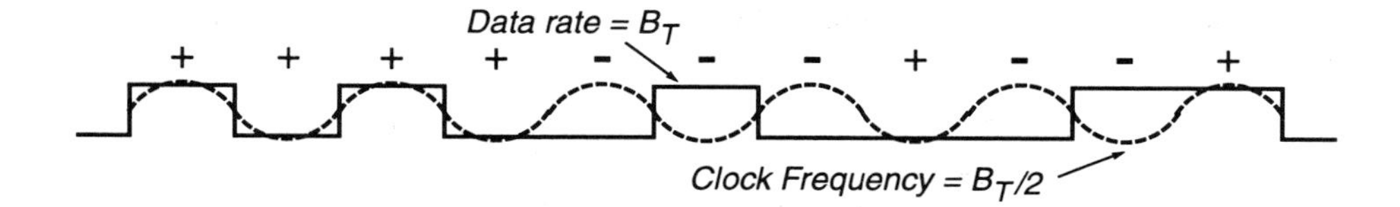

Figure 2.6 Random NRZ data and a tone with a frequency of one-half of the bit-rate.

This operation implies that we need to find the difference between the positive and negative pulses for all possible sequences and average the square. This seems an ominous task and it illustrates the power of the using the expectation operator and the statistical correlation between bits. $\overline{P_0}$ can be obtained as follows;

$$\overline{P_0} = E\left[\frac{1}{N}\sum_{n=0}^{N-1} r_n(\cdot)\right]^2 \tag{2.44}$$

$$= \frac{1}{N^2}\sum_{n=0}^{N-1}\sum_{k=0}^{N-1} E\left[r_n(\cdot)r_k(\cdot)\right] \tag{2.45}$$

$$= 1/N, \tag{2.46}$$

which is the same result obtained in the frequency domain. The result also shows that the dc value does approach zero, for equally likely ± 1 signals, as the length of the period NT is increased.

The reason for devoting a lengthy discussion to the power in the dc component is because the expected value of all harmonic power components $\overline{P_m}$ have a similar interpretation. The mean of the random Fourier coefficients $a_m(\cdot)$ and $b_m(\cdot)$ are in fact all equal to zero as can be seen from (2.14) and (2.19). Fig. 2.6 illustrates how a tone at half of the bit-rate lines up in phase with the data half of the time and is out of phase with the data the other half. Hence, there is no frequency component at any of the harmonics on average. However, any individual sample function will have bits that line-up in the appropriate order such that there is a residual component at, for instance, the 3rd, or the 7th, harmonic. When these values are squared and averaged over all possible sample functions, a non-zero average power $\overline{P_m}$ is obtained.

This has been a rather long journey to confirm the result that was anticipated in Fig. 2.4, that the PSD is a sinc^2 function in the ensemble average. In the sections to follow, these results will be considered within a larger framework so that similar results can be obtained much more readily.

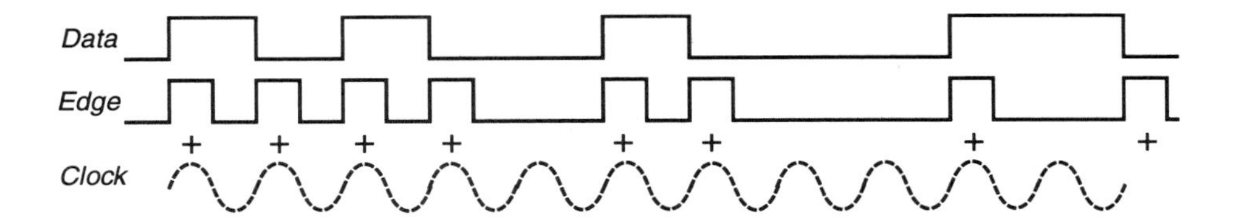

Figure 2.7 NRZ data, edge-detected data, and a tone at the bit-rate that is in phase-alignment with the edge-detected signal.

2.2.2 Fourier Series Representation of Edge-Detected Data

It was determined in the previous section that the PSD of NRZ data has a null at the bit-rate, but clearly, if bits are being transmitted at a rate of B_T, then there must be information about the bit-rate contained within the data signal itself. Since any linear operation on the data will always contain a null at the bit-rate, it is appropriate to consider what types of nonlinear operation can be performed on the data that will generate a tone at either B_T or a multiple of B_T. One such nonlinear operation is edge-detection. Since the data does not return to zero in the bit-interval, there is no discernible timing information contained in the data, unless adjacent bits are different. Therefore the timing information is contained in the transitions between adjacent, non-identical, bits. However, since this transition is equally likely to be positive as negative, at any time $t = nT$ for $[n = 1, 2, 3 \ldots]$, this random phase reversal, as was illustrated in Fig. 2.6, prevents the accumulation of a steady-state resonance if the data signal were applied directly to the input of a bandpass filter. It was found that a residual signal at half the bit-rate exists simply due the the random placement of bits in the proper phase, but this frequency component is small and indistinguishable from all of the other frequency components in the data signal. Noticing that the random phase reversals are what prevents a resonant circuit at $B_T/2$ from sustaining oscillation, we can devise an operation that removes the random phase reversals, and a signal with a strong clock component can be derived. Consider generating a positive pulse for a fraction of the bit period every time that a transition of the data is encountered. This operation is illustrated in Fig. 2.7. It can be seen that the edge-detected pulses $e_N(t, \cdot)$ are always in-phase with a resonant signal at the bit-rate; $e_N(t, \cdot)$ could be used as the input to a resonant circuit, and a sustained oscillation at a frequency of B_T would result. Therefore, one would expect to find a strong component in the frequency domain description of $e_N(t, \cdot)$ at B_T. The task at hand is to determine the PSD of $e_N(t, \cdot)$ and predict the power in the derived clock signal.

Functional Form of the Edge-Detected Signal The signal $e_N(t, \cdot)$ can be expressed in the same manner as the NRZ data. The fundamental pulse shape $e_T(t)$ is shown in

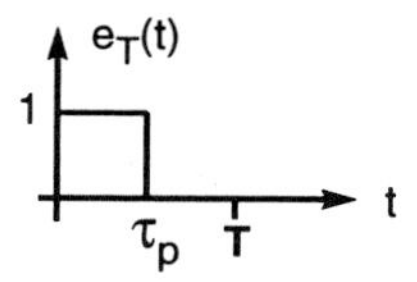

Figure 2.8 Pulse shape for edge-detected data

Fig. 2.8, where

$$e_T(t) = \begin{cases} 1 & \text{for } 0 \leq t \leq \tau_p \\ 0 & \text{elsewhere.} \end{cases} \tag{2.47}$$

A time limited signal derived from an N-bit data sequence is then given by

$$e(t, \cdot) = \sum_{n=0}^{N-1} s_n(\cdot) e_T(t - nT), \tag{2.48}$$

where $s_n(\cdot)$ is a random variable derived from two adjacent data bits. If the data bits are identical then $s_n(\cdot) = 0$, and $s_n(\cdot) = 1$ when the adjacent bits are different. It is clear that adjacent data bits are equally likely to be identical as they are different so that

$$s_n(\cdot) = \begin{cases} 1 & \text{with Probability } 1/2 \\ 0 & \text{with Probability } 1/2 \end{cases} \tag{2.49}$$

The random variable $s_n(\cdot)$ can be written in terms of a random variable $q_n(\cdot)$.

$$s_n(\cdot) = 1/2(1 + q_n(\cdot)), \tag{2.50}$$

where $q_n(\cdot)$ has statistics that are identical with the original data polarity random variable $r_n(\cdot)$.

$$q_n(\cdot) = \begin{cases} +1 & \text{with Probability } 1/2 \\ -1 & \text{with Probability } 1/2, \end{cases} \tag{2.51}$$

and

$$E[q_n(\cdot)] = 0 \tag{2.52}$$

$$E[q_n(\cdot) q_k(\cdot)] = \begin{cases} 1 & \text{for } n = k \\ 0 & \text{for } n \neq k. \end{cases} \tag{2.53}$$

Therefore, $e(t, \cdot)$ can be written as the sum of a deterministic signal and a random signal. This decomposition into two parts is illustrated in Fig. 2.9, and the signal can be expressed analytically as,

$$e(t, \cdot) = \frac{1}{2} \sum_{n=0}^{N-1} e_T(t - nT) + \frac{1}{2} \sum_{n=0}^{N-1} q_n(\cdot) e_T(t - nT). \tag{2.54}$$

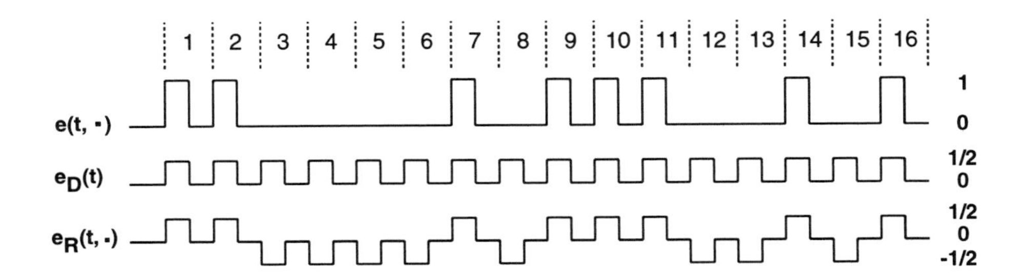

Figure 2.9 Decomposition of a edge-detected NRZ random signal into a deterministic and random parts.

The signal $e_N(t, \cdot)$ is now derived by repeating $e(t, \cdot)$ indefinitely.

Power Spectrum of the Edge-Detected Signal Since $e_N(t, \cdot)$ is periodic with a period of NT, it can be represented by a Fourier series with a fundamental frequency of B_T/N. Obtaining the Fourier series coefficients is a linear operation, so we can utilize superposition to determine the coefficients due to the deterministic part and random parts of the signal separately. We can write

$$e_N(t, \cdot) = e_{ND}(t) + e_{NR}(t, \cdot), \tag{2.55}$$

and the deterministic portion can be written as

$$e_{ND}(t) = \frac{a_{D0}}{2} + \sum_{M=1}^{\infty} a_{DM} \cos\left(\frac{2\pi M t}{T}\right) + \sum_{M=1}^{\infty} b_{DM} \sin\left(\frac{2\pi M t}{T}\right). \tag{2.56}$$

Evaluated a_{DM} directly from (2.6)

$$a_{DM} = \frac{2}{T}\int_0^T \frac{1}{2} e_T(t) \cos\left(\frac{2\pi M t}{T}\right) dt \tag{2.57a}$$

$$= \frac{1}{T}\int_0^{\tau_p} \cos\left(\frac{2\pi M t}{T}\right) dt \tag{2.57b}$$

$$= \frac{1}{2\pi M} \sin\left(\frac{2\pi M \tau_p}{T}\right) \tag{2.57c}$$

$$= \frac{1}{\pi M} \sin\left(\frac{\pi M \tau_p}{T}\right) \cos\left(\frac{\pi M \tau_p}{T}\right), \tag{2.57d}$$

and defining a pulse-width parameter

$$p \triangleq \tau_p/T, \tag{2.58}$$

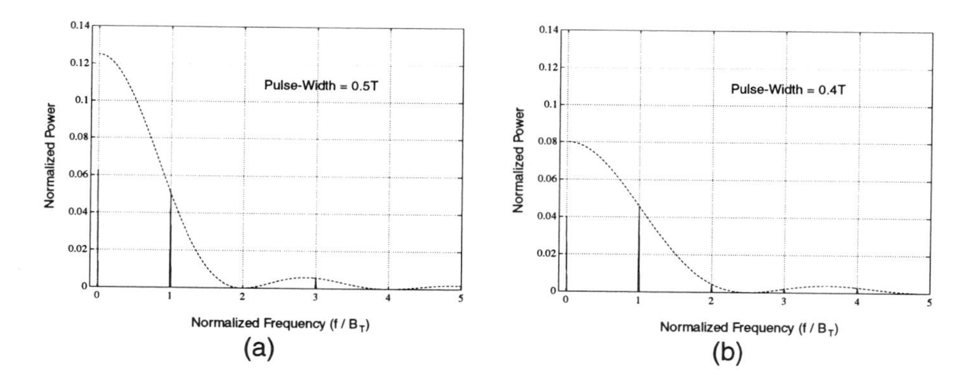

Figure 2.10 The power in harmonics of the deterministic portion of an edge-detected NRZ data signal: (a) $p = 0.5$, (b) $p = 0.4$. The dotted line is the sinc function envelope that is controlled by the shape of the edge-detected pulses.

the Fourier coefficients a_{DM} for the deterministic part are

$$a_{DM} = p\text{sinc}(Mp)\cos(\pi Mp). \tag{2.59a}$$

The quadrature coefficients b_{DM} are found similarly to be

$$b_{DM} = p\text{sinc}(Mp)\sin(\pi Mp). \tag{2.59b}$$

Clearly the power in each harmonic for the deterministic portion is

$$\overline{P_{DM}} = \begin{cases} \frac{p^2}{4} & \text{for } M = 0 \\ \frac{p^2}{2}\text{sinc}^2(Mp) & \text{for } M \neq 0 \end{cases} \tag{2.60}$$

where the frequency increment between harmonics is now B_T instead of B_T/N as in the previous section. Comparing the average power in the time and frequency domains,

$$\sum_{M=0}^{\infty} \overline{P_{DM}} = \frac{p}{4}\left[\sum_{M=1}^{\infty} 2p\text{sinc}^2(Mp) + p\right] = \frac{p}{4}. \tag{2.61}$$

This is equivalent to the time-domain result obtained by averaging the square of the deterministic pulse over one period T, which has a magnitude of $1/2$ over the time interval $[0, pT]$. The average power in the harmonics is plotted in Fig. 2.10 for $p = 0.5$ and $p = 0.4$. Notice that particular harmonics of the bit-rate can be nulled by choosing the value of p appropriately. For $p = 1/2.5 = 0.4$, there is a spectral null at multiples of $2.5B_T$ so that every 5th harmonic is zero. For the case of $p = 1/2$, all even harmonic are nulled.

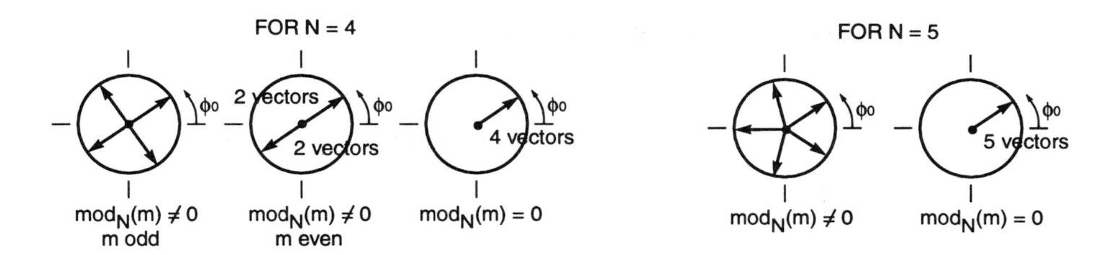

Figure 2.11 Graphical illustration showing the sum of cosines and sines as projects onto "x" and "y" axes of unit magnitude vectors.

We now want to add these deterministic Fourier coefficients with those obtained from the random portion of the signal. A problem arises in that the the random portion has a period that is N-times longer than the deterministic part. This can be fixed by stuffing zeros in the coefficients for the deterministic part at frequencies that are not a multiple of B_T. Alternatively we could have obtained this result directly by using a period of NT instead of T. In this case the Fourier series coefficients for the deterministic part are

$$a_{Dm} = \frac{p}{N} \text{sinc}\left(\frac{pm}{N}\right) \sum_{n=0}^{N-1} \cos\left(\frac{2\pi m}{N}(n + p/2)\right) \tag{2.62a}$$

$$b_{Dm} = \frac{p}{N} \text{sinc}\left(\frac{pm}{N}\right) \sum_{n=0}^{N-1} \sin\left(\frac{2\pi m}{N}(n + p/2)\right) \tag{2.62b}$$

The two summation terms in (2.62) are the sums of the projections onto the "x" and "y" axes respectively of N equally spaced vectors around the unit circle with an initial phase offset of $\phi_0(m) = \pi pm/N$ radians. This sum can be illustrated graphically, as in Fig. 2.11. Since the total phase between successive vectors is $2\pi m/N$, each new vector is traced by traversing the unit circle Integer(m/N) times, and then adding a phase increment of 2π times the remainder of m/N. For the case of N-odd, there are always N equally spaced vectors, and for N-even, the vectors continue to double-up, depending on whether N is divisible by a power of 2. In both cases the distribution of the vectors is symmetric, so that if these vectors were the spokes of a bicycle wheel that were balanced horizontally on its hub, then hanging an equal weight from from each of the spokes will keep the wheel in a level position. Due to this symmetry, the vector sum is equal to zero in these cases. If the spokes were distributed in an unsymmetrical way, then the wheel would tip over in the direction of the vector sum of the spokes. For the above summations, only in the case where m is a multiple of N ($\text{mod}_N(m) = 0$) do all the vectors line up on the offset angle and their sum accumulates to a non-zero value. Therefore the Fourier coefficients are given by

$$a_{Dm} = \begin{cases} \frac{p}{N}\text{sinc}\left(\frac{pm}{N}\right) N \cos\left(\frac{\pi pm}{N}\right) & \text{for mod}_N(m) = 0 \\ 0 & \text{for mod}_N(m) \neq 0 \end{cases} \tag{2.63a}$$

$$b_{Dm} = \begin{cases} \frac{p}{N}\text{sinc}\left(\frac{pm}{N}\right) N \sin\left(\frac{\pi pm}{N}\right) & \text{for mod}_N(m) = 0 \\ 0 & \text{for mod}_N(m) \neq 0. \end{cases} \tag{2.63b}$$

Clearly the average power $\overline{P_{Dm}}$ is the same as $\overline{P_{DM}}$ given in (2.60), where $M = m/N$.

Now we need to determine the Fourier coefficients for the random portion of $e_N(t, \cdot)$. This will be simple, however, since the form of the signal is very similar to the form of the random NRZ data itself. Recall that $e_{NR}(t, \cdot)$ is given by,

$$e_{NR}(t, \cdot) = \frac{1}{2} \sum_{n=0}^{N-1} q_n(\cdot) e_T(t - nT). \tag{2.64}$$

whereas the data itself is

$$d_N(t, \cdot) = \sum_{n=0}^{N-1} r_n(\cdot) p_T(t - nT). \tag{2.65}$$

It was shown in (2.11) that the Fourier coefficient $a_m(\cdot)$ for NRZ data is

$$a_m(\cdot) = \frac{2}{NT} \sum_{n=0}^{N-1} r_n(\cdot) \int_{nT}^{(n+1)T} \cos\left(\frac{2\pi mt}{NT}\right) dt, \tag{2.66}$$

whereas $a_{Rm}(\cdot)$ for the random part of the edge detected signal is

$$a_{Rm}(\cdot) = \frac{1}{NT} \sum_{n=0}^{N-1} q_n(\cdot) \int_{nT}^{(n+p)T} \cos\left(\frac{2\pi mt}{NT}\right) dt. \tag{2.67}$$

It was also shown in (2.13e) that $a_m(\cdot)$ could be simplified to

$$a_m(\cdot) = \frac{2\sin(\phi)}{\pi m} \sum_{n=0}^{N-1} r_n(\cdot) \cos(\theta_n + \phi), \tag{2.68}$$

where $\phi = \pi m/N$. A similar result can be obtained for $a_{Rm}(\cdot)$. Defining $\phi_p \triangleq \pi pm/N$, then

$$a_{Rm}(\cdot) = \frac{\sin(\phi_p)}{\pi m} \sum_{n=0}^{N-1} q_n(\cdot) \cos(\theta_n + \phi_p). \tag{2.69}$$

Therefore the expressions for the Fourier coefficient of $e_{NR}(t, \cdot)$ are

$$a_{Rm}(\cdot) = \frac{p}{N}\text{sinc}(pm/N) \sum_{n=0}^{N-1} q_n(\cdot) \cos\left(\frac{2\pi m}{N}(n + p/2)\right). \tag{2.70a}$$

$$b_{Rm}(\cdot) = \frac{p}{N}\text{sinc}(pm/N) \sum_{n=0}^{N-1} q_n(\cdot) \sin\left(\frac{2\pi m}{N}(n + p/2)\right), \tag{2.70b}$$

and since $q_n(\cdot)$ has identical statistics of $r_n(\cdot)$, then the expected value of the time averaged power is

$$\overline{P_{Rm}} = \begin{cases} \frac{p^2}{4N} & \text{for } m = 0 \\ \frac{p^2}{2N}\text{sinc}^2(pm/N) & \text{for } m \neq 0. \end{cases} \tag{2.71}$$

Now we have derived the power in the deterministic and random parts. All that remains is to find the average power in the total signal. From superposition we know that

$$\begin{aligned} a_m(\cdot) &= a_{Dm} + a_{Rm}(\cdot) \\ b_m(\cdot) &= b_{Dm} + b_{Rm}(\cdot). \end{aligned} \tag{2.72}$$

The time averaged power is then

$$\begin{aligned} \overline{P_m(\cdot)} = {} & \frac{1}{2}(a_{Dm}^2 + 2a_{Dm}a_{Rm}(\cdot) + a_{Rm}^2(\cdot)) \\ & + \frac{1}{2}(b_{Dm}^2 + 2b_{Dm}b_{Rm}(\cdot) + b_{Rm}^2(\cdot)). \end{aligned} \tag{2.73}$$

Taking the expected value gives

$$\overline{P_m} = \overline{P_{Dm}} + \overline{P_{Rm}} + a_{Dm}E[a_{Rm}(\cdot)] + b_{Dm}E[b_{Rm}(\cdot)],$$

and since the expected values of $a_{Rm}(\cdot)$ and $b_{Rm}(\cdot)$ are zero, then the total power is just the sum of the two individual powers,

$$\overline{P_m} = \overline{P_{Dm}} + \overline{P_{Rm}}. \tag{2.74}$$

Therefore, the expected value of the time averaged power per harmonic of the edge detected signal is

$$\boxed{\overline{P_m} = \begin{cases} \frac{p^2}{4}[1 + 1/N] & \text{for } m = 0 \\ \frac{p^2}{2}[1/N]\text{sinc}^2(pm/N) & \text{for } m \neq 0, \quad \text{mod}_N(m) \neq 0 \\ \frac{p^2}{2}[1 + 1/N]\text{sinc}^2(pm/N) & \text{for } m \neq 0 \quad \text{mod}_N(m) = 0. \end{cases}} \tag{2.75}$$

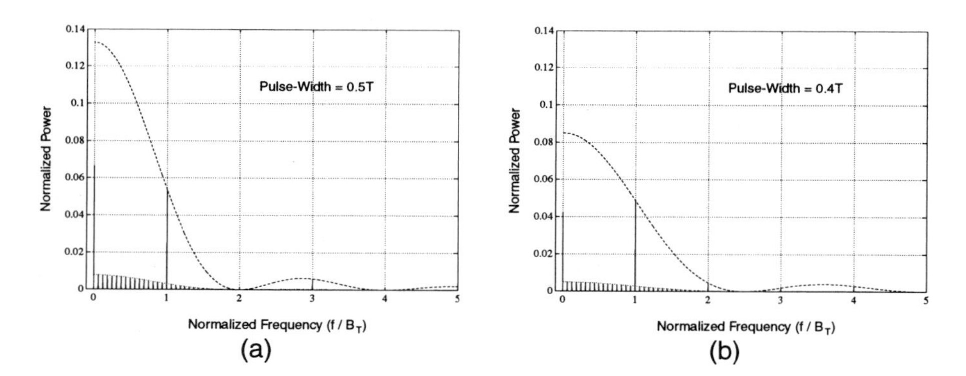

Figure 2.12 The power in harmonics of an edge-detected NRZ data signal for: (a) $p = 0.5$, (b) $p = 0.4$.

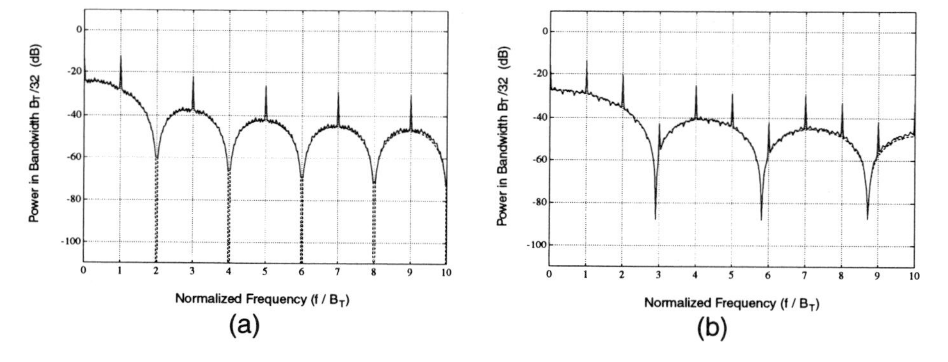

Figure 2.13 Simulated and calculated power in harmonics of an edge-detected NRZ data signal for: (a) $p = 0.5$, (b) $p = 0.3438$.

This power spectrum is plotted in Fig. 2.12 for $N = 16$. It can be seen that the large spikes at multiples of the bit-rate are due to the deterministic part, and the power in the random part is spread more uniformly over all frequencies. This analytical expression can be verified in simulation. A discrete-time rectangular NRZ data sequence was generated using a a sampling interval of 32 samples-per-bit. A pulse of width pT was generated whenever a transition in the data occurred. A Discrete Fourier Transform DFT was taken from the edge-detected data. Since the frequency interval for the DFT is $\Delta f = B_T/32$, then the PSD from the simulated data was compared to the calculated value for $N = 32$. The results are plotted in Fig. 2.13, where it can be seen that the simulated value is coincident with the calculated value. This result is consistent with expectations based on arguments about resonant circuits. Once the random phase reversals have been removed from the data, a sustained oscillation can appear at the output of a bandpass filter tuned to the data-rate. The edge-detected signal must, therefore, have a strong spectral component at B_T. This was indeed found to be true, and it was also found that by varying the pulse-width pT, the relative magnitudes of the harmonics of the clock could be altered. Specifically, for $p = 1/2$, all even clock harmonics are nulled.

2.2.3 Summary of Fourier Series Analysis

Thus far a Fourier series representation of rectangular NRZ data, and an edge-detected signal derived from this data have been found. It was shown that the coefficients of the Fourier series are random variables, and are linear combinations of the random data. The ensemble average of each of these coefficients is zero, because it was assumed the data was equally likely to be positive as negative. Meaningful results of the frequency content of the random signal were obtained by finding the time-averaged power in each harmonic for a given sample function of the random process, and then taking the expected value of this power over all possible sample functions of the ensemble. The result of this operation is analogous to what one would observe in the laboratory, if a long random sequence is input to a spectrum analyzer, and the display of the spectrum analyzer is set to average several sweeps.

Although the calculations in the previous section were straightforward, with unambiguous interpretations, they were also quite cumbersome. The results have been obtained directly without introducing negative frequencies, impulse functions, or several subtle concepts from the general theory of random signals such as: autocorrelation functions, stationarity, cyclo-stationarity and ergodicity. However each of these concepts have been referenced implicitly. In the following section the above concepts will be introduced, and the results obtained thus far will be placed in a more general framework so that the effect of further processing can be determined quickly. The goal of this chapter is to develop an intuition about random data and data-derived signals, so that the frequency content of such a random signal can be determined almost by inspection. As with any useful intuition, it must be based on a solid grasp of fundamental concepts. In this section the foundations have been laid for more sophisticated analysis to come.

2.3 FOURIER TRANSFORM FREQUENCY DOMAIN REPRESENTATIONS

In the previous section we used the Fourier series to represent a pseudo-random rectangular NRZ data sequence that repeated every N bits. It was found that the signal had a discrete power spectrum with power only at frequencies of $\Delta f = mB_T/N = m/NT$ where $m = [1, 2, 3, \ldots]$. As the length of the period NT is increased, the frequency increments get closer together. The amplitude of the power in any given harmonic is reduced by N as the average power per period becomes distributed over more and more frequencies. If the power coefficients are divided by the frequency interval, then the amplitude of the coefficient is independent of N, and gives the power normalized to a one Hertz bandwidth. Dividing by the frequency interval is equivalent to multiplying by NT, so the resulting coefficients can also be considered as the total energy in N bits of the pseudo-random sequence.

In the previous section it was also found that using complex numbers to represent the in-phase and quadrature components of the coefficients simplified the analysis. The Fourier series was defined such that

$$f(t) = \frac{a_0}{2} + \sum_{m=1}^{\infty} a_m \cos\left(\frac{2\pi mt}{T}\right) + \sum_{m=1}^{\infty} b_m \sin\left(\frac{2\pi mt}{T}\right). \tag{2.76}$$

Letting

$$\omega_m \triangleq \frac{2\pi m}{T}, \tag{2.77}$$

and substituting the following identities

$$\cos(\omega_m t) \equiv \frac{e^{j\omega_m t} + e^{-j\omega_m t}}{2} \tag{2.78a}$$

$$\sin(\omega_m t) \equiv \frac{e^{j\omega_m t} - e^{-j\omega_m t}}{2j}, \tag{2.78b}$$

then the Fourier series becomes

$$f(t) = \frac{a_0}{2} + \frac{1}{2}\sum_{m=1}^{\infty} (a_m - jb_m)\, e^{j\omega_m t} + \frac{1}{2}\sum_{m=1}^{\infty} (a_m + jb_m)\, e^{-j\omega_m t}. \tag{2.79}$$

A complex coefficient can be defined as in (2.25), such that

$$c_m = \frac{a_m - jb_m}{2}. \tag{2.80}$$

Using this complex coefficient, the Fourier series can be expressed as

$$f(t) = \frac{a_0}{2} + \sum_{m=1}^{\infty} c_m e^{j\omega_m t} + \sum_{m=1}^{\infty} c_m^* e^{-j\omega_m t}. \tag{2.81}$$

The complex coefficient c_m can be extracted from the defining equations for a_m and b_m.

$$a_m - jb_m = \frac{2}{T}\int_0^T f(t)\left[\cos(\omega_m t) - j\sin(\omega_m t)\right] dt \tag{2.82}$$

$$c_m = \frac{1}{T}\int_0^T f(t)e^{-j\omega_m t}. \tag{2.83}$$

For m negative we can write

$$c_{-m} = \frac{1}{T}\int_0^T f(t)e^{j\omega_m t} \tag{2.84}$$

$$c_{-m} = \frac{1}{T}\int_0^T f(t)\left[\cos(\omega_m t) + j\sin(\omega_m t)\right] dt \tag{2.85}$$

$$c_{-m} = \frac{a_m + jb_m}{2} = c_m^*. \tag{2.86}$$

So the complex coefficients display conjugate symmetry. Therefore, summing $c_m^* e^{-j\omega_m t}$ over positive frequencies is the same as summing $c_m e^{j\omega_m t}$ over negative frequencies, and

$$f(t) = \frac{a_0}{2} + \sum_{m=1}^{\infty} c_m e^{j\omega_m t} + \sum_{m=-\infty}^{-1} c_m e^{j\omega_m t}. \tag{2.87}$$

Since the dc coefficient $c_0 = a_0/2$, then the Fourier series can be written compactly in complex form as

$$\boxed{\begin{aligned} f(t) &= \sum_{m=-\infty}^{\infty} c_m e^{j\omega_m t} \\ c_m &= \frac{1}{T}\int_{-T/2}^{T/2} f(t)e^{-j\omega_m t} dt, \end{aligned}} \tag{2.88}$$

where

$$c_m = \frac{a_m - jb_m}{2} \quad \text{and} \quad |c_m|^2 = \frac{a_m^2 + b_m^2}{4}. \tag{2.89}$$

The power in a real signal at the mth harmonic is

$$\overline{P_m} = |c_m|^2 + |c_m^*|^2 \quad = |c_m|^2 + |c_{-m}|^2 \quad = 2|c_m|^2 \quad \text{for} \quad m \neq 0, \tag{2.90a}$$

and the dc power is just

$$\overline{P_0} = |c_0|^2. \tag{2.90b}$$

Parseval's theorem relating the power in the time and frequency domains is also expressed compactly using the complex coefficient.

$$\overline{P} = \sum_{m=-\infty}^{\infty} |c_m|^2 = \frac{1}{T}\int_{-T/2}^{T/2} |f(t)|^2 dt. \tag{2.91}$$

The frequency interval between successive harmonics is $\Delta f = 1/T$, and the periodic time function can be written as

$$f(t) = T \sum_{m=-\infty}^{\infty} c_m e^{j2\pi f_m t} \Delta f. \tag{2.92}$$

An energy spectral density coefficient can be defined as

$$e_m \triangleq T c_m \tag{2.93}$$

so that the interpretation of this coefficient is that $2|e_m|^2 \Delta f = T P_m$ which is the total energy over the time T of the mth harmonic, and $2|e_m|^2$ is the energy per unit Hertz. In terms of the energy spectral density coefficients, the Fourier series can be expressed as

$$e_m = \int_{-T/2}^{T/2} f(t) e^{-j2\pi f_m t} dt \tag{2.94a}$$

$$f(t) = \sum_{m=-\infty}^{\infty} e_m e^{j2\pi f_m t} \Delta f \tag{2.94b}$$

Fourier Transform as Limiting Case of Fourier Series In the limit as $T \to \infty$ the frequency interval $\Delta f \to 0$, and e_m becomes a continuous function of frequency $F(j2\pi f)$ known as the Fourier Transform of $f(t)$, where the defining relationships are

$$\boxed{\begin{aligned} F(j2\pi f) &= \int_{-\infty}^{\infty} f(t) e^{-j2\pi f t} dt \\ f(t) &= \int_{-\infty}^{\infty} F(j2\pi f) e^{j2\pi f t} df \end{aligned}} \tag{2.95}$$

The portion of the time averaged signal energy contained within the bandwidth from f_1 to f_2 is

$$\overline{E_{f_1 f_2}} = 2\int_{f_1}^{f_2} |F(j2\pi f)|^2 df. \tag{2.96}$$

$f(t)$	$\leftrightarrow$	$F(j2\pi f)$
$f(t-t_0)$	$\leftrightarrow$	$e^{-j2\pi f t_0}F(j2\pi f)$
$f(t)e^{j2\pi f_0 t}$	$\leftrightarrow$	$F(j2\pi(f-f_0))$
$f(at)$	$\leftrightarrow$	$\frac{1}{\lvert a\rvert}F(j2\pi f/a)$
$\frac{df(t)}{dt}$	$\leftrightarrow$	$j2\pi fF(j2\pi f)$
$\int_{-\infty}^{t} f(\tau)d\tau$	$\leftrightarrow$	$\frac{F(j2\pi f)}{j2\pi f}+\frac{1}{2}F(0)\delta(f)$
$f_1(t)*f_2(t)$	$\leftrightarrow$	$F_1(j2\pi f)F_2(j2\pi f)$
$f_1(t)f_2(t)$	$\leftrightarrow$	$F_1(j2\pi f)*F_2(j2\pi f)$
$f^*(t)$	$\leftrightarrow$	$F^*(-j2\pi f)$
$\int_{-\infty}^{\infty}\lvert f(t)\rvert^2dt$	$=$	$\int_{-\infty}^{\infty}\lvert F(j2\pi f)\rvert^2df$

Table 2.1 Properties of the Fourier transform.

Certainly this information is nothing new. Excellent treatments of Fourier analysis can be found in [7, 8, 9, 10] and countless other texts. However, various authors define the Fourier transform and Fourier series coefficients differently with constants of 2π and T popping in and out unexpectedly like unwanted guests. The previous discussion has provided a unified development of the Fourier series and Fourier transform with clear connections between $F(j2\pi f)$ and c_m. These connections are extremely important when interpreting spectral density results for random data. Table 2.1 lists several properties of the Fourier transform that will be used extensively. Table 2.2 lists some commonly used transform pairs. The last transform pair is particularly interesting. It shows that the Gaussian is an eigenfunction of the Fourier transform. In addition to the above tables it is also useful to know that the integral of a sinc function and a sinc^2 function are both equal to unity according to

$$T\int_{-\infty}^{\infty}\text{sinc}(fT)df = 1 \tag{2.97}$$

$$T\int_{-\infty}^{\infty}\text{sinc}^2(fT)df = 1. \tag{2.98}$$

$\delta(t)$	$\leftrightarrow$	1
1	$\leftrightarrow$	$\delta(f)$
$u(t)$	$\leftrightarrow$	$\frac{1}{j2\pi f} + \frac{1}{2}\delta(f)$
$\text{sgn}(t)$	$\leftrightarrow$	$\frac{1}{j\pi f}$
$\text{rect}(t/T)$	$\leftrightarrow$	$T\text{sinc}(fT)$
$\text{sinc}(Ft)$	$\leftrightarrow$	$\frac{1}{F}\text{rect}(f/F)$
$e^{j2\pi f_0 t}$	$\leftrightarrow$	$\delta(f - f_0)$
$\cos(2\pi f_0 t)$	$\leftrightarrow$	$\frac{1}{2}\delta(f - f_0) + \frac{1}{2}\delta(f + f_0)$
$\sin(2\pi f_0 t)$	$\leftrightarrow$	$\frac{1}{2j}\delta(f - f_0) - \frac{1}{2j}\delta(f + f_0)$
$e^{-\|t\|/\tau}$	$\leftrightarrow$	$\frac{2\tau}{1+(2\pi f\tau)^2}$ for $\tau > 0$
$\frac{1}{\sqrt{2\pi}\sigma}e^{-\frac{1}{2}\left(\frac{t-\mu}{\sigma}\right)^2}$	$\leftrightarrow$	$e^{-j2\pi f\mu}e^{-\frac{1}{2}(2\pi f\sigma)^2}$

Table 2.2 Fourier transform pairs.

$\lim_{T\to\infty} \mathrm{rect}(t/T)$	$\leftrightarrow$	$\lim_{T\to\infty} T\mathrm{sinc}(fT) = \delta(f)$
$\lim_{T\to\infty} \mathrm{rect}(t/T) * \frac{1}{T}\mathrm{rect}(t/T)$	$\leftrightarrow$	$\lim_{T\to\infty} T\mathrm{sinc}^2(fT) = \delta(f)$
$\lim_{F\to 0} \mathrm{sinc}(Ft)$	$\leftrightarrow$	$\lim_{F\to 0} \frac{1}{F}\mathrm{rect}(f/F) = \delta(f)$
$\lim_{\sigma\to\infty} e^{-\frac{1}{2}\left(\frac{t}{\sigma}\right)^2}$	$\leftrightarrow$	$\lim_{\sigma\to\infty} \sqrt{2\pi}\sigma e^{-\frac{1}{2}(2\pi f\sigma)^2} = \delta(f)$

Table 2.3 Various equivalent forms of the impulse function.

These and other unit integrals can be used in the limit to represent an impulse function, where the impulse $\delta(t)$ is defined by

$$\delta(t) = \begin{cases} 0 & \text{for } t \neq 0 \\ \text{undefined} & \text{for } t = 0 \end{cases} \tag{2.99a}$$

$$\int_{0^-}^{0^+} \delta(t)dt = 1. \tag{2.99b}$$

Several equivalent representations of an impulse function are given in table 2.3. The time domain functions begin with: a rectangular pulse, a triangular pulse, a sinc pulse, and a Gaussian pulse. Each of these pulse are stretched wider and wider in time so that in the limit, the result is a dc value of unity. In the frequency domain the Fourier transforms are: a sinc pulse, a sinc^2 pulse, a rectangle, and a Gaussian, respectively. Each of these functions get narrower in frequency and approach an impulse in the limit.

2.3.1 Fourier Transform of NRZ data

Now we can return to the problem of finding the energy spectrum of random rectangular NRZ data, taking advantage of the properties of the Fourier transform to simplify the analysis. In the previous section a random data segment N-bits long was repeated indefinitely. For the the Fourier transform to exist the signal must have finite energy, so we will deal only with one period of the pseudo-random sequence, but we can let the period grow arbitrarily large. The NRZ data is given by

$$d(t,\cdot) = \sum_{n=0}^{N-1} r_n(\cdot)p_T(t-nT). \tag{2.100}$$

The Fourier transform of this time-limited data sequence is

$$F_d(j2\pi f, \cdot) = \int_{-\infty}^{\infty} \sum_{n=0}^{N-1} r_n(\cdot) p_T(t - nT) e^{-j2\pi ft} dt, \tag{2.101}$$

and this can be written as the sum of integrals just as (2.11)

$$F_d(j2\pi f, \cdot) = \sum_{n=0}^{N-1} r_n(\cdot) \int_{nT}^{(n+1)T} e^{-j2\pi ft} dt, \tag{2.102}$$

and after evaluating the integral

$$F_d(j2\pi f, \cdot) = T\text{sinc}(fT) \sum_{n=0}^{N-1} r_n(\cdot) e^{-j2\pi f(n+1/2)T} dt. \tag{2.103}$$

This is the same functional form obtained for the Fourier series coefficients in (2.26), except that now the sinc function is continuous in frequency and not simply the envelope of discrete coefficients. Since $F_d(j2\pi f, \cdot)$ is a random variable, we can take the expected value by ensemble averaging as was done for the Fourier series coefficients. Defining the energy spectral density of the data sequence as

$$S_d(f) \triangleq E\left[|F_d(j2\pi f, \cdot)|^2\right] = E\left[F_d(j2\pi f, \cdot) F_d^*(j2\pi f, \cdot)\right], \tag{2.104}$$

then

$$S_d(f) = T^2 \text{sinc}^2(fT) \sum_{n=0}^{N-1} \sum_{m=0}^{N-1} E[r_n(\cdot) r_m(\cdot)] e^{-j2\pi f(n-m)T}. \tag{2.105}$$

Since the data is uncorrelated, the double sum becomes

$$\sum_{n=0}^{N-1} \sum_{m=0}^{N-1} E[r_n(\cdot) r_m(\cdot)] e^{-j2\pi f(n-m)T} = \sum_{n=0}^{N-1} e^{j0} = N. \tag{2.106}$$

Therefore, the energy spectral density for the random NRZ data is

$$\boxed{S_d(f) \quad = \quad NT^2 \text{sinc}^2(fT).} \tag{2.107}$$

This result is consistent with the total energy in the time domain

$$\int_{-\infty}^{\infty} d^2(t, \cdot) dt = NT \tag{2.108a}$$

$$NT \int_{-\infty}^{\infty} T\text{sinc}^2(fT) df = NT \tag{2.108b}$$

It is useful to define an energy spectral density that is normalized to give the average energy in each bit.

$$\boxed{S_{Bd}(f) \quad = \quad T^2\text{sinc}^2(fT).} \tag{2.109}$$

This is simply $S_d(f)/N$. As expected the result depends of T, but is independent of the length of the data sequence N.

Comparison of Results with the Periodic Case We can compare this result with the periodic case of section 2.2. Recall from (2.37) that the power in the mth harmonic is

$$\overline{P_m} = \begin{cases} \frac{1}{N} & \text{for } m = 0 \\ \frac{2}{N}\text{sinc}^2(m/N) & \text{for } m \neq 0 \end{cases} \tag{2.110}$$

where the frequency interval is $\Delta f = 1/NT$, and the frequency $f_m = m/NT$; the energy in one fundamental period NT is $NT\overline{P_m}$, and the average-energy-per-bit $\overline{E_{Bm}}$ in a time NT is simply $T\overline{P_m}$, so that

$$\overline{E_{Bm}} = \begin{cases} \frac{T}{N} & \text{for } m = 0 \\ \frac{2T}{N}\text{sinc}^2(f_m T) & \text{for } m \neq 0 \end{cases} \tag{2.111}$$

The energy in a bandwidth of Δf in the signal $d(t,\cdot)$ can be found by integrating $S_d(f)$.

$$\overline{E_{Bd}}\Big|_{f_m-\frac{\Delta f}{2}}^{f_m+\frac{\Delta f}{2}} = 2\int_{f_m-\Delta f/2}^{f_m+\Delta f/2} T^2\text{sinc}^2(fT)df \tag{2.112a}$$

$$\simeq 2T^2\text{sinc}^2(f_m T)\Delta f \quad \text{for } \Delta f \text{ small} \tag{2.112b}$$

$$= \frac{2T}{N}\text{sinc}^2(f_m T) \qquad \text{for} \quad m \neq 0 \tag{2.112c}$$

$$= \frac{T}{N} \qquad \text{for} \quad m = 0. \tag{2.112d}$$

Therefore, the continuous energy spectrum per bit $S_{Bd}(f)$ multiplied by the incremental bandwidth is equivalent to the discrete Fourier series coefficients multiplied by the bit-period. It is also useful to define an average power spectral density $P_d(f)$ for the time-limited data sequence as the average-energy-per-bit divided by T. Therefore,

$$\boxed{P_d(f) \quad = \quad T\text{sinc}^2(fT)} \tag{2.113}$$

which is the envelope of the Fourier coefficient $E[|c_m(\cdot)|^2]/\Delta f$

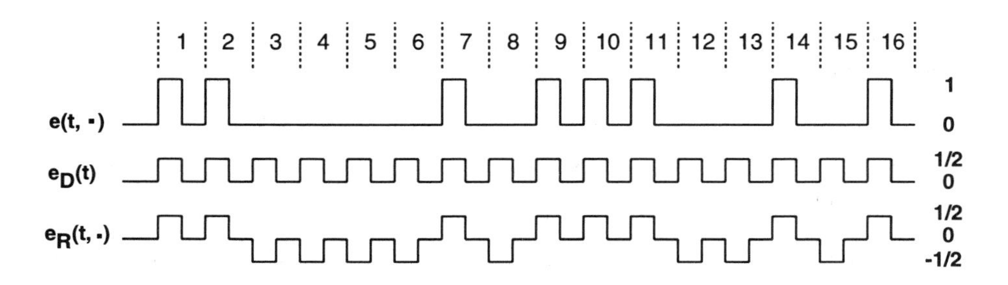

Figure 2.14 Decomposition of a edge-detected NRZ random signal into a deterministic and random parts.

2.3.2 Fourier Transform of Edge-Detected Data

Energy Spectrum of the Deterministic Part of Edge-Detected Data Now we can use the Fourier transform to find the energy spectral density of the edge-detected data analyzed in section 2.2.2. The random pulses $e(t, \cdot)$ were separated into the sum of a random and deterministic part. This separation was illustrated in Fig. 2.9 and is repeated here in Fig 2.14 for convenience. It was found that the deterministic part had a Fourier series representation. Therefore the time limited signal $e_D(t)$ can be written as the product of the periodic Fourier series and a rectangular windowing function $W_{NT}(t)$. From (2.59)

$$e_D(t) = W_{NT}(T) \sum_{M=-\infty}^{\infty} c_M e^{\frac{j2\pi Mt}{T}} \tag{2.114a}$$

$$c_M = \frac{p}{2}\text{sinc}(Mp)e^{-j\pi Mp}, \tag{2.114b}$$

where,

$$W_{NT}(t) = \text{rect}\left(\frac{t - NT/2}{NT}\right) = \begin{cases} 1 & \text{for } 0 \le t \le NT \\ 0 & \text{elsewhere.} \end{cases} \tag{2.114c}$$

Since the Fourier transform is a linear operator, the Fourier transform $F_{e_{DN}}(j2\pi f)$ of the periodic pulse stream is the sum of the Fourier transforms of $\exp(j2\pi Mt/T)$ weighted by the coefficients c_M. Therefore,

$$F_{e_{DN}}(j2\pi f) = \sum_{M=-\infty}^{\infty} \frac{p}{2}\text{sinc}(Mp)e^{-j\pi Mp}\delta\left(f - \frac{M}{T}\right), \tag{2.115}$$

and since the impulses are non-zero only at one point, the sinc function can be expressed as a continuous envelope.

$$F_{e_{DN}}(j2\pi f) = \frac{p}{2}\text{sinc}(fpT)e^{-j\pi fpT} \sum_{M=-\infty}^{\infty} \delta\left(f - \frac{M}{T}\right), \tag{2.116}$$

The Fourier transform of the windowing function can be found from the rectangular function entry of table 2.2. Using this result, together with the time shifting property of table 2.1;

$$F_{W_{NT}}(j2\pi f) = NT\text{sinc}(fNT)e^{j\pi fNT}. \tag{2.117}$$

The Fourier transform of $e_D(t)$ is then found using the multiplication property of table 2.1, such that

$$F_{e_D}(j2\pi f) = \mathcal{F}[e_{DN}(t)W_{NT}(t)] = F_{e_{DN}}(j2\pi f) * F_{W_{NT}}(j2\pi f). \tag{2.118}$$

Recalling that convolving a function with an impulse just shifts that function to the center of the impulse,

$$f(t) * \delta(t - t_0) = f(t - t_0), \tag{2.119}$$

then

$$F_{e_D}(j2\pi f) = \frac{p}{2}\text{sinc}(fpT)e^{-j\pi fpT} \sum_{M=-\infty}^{\infty} NT\text{sinc}\left(\left(f - \frac{M}{T}\right)NT\right) e^{j\pi\left(f - \frac{M}{T}\right)NT}. \tag{2.120}$$

This shows that the Fourier transform of the time limited pulses is the same as for the periodic function except the impulse functions have been replaced with narrow sinc functions. The width of the envelope sinc pulse is determined by the pulse width pT and the narrow sinc pulse width is controlled by the length of the sequence N. When N is large the narrow since pulses have most of their energy concentrated in a small bandwidth around harmonics of the bit-rate, and there is minimal interaction between adjacent sinc pulses. Therefore the energy spectral density

$$S_{e_D}(f) = \left[\frac{p}{2}\text{sinc}(fpT)\right]^2 \left| \sum_{M=-\infty}^{\infty} NT\text{sinc}\left(\left(f - \frac{M}{T}\right)NT\right) e^{j\pi\left(f - \frac{M}{T}\right)NT} \right|^2 \tag{2.121}$$

can be approximated for large N as

$$S_{e_D}(f) = \left[\frac{p}{2}\text{sinc}(fpT)\right]^2 \sum_{M=-\infty}^{\infty} N^2T^2\text{sinc}^2\left(\left(f - \frac{M}{T}\right)NT\right). \tag{2.122}$$

This energy spectrum is plotted in Fig. 2.15 for $p = 0.4$, and $N = 16$.

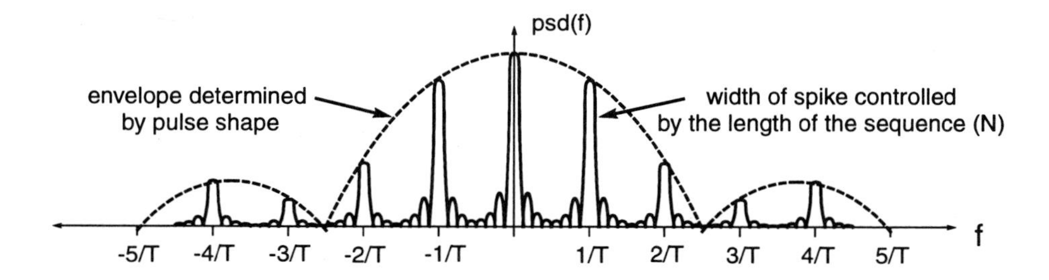

Figure 2.15 The energy spectral density of the deterministic part of an edge-detected 16-bit data stream, where detected pulse-width is $T/2.5$.

Energy Spectrum of the Random Part of Edge-Detected Data For the random part $e_R(t,\cdot)$ the energy spectrum is the same as for the NRZ data with T replaced by pT and the amplitude reduced by $1/2$. Whereas the Fourier transform for the NRZ data in (2.103) is

$$F_d(j2\pi f,\cdot) = T\text{sinc}(fT)\sum_{n=0}^{N-1} r_n(\cdot)e^{-j2\pi f(n+1/2)T}dt, \tag{2.123}$$

the Fourier transform of $e_R(t,\cdot)$ is

$$F_{e_R}(j2\pi f) = \frac{pT}{2}\text{sinc}(fpT)\sum_{n=0}^{N-1} q_n(\cdot)e^{-j2\pi f(n+p/2)T}dt. \tag{2.124}$$

The energy spectrum of $e_R(t,\cdot)$ is then

$$S_{e_R}(f) = N\left[\frac{pT}{2}\text{sinc}(fpT)\right]^2. \tag{2.125}$$

The spectrum of the random part of the signal is a factor of N less than the envelope of the spectrum for the deterministic part.

Using the principle of superposition to find the energy spectrum of the total signal $e(t,\cdot)$

$$F_e(j2\pi f,\cdot) = F_{e_D}(j2\pi f) + F_{e_R}(j2\pi f,\cdot), \tag{2.126}$$

and the expected value of the energy spectrum is then

$$S_e(f) = E\left[\left[F_{e_D}(j2\pi f) + F_{e_R}(j2\pi f,\cdot)\right]\left[F^*_{e_D}(j2\pi f) + F^*_{e_R}(j2\pi f,\cdot)\right]\right] \tag{2.127}$$

Expanding this we obtain

$$S_e(f) = S_{e_D}(f) + S_{e_R}(f) + F_{e_D}(j2\pi f)E\left[F^*_{e_R}(j2\pi f,\cdot)\right] + F^*_{e_D}(j2\pi f)E\left[F_{e_R}(j2\pi f,\cdot)\right], \tag{2.128}$$

and since the random variable $F_{e_R}(j2\pi f,\cdot)$ is obtained from a linear combination of zero-mean random variables, then

$$E\left[F_{e_R}(j2\pi f,\cdot)\right] = \frac{pT}{2}\text{sinc}(fpT)\sum_{n=0}^{N-1} E[q_n(\cdot)]e^{-j2\pi f(n+p/2)T}dt = 0, \tag{2.129}$$

and the total energy spectrum is just the sum of the individual energy spectrums

$$S_e(f) = S_{e_D}(f) + S_{e_R}(f). \tag{2.130}$$

Therefore the energy spectral density of $e(t,\cdot)$ for large N is given by

$$\boxed{S_e(f) = N\left[\frac{pT}{2}\text{sinc}(fpT)\right]^2\left[1 + N\sum_{M=-\infty}^{\infty}\text{sinc}^2\left(\left(f-\frac{M}{T}\right)NT\right)\right].} \tag{2.131}$$

For very large N the energy in the narrow sinc pulses will be concentrated in a very small bandwidth and can be approximated as an impulse function with all of its energy concentrated at one frequency. From (2.97)

$$\int_{-\infty}^{\infty}\text{sinc}^2(fNT)df = \frac{1}{NT} \tag{2.132}$$

therefore,

$$N\sum_{M=-\infty}^{\infty}\text{sinc}^2\left(\left(f-\frac{M}{T}\right)NT\right) \simeq \frac{1}{T}\sum_{M=-\infty}^{\infty}\delta\left(f-\frac{M}{T}\right). \tag{2.133}$$

The energy spectral density per bit is obtained by dividing $S_e(f)$ by the number of bits. The final results is then

$$\boxed{S_{Be}(f) = \left[\frac{pT}{2}\text{sinc}(fpT)\right]^2\left[1 + \frac{1}{T}\sum_{M=-\infty}^{\infty}\delta\left(f-\frac{M}{T}\right)\right],} \tag{2.134}$$

and dividing by T gives the power spectral density

$$\boxed{P_e(f) = \left[\frac{p}{2}\text{sinc}(fpT)\right]^2\left[T + \sum_{M=-\infty}^{\infty}\delta\left(f-\frac{M}{T}\right)\right].} \tag{2.135}$$

2.3.3 Power-Spectral Densities of Various PCM Signals

A general expression for the power spectra of signals modulated by random, and pseudorandom data sequences can be derived by representing the random data sequence as a Markov process with a known transition probability matrix. This analysis was first reported in 1961 by Titsworth and Welch in a Jet Propulsions Laboratory Technical Report [6], and was summarized later in a book by Lindsey and Simon [11, sec. 1-5]. The general expression is rather complicated, and requires several definitions that will not be discussed here. For the special case of binary, symmetric, equally likely signals, ($s_1(t) = -s_0(t) = s(t)$) the general expression for the PSD reduces to to the simple result

$$P(f) = \frac{1}{T}|F_s(j2\pi f)|^2, \tag{2.136}$$

where $F_s(j2\pi f)$ is the Fourier transform of the data pulse $s(t)$. When the data pulse is rectangular,

$$s(t) = \text{rect}(t/T), \tag{2.137}$$

the Fourier transform is given by

$$F_s(j2\pi f) = T\text{sinc}(fT). \tag{2.138}$$

Therefore, the PSD, using the method of Titsworth and Welch, is

$$P(f) = T\text{sinc}^2(fT), \tag{2.139}$$

which is the same result that we derived by applying the definition of the Fourier transform directly to the signal. In addition to NRZ data, Lindsey and Simon give results for various pulse-code-modulation (PCM) formats. These formats are illustrated in Fig. 1.5 of [11], and are summarized here in Fig. 2.16. Although we will be dealing with NRZ data in the remainder of this dissertation, before moving on, it is instructive to consider the spectra of other data formats.

Return-to-Zero (RZ) Signaling Format PSD The RZ format has a dc value, and also has spectral lines at harmonics of the bit-rate. For equiprobable data, the PSD as given in (1-23) of [11] is

$$P(f) = \underbrace{\frac{1}{16}\delta(f)}_{\text{(dc value)}} + \underbrace{\frac{1}{16}\sum_{\substack{n=-\infty \\ n\neq 0}}^{\infty}\left(\frac{2}{n\pi}\right)^2 \delta(f - nB_T) +}_{\text{(clock tone harmonics)}} \underbrace{\frac{T}{16}\text{sinc}^2(fT/2)}_{\text{(continuous spectrum)}} . \tag{2.140}$$

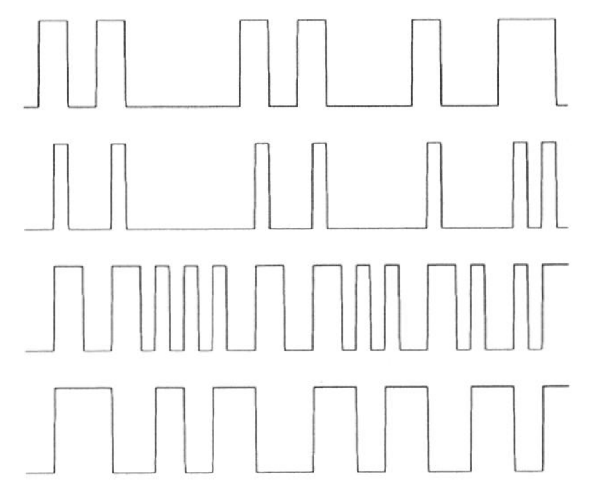

Figure 2.16 Various pulse-code-modulation (PCM) formats for transmission of binary data.

Non-Return-to-Zero (NRZ) Signaling Format PSD We have already shown that the PSD for NRZ data is given by

$$P(f) = T\mathrm{sinc}^2(fT). \tag{2.141}$$

Bi-Phase or Manchester Coding PSD Bi-phase, or Manchester coded waveforms are obtained by dithering an NRZ bit-stream with the system clock, and [11] gives the PSD in (1-25) as

$$P(f) = T\mathrm{sinc}^2(fT/2)\sin^2(\pi fT/2). \tag{2.142}$$

Delay Modulation or Miller Coding PSD The PSD for delay modulation is given in (1-31) in [11]. If we define a parameter θ such that

$$\theta \triangleq \pi fT, \tag{2.143}$$

and two vectors $\boldsymbol{a}$ and $\boldsymbol{b}$ as

$$\boldsymbol{a} = \begin{bmatrix} 23 \\ -2 \\ -22 \\ -12 \\ 5 \\ 12 \\ 2 \\ -8 \\ 2 \end{bmatrix}, \qquad \boldsymbol{b} = \begin{bmatrix} \cos(0) \\ \cos(\theta) \\ \cos(2\theta) \\ \cos(3\theta) \\ \cos(4\theta) \\ \cos(5\theta) \\ \cos(6\theta) \\ \cos(7\theta) \\ \cos(8\theta) \end{bmatrix}, \tag{2.144}$$

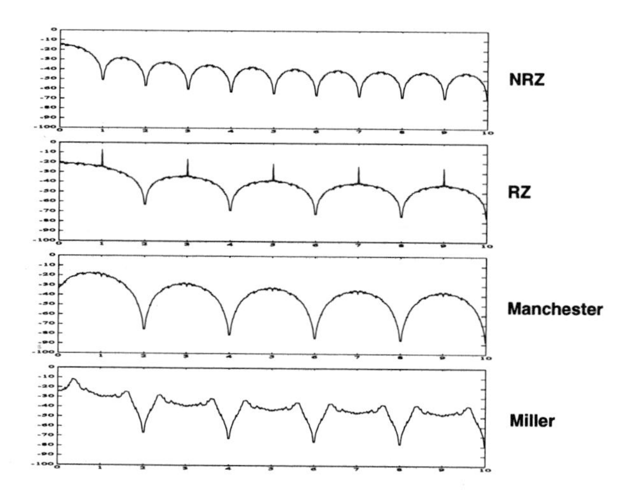

Figure 2.17 Power spectral Densities for RZ, NRZ, Manchester coded, and Miller coded, binary signaling formats.

then the PSD can be written as

$$P(f) = \frac{T}{2\theta^2(17 + 8\cos(8\theta))}[\boldsymbol{a} \cdot \boldsymbol{b}]. \tag{2.145}$$

Comparison of Spectra for Various PCM Formats The power spectra for the above PCM signaling formats are plotted in Fig. 2.17. We notice that the PSD for RZ data has the same functional form as for NRZ data, except that the bandwidth is doubled, and there are spectral lines in the RZ spectrum. The spectral lines arise because the random phase reversals that we saw in NRZ data are no longer present. Since the RZ data is always forced to return to zero, there is no ambiguity about the starting point of a transition. In other words, falling edges only occur at the start of a bit period, and rising edges only occur in the middle of a bit-period. Since spectral lines are present in RZ data, we could extract the clock directly from the data signal without using edge-detection circuits. However, the penalty in terms of increased bandwidth required, is most often too high a cost to pay for this convenience.

Manchester coded data also has its power spread over a larger bandwidth than NRZ data. However, due to the presence of at least one transition per bit-period, there is little dc energy in this signal. This can be important for practical circuit design. For example, when the data is detected with an optical transducer, there will be indeterminate dc offsets. Further, there will be an unknown dark current present in the photodiode detector, also giving rise to an unknown dc value in the final data steam. Often the data processing circuitry that follows the optical transducer requires a well defined dc value, necessitating a restoration of the dc value of the data. A common technique for restoring the dc value is to average the data, compare it to a reference, and add the difference back to the data. The problem with this technique is that it performs a highpass function on the data, and any dc components of the data will be filtered out. This is a serious problem in dealing with NRZ data which has most of its energy concentrated at low frequencies. However, with Manchester coded data, the problem is averted.

Miller Coding (delay modulation) offers desirable time-domain and frequency domain properties. In the time domain there is an average of one transition per bit-period as opposed to 1/2 for NRZ data. We will see in chapter 4 that the accuracy of the recovered clock is proportional to the square-root of the average number of transitions per bit-period. Miller coding also has desirable frequency-domain properties. As in the case of Manchester coding, the power at dc is also zero, so that we can avoid problems with restoring the dc value. The primary benefit is that most of the power is concentrated in a much narrower frequency band than for RZ, NRZ, or Manchester coding. This means that a narrowband filter can pass the majority of the signal power, while reducing the contribution of additive broadband noise in comparison with the other signaling formats.

2.3.4 Summary of Fourier Transform Analysis

The Fourier transform was introduced as a limiting case of the Fourier series coefficients normalized to give the energy in one fundamental period per unit bandwidth. Frequency analysis was simplified using the Fourier transform by taking advantage of several useful properties listed in table 2.1. The results obtained using Fourier transform analysis can be related to laboratory measurements via the PSD defined as $E|F(j2\pi f, \cdot)|^2/NT$. This is the energy spectral density divided by the time-interval of the data sequence. The time averaged process can be related to the ensemble expectation if it is assumed that small time segments of length n of a long N-bit data stream are identical to ensemble sample functions of an n-bit data signal. [3]

[3] A random process is said to be ergodic if the averages over a long period of time converge to ensemble averages. This assumption is typically made when relating results obtained by using ensemble averaging to laboratory measurements. In the lab we only have one sample function $d(\cdot, \xi_1)$ of the random process $d(\cdot, \cdot)$, so the only way to obtain any statistical data is to time-average this one sample function.

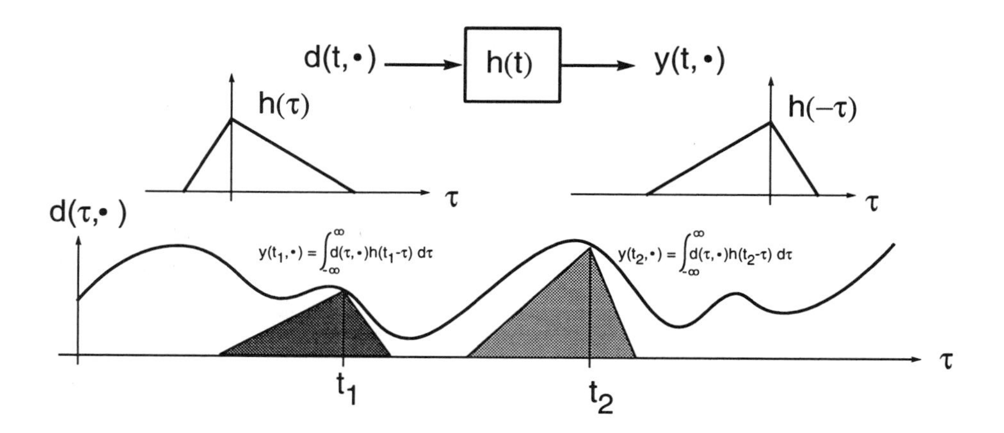

Figure 2.18 Illustration of the convolution integral.

2.4 LINEAR FILTERING OF RANDOM DATA

The primary reason for going to all the trouble to find the Fourier transform of random data is that the effect of linear filtering can be determined simply and intuitively in the frequency domain. Frequency domain analysis can be used to determine the optimal shape of the transitions of NRZ data, and the optimal pulse-shape of the of edge-detected data. We will also use frequency domain analysis in chapter 3 to find the optimal shaped weighting function used to average the noise in the detector circuit. Parseval's theorem also provides a method for determining the average energy per bit by integrating the energy spectral density function in the frequency domain. This is often simpler than performing the equivalent operation in the time domain.

Linear Filtering as a Convolution Integral An input to a linear network, such as a random data signal $d(t,\cdot)$, can be considered as an impulse of magnitude $d(n\Delta t,\cdot)\Delta t$ for each time interval Δt. If a network has an impulse response $h(t)$, and if the network is linear, then the output of the network, via superposition, is a weighted sum of impulse responses shifted in time. In the limit as $\Delta t \to 0$, the superposition sum becomes a convolution integral, and the output $y(t,\cdot)$ is given by

$$y(t,\cdot) = d(t,\cdot) * h(t) = \int_{-\infty}^{\infty} d(\tau,\cdot)h(t-\tau)d\tau \tag{2.146}$$

This convolution operation is illustrated in Fig. 2.18 The power of using the Fourier transform for analysis of linear filtering is provided by the convolution property of table 2.1. Taking the Fourier transform of both sides of (2.146) gives

$$F_y(j2\pi f,\cdot) = F_d(j2\pi f,\cdot)H(j2\pi f), \tag{2.147}$$

and the energy spectrum of the output signal $y(t,\cdot)$ is simply

$$S_y(f) = S_d(f)|H(j2\pi f)|^2. \tag{2.148}$$

$|H(j2\pi f)|^2$ is known as the "frequency response" of the filter, since it scales the input signal spectrum; $H(j2\pi f)$ is known as the "transfer function" because the input Fourier transform is transferred to the output through multiplication by $H(j2\pi f)$. It should be pointed out that $F_y(j2\pi f,\cdot)$ is the "steady-state" output after all transients have died out; this is an artifact of starting the convolution integral at $-\infty$, which implies that the input was applied to the system just prior to the "Big-Bang."

2.4.1 Bandlimited NRZ Data

Thus far we have been dealing only with rectangular NRZ data because it is easy to represent analytically as the sum of square pulses. However, real data will have non-zero rise and fall times. It was shown in section 2.1 that bandlimited data can be obtained by passing rectangular data through a linear filter, such that

$$y(t,\cdot) = h(t) * \left[\sum_{n=0}^{N-1} r_n(\cdot)p_T(t-nT)\right]. \tag{2.149}$$

The filtering operation introduces memory into the signal, so that $y(t,\cdot)$ is a linear combination of the original data over a time interval T_H, where T_H is the time over which $h(t)$ is non-zero. If T_H is longer than the bit-period T, then intersymbol interference (ISI) will exist. If T_H is less than T, then there will be a time in an interval $nT \leq t \leq (n+1)T$ such that $y(t,\cdot)$ is a linear combination of the data only over one bit.[4]

NRZ Data with Sinusoidal Transitions This discussion is more clearly illustrated by an example. Consider a filter with an impulse response of a half-cosine as shown in Fig. 2.19. This impulse response can be written as

$$h(t) = \frac{\pi}{2T}\cos\left(\frac{\pi t}{T}\right)\mathrm{rect}(t/T). \tag{2.150}$$

The constant multiplier of $\pi/2T$ is for normalization. The Fourier transform can be obtained easily using the multiplication property. The transform of the cosine wave is

$$F_c(j2\pi f) = \frac{\pi}{4T}\delta\left(f \pm B_T/2\right), \tag{2.151}$$

[4] It is possible to to obtain no ISI with T_H longer than T. This requires that the filters impulse response be orthogonal to shifted data bits at discrete sampling instances. This technique is encountered frequently in bandlimited channels, where the actual data pulse may extend over several bit periods. The pulses are designed to have zero-crossings at the center of each bit-period so that at one particular time instance in each bit-period, the data signal amplitude is due only to the current bit.

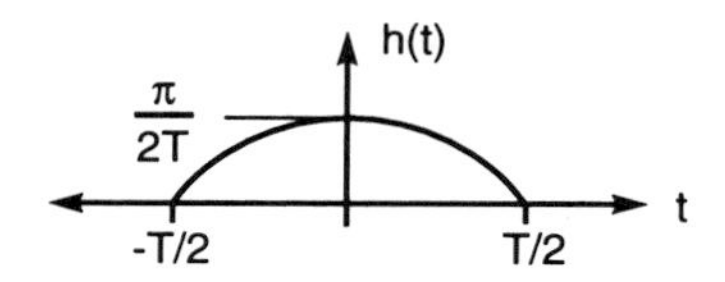

Figure 2.19 Half-cosine impulse response of a low-pass filter.

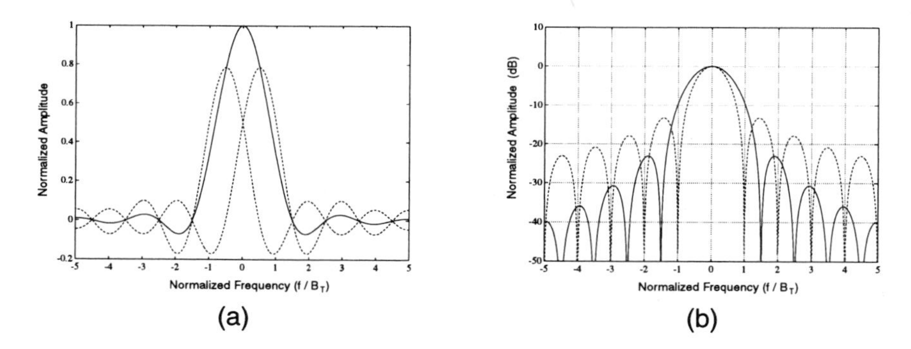

Figure 2.20 Magnitude squared of the transfer function of a filter with a half-cosine impulse response: (a) linear squared magnitude, (b) magnitude in dB compared to a sinc function.

where B_T is the bit-rate $= 1/T$. The Fourier transform of the rectangular window is found from table 2.2 to be the familiar sinc function.

$$F_r(j2\pi f) = T\text{sinc}(fT) \tag{2.152}$$

The Fourier transform of the product is the convolution in the frequency domain of the individual Fourier transforms. Therefore,

$$H(j2\pi f) = \frac{\pi}{4}\left[\text{sinc}\left((f - B_T/2)\,T\right)\right) + \text{sinc}\left((f + B_T/2)\,T\right))\right]. \tag{2.153}$$

The filter frequency response is the sum of two sinc function of equal magnitude shifted so that the centers are at $\pm B_T/2$. The squared magnitude of this transfer function is plotted in Fig. 2.20a; this is compared with the magnitude of the sinc function itself in Fig. 2.20b, where it can be seen that $|H(j2\pi f)|^2$ provides better attenuation of high-frequencies. When rectangular NRZ data is applied to the input of this filter, a little thought will indicate that the data will have sinusoidal transitions. If the data were a square-wave, then the output would be a single tone at half the data rate. When the data is random, the output will be constant when the data doesn't change, and the output will follow a sinusoidal path in its transition from a high-to-low value or visa-versa. A typical NRZ data sequence is shown in Fig. 2.21a, and the data filtered

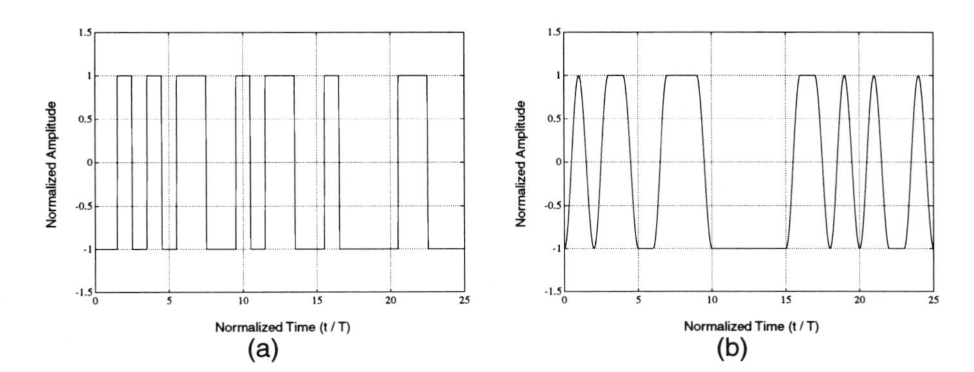

Figure 2.21 Typical random data sequences for: (a) rectangular NRZ data, (b) rectangular NRZ data passed through a filter with a half-cosine impulse response.

by a half-cosine impulse response is shown in Fig. 2.21b. It was shown in (2.109) that the energy-spectral-density-per-bit (ESDB) of rectangular NRZ data is

$$S_{Bd}(f) = T^2 \mathrm{sinc}^2(fT). \tag{2.154}$$

Therefore, the ESDB of half-cosine filtered data is

$$S_{By}(f) = |H(j2\pi f)|^2 T^2 \mathrm{sinc}^2(fT), \tag{2.155}$$

and substituting (2.153), $S_{By}(f)$ is given by

$$\boxed{S_{By}(f) \quad = \quad \left[\tfrac{\pi}{4}\left[\mathrm{sinc}\left((f - B_T/2)T\right)\right) + \mathrm{sinc}\left((f + B_T/2)T\right))\right]\right]^2 T^2 \mathrm{sinc}^2(fT)} \tag{2.156}$$

This energy spectrum is plotted in Fig. 2.22b, which shows (2.156) plotted in dashed lines compared to numerical simulation plotted in a solid line. The discrepancy between the results is due to the the discrete time nature of the simulation used. In order to force $h(t)$ to zero in the discrete time simulation at $t = 0$ and at $t = T$, a frequency of

$$f\mathrm{sim} = \frac{B_T}{2}\frac{N_s}{N_s - 1} \tag{2.157}$$

had to be used, where $N_s = 32$ is the number of samples per bit. The simulated and calculated spectrums for the input rectangular NRZ data is shown in Fig. 2.22a for comparison. The simulation was performed using 32 samples per bit. The results of Fig. 2.22 show the integral of the one-sided PSD over a bandwidth increments of $B_T/32$ where B_T has been normalized to unity. Therefore, to get the PSD from these plots simply divide by Δf, which is equivalent to multiplying by 32 or adding $10 \log 32 = 15.05$ dB.

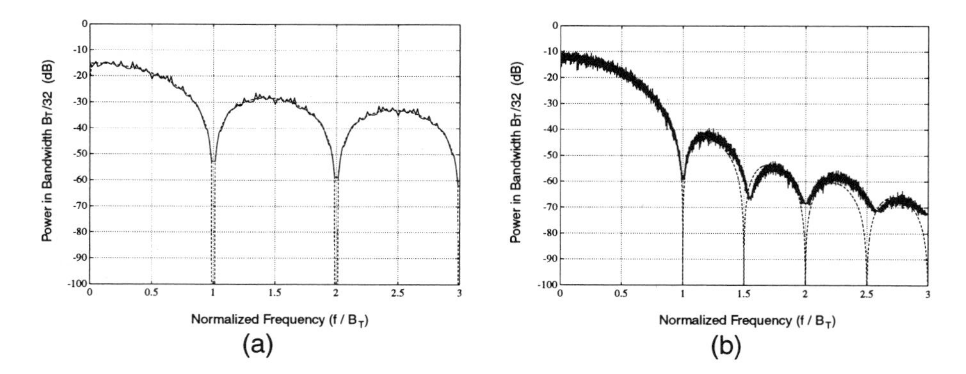

Figure 2.22 Simulated and calculated power spectrums for: (a) rectangular NRZ data, (b) rectangular NRZ data passed through a filter with a half-cosine impulse response.

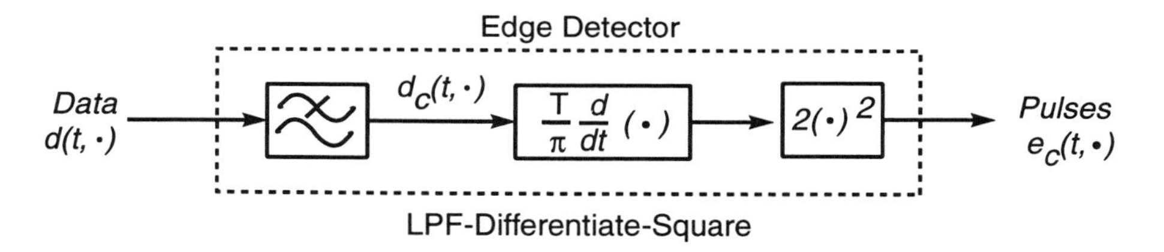

Figure 2.23 Block diagram of a circuit used to detect transition in random NRZ data.

2.4.2 Bandlimited Edge-Detected Data

We can now make use of the linear filtering properties of random data signals to find the ESDB of edge-detected data where the pulses are no longer rectangular. Often the data transitions are detected using the circuit of Fig. 2.23. If the lowpass-filter in Fig. 2.23 has a half-cosine impulse function, then the transitions will be sinusoidal, and of the form

$$\begin{cases} -\sin\left(2\pi(B_T/2)t\right) & \text{for a negative transition } \frac{-T}{2} \leq t \leq \frac{T}{2} \\ \sin\left(2\pi(B_T/2)t\right) & \text{for a positive transition } \frac{-T}{2} \leq t \leq \frac{T}{2}. \end{cases} \tag{2.158}$$

T/π times the derivative of the data is equal to zero when there is no transition and, is equal to

$$\begin{cases} -\cos\left(\pi B_T t\right) & \text{for a negative transition } \frac{-T}{2} \leq t \leq \frac{T}{2} \\ \cos\left(\pi B_T t\right) & \text{for a positive transition } \frac{-T}{2} \leq t \leq \frac{T}{2}. \end{cases} \tag{2.159}$$

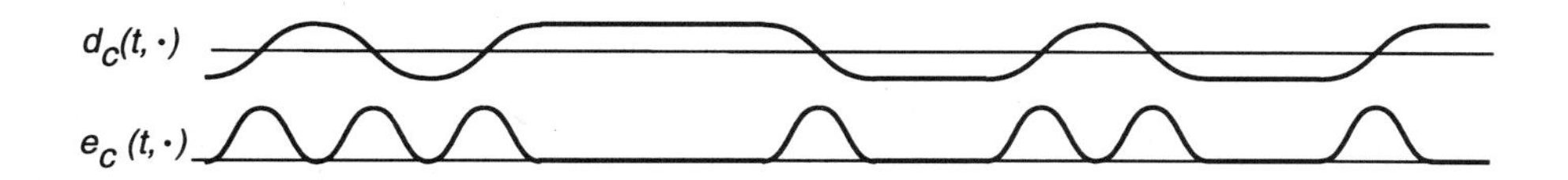

Figure 2.24 NRZ data with sinusoidal transitions and raised cosine pulses at each transition.

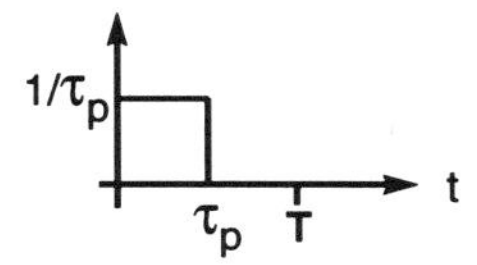

Figure 2.25 Pulse shape for edge-detected data normalized to have unit area.

After squaring and multiplying by 2, the result is that the edge-detected data is zero for no transition, and for both positive and negative transitions the signal is

$$e_c(t, \cdot) = 2\cos^2(\pi B_T t) \quad \text{(for} \quad \pm \quad \text{transitions)} \tag{2.160}$$

$$= 1 + \cos(2\pi B_T t) \quad \text{for} \quad -T/2 \leq t \leq T/2. \tag{2.161}$$

The resulting signal gives a raised cosine pulse when a transition occurs, as illustrated in Fig. 2.24. If the data signal were alternating every bit, then $e_c(t, \cdot)$ would be a single tone at the bit-rate.

Derivation of the Energy Spectral Density Based on Rectangular Pulses Results To find the ESDB of $e_c(t, \cdot)$ the Fourier transform could be obtained directly from the definition. However, it is simpler to apply the results already obtained for the rectangular edge-detected data. If the fundamental pulse shape $e_T(t)$ from Fig. 2.8, is normalized to have unit area as shown in Fig. 2.25, then the new pulse $u_T(t)$ is given by

$$u_T(t) = \begin{cases} 1/pT & \text{for } 0 \leq t \leq pT \\ 0 & \text{elsewhere,} \end{cases} \tag{2.162}$$

and the ESDB from (2.134) is

$$\widehat{S_{Beu}}(f) = \frac{S_{Be}(f)}{(pT)^2} = \left[\frac{1}{2}\text{sinc}(fpT)\right]^2 \left[1 + N \sum_{M=-\infty}^{\infty} \text{sinc}^2\left(\left(f - \frac{M}{T}\right) NT\right)\right]. \tag{2.163}$$

As p tends toward zero, the envelope gets broader, until in the limit it approaches a constant of $1/4$, and the normalized energy spectrum is shown if Fig. 2.26 In the time

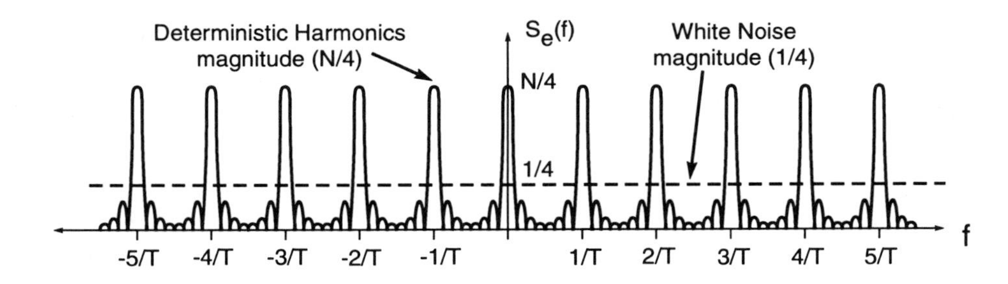

Figure 2.26 Normalized energy spectrum for edge detected data with pulses of unit area.

domain, as p approaches zero, then the train of unit area pulses $u_T(t)$ become a train of impulse functions. The signal that we desire can now be represented as a convolution of a kernel raised-cosine pulse with this train of random impulses.

$$e_c(t,\cdot) = \lim_{p\to 0} e_u(t,\cdot) * [1+\cos(2\pi B_T t)]\,\mathrm{rect}(t/T) \tag{2.164}$$

Defining a normalized transfer function $G(j2\pi f)$ such that

$$G(j2\pi f) \triangleq \mathcal{F}\left\{\frac{1}{T}[1+\cos(2\pi B_T t)]\mathrm{rect}(t/T)\right\}, \tag{2.165}$$

we can easily recognize that $G(j2\pi f)$ as the Fourier transform of $\mathrm{rect}(t/T)$ convolved with impulses of magnitude $1/T$ at $f = 0$, and impulses of magnitude $1/2T$ at $f = \pm B_T$, so that $G(j2\pi f)$ is simply expressed as the superposition of three sinc functions.

$$G(j2\pi f) = \sum_{m=-1}^{1}\left(\frac{1}{2}\right)^{|m|} \mathrm{sinc}\left((f - mB_T)T\right) \tag{2.166}$$

$|G(j2\pi f)|^2$ is plotted if Fig. 2.27a, and is compared to a sinc^2 function in Fig. 2.27b. The ESDB for the pulses $e_c(t,\cdot)$ is then given by

$$S_{Bec}(f) = T^2|G(j2\pi f)|^2 \lim_{p\to 0}\widehat{S_{Beu}}(f), \tag{2.167}$$

or

$$S_{Bec}(f) = \frac{T^2}{4}|G(j2\pi f)|^2\left[1+\sum_{M=-\infty}^{\infty}\mathrm{sinc}^2\left(\frac{f-MB_T}{B_T/N}\right)\right]. \tag{2.168}$$

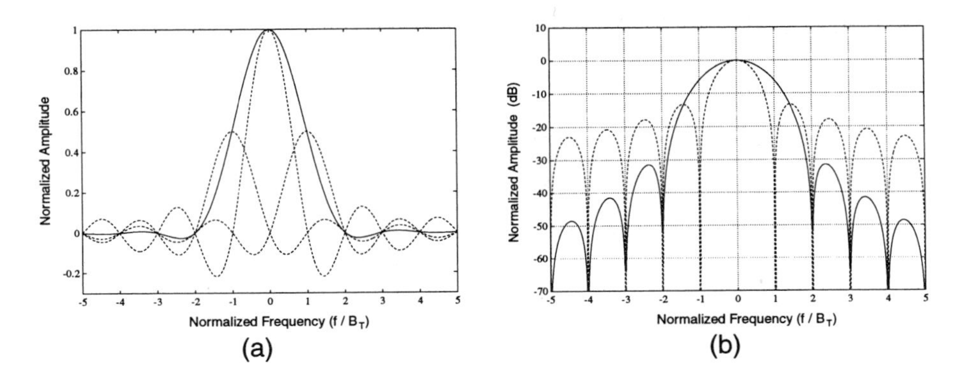

Figure 2.27 Squared magnitude response of a filter with a raised-cosine impulse response: (a) linear plot, (b) magnitude in dB compared to a sinc^2 function.

As the number of bits N grows the narrow sinc^2 pulses can be replaced by impulses with equal area as in (2.134), so that

$$S_{Bec}(f) = \frac{T^2}{4}\left[\sum_{m=-1}^{1}\left(\frac{1}{2}\right)^{|m|}\mathrm{sinc}\left(\frac{f - mB_T}{B_T}\right)\right]^2\left[1 + \frac{1}{T}\sum_{M=-\infty}^{\infty}\delta(f - MB_T)\right] \tag{2.169}$$

Since the envelope of $S_{Bec}(f)$ is $|G(j2\pi f)|^2$, then this energy spectrum will have the desirable property that all harmonics of the signal at multiples of the bit-rate are nulled. This property results from having a kernel-pulse that is non-zero in the interval $t \in [0, T]$, whereas the rectangular pulses were only non-zero for $t \in [0, pT]$.

Discrete Power Spectrum for Comparison with Simulation The ESDB from (2.169) can be converted to energy dissipated in a 1Ω resistor by integrating $S_{Bec}(f)$ over the appropriate frequency intervals. If $e_c(t, \cdot)$ is input to a spectrum analyzer with bandwidth intervals of $\Delta f = B_T/N_s$, where $f_n = n\Delta f$, then the average two-sided energy-per-bit of the signal in the nth frequency bin is

$$\begin{aligned}\overline{E_{Bec}}(f_n) = \frac{T^2}{4}\int_{f_n - B_T/2N_s}^{f_n + B_T/2N_s} & |G(j2\pi f_n)|^2 df + \\ & \left[\frac{T}{4}|G(j2\pi f_n)|^2 \quad \text{for} \quad \mathrm{mod}_{B_T}(f_n) = 0\right].\end{aligned} \tag{2.170}$$

For N_s large, the integral can be approximated by $|G(j2\pi f_n)|^2\Delta f$. Therefore,

$$\overline{E_{Bec}}\left(\frac{nB_T}{N_s}\right) = \underbrace{\frac{1}{4}\frac{T}{N_s}\left|G\left(j2\pi\frac{nB_T}{N_s}\right)\right|^2}_{\text{mod}_{N_s}(n)\neq 0} + \frac{1}{4}T|G(0)|^2\hat{\delta}(n)$$
$$+\frac{1}{4}T|G(j2\pi B_T)|^2\hat{\delta}\left(\frac{n}{N_s}-1\right)$$
$$+\frac{1}{4}T|G(-j2\pi B_T)|^2\hat{\delta}\left(\frac{n}{N_s}+1\right), \tag{2.171}$$

where

$$\hat{\delta}(n) = \begin{cases} 1 & \text{for } n = 0 \\ 0 & \text{for } n \neq 0. \end{cases} \tag{2.172}$$

The average power is obtained by dividing the energy by the time interval T. Considering positive frequencies, and remembering that the dc component doesn't get doubled, then

$$\overline{P_{ec}}\left(\frac{nB_T}{N_s}\right) = \frac{1}{2}\left[\underbrace{\frac{1}{N_s}\left|G\left(j2\pi\frac{nB_T}{N_s}\right)\right|^2}_{\text{mod}_{N_s}(n)\neq 0} + \frac{1}{2}|G(0)|^2\hat{\delta}(n)\right.$$
$$\left. + |G(j2\pi B_T)|^2\hat{\delta}\left(\tfrac{n}{N_s}-1\right)\right]. \tag{2.173}$$

(2.173) gives the power in N_s equally spaced frequency bins; this can be compared directly with simulation results. First, however, we realize that the dc value due to the deterministic part is $1/4|G(0)|^2 = (1/2)^2$, so the dc term can be removed by subtracting $1/2$ from the original signal. It is clear that the average value of $e_c(t,\cdot)$ is zero when no pulse occurs, and unity when there is a pulse. Since the probability of a pulse is $1/2$, then the expected value of the signal is $1/2$, so that by subtracting $1/2$ from $e_c(t,\cdot)$ produces a zero-mean random process. A plot of this signal $\hat{e_c}(t,\cdot)$ is shown in Fig. 2.28a with the random NRZ data $d_c(t,\cdot)$. After removal of the mean, the power spectrum is shown plotted in Fig. 2.28b. The calculated spectrum for $N_s = 32$ is shown in dashed line and a simulation using 32 samples per bit is plotted with a solid line. The simulated curves shows small variations around the calculated curve. These variation can be reduced by averaging over even more data segments.

2.4.3 Bandpass Filtering of Edge-Detected Data

A clock at the receiver is often extracted from the data by passing the edge-detected signal through a bandpass filter tuned to the data rate. This operation is illustrated

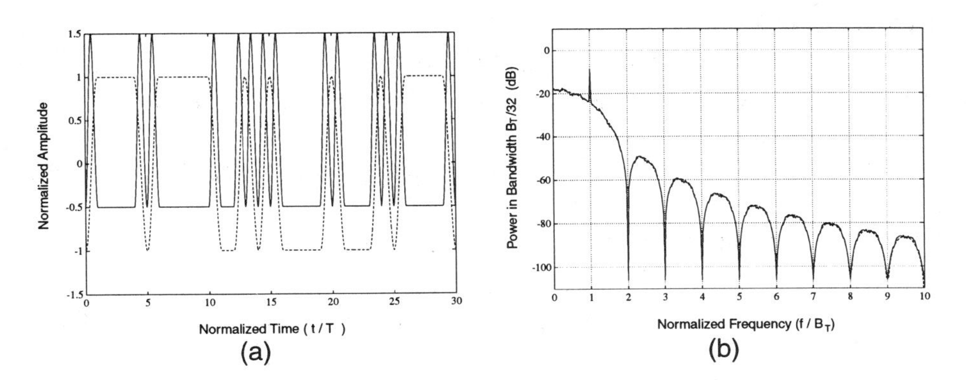

Figure 2.28 Transition detected pulses using a raised-cosine kernel function: (a) the zero-mean pulse stream in the time domain and the NRZ random data, (b) calculated and simulated normalized power in a bandwidth of $B_T/32$.

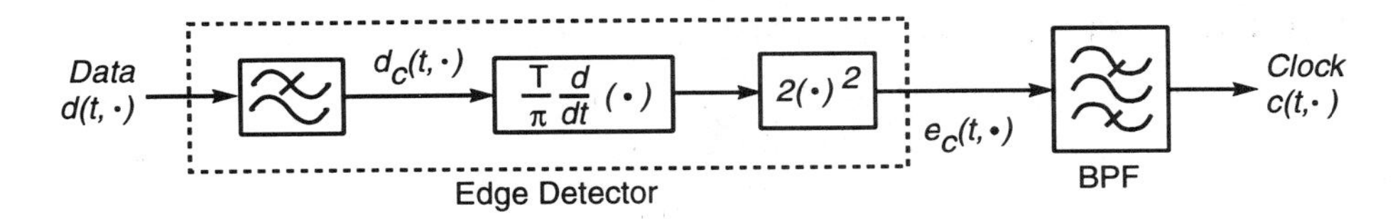

Figure 2.29 Block diagram showing the extraction of a clock by bandpass filtering the edge-detected data.

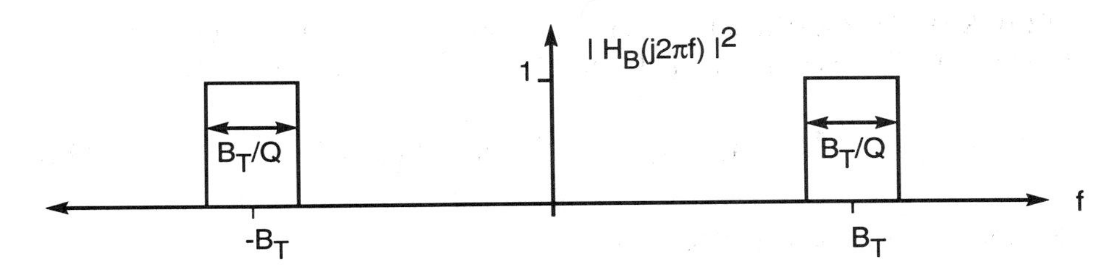

Figure 2.30 Magnitude response of an ideal bandpass filter.

in Fig. 2.29. An important figure of merit for an oscillator is the ratio of the power in the pure tone, to the power in the side-bands. We can determine this ratio for the clock $c(t, \cdot)$ if the transfer function of the bandpass filter (BPF) is known. For a first order analysis we'll consider an ideal bandpass filter with a magnitude response $|H_B(j2\pi f)|^2$ shown in Fig. 2.30. The Q-factor of the filter is defined as the ratio of the center-frequency to the bandwidth. Therefore, $Q = B_T/B$. The ESDB at the output of the BPF is

$$S_{Bc}(f) = S_{Bec}(f)|H_B(j2\pi f)|^2, \tag{2.174}$$

where $S_{Bec}(f)$ is given in (2.169). The average energy-per-bit in the tone is

$$\overline{E_{\text{tone}}} = \frac{T^2}{4}|G(j2\pi f)|^2 \frac{1}{T}\int_{-\infty}^{\infty} \delta(f \pm B_t)df = \frac{T}{2}|G(j2\pi B_T)|^2. \tag{2.175}$$

The average energy-per-bit in the side-band around the tone is

$$\overline{E_{\text{sb}}} = 2\left(\frac{T^2}{4}\right)\int_{B_T - B_T/2Q}^{B_T + B_T/2Q} |G(j2\pi f)|^2 df, \tag{2.176a}$$

and for large Q this can be approximated by

$$\overline{E_{\text{sb}}} \simeq \frac{T^2}{2}|G(j2\pi B_T)|^2 \frac{B_T}{Q} = \frac{T}{2}|G(j2\pi B_T)|^2 \frac{1}{Q}. \tag{2.176b}$$

Therefore, the energy ratio is simply equal to the selectivity of a bandpass filter:

$$\boxed{\frac{\overline{E_{\text{tone}}}}{\overline{E_{\text{sb}}}} = Q} \tag{2.177}$$

This quantity Q will play an important role in section 2.5, where relationships between the energy spectrum of a random signal and its time-domain statistics will be developed.

Energy Ratio for Rectangular Pulses The previous result can be compared to an edge-detected signal using rectangular pulses. $S_{Be}(f)$ from (2.134) is given by

$$S_{Be}(f) = \left[\frac{pT}{2}\text{sinc}(fpT)\right]^2 \left[1 + \frac{1}{T}\sum_{M=-\infty}^{\infty} \delta\left(f - \frac{M}{T}\right)\right], \tag{2.178}$$

The average energy-per-bit in the tone is

$$\overline{E_{\text{tone}}} = \frac{T}{2}[p\text{sinc}(p)]^2, \tag{2.179}$$

and the in the side-bands

$$\overline{E_{\text{sb}}} = \frac{T}{2}[p\text{sinc}(p)]^2 \frac{1}{Q}, \tag{2.180}$$

so the energy ratio is also equal to the filter selectivity:

$$\boxed{\frac{\overline{E_{\text{tone}}}}{\overline{E_{\text{sb}}}} = Q} \tag{2.181}$$

This result holds for all values of p, however, the absolute power in the tone varies with p. Other sources of noise in the system will raise the noise-floor above $1/Q$ times the tone level. In this case it is important to maximize the power in the tone for a given peak-to-peak signal level. The raised-cosine pulses were normalized to have a peak-to-peak value of 2, and the rectangular pulses had a peak-to-peak value of 1. For an equal comparison the rectangular pulses will be multiplied by 2 which multiplies the energy by 4. Therefore

$$\overline{E\mathrm{rec}}/\overline{E\mathrm{rcos}} = 2T[p\mathrm{sinc}(p)]^2/\frac{T}{2}|G(j2\pi B_T)|^2 \tag{2.182a}$$

$$= \left[\frac{4}{\pi}\sin(\pi p)\right]^2, \tag{2.182b}$$

where $G(j2\pi B_T) = 1/2$ was substituted. The pulse width parameter varies from $0 \leq p \leq 1$. It is easy to see that the value of p that maximizes the energy ratio is $p = 1/2$, and at the maximum value

$$\overline{E\mathrm{rec}}/\overline{E\mathrm{rcos}} = \left(\frac{4}{\pi}\right)^2 = 1.621. \tag{2.183}$$

This shows that for signals of equal amplitude, the raised-cosine signal has only 61.7% of the tone power as a rectangular signal with a pulse width of $T/2$. Since the tone power is due to the deterministic part of the signal, then this term $(4/\pi)$ is just the Fourier series coefficient of the fundamental tone for a square-wave. This comparison of the tone powers is interesting, and has been done to illustrate the analytical techniques. As a practical matter, however, we realize at high-speeds there are no such things as rectangular pulses.

2.4.4 Summary of Linear Filtering Results

Thus far we have been dealing with the energy, or power spectral densities of a random signal. Frequency domain analysis provides a useful tool for analyzing the effect of linear filtering. Insight is also gained that aids in signal design. Pulse-shapes can be tailored to null specific harmonics in the edge-detected signal. However, the PSD does not provide a unique description of a signal. Since phase information has been ignored in obtaining the PSD, several signals with various phase-shifts can have the same power spectrum. Often the phase information is crucial in predicting performance. For example the phase-jitter in the recovered clock is a key parameter that effects the overall bit-error rate (BER) of the recovered data. The phase response of a filter can convert excess side-band power into either amplitude-modulation, or phase-jitter. This topic will be taken up in sections 2.6 and 2.7, but before moving on, we need to develop a more general connection between the PSD and time domain statistics. This connection is provided by Parseval's theorem , which allows us to find the average power in the time domain by integrating the PSD over a given bandwidth. The following section will show how the average power of the random signal in the frequency domain is related to the variance of the random-process.

2.5 REVIEW OF GENERAL THEORY OF RANDOM SIGNALS

Thus far when we wanted to find the power-spectral density of a random process we first found an explicit representation of the signal in the time-domain, and transformed the signal into the frequency domain, where the frequency domain representation was itself a random process. Then the expected value of the time averaged energy was determined over the ensemble of random sample functions. If the frequency transformations, and the expected value are linear operators, then the order of expectation and frequency domain transformation can be reversed. Doing these operations in reverse order can be extremely useful. In most cases an explicit form of the random signal cannot be written, but the statistics of the signal are known. Therefore the PSD can be found directly from the time-domain statistics.

2.5.1 Autocorrelation functions: Time and Ensemble

An important function derived from a random process $r(\cdot,\cdot)$ is the time-autocorrelation function defined by[5]

$$a_r(\tau,\cdot) \triangleq \lim_{t_r\to\infty}\int_{-t_r/2}^{t_r/2} r(t,\cdot)r(t+\tau,\cdot)dt \tag{2.184}$$

This function gives an indication of the speed in which the signal $r(t,\cdot)$ varies with time. For τ large, one would expect that the correlation goes to zero, and for τ small, the correlation will be a maximum. The function $a_r(\tau,\cdot)$ is also a random variable,

[5]This in provided that the integral converges. One condition can be artificially imposed is to consider only time-limited data sequence. This becomes a problem when the data sequence is passed through a linear filter with an infinite duration impulse response, then the output will no longer be time limited. However, as long as the filter is "well behaved" and is the type of filter that one finds in practice, then provided that the input data has finite energy, then the output data will also have finite energy.

and the expected value can be taken such that

$$A_r(\tau) \triangleq E[a_r(\tau,\cdot)] = \lim_{t_r\to\infty}\int_{-t_r/2}^{t_r/2} E[r(t,\cdot)r(t+\tau,\cdot)]dt \tag{2.185}$$

The expression inside the integral is defined as the the ensemble auto-correlation function.

$$R_r(t,\tau) \triangleq E[r(t,\cdot)r(t+\tau,\cdot)] \tag{2.186}$$

For a stationary process the signal statistics are independent of time so that $R_r(t,\tau)$ depends only on the time offset τ.

$$R_r(t,\tau) = R_r(\tau) \quad \text{for } r(t,\cdot) \text{ stationary}, \tag{2.187}$$

If in addition to being stationary, $R_r(\tau)$ is non-zero only over a finite interval T_r, then $A_r(\tau)$ is proportional to $R_r(\tau)$ according to

$$A_r(\tau) = T_r R_r(\tau), \tag{2.188}$$

In the general case, when the random signal is non-stationary, then

$$A_r(\tau) = \lim_{t_r\to\infty}\int_{-t_r/2}^{t_r/2} R_r(t,\tau)dt \tag{2.189}$$

Fourier Transform of $A_r(\tau)$ The expected value of the time-autocorrelation function is a deterministic signal that depends only on the offset τ. The Fourier transform can be obtained for this signal such that

$$F_A(j2\pi f) = \int_{-\infty}^{\infty} A_r(\tau)e^{-j2\pi f\tau}d\tau. \tag{2.190a}$$

Substituting (2.185) for $A_r(\tau)$

$$F_A(j2\pi f) = \int_{-\infty}^{\infty} \lim_{t_r\to\infty}\int_{-t_r/2}^{t_r/2} E[r(t,\cdot)r(t+\tau,\cdot)]dt \quad e^{-j2\pi f\tau}d\tau. \tag{2.190b}$$

Letting $\tau = \hat{t} - t$, and moving the expectation outside both integrals, then

$$F_A(j2\pi f) = E\left[\int_{\hat{t}=-\infty+t}^{\infty+t} \lim_{t_r\to\infty}\int_{t=-t_r/2}^{t_r/2} r(t,\cdot)r(\hat{t},\cdot)e^{-j2\pi f(\hat{t}-t)}dtd\hat{t}\right] \tag{2.190c}$$

In the limit[6] both integrals span $[-\infty, +\infty]$ and the result is

$$F_A(j2\pi f) = E\left[\int_{-\infty}^{\infty} r(\hat{t},\cdot)e^{-j2\pi f\hat{t}}d\hat{t}\int_{-\infty}^{\infty} r(t,\cdot)e^{j2\pi ft}dt\right]. \tag{2.190d}$$

We recognize that each integral is a Fourier transform, and since $r(t,\cdot)$ is a real signal, then

$$F_A(j2\pi f) = E\left[F_r(j2\pi f,\cdot)F_r^*(j2\pi f,\cdot)\right] = S_r(f). \tag{2.190e}$$

Therefore, the energy-spectral density $S_r(f)$ can be found directly by taking the Fourier transform of the expected value of the auto-correlation function $A_r(\tau)$. The autocorrelation function evaluated at 0 gives the expected value of the total energy in the signal.

$$A_r(0) = \int_{-\infty}^{\infty} E[|r(t,\cdot)|^2]dt = \int_{-\infty}^{\infty} S_r(f)df \tag{2.191}$$

2.5.2 NRZ Data Revisited

The time-autocorrelation function can be used to determine the ESD of random NRZ data. For an N-bit sequence, $a_d(0,\cdot)$ will equal the total energy in the random sample function. As the time offset τ is shifted slightly away from zero, only a fraction of identical bits will line up in the correlation. The remaining portion of the integral will be over two different and uncorrelated bits. As τ is increased to the point where no identical bits line up, then the average correlation will be zero. Therefore, the expected value, $A_d(\tau)$ will start at a maximum of NT at $\tau = 0$ and fall off linearly to zero at $\tau = \pm T$, and will be zero for any value of $|\tau| > T$. A plot of $A_d(\tau)$ is shown in Fig. 2.31. $A_d(\tau)$ can be easily recognized as the convolution of two identical rectangular pulses.

$$A_d(\tau) = N[\text{rect}(t/T) * \text{rect}(t/T)] \tag{2.192}$$

Therefore the ESD of the random data is the Fourier transform of $A_d(\tau)$ and is simply given by

$$S_d(f) = N\mathcal{F}\left\{\text{rect}(t/T)\right\}^2 = NT^2\text{sinc}^2(fT). \tag{2.193}$$

This is identical to (2.107), but the result was obtained with much less effort. Using the autocorrelation $A_d(\tau)$ also provides additional insight into the spectral content, by explicitly showing a time domain description of the how fast the signal varies on average.

---

[6] Taking the limit of the term $\lim_{t\to\infty}(-\infty + t)$ can be problematic in the strictest mathematical sense. However if $r(t,\cdot)$ is assumed to be time limited to $[-T_r, T_r]$, then problems can be circumvented. For the this case the second integral has finite limits $\int_{-T_r}^{T_r}$. With $|t|$ limited to $|T_r|$, then taking the limit is no problem, and the first integral is then integrated over $\int_{-\infty}^{\infty}$. We will develop analytical tools for time-limited data, and then apply the results to data that is not time-limited, but has "tails" that die out much faster than the length of the "main-part" of the data sequence.

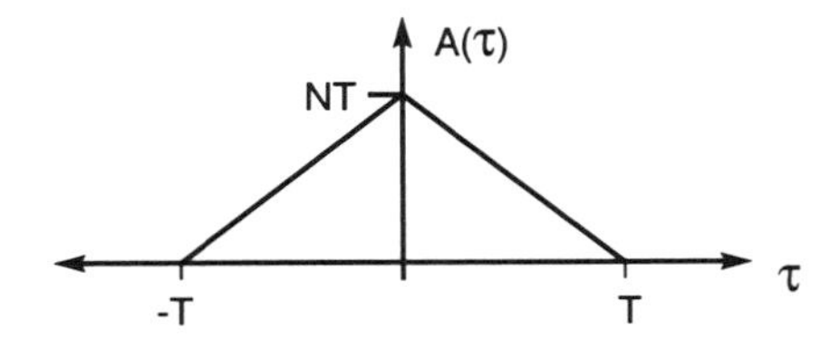

Figure 2.31 Expected value of the time auto-correlation function for random rectangular NRZ data.

2.5.3 Theory of Stationary Random Processes Applied to Non-stationary Signals

There exists a complete theory of spectral analysis of random processes, where frequency domain techniques are used when the random process is stationary or at least wide-sense stationary. [1, 2, 3, 4]. A stationary random process is one in which all of the statistics of the process are independent of time. A process is said to be stationary in the wide-sense if all of the first and second order statistics are independent of time. Wide-sense stationarity implies that the ensemble autocorrelation function $R_d(t, \tau)$ is only a function of the time offset τ.

Basic Results for Stationary Random Processes For a wide-sense stationary random process the power spectral density is defined as the Fourier transform of $R_r(\tau)$.

$$P_r(f) = \int_{-\infty}^{\infty} R_r(\tau) e^{-j2\pi f \tau} d\tau \tag{2.194}$$

The inverse Fourier transform gives $R_r(\tau)$ when the power spectral density is known.

$$R_r(\tau) = \int_{-\infty}^{\infty} P_r(f) e^{j2\pi f \tau} df \tag{2.195}$$

Setting the time offset to zero gives

$$R_r(0) = E[r(t, \cdot) r(t, \cdot)] = \sigma_r^2 = \int_{-\infty}^{\infty} P_r(f) df. \tag{2.196}$$

Therefore, the variance of the random processes in the time domain is obtained by integrating $P_r(f)$ over frequency. The power spectral density at the output of a linear filter with an impulse response of $h(t)$ is the same as the results obtained in section 2.4.

$$P_y(f) = P_r(f) |H(j2\pi f)|^2. \tag{2.197}$$

The following section will demonstrate how there results can be applied to cyclo-stationary random processes in certain instances.

Cyclo-Stationary Random Data Although it has not been explicitly stated, the NRZ data signal is a non-stationary process. However, there is a definite structure embedded in the randomness, and the statistics of the signal are periodic with a period equal to the data bit-period T. This type of random process is known as a cyclo-stationary process, because the statistics are periodic. To see that the statistics are periodic we'll consider the auto-correlation function of rectangular NRZ data $R_d(t, \tau)$ given by

$$R_d(t, \tau) = E[d(t, \cdot)d(t + \tau, \cdot)] \tag{2.198}$$

For rectangular data the function $R_d(t, \tau)$ is illustrated in Fig. 2.32, where it can be seen that the shape of $R_d(t, \tau)$ is unchanged, but is shifted depending on where t lies in the interval $[nT, (n+1)T]$. For the first period centered around $t = 0$, this function can be written as

$$R_d(t, \tau) = \text{rect}\left(\frac{t+\tau}{T}\right). \tag{2.199}$$

In section 2.5.1 it was shown that the expected value of the time-autocorrelation function is

$$A_d(\tau) = \int_{-\infty}^{\infty} R_d(t, \tau)dt \tag{2.200a}$$

For a time limited data sequence of N-bits, then the integral can be replaced by the sum of N integrals, each integrated over one period, and since the statistics are periodic, then $A_d(\tau)$ is just N times the integral over one period.

$$A_d(\tau) = N \int_{-T/2}^{T/2} \text{rect}\left(\frac{t+\tau}{T}\right) dt \tag{2.200b}$$

The integral of $R_d(t, \tau)$ over each period is plotted in Fig. 2.33. Multiplying the integrand by $\text{rect}(t/T)$ doesn't change the integral since the the function is unity over exactly the limits of integration. Therefore,

$$A_d(\tau) = N \int_{-T/2}^{T/2} \text{rect}(t/T)\text{rect}\left(\frac{t+\tau}{T}\right) dt \tag{2.200c}$$

$$= N \int_{-\infty}^{\infty} \text{rect}(t/T)\text{rect}\left(\frac{t+\tau}{T}\right) dt \tag{2.200d}$$

$$= N \int_{-\infty}^{\infty} \text{rect}(-t/T)\text{rect}\left(\frac{\tau-t}{T}\right) dt, \tag{2.200e}$$

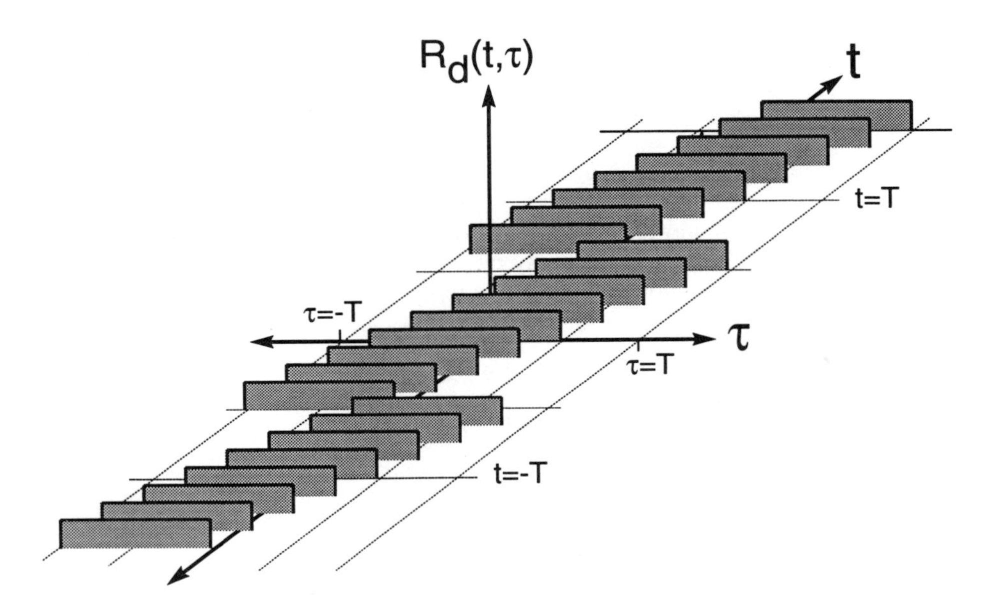

Figure 2.32 Periodic autocorrelation function for rectangular NRZ data.

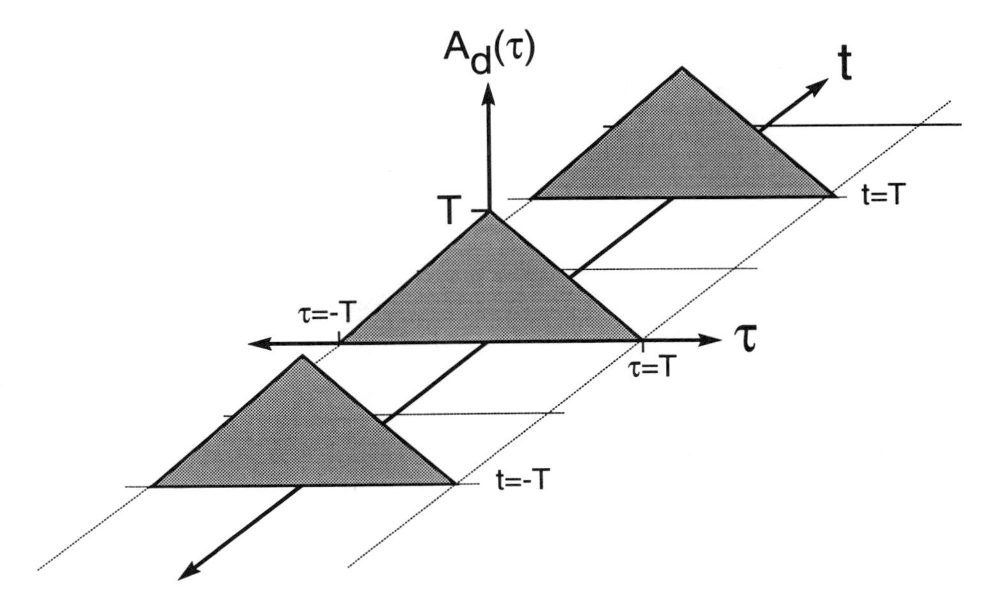

Figure 2.33 Integral per period of the autocorrelation function of rectangular NRZ data.

and since the rect function is symmetric $\text{rect}(t/T) = \text{rect}(-t/T)$, then

$$A_d(\tau) = N \int_{-\infty}^{\infty} \text{rect}(t/T)\text{rect}\left(\frac{\tau - t}{T}\right) dt \tag{2.200f}$$

$$= N[\text{rect}(t/T) * \text{rect}(t/T)]. \tag{2.200g}$$

This is the same function plotted in Fig 2.31. However, we can now make use of the results from the theory of stationary random processes to relate $A_d(\tau)$ to signal statistics. For a cyclo-stationary random process the variance is periodic with period T. From the definition of $A_d(\tau)$ for an N-bit sequence,

$$A_d(\tau) = NT\left[\frac{1}{T}\int_{-T/2}^{T/2} R_d(t,\tau)dt\right] \tag{2.201a}$$

$$A_d(0) = NT\left[\frac{1}{T}\int_{-T/2}^{T/2} \sigma_d^2(t)dt\right] \tag{2.201b}$$

$$A_d(0)/NT = \overline{\sigma_d^2(t)} \tag{2.201c}$$

This result states that the average variance over one period is equal to the time auto-correlation function divided by the time interval NT. For the case of rectangular NRZ data the variance can be determined easily in the time domain. The data is either +1 or -1, so the variance is equal to 1. In the general case the variance will be periodic, but in this special cases it is a constant. The variance is equal to $R_d(t, 0)$, and from Fig. 2.32 it can be seen that $R_d(t, 0)$ is always unity. It is also clear from Fig. 2.33 that $A_d(0)/NT = 1$. The results are summarized in table 2.4 for a cyclo-stationary random data sequence that is limited in time.

Time-Limited Random Signals Passed Through Linear Filters As we alluded to earlier, when time-limited data is passed through a filter with an infinite impulse response, the data will no longer be time limited. This raises the question of how to deal with such a situation. For the time limited case it was shown that the expected value of the energy in the signal is given by

$$\overline{E_d} = A_d(0) = E\left[\int_0^{NT} d^2(t,\cdot)dt\right] \tag{2.202}$$

$$= \int_{-\infty}^{\infty} S_d(f)df. \tag{2.203}$$

If the time limited data is passed through a filter, then there will be a transient response at the beginning and end of the data sequence and a steady-state response for most of

$$
\begin{aligned}
R_d(t,\tau) &= E[d(t,\cdot)d(t+\tau,\cdot)] \\
R_d(t,0) &= E[|d(t,\cdot)|^2] &&= \sigma_d^2(t) \\
\overline{\sigma_d^2} &= \frac{1}{NT}\int_{-NT/2}^{NT/2}\sigma_d^2(t)dt &&= \frac{1}{T}\int_{-T/2}^{T/2}\sigma_d^2(t)dt \\
a_d(\tau,\cdot) &= \lim_{t_d\to\infty}\int_{-t_d/2}^{t_d/2} d(t,\cdot)d(t+\tau,\cdot)dt &&= \int_{-NT/2}^{NT/2} d(t,\cdot)d(t+\tau,\cdot)dt \\
A_d(\tau) &= E[a_d(\tau,\cdot)] &&= \int_{-NT/2}^{NT/2} R_d(t,\tau)dt \\
S_d(f) &= \mathcal{F}\{A_d(\tau)\} \\
P_d(f) &= S_d(f)/NT \\
A_d(0) &= \int_{-NT/2}^{NT/2}\sigma_d^2(t)dt &&= NT\overline{\sigma_d^2} \\
A_d(0) &= \int_{-\infty}^{\infty} S_d(f)df &&= NT\overline{\sigma_d^2} \\
\overline{\sigma_d^2} &= \int_{-\infty}^{\infty} P_d(f)df &&= A_d(0)/NT
\end{aligned}
$$

Table 2.4 Summary of Relations for Time-Limited Cyclo-Stationary Random Signals.

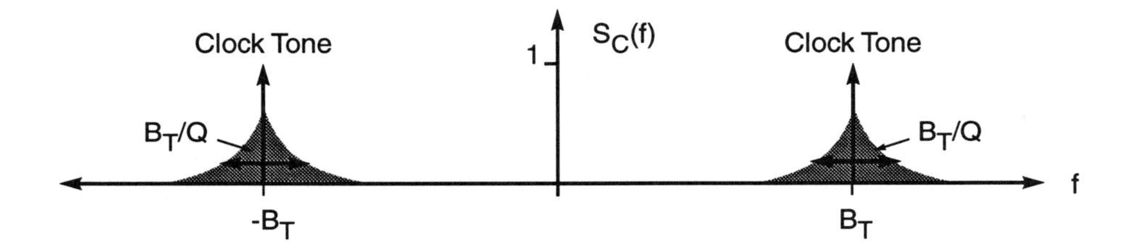

Figure 2.34 Energy spectrum of a clock recovered by passing an edge-detected data signal through a BPF.

the duration of the data. Therefore, the energy can be written as

$$A_d(0) = \int_0^{t_1} + \int_{t_1}^{NT+t_2} + \int_{NT+t_2}^{NT+t_3}, \tag{2.204}$$

where transient behavior occurs in the time intervals $[0, t_1]$ and $[NT + t_2, NT + t_3]$. As the number of bits becomes very large, the first and third integrals are negligible compared to the middle integral, provided that the filter is stable and the transient response dies out over time. As a practical consideration, when the length of the random sequence NT is very large compared with the time constant of a filter that operates on the data, then the output energy of the filter can be approximately obtained by integrating over the interval $[0, NT]$, and ignoring the transient behavior. Therefore, the results of table 2.4 can be applied to data that is not time limited, provided that N is large compared to the normalized transient time t_{tran}/T.

2.6 RANDOM AMPLITUDE MODULATION

Now we're finally in a position to use the theory for a practical problem. We would like to predict the rms value of the amplitude modulation in a clock signal that was recovered from the data by passing the edge-detected signal through a bandpass filter (BPF). The energy spectrum of an edge-detected signal at the output of a BPF is shown in Fig. 2.34. It will be assumed that the side-band energy in the recovered clock is due entirely to amplitude modulation, and the recovered clock signal will be assumed to be limited in time to $[-NT/2, NT/2]$. If the clock is not time limited, then the analysis will still hold if the energy in the transient tails of the clock are negligible compared to the energy in the interval $[-NT/2, NT/2]$. Under these assumptions the clock signal can be expressed as

$$c(t, \cdot) = [(1 + a_m(t, \cdot)) \cos(2\pi B_T t)] \operatorname{rect}(t/NT), \tag{2.205}$$

where $a_m(t,\cdot)$ is a zero-mean random variable that represents the normalized amplitude modulation. It is desired to find the power spectral density for this clock signal. We start by finding the Fourier transform of $c(t,\cdot)$.

$$\begin{aligned} F_c(j2\pi f,\cdot) &= \left[\tfrac{1}{2}\delta(f\pm B_T)+\tfrac{1}{2}F_{am}(j2\pi(f\pm B_T)\right] * NT\text{sinc}(fNT) \\ &= \tfrac{1}{2}\left[NT\text{sinc}(NT\,(f\pm B_T))\right]+ \\ &\quad \tfrac{1}{2}\left[F_{am}(j2\pi f) * NT\text{sinc}(NT\,(f\pm B_T))\right]. \end{aligned} \tag{2.206}$$

Since the number of bits is assumed to be large, the sinc function can be approximated as an impulse.

$$NT\text{sinc}(fNT)\simeq\delta(f).$$

Replacing the second sinc function with an impulse gives

$$F_c(j2\pi f,\cdot)=\frac{1}{2}\left[NT\text{sinc}(NT\,(f\pm B_T))+F_{am}\left(j2\pi(f\pm B_T)\right)\right] \tag{2.207}$$

We delay replacing the first sinc function with an impulse until after the magnitude is squared to avoid mathematical difficulties associated with squaring an impulse function. Since the signal is also narrow-band, the sidebands don't overlap and the squared magnitude is

$$\begin{aligned} |F_c(j2\pi f,\cdot)|^2 &= \tfrac{1}{4}\left[NT\text{sinc}(NT(f\pm B_T)\right]^2+ \\ &\quad \tfrac{1}{4}\left[|F_{am}\left(j2\pi(f-B_T)\right)|^2+|F_{am}\left(j2\pi(f+B_T)\right)|^2\right]+ \\ &\quad \text{cross-terms}(f,\cdot). \end{aligned} \tag{2.208}$$

The ESD is found by taking the expected value of $|F_c(j2\pi f,\cdot)|^2$. Since the expected value of $F_c(j2\pi f,\cdot)=0$, then the expected value of the cross-terms are also zero. The sinc^2 function can now be replaced by an impulse function of equal area. Therefore,

$$S_c(f)=E\left[|F_c(j2\pi f,\cdot)|^2\right]=\frac{1}{4}\left[NT\delta(f\pm B_T)+S_{am}(f\pm B_T)\right], \tag{2.209}$$

where $S_{am}(f)$ is the ESD for the baseband amplitude modulation. The variance of the baseband amplitude modulation is obtained by integrating the ESD over all frequencies.

$$NT\overline{\sigma_{am}^2}=\int_{-\infty}^{\infty}S_{am}(f)df \tag{2.210}$$

The total energy in the side-bands of the clock is then

$$\overline{E_{sb}}=\frac{1}{4}\int_{-\infty}^{\infty}\left[S_{am}(f-B_T)+S_{am}(f+B_T)\right]df \tag{2.211a}$$

$$=\frac{NT}{2}\overline{\sigma_{am}^2}, \tag{2.211b}$$

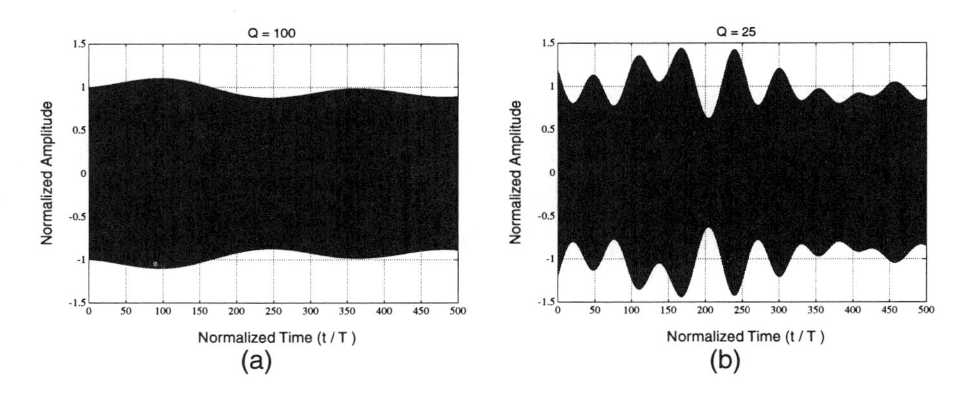

Figure 2.35 Clock recovered by passing edge-detected NRZ data through, linear-phase bandpass filter with approximate selectivities of: (a) $Q = 100$, (b) $Q = 25$.

and the energy in the pure tone is

$$\overline{E_{\text{tone}}} = \frac{1}{4}NT \int_{-\infty}^{\infty} [\delta(f - B_T) + \delta(f + B_T)]\, df = \frac{1}{2}NT. \tag{2.212}$$

Therefore, the variance of the amplitude modulation is given by the energy ratio

$$\boxed{\overline{\sigma_{am}^2} = \frac{\overline{E_{sb}}}{\overline{E_{\text{tone}}}}} \tag{2.213}$$

It was shown in (2.177) and (2.181) that this energy ratio is approximately equal to $1/Q$ for a narrow-band ideal BPF. Therefore, we have finally arrived at the simple and useful result, that the rms deviation in the envelope of the clock signal derived by passing an edge-detected signal through and ideal, linear-phase BPF with a selectivity of Q.

$$\boxed{\overline{\sigma_{am}} = \frac{1}{\sqrt{Q}}} \tag{2.214}$$

Simulations Results: Amplitude Modulations of Extracted Clock Simulation results of clock extraction using BPFs with approximate Q values of 100 and 25 are shown in Fig. 2.35. For $Q = 100$ the predicted rms amplitude modulation is 10%, and is 20% for $Q = 25$. The normalized PSD of the clock signal for $Q \sim 25$ is shown in Fig. 2.36. For any non-ideal filter shape an effective Q can be defined in terms of the sideband power according to

$$P_{sb} = 2Q_{eq}B_T|H(j2\pi B_T)|^2, \tag{2.215}$$

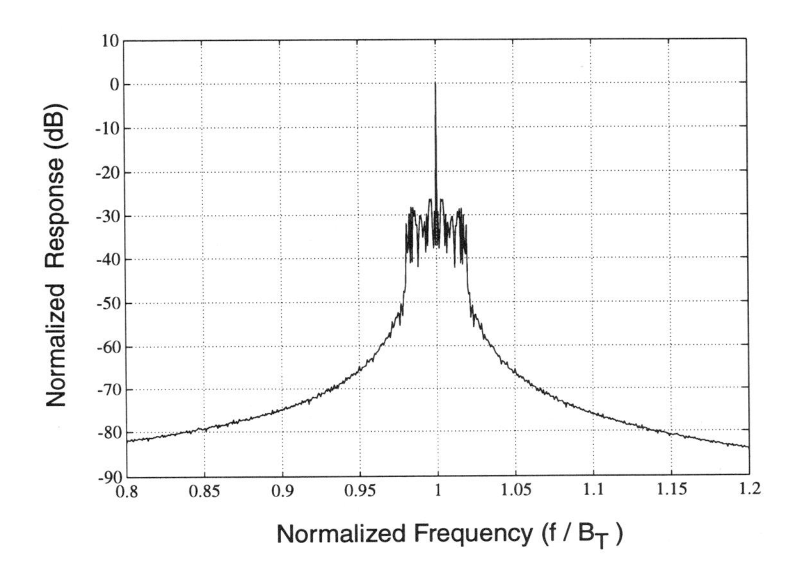

Figure 2.36 Normalized narrowband spectrum of a bandpass filter with $Q_{eq} = 20.75$.

and the equivalent selectivity is then

$$Q_{eq} = \frac{1}{B_T |H(j2\pi B_T)|^2} \int_0^\infty |H(j2\pi f)|^2 df. \tag{2.216}$$

With this definition of Q_{eq} we can express the rms envelope deviation for any arbitrarily shaped bandpass filter as

$$\boxed{\overline{\sigma_{am}} = \frac{1}{\sqrt{Q_{eq}}}} \tag{2.217}$$

Using (2.216) the equivalent Q values for the simulated filters were found to be 92.63 and 20.75 respectively. The theory predicts that the standard deviation in the the envelope is $1/\sqrt{Q_{eq}}$, which gives us values for σ_{am} of 0.1039 and 0.2198 respectively. These results can be compared with the simulation by extracting the envelope of the clock signals and determining the envelopes statistics. Fig. 2.37 shows histograms of the deviation in the envelopes of the simulated recovered clocks. The simulated rms envelope deviation was $\sigma_{am} = 0.1038$ for $Q_{eq} = 92.63$, and $\sigma_{am} = 0.2195$ for $Q_{eq} = 20.75$, which is within 0.14% of the results predicted using the frequency domain power ratios. One final remark can be made about the envelope. The narrowband filter has a bandwidth of approximately B_T/Q_{eq}. It was shown that the energy in this band is just the baseband amplitude modulation shifted to the

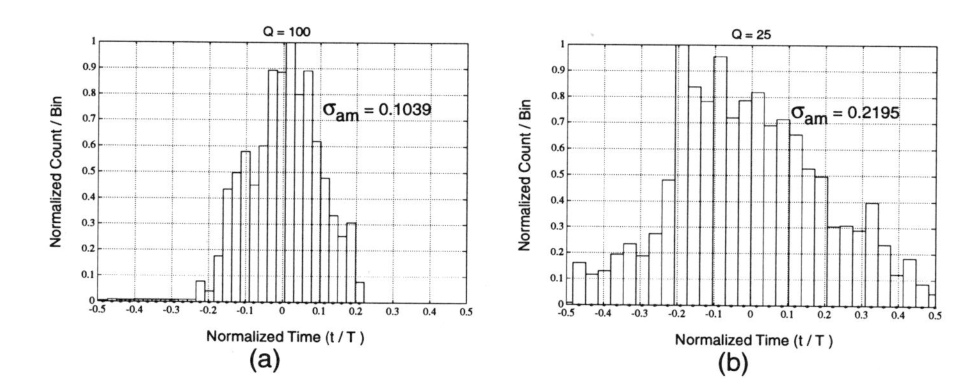

Figure 2.37 Histograms of the random envelope deviation in simulated edge-detected NRZ data passing through an ideal, linear- phase BPF with: (a) $Q = 100$, (b) $Q = 25$.

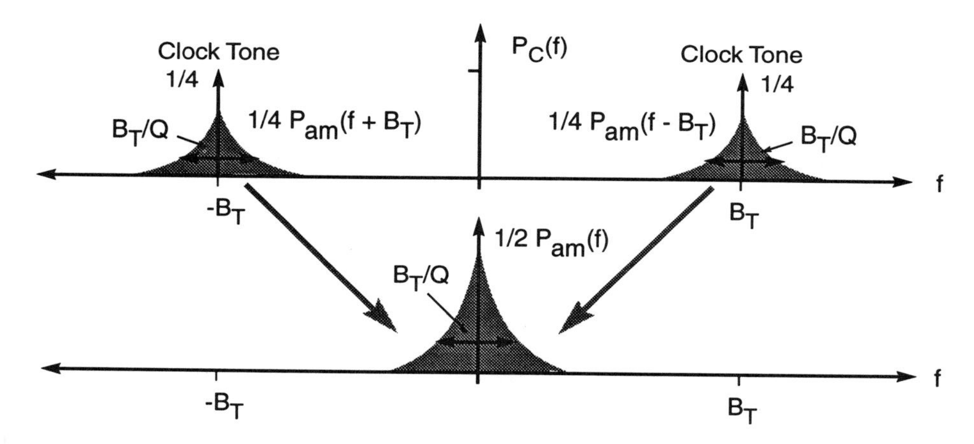

Figure 2.38 Illustration of the narrowband spectrum being down converted to be centered around dc.

clock frequency. Therefore, by down converting the narrowband spectrum to dc as shown in Fig. 2.38 it is clearly seen that the amplitude has a maximum frequency of approximately $B_T/2Q_{eq}$. Therefore, we should expect on average that the envelope will have one random cycle in every $2Q_{eq}$ clock periods.

Energy Spectral Densities are not Unique Mappings As stated previously, the energy spectral densities ignore phase information, and the the mapping of a signal from the time domain to the ESD is not unique. In fact several different signals can have the same ESD. In this section the narrow-band spectrum was assumed to have been the result of only amplitude modulation. This is the case when the signal was passed through a linear-phase filter. In the following section we will show how the identical ESD could have resulted from phase-modulation.

2.7 PHASE-JITTER

The narrow-band spectrum of Fig. 2.34 could have resulted from a clock signal with phase-modulation only. Such a signal can be written as

$$c(t,\cdot) = [\sin(2\pi B_T t + \phi(t,\cdot))]\,\mathrm{rect}(t/NT) \tag{2.218}$$

Using the identity

$$\sin(A+B) = \sin B \cos A + \cos B \sin A,$$

then the phase-noise can be separated from the center frequency.

$$c(t,\cdot) = [\sin(\phi(t,\cdot))\cos(2\pi B_T t) + \cos(\phi(t,\cdot))\sin(2\pi B_T t)]\,\mathrm{rect}(t/NT). \tag{2.219}$$

When the phase modulation is small, then we can make the following first-order approximations,

$$\begin{aligned} \sin(\phi(t,\cdot)) &\simeq \phi(t,\cdot) \\ \cos(\phi(t,\cdot)) &\simeq 1. \end{aligned}$$

Under the small-signal approximation, the clock signal can be written as

$$c(t,\cdot) \simeq [\phi(t,\cdot)\cos(2\pi B_T t) + \sin(2\pi B_T t)]\,\mathrm{rect}(t/NT), \tag{2.220}$$

and we see that for small angle deviations, the phase-modulation is approximately equal to amplitude modulation of a carrier in quadrature with the main clock tone.

To obtain the power spectrum, we start by taking the Fourier transform of the random clock signal of (2.220).

$$F_c(j2\pi f,\cdot) = \frac{1}{2}F_\phi(j2\pi(f \pm B_T),\cdot) + \frac{1}{2j}\delta(f - B_T) - \frac{1}{2j}\delta(f + B_T) \tag{2.221}$$

The expected value of the squared magnitude of F_c divided by the time interval NT, gives us the desired PSD;

$$P_c(f) = \frac{1}{4}P_\phi(f \pm B_T) + \frac{1}{4}\delta(f \pm B_T) \tag{2.222}$$

The phase-noise variance can be found from the baseband PSD by integrating over all frequencies.

$$\sigma_\phi^2 = \int_{\infty}^{-\infty} P_\phi(f)df \tag{2.223}$$

Using the narrowband assumption that the overlap of the positive and negative parts of the frequency spectrum is insignificant, then the baseband power can be expressed as an integral over the bandpass spectrum. The sideband power is therefore

$$\overline{P_{\text{sb}}} = \frac{1}{4}\int_{-\infty}^{\infty} P_\phi(f \pm B_T)df = \frac{1}{2}\int_{-\infty}^{\infty} P_\phi(f)df = \frac{\sigma_\phi^2}{2}, \tag{2.224}$$

and the power in the tone is

$$\overline{P_{\text{tone}}} = \frac{1}{4}\int_{\infty}^{-\infty} \delta(f \pm B_T)df = \frac{1}{2}. \tag{2.225}$$

Therefore, the noise variance, in radians squared, is just given by the ratio

$$\sigma_\phi^2 = \frac{\overline{P_{\text{sb}}}}{\overline{P_{\text{tone}}}} \tag{2.226}$$

For a bandpass filter with Q_{eq} defined by (2.216), then we obtain the simple result

$$\boxed{\sigma_\phi = \frac{1}{\sqrt{Q_{eq}}}.} \tag{2.227}$$

This is the same result that we obtained for amplitude modulation which is not surprising since the small-signal linearization converted the phase-noise to amplitude noise. In the next few pages the same result will be derived more rigorously, and it will be shown that a second-order approximation of the rms phase noise can be given by

$$\boxed{\sigma_\phi = \frac{1}{\sqrt{1 + Q_{eq}}}.} \tag{2.228}$$

An illustration of the narrowband spectrum to be derived that will give rise to the above second-order approximation is shown in Fig. 2.39. The reader wishing to skip the following derivation, may wish to proceed directly to the simulation results on page 96.

2.7.1 Second-Order Estimate of Phase-Noise Variance

The goal of this analysis is to find an expression for the narrowband power spectrum $P_c(f)$ in terms of the baseband PSD $P_\phi(f)$ of the random phase modulation. We can

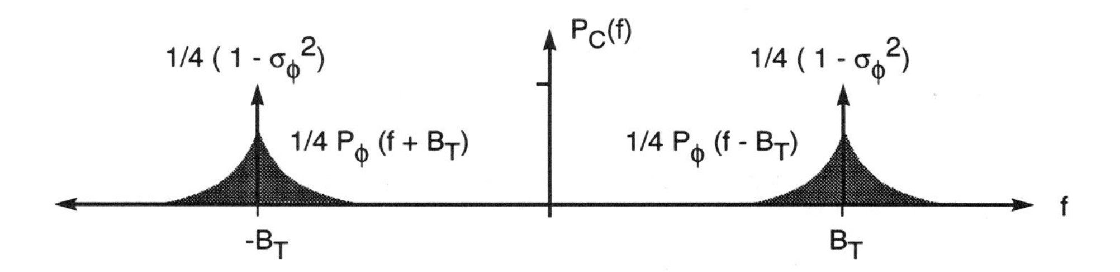

Figure 2.39 Narrowband power spectral density of a signal due to random phase modulation.

find the ESD $S_c(f)$ of this clock signal using the techniques of section 2.5, by first finding the time-autocorrelation function.

$$a_c(\tau, \cdot) = \int_{-NT/2}^{NT/2} c(t, \cdot)c(t + \tau, \cdot)dt \tag{2.229}$$

The expected value of $a_c(\tau, \cdot)$ is the integral of the ensemble autocorrelation function.

$$A_c(\tau) = E[a_c(\tau, \cdot)] = \int_{-NT/2}^{NT/2} R_c(t, \tau)dt \tag{2.230}$$

The narrow-band spectrum that we are looking for is the ESD of $c(t, \cdot)$, which is the Fourier transform of $A_c(\tau)$.

$$S_c(f) = \mathcal{F}\{A_c(\tau)\} \tag{2.231}$$

Explicit Expression for $R_c(t, \tau)$ The above outlined analysis can be carried out by first finding an explicit expression for the ensemble autocorrelation function $R_c(t, \tau)$ of the clock signal. By definition

$$\begin{aligned} R_c(t, \tau) = E\Big[& [\sin(2\pi B_T t + \phi(t, \cdot))] \operatorname{rect}(t/NT) \times \\ & [\sin(2\pi B_T(t + \tau) + \phi(t + \tau, \cdot))] \operatorname{rect}((t + \tau)/NT)\Big] \end{aligned} \tag{2.232}$$

Now we can make make use of some trigonometric manipulations to separate terms into sum and difference frequencies. Recalling

$$\sin A \sin B = \frac{1}{2}\cos(A - B) - \frac{1}{2}\cos(A + B),$$

then

$$\begin{aligned} R_c(t,\tau) = E\Big[& \frac{1}{2}\cos\left(2\pi B_T\tau + \phi(t+\tau,\cdot) - \phi(t,\cdot)\right) \\ & - \frac{1}{2}\cos\left(2\pi(2B_T)t + 2\pi B_T\tau + \phi(t+\tau,\cdot) + \phi(t,\cdot)\right) \\ & \Big]\text{rect}(t/NT)\text{rect}\left((t+\tau)/NT\right), \end{aligned} \tag{2.233}$$

and recalling also that

$$\cos(A+B) = \cos A\cos B - \sin A\sin B,$$

then the fast varying center-frequency terms can be separated from the random phase-noise terms.

$$\begin{aligned} R_c(t,\tau) = \Big\{ & \frac{1}{2}\cos(2\pi B_T\tau)E[\cos\left(\phi(t+\tau,\cdot) - \phi(t,\cdot)\right)] \\ & - \frac{1}{2}\sin(2\pi B_T\tau)E[\sin\left(\phi(t+\tau,\cdot) - \phi(t,\cdot)\right)] \\ & - \frac{1}{2}\cos\left(2\pi(2B_T)t + 2\pi B_T\tau\right) E[\cos\left(\phi(t+\tau,\cdot) + \phi(t,\cdot)\right)] \\ & + \frac{1}{2}\sin\left(2\pi(2B_T)t + 2\pi B_T\tau\right) E[\sin\left(\phi(t+\tau,\cdot) + \phi(t,\cdot)\right)] \\ & \Big\}\text{rect}(t/NT)\text{rect}\left((t+\tau)/NT\right). \end{aligned} \tag{2.234}$$

With one more application of the previous trig identity and

$$\sin(A+B) = \sin A\cos B + \cos A\sin B,$$

then terms involving t, can be separated from terms involving τ, and we finally get the desired form of the auto-correlation function:

$$\begin{aligned} R_c(t,\tau) = \Big\{ & \frac{1}{2}\cos(2\pi B_T\tau)E[\cos(\phi(t+\tau,\cdot)-\phi(t,\cdot))] \\ & -\frac{1}{2}\sin(2\pi B_T\tau)E[\sin(\phi(t+\tau,\cdot)-\phi(t,\cdot))] \\ & -\frac{1}{2}\cos(2\pi(2B_T)t)\cos(2\pi B_T\tau)E[\cos(\phi(t+\tau,\cdot)+\phi(t,\cdot))] \\ & +\frac{1}{2}\sin(2\pi(2B_T)t)\sin(2\pi B_T\tau)E[\cos(\phi(t+\tau,\cdot)+\phi(t,\cdot))] \\ & +\frac{1}{2}\sin(2\pi(2B_T)t)\cos(2\pi B_T\tau)E[\sin(\phi(t+\tau,\cdot)+\phi(t,\cdot))] \\ & +\frac{1}{2}\cos(2\pi(2B_T)t)\sin(2\pi B_T\tau)E[\sin(\phi(t+\tau,\cdot)+\phi(t,\cdot))] \\ & \Big\}\mathrm{rect}(t/NT)\mathrm{rect}((t+\tau)/NT)\,. \end{aligned} \tag{2.235}$$

Since we are dealing with a narrow-band signal, then the baseband modulation, by definition, varies much slower than the tone. Therefore, the terms involving

$$\begin{gathered} E[\cos(\phi(t+\tau,\cdot)+\phi(t,\cdot))] \\ E[\sin(\phi(t+\tau,\cdot)+\phi(t,\cdot))] \end{gathered}$$

are expectations of a slowly varying signal, and these terms remain essentially constant over several periods of the double-frequency $(2B_T)$ signal. Hence, when (2.235) is integrated over time, the last four terms will vanish.

Approximations for Small Angles To continue the analysis it is helpful at this point to make some approximations assuming that the phase modulation is small. This is a valid assumption, because any clock signal that has a large cycle-to-cycle phase jitter $\phi(t,\cdot)$ is of no use to us, so there is no need to analyze it. Instead we will be considering a clock signal with small phase deviations. Recalling the series expansions for sine and cosine around zero

$$\begin{aligned} \sin\theta &= \theta - \tfrac{\theta^3}{3!} + \tfrac{\theta^5}{5!} - \cdots \\ \cos\theta &= 1 - \tfrac{\theta^2}{2!} + \tfrac{\theta^4}{4!} - \cdots, \end{aligned}$$

and ignoring any terms of 3rd order or greater, then

$$E\left[\sin(\phi(t+\tau,\cdot)-\phi(t,\cdot))\right] \simeq E[\phi(t+\tau,\cdot)] - E[\phi(t,\cdot)], \tag{2.236}$$

and for a zero-mean phase-noise process

$$E[\phi(t+\tau,\cdot)] - E[\phi(t,\cdot)] = 0 - 0 = 0. \tag{2.237}$$

Using the small-angle approximation for the cosine function

$$\begin{aligned} E\left[\cos\left(\phi(t+\tau,\cdot) - \phi(t,\cdot)\right)\right] \simeq 1 - \frac{1}{2}\Big[E[\phi^2(t+\tau,\cdot)] + \; E[\phi^2(\tau,\cdot)] \\ - 2E[\phi(t+\tau,\cdot)\phi(t,\cdot)]\Big]. \end{aligned} \tag{2.238}$$

If the base-band phase-noise process is assumed to by wide-sense stationary with a variance of σ_ϕ^2 and an auto-correlation function $R_\phi(\tau)$, then

$$E\left[\cos\left(\phi(t+\tau,\cdot) - \phi(t,\cdot)\right)\right] \simeq 1 - \sigma_\phi^2 + R_\phi(\tau). \tag{2.239}$$

Energy Spectral Density Now the expected value of the time-autocorrelation function $A_c(\tau)$ can be found by integrating $R_c(t,\tau)$. Define an effective ensemble auto-correlation function $\widehat{R_c}(\tau)$ as,

$$\widehat{R_c}(\tau) \triangleq \frac{1}{2}\cos(2\pi B_T \tau)\left[1 - \sigma_\phi^2 + R_\phi(\tau)\right]. \tag{2.240}$$

Integrating the horrendous expression in (2.235) reduces to

$$\begin{aligned} A_c(\tau) &= \int_{-NT/2}^{NT/2} R_c(t,\tau)dt \\ &= \widehat{R_c}(\tau) \int_{-NT/2}^{NT/2} \text{rect}(t/NT)\text{rect}\left((t+\tau)/NT\right)dt \\ &= \widehat{R_c}(\tau)\left[\text{rect}(\tau/NT) * \text{rect}(\tau/NT)\right] \end{aligned} \tag{2.241}$$

Taking the Fourier transform of $A_c(\tau)$ will finally give us the energy spectral density function of the narrowband signal in terms of the baseband ESD. In anticipation of the final result we'll define a power spectral density $P_c(f)$ as the Fourier transform of $\widehat{R_c}(\tau)$.

$$P_c(f) \triangleq \mathcal{F}\left\{\widehat{R_c}(\tau)\right\} \tag{2.242a}$$

$$P_c(f) = \frac{1}{4}(1 - \sigma_\phi^2)\delta(f \pm B_T) + \frac{1}{4}P_\phi(f \pm B_T), \tag{2.242b}$$

where $P_\phi(f)$ is the PSD of the baseband phase noise. Therefore,

$$S_c(f) = \mathcal{F}\left\{\widehat{R_c}(\tau)\right\} * \mathcal{F}\left\{\text{rect}(\tau/NT) * \text{rect}(\tau/NT)\right\} \tag{2.243a}$$

$$S_c(f) = P_c(f) * [NT\text{sinc}(fNT)]^2, \tag{2.243b}$$

and using the now familiar approximation of the sinc^2 function with an impulse of equal area we obtain

$$S_c(f) = P_c(f) * NT\delta(f) \tag{2.243c}$$

$$S_c(f) = NTP_c(f), \tag{2.243d}$$

dividing the ESD by the time interval NT the power spectral density of the phase-modulated signal is as anticipated

$$\boxed{P_c(f) = \frac{1}{4}(1-\sigma_\phi^2)\delta(f \pm B_T) + \frac{1}{4}P_\phi(f \pm B_T).} \tag{2.244}$$

Determining the Phase Noise Variance from $P_c(f)$ It was assumed that the time domain signal corresponding to the energy spectrum was a constant amplitude tone with small-signal phase modulation. The expression for the narrowband PSD was expressed in terms of the PSD of the baseband phase-noise as in (2.244). The result is that the PSD consists of a pure tone plus the baseband noise PSD shifted to $\pm B_T$. This is illustrated in Fig. 2.39. The phase-noise variance can be found by taking the ratio of the tone power and the sideband power. From (2.244) the tone power is

$$\overline{P_{\text{tone}}} = \frac{1}{4}\int_{-\infty}^{\infty}(1-\sigma_\phi^2)\delta(f \pm B_T)df = \frac{1}{2}(1-\sigma_\phi^2), \tag{2.245}$$

and the sideband power is

$$\overline{P_{\text{sb}}} = \frac{1}{4}\int_{-\infty}^{\infty} P_\phi(f \pm B_T)df \tag{2.246a}$$

$$= \frac{1}{2}\int_{-\infty}^{\infty} P_\phi(f)df \tag{2.246b}$$

$$= \frac{\sigma_\phi^2}{2}. \tag{2.246c}$$

Therefore the ratio of the two powers is simply related to the phase-noise variance by

$$\frac{\overline{P_{\text{sb}}}}{\overline{P_{\text{tone}}}} = \frac{\sigma_\phi^2}{1-\sigma_\phi^2}. \tag{2.247}$$

Alternatively, we can express the noise variance in terms of the power ratio

$$\boxed{\sigma_\phi^2 = \frac{\overline{P_{\text{sb}}}/\overline{P_{\text{tone}}}}{1 + \overline{P_{\text{sb}}}/\overline{P_{\text{tone}}}}.} \quad (2.248)$$

Returning to the example of edge-detected NRZ data passing through an ideal bandpass filter of selectivity Q, the above power ratio, which is the same as the energy ratio, is just equal to $1/Q$. The noise variance of the recovered clock signal is therefore,

$$\sigma_\phi^2 = \frac{1}{1+Q}. \quad (2.249)$$

For an arbitrary filter we use the equivalent selectivity, so that the general result is

$$\boxed{\sigma_\phi = \frac{1}{\sqrt{1+Q_{eq}}} \qquad \text{radians}} \quad (2.250)$$

This result, however, assumes that all of the sideband energy is converted to phase-noise, and there is no contribution to the envelope deviation. Therefore, (2.250) gives an upper-bound on the phase-noise obtained by filtering random edge-detected data. In the following section we will show how a nonlinear phase filter distributes the noise power between amplitude and phase modulation. Before, ending this section, however, we will give some simulation results that verify the theory.

2.7.2 Simulation of Narrowband Phase-Noise

To illustrate the application of the above theory for predicting phase noise, the following signal was simulated for various values of rms phase-noise $\theta(t,\cdot)$

$$c(t,\cdot) = \cos(2\pi f_0 t + \theta(t,\cdot)). \quad (2.251)$$

The simulated eye-diagrams for two cases are shown in Fig. 2.40. The noise variance can be calculated in the time domain because we have explicit control of the phase-noise in the simulation. The normalized histograms for the phase noise are shown in Fig. 2.41, where the rms phase deviation was calculated to be 0.0994 and 0.1962 radians respectively. The power-spectral density of the clock signals with random phase jitter was obtained by taking a discrete Fourier transform of the signal. The signal was broken into 128 intervals of 32 periods each, and the power spectrum of each short segment was averaged to obtain the estimates of the PSDs shown in Fig. 2.42. The theory predicts that the phase-noise variance is just the ratio of the tone power to the

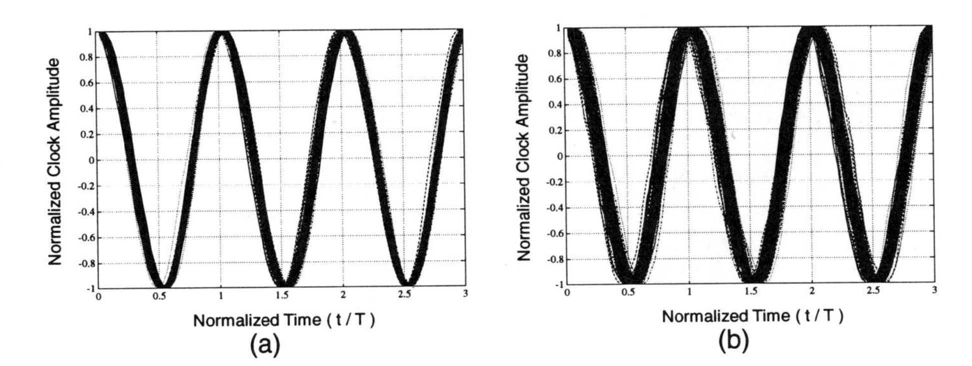

Figure 2.40 Eye-diagrams of a clock signal with random phase-noise: (a) $\sigma_\theta \simeq 0.1$, (b) $\sigma_\theta \simeq 0.2$.

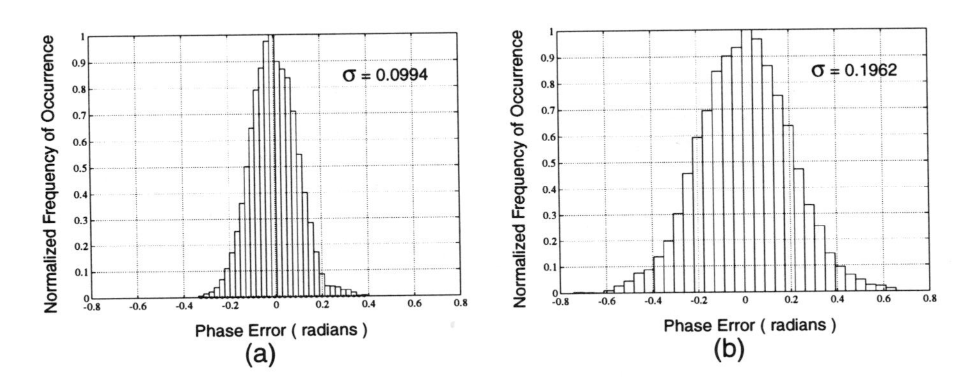

Figure 2.41 Normalized phase-noise histograms: (a) $\sigma_\theta = 0.0994$, (b) $\sigma_\theta = 0.1962$.

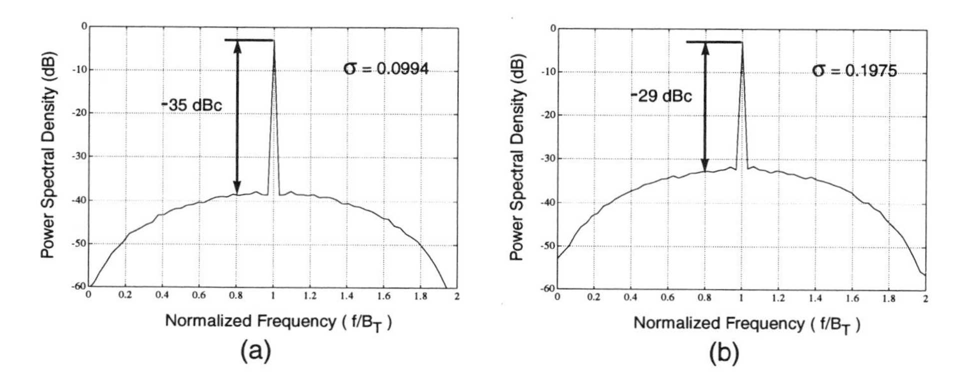

Figure 2.42 Average power in a bandwidth of $B_T/32$. The rms phase-noise calculated from the frequency domain is: (a) $\sigma_\theta = 0.0994$, (b) $\sigma_\theta = 0.1975$.

sideband power. These ratios are $(101.1)^{-1}$ and $(25.64)^{-1}$ respectively from which we calculate the rms noise as

$$\begin{aligned} \sigma_\theta &= \frac{1}{\sqrt{101.1}} = 0.0994 = \quad 5.70^\circ \qquad \text{for (a)} \\ \sigma_\theta &= \frac{1}{\sqrt{25.64}} = 0.1975 = 11.32^\circ \qquad \text{for (b)} \end{aligned} \tag{2.252}$$

For a gaussian random variable, the peak-to-peak deviation is approximately 6σ. We can see from the eye-diagrams that the phase deviation is approximately 34.2° and 67.9° in Figs. 2.40a and b respectively.

Since the simulation was performed with 32 samples per period. The bandwidth is therefore $B_T/32$. In order to double the rms phase noise the sideband power has to increase by a factor of 4. Therefore we see that the noise power is 6 dB higher in the second simulation. It is shown that the noise power is -35 dBc in Fig. 2.42a, and -29 dBc in Fig. 2.42b at an offset of 20% of the clock frequency. To convert these numbers to dBc/Hz, we subtract $10\log(B_T/32)$. For the example of a 10 GHz clock $10\log(B_T/32) = 85$ dB. This results in the noise power being down by -120 dBc/Hz and -114 dBc/Hz at 2 GHz offset, in Figs. 2.42a and b respectively.

These simulation results verify that the theory can be used to predict the rms noise in the time domain from the PSD provided that the assumption of no amplitude modulation has not been violated. In this example the error in the estimate is 0.66% in case (b), and various other simulations have shown that the error is typically on this order for the length of the simulation and number of samples per period chosen. In the case of frequency noise in a voltage controlled oscillator, the phase is the integral of the frequency noise, and the phase variance goes to infinity as time increases.[7] We will defer discussion of this important and practical case until Chapter 8. In the following section, all of the results of this chapter will be tied together to obtain the connection of how random amplitude modulation, through nonlinear phase shifts, can be converted to phase noise. With this information we will be able to predict the minimum obtainable phase jitter in a clock recovered from NRZ data by using a bandpass filter.

2.8 EFFECT OF BPF PHASE RESPONSE ON ANGLE AND AMPLITUDE MODULATION

It was found that for narrowband amplitude modulation, the rms deviation in the envelope is $1/\sqrt{Q}$, which is the same result obtained for the rms phase deviation. Therefore, it has been shown that a BPF with an ideal magnitude response can generate a clock signal that contains only amplitude modulation, or only phase modulation. The actual distribution of the sideband energy between envelope deviations and phase-jitter will be determined by the phase response of the filter. A filter with linear phase (constant group delay) will give rise to a clock signal with only amplitude modulation. However, a real analog filter can only approximate a constant group delay. Any deviation from a constant delay will result in a conversion of envelope deviation energy into phase-jitter. Even if a BPF has linear phase, other non-ideal circuit elements will convert amplitude modulation to phase-jitter. Since the recovered clock signal has a randomly varying envelope, induced by random, data-dependent amplitude modulation, then a limiter, or automatic-gain-control (AGC) amplifier needs to be used to produce a constant amplitude clock. The phase-response of such a limiter circuit will also contribute to clock jitter. The nonlinear phase response of the limiter provides different delays through the circuit at different frequencies. In addition, nonlinear, voltage dependent parasitic capacitance in integrated circuits cause the delay time through the circuit to be amplitude dependent, adding further to the phase-jitter. It is a complicated matter to determine the relative contributions of these nonlinear delays to the phase-jitter, and envelope deviation respectively. However, assuming that the rms phase-jitter in the clock due the the random data itself is $1/\sqrt{Q}$ is a reasonable first order approximation for the final clock after the amplitude modulation has been removed: Likewise it is a

[7] This is a practical example of the random walk problem, where a man (presumably drunk, or extremely dizzy) takes steps in random directions. The distance the man is away from the starting point is a random variable. It is equally likely that the man will walk in any direction; one might therefore expect to find the man at the starting point as time increases since each random movement will be balanced by one in the opposite direction. However, this is not the case, and the variance of the distance from the starting point approaches infinity as the time is increased to infinity. This process is also known as Brownian motion. A little thought will reveal that the variance must go to infinity, otherwise no gases would ever diffuse.

reasonable approximation to assume the rms envelope deviation at the output of the BPF is also $1/\sqrt{Q}$.

2.9 SUMMARY

If you look at the literature concerning cyclo-stationary random processes [5, 6, 11], and compare that with the seat-of-the-pants analysis used by most circuit hackers, you'll find an enormous gap. In this chapter we have tried to fill this gap by explaining some of the concepts from random signal theory in the language of circuit design such as, Q-factors, transfer functions, and impulse responses. As far as signals go, pulse-amplitude-modulated (PAM) baseband signals are not very complex; yet a mastery of the theory required to make even the simplest of calculations becomes too involved to make it worthwhile. Also a complete theory can often become formalism for formalism sake. The reason is because signals will be processed by nonlinear operations such as edge-detectors and limiters; following formal analysis through such a systems becomes intractable. In this chapter we have aimed at developing an intuition about such signals so that intelligent systems can be designed even when closed-form mathematical descriptions escape us.

This chapter certainly is not a rigorous mathematical tour de force, but it does help us to reach our goal of designing better receiver ICs. We have been primarily concerned with the following questions.

- What is the spectrum of random PAM data for various signaling formats?
- How is the signal affected by linear filtering, both in the time and frequency domains?
- How is the spectrum affected by nonlinear operations?
- How does prefiltering the data before nonlinear processing affect the harmonic content of the clock tone?
- What is the ratio of the clock-tone energy to the random sideband energy, and how can this ratio be maximized?
- How much amplitude modulation will result when a clock is extracted from random data using a bandpass filter of a given Q?
- How much phase-jitter is present in a clock extracted from random data, where does it come from, and how can it be reduced?

By the end of this chapter, the reader should have the analytical tools to answer these questions. More importantly, the reader should develop a *feel* for the characteristics of random data, and be able to predict the basic behavior of certain circuits by inspection.

REFERENCES

[1] Wilbur B. Davenport, Jr. and William L. Root. *An Introduction to the Theory of Random Signals and Noise*. IEEE Press, New York, 1987. IEEE PRESS edition of a book published by McGraw Hill Book Company in 1958 under the same title.

[2] Paul G. Hoel, Sidney C. Port, and Charles J. Stone. *Introduction to Stochastic Processes*. Houghton Mifflin, Boston, 1972.

[3] Athanasios Papoulis. *Probability, Random Variables, and Stochastic Processes*. McGraw Hill, New York, 1965.

[4] Richard E. Mortensen. *Random Signals and Systems*. John Wiley & Sons, New York, 1987.

[5] W. R. Bennett. Statistics of regenerative digital transmission. *Bell Syst. Tech. J.*, 37(6):1501–1542, November 1958.

[6] R. C. Titsworth and L. R. Welch. Power spectra of signals modulated by random and pseudorandom sequences. Technical Report 32-140, Jet Propulsions Laboratory, Pasadena, Calif., October 1961.

[7] William H. Hayt, Jr. and Jack E. Kemmerly. *Engineering Circuit Analysis*. McGraw Hill, Inc., New York, third edition, 1978.

[8] Mischa Schwartz. *Information Transmission, Modulation, and Noise*. McGraw Hill, Inc., New York, third edition, 1980.

[9] Frederic de Coulon. *Signal Theory and Processing*. Artech House, Inc., Dedham MA, 1986. Translation of *Theorie et Traitement des Signaux*, originally published in French as volume VI of the *Traité d'Electricité* by The Presses Polytechniques Romandes, Lausanne, Switzerland. ©1984.

[10] K. Sam Shanmugam. *Digital and Analog Communication Systems*. John Wiley & Sons, New York, 1979.

[11] William C. Lindsey and Marvin K. Simon. *Telecommunication Systems Engineering*. Dover Publications Inc., New York, 1991. Dover edition first published in 1991 is an unabridged, unaltered republication of the work first published by Prentice-Hall, Inc., Englewood Cliffs, N.J., 1973 in its "Prentice-Hall Information and System Science Series.".

3

OPTIMAL DECISION THEORY APPLIED TO HIGH-SPEED IC RECEIVER DESIGN

The purpose of a telecommunication system is to convey a message, as accurately as possible, from a source to a destination. A model for a typical system is shown in Fig. 3.1. Along the way, the transmitted message can be corrupted by noise and distortion as it travels to its final destination. The purpose of a receiver is to observe the corrupted received signal, and estimate what the original message should have been.

Multilayered Sources of Errors in Communication A receiver is considered "optimal" if it provides the "best" performance relative to some quantitative performance measure, under a given set of assumptions. Errors in communication can occur at several levels; choosing a criteria for optimality in the context of the overall system is non-trivial, because the quality of the received message is generally a subjective judgment. To illustrate this hierarchy of communication, we could represent a conversation between two people using the block diagram of Fig. 3.1. In this example, the source will be considered as a thought, or an idea in the brain of the speaker. This thought can be pictured as roaming around in a multidimensional vector-space of all thoughts. The speaker then maps only a shadow of this idea onto a lower dimensional vector space of thoughts that can be expressed by words. This mapping is analogous to quantization, where an infinite dimensional signal is mapped to a finite number of discrete levels. At this point, much information may have already been lost. The speaker may realize that the idea he is about to express is difficult to understand, so he may repeat key phrases, or offer an analogy to reduce confusion. This is similar to channel coding, where a communication system will purposely add redundancy to reduce errors. Before sending the message, the speaker evaluates the conditions of the communication channel. If the room is noisy, then the speaker might adjust his volume to keep the signal-to-noise ratio at an acceptable level as he modulates his vocal chords and transmits a sound wave in the direction of the listener's ears. If the listener has a good idea of what the speakers voice sounds like, and knows that English

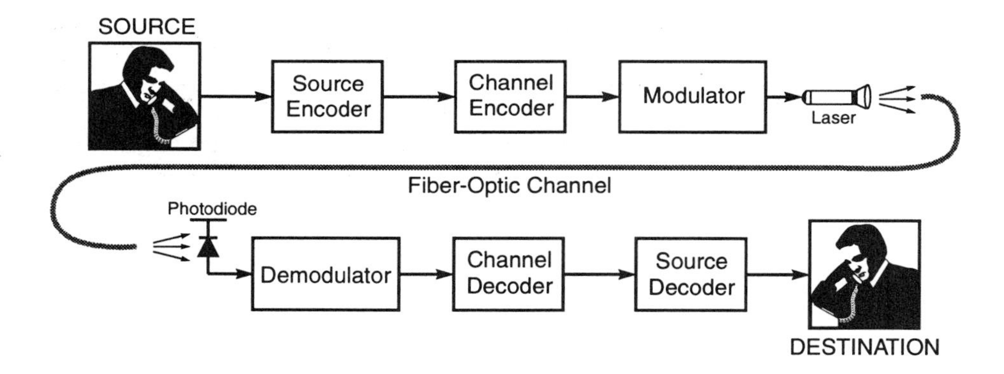

Figure 3.1 Block diagram of a fiber-optic telecommunications link.

words are being spoken, then he can "tune-in" to the speaker, and the soundwaves can be converted to words by the listener, even when the noise is larger than the signal. Electronic receivers perform similar demodulations in the presence of noise by only looking for a given frequency, phase, or pulse shape, and ignoring large background noise. The listener then takes the sequence of words that he has understood, and does the channel decoding. Perhaps one or two words were not clear, but the listener waits for the completion of the sentence, then the missing word can often be filled in by the context of the sentence. Finally, the listener maps the corrected sequence of words to a thought in his own brain. Clearly there are several things that can prevent the errorless communication of ideas. The largest source of error in this example is undoubtedly the mappings of ideas to words, and visa-versa. Not only is this an approximation at best, but to make matters worse, there is no guarantee that the two people talking are using identical mappings. The same words can mean different things to different people, especially when the two speakers are of the opposite sex!

The fascinating subject of human communication has often been left to neurophysiologists, and psychologists. However, engineers have recently utilized biological models to implement neural-network sensors for machine perception, and artificial intelligence. Likewise, utilization of results from linguistics has resulted in improved systems for speech synthesis and recognition. Human perception is often ignored by engineers doing quantitative analysis of communication systems, but ultimately for voice, video, fax, etc., it is the final human perception that determines the quantitative performance criteria that must be met. As interesting as this subject is, we will leave it behind, and concentrate only on errors in the transmission of the words themselves, and not in the interpretation of these words by humans.

Video Telecommunications Example Towards developing a quantitative measure of receiver performance, we will consider video communication, and elaborate further on each of the functional blocks in Fig. 3.1. The source encoder in a video system performs the very important task of data rate compression by removing the redundancy in the input signal. Its goal is to reduce the data rate to such an extent that the statistics of the final quantized data are random and uncorrelated. Such techniques include: differential coding, linear predictive coding, subband coding, and vector quantization. Compression of the data is performed both within and between frames. Motion compensation is utilized for inter-frame data reduction, where only the errors between the image, and the translated portion of the previous image are quantized.

Channel Coding The channel coder now takes the random, uncorrelated data from the source coder and adds redundancy in an efficient and predictable manner, so that the receiver will be able to correct for errors in transmission. A spell-checker programs operates on this principle. Errors can be corrected in misspelled words only because all possible combinations of letters in the alphabet are not allowed, only those that constitute words in the English language. Often when a word is misspelled, it can be corrected by finding the nearest legal spelling that most closely matches the incorrect word, provided that there are not too many initial errors. If the minimum-distance vector search is not adequate to correct the errors, then we must look to the context of the sentence. Errors can also be corrected in this manner because all sequences of words are not allowed to be strung together. We only allow sequences that obey a certain grammatical structure. For a video system, the spell-checker is analogous to a convolutional, or Viterbi code that is used to structure the order of short bit sequences, or words. The Viterbi decoder is used to remove bit errors, provided that they don't occur in clusters. The convolutional coded bits, or words, can then be organized into blocks resembling sentences that obey a fixed structure. These block codes can remove errors in whole words.

Inter-Bit Correlation of Coded Data Sequences In the previous discussion it was implicitly assumed that a device at the receiver exists that can detect the individual letters of the alphabet of possible transmitted signals, and make a decision as to which one was sent. In the example of spoken english, the symbol alphabet is the familiar 26-letter alphabet, plus the digits 0–9, and various punctuation symbols. For a binary system, the alphabet consists of only two symbols, generally referred to as "one" and "zero," each with an equal probability of being sent. The previous discussion of coding in telecommunication is intended to illustrate that there exists a structure within the transmitted data so that *the actual transmitted bits are indeed correlated.* Therefore, if we wanted to develop an optimal decision rule for deciding the polarity of each binary bit, the decision rule would have to include all of the correlation information about the random data, and the data's past history. For example, if a coder is used that forces a transition every 8-bits, and the receiver has just detected 7 "zeros" in a row, then this information needs to be given to the decision circuit so that the next decision is biased in favor of detecting a "one." However, the decoding of the receiver is usually implemented hierarchically. First, the bits are detected assuming no prior knowledge of the statistics that the coding imposes on the data. Next, bit-errors are corrected by convolutional decoding of the detected bit. Finally, word errors can be corrected by the block decoder. This separation of tasks makes the implementation simpler, however, the performance is degraded compared to a conceivable system that takes into account all of the structure in the data in every decision.[1]

Concept of a Receiver The concept of a receiver for telecommunications is quite broad. Issues that determine performance span the range from human perception, to quantitative measures, such as distortion, signal-to-noise-ratio, and probability of error. In the remainder of this work we will use the term receiver in a much more limited sense to refer to a circuit that looks at the received signal over one bit period, and decides which bit was sent, using no knowledge of previous bits. This type of receiver assumes that the binary data is random, so that all bits, and all sequences of bits of arbitrary length, are mutually uncorrelated. We realize that this assumption is violated when a channel coder is used, but the data can be made to "look" uncorrelated, especially when time-division multiplexing is used, so we will adopt this model and use it from now on. As a further rationale for adopting this model, the penalty incurred, in terms of increased signal-power required in the simple receiver to reach the same level of performance as the "all-in-one" decision circuit, is only about 0.5dB, or 12%. Therefore, the reduction in receiver complexity afforded by using the uncorrelated model far outweighs the power penalty suffered. The performance criteria most applicable to this type of receiver is the probability of error, or the bit-error-rate (BER). Therefore, we seek to find the receiver that minimizes the BER for a given set of assumptions. To obtain a mathematical description of the receiver we turn to the theory of hypothesis testing. Before jumping straight into the theory, it is helpful to take a moment to reflect, qualitatively, on the operations that the receiver needs to perform, and to obtain an intuitive feel for the type of processing required. In this manner we can develop a list of ideas, that seem like reasonable things to do, and then compare this list with the theoretical results.

[1] Trellis-coded modulation is an example of a technique that combines the modulation and coding of the signal into a single step for improved performance.

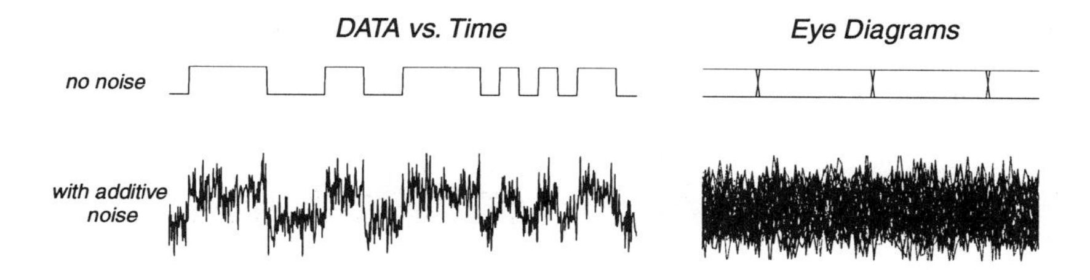

Figure 3.2 Rectangular NRZ data and eye-diagrams both without and with additive noise.

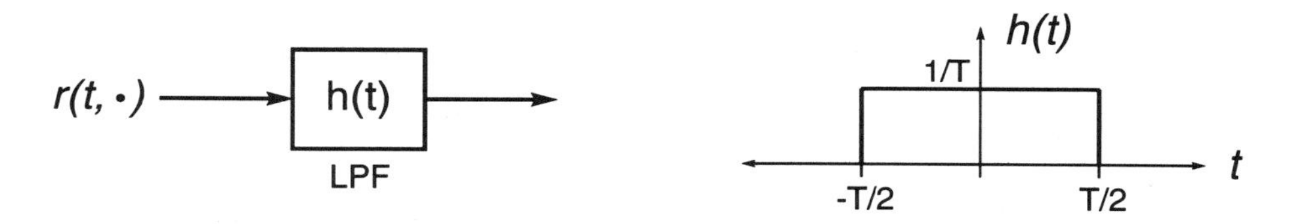

Figure 3.3 Impulse response of a moving-average lowpass filter.

3.1 QUALITATIVE DETECTION OF INDEPENDENT BINARY PULSES

The task of a binary receiver is to determine which of two possible signals were transmitted. For the case of zero-mean rectangular NRZ data, the symbols are either $+V$ or $-V$ in an interval of T seconds. An NRZ data stream is shown in Fig. 3.2 together with its eye-diagram, both for data with and without added noise. From the eye-diagram of the noisy data it can be seen that there is virtually no opening in which to obtain a valid data sample. If a decision were to be made about the polarity of the signal by using only one sample, then several errors will result. A better approach is to average the signal over the bit-period. Since the data signal is constant over this interval, and the noise is essentially uncorrelated with zero-mean, the signal will add coherently to the average, and the noise contribution to the average will tend towards zero — if the bit-period is long enough. Averaging the signal over each bit period and sampling at intervals of T, is equivalent to passing the signal through a "moving-average" lowpass filter and sampling also at intervals of T, as illustrated in Fig. 3.3. The "moving average" filter has a rectangular impulse response of

$$h_r(t) = \frac{1}{T}\text{rect}(t/T) \tag{3.1}$$

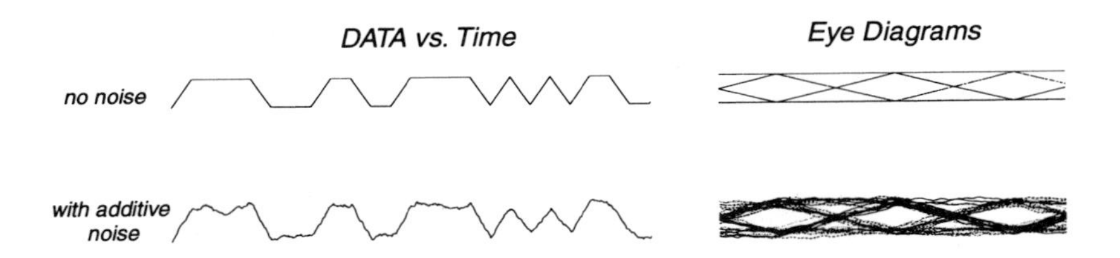

Figure 3.4 Rectangular NRZ data and eye-diagrams with and without additive noise after passing through a lowpass filter with a rectangular impulse response.

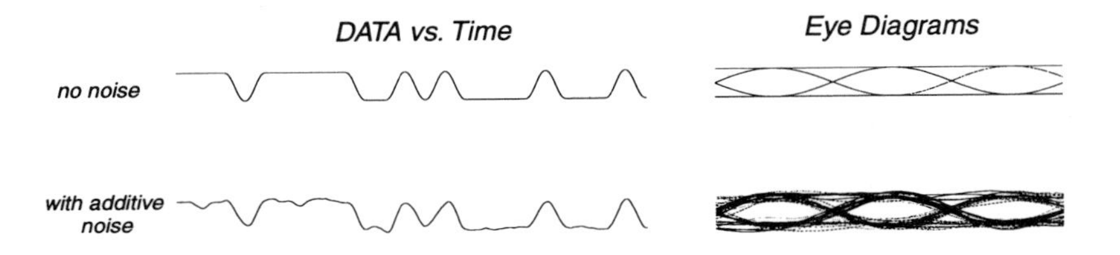

Figure 3.5 Rectangular NRZ data and eye-diagrams with and without additive noise after passing through a lowpass filter with a half-cosine impulse response.

The data signals of Fig. 3.2 are shown in Fig. 3.4 after being filtered by a moving-average lowpass filter. It can be seen that the averaging operation reduces the noise and provides a wider opening in the eye-diagram.

Clock Jitter Tolerance Instead of using a strict average, we could also have used a weighted average, and this may give desirable results in some instances. Using the half-cosine weighting function

$$h_c(t) = \frac{\pi}{2T} \cos\left(\frac{\pi t}{T}\right) \mathrm{rect}(t/T), \tag{3.2}$$

the data signals of Fig. 3.5 will result. In this case the vertical eye-opening is not as high as in the case for a rectangular impulse response. However, the horizontal eye-opening is wider. Therefore, we would expect to find that using a weighted average filter, that concentrates most of the energy in the center of the data pulse, will be less sensitive to clock jitter. In other words when the data is sampled at a point that is offset from the center of the eye, the SNR penalty is not as severe as when a rectangular impulse response filter is used.

Colored Noise In the general case the additive noise will be non-white or colored. The lowpass filter that averages the data can be tailored to reduce the noise by biasing the filter's frequency response away from where the noise has the most power. As an example, if the noise PSD increases with the square of frequency, then the desired lowpass noise filter should have good suppression of high-frequency content. The frequency response of the rectangular and half-cosine filters were given in Fig. 2.20(b), where it was noted that the half-cosine filter provided better high-frequency attenuation compared to a simple moving average filter. The optimal filter in the presence of colored noise will be the one that produces the largest SNR at a given sampling point. We would expect the optimal filter to have its energy concentrated in frequency bands where the signal power is the strongest, and have small amounts of energy where the noise is the strongest.

In sections to follow, a receiver will be derived that is optimal in the sense that the BER is minimized provided that a given set of assumptions are satisfied. From the previous discussion we realize that the receiver should perform the following functions.

- The receiver should perform filtering to average the noise.
- A weighted average filter can be implemented by using a windowing function which might improve receiver performance in the presence of clock jitter.
- The frequency response of the filter can be altered to improve receiver performance in colored noise.

In the following sections a mathematical framework will be introduced so that the above statements about receiver performance can be quantified.

3.2 HYPOTHESIS TESTING

Given the assumptions that each data symbol is independent of all previous data symbols, and that the a priori probabilities for the symbols are known, the problem of determining an optimal decision criteria for the receiver can be posed mathematically as a hypothesis test [1, 2]. The receiver observes a signal over a bit interval, and determines the probability that the received signal resulted from each one of the pulses in the alphabet. A cost is associated with an incorrect decision, and the optimum receiver is one which minimizes the expected cost. Fig. 3.6 illustrates a channel for a binary communication system. The receiver's task is to observe the received signal $r(t, \cdot)$, over an interval T, and determine whether $s_0(t)$ or $s_1(t)$ was the transmitted signal in that interval. At the receiver two separate hypotheses can be formulated:

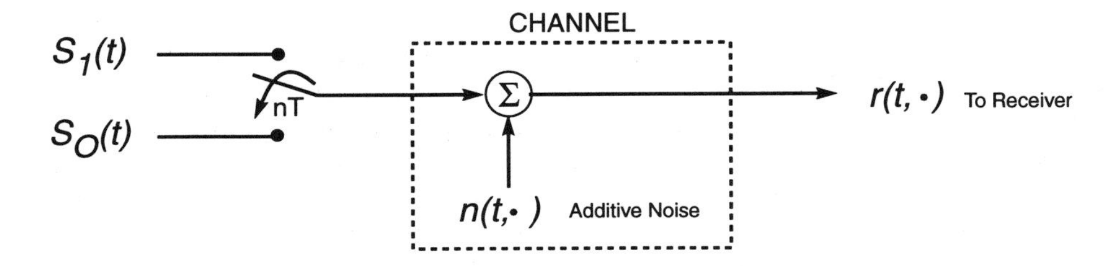

Figure 3.6 Block diagram of a channel for communication of binary data.

- $H_0 \Rightarrow$ Hypothesis that $s_0(t)$ was sent,
- $H_1 \Rightarrow$ Hypothesis that $s_1(t)$ was sent.

Without loss of generality we can consider the received signal to be a series of samples. Later the number of samples can be made to approach infinity. Therefore, the received signal $r(t,\cdot)$ can be represented as a K dimensional vector $\boldsymbol{r}(\cdot)$, where K is the number of samples in an interval of length T.[2] Based on the two hypotheses, a pair of multidimensional probability density functions (pdfs) can be defined.

$$P_0(\boldsymbol{r}) \triangleq P(\boldsymbol{r}|H_0), \tag{3.3}$$

where $P_0(\boldsymbol{r})$ is the conditional pdf as a function of $\boldsymbol{r}$ given that $s_0(t)$ was sent. Likewise

$$P_1(\boldsymbol{r}) \triangleq P(\boldsymbol{r}|H_1) \tag{3.4}$$

is the conditional pdf given that $s_1(t)$ was sent. The noise can always be considered to be additive by definition, so that the received signals under the two hypotheses are

$$\begin{aligned} \boldsymbol{r}(\cdot)|H_0 &= \boldsymbol{s}_0 + \boldsymbol{n}(\cdot) \\ \boldsymbol{r}(\cdot)|H_1 &= \boldsymbol{s}_1 + \boldsymbol{n}(\cdot). \end{aligned} \tag{3.5}$$

For a binary decision, the vector spaces spanned by $\boldsymbol{r}(\cdot)$ can be divided into two regions. In the region $\mathcal{R}_0$ the receiver chooses that $s_0(t)$ was sent, and in $\mathcal{R}_1$, $s_1(t)$ is chosen. A decision function $d(\boldsymbol{r})$ can also be defined in these regions such that

$$\begin{aligned} d(\boldsymbol{r}) &= 0 \qquad \text{for} \quad \boldsymbol{r} \in \mathcal{R}_0 \\ d(\boldsymbol{r}) &= 1 \qquad \text{for} \quad \boldsymbol{r} \in \mathcal{R}_1 \end{aligned} \tag{3.6}$$

[2] A comment about the slightly sloppy notation used may be helpful here. The random received vector is denoted as $r(\cdot)$, whereas any vector in the space spanned by $r(\cdot)$ is noted as r. The vector r is not a random vector, but is only a dummy variable used to specify the coordinates in a vector space.

Finally a cost matrix can be defined such that c_{ij} is the cost of choosing i when in fact j was actually sent. The expected cost is now written as

$$\begin{aligned} \bar{c} \;=\; & c_{00} \cdot \Pr[d(\boldsymbol{r}(\cdot)) = 0 \;\; |H_0] \;+\; c_{01} \cdot \Pr[d(\boldsymbol{r}(\cdot)) = 0 \;\; |H_1] + \\ & c_{10} \cdot \Pr[d(\boldsymbol{r}(\cdot)) = 1 \;\; |H_0] \;+\; c_{11} \cdot \Pr[d(\boldsymbol{r}(\cdot)) = 1 \;\; |H_1]. \end{aligned} \tag{3.7}$$

With the assumption that the a priori probabilities are known:

$$\begin{aligned} \pi_0 &= \Pr[H_0], \\ \pi_1 &= \Pr[H_1], \end{aligned} \tag{3.8}$$

the average cost can be written as an integral of the conditional pdfs over the two decision regions.

$$\begin{aligned} \bar{c}(d,\pi) = & c_{00}\pi_0 \int_{\mathcal{R}_0} P_0(\boldsymbol{r})d\boldsymbol{r} + c_{01}\pi_1 \int_{\mathcal{R}_0} P_1(\boldsymbol{r})d\boldsymbol{r} + \\ & c_{10}\pi_0 \int_{\mathcal{R}_1} P_0(\boldsymbol{r})d\boldsymbol{r} + c_{11}\pi_1 \int_{\mathcal{R}_1} P_1(\boldsymbol{r})d\boldsymbol{r}. \end{aligned} \tag{3.9}$$

The problem now stated in mathematical terms is to choose the regions $\mathcal{R}_0$ and $\mathcal{R}_1$ such that the above average cost is minimized.

Determination of Optimal Decision Regions The integral of a probability density function, by definition, must equal unity. Since the two decision regions are mutually exclusive, and span the entire vector space,

$$\begin{aligned} \int_{\mathcal{R}_0} P_0(\boldsymbol{r})d\boldsymbol{r} + \int_{\mathcal{R}_1} P_0(\boldsymbol{r})d\boldsymbol{r} &= 1 \\ \int_{\mathcal{R}_0} P_1(\boldsymbol{r})d\boldsymbol{r} + \int_{\mathcal{R}_1} P_1(\boldsymbol{r})d\boldsymbol{r} &= 1. \end{aligned} \tag{3.10}$$

As a result, the average cost can be written as an integral over only one of either of the decision regions $\mathcal{R}_0$ and $\mathcal{R}_1$. Therefore, integrating over $\mathcal{R}_1$, the average cost is

$$\begin{aligned} \bar{c}(d,\pi) \;=\; & c_{00}\pi_0 \left[1 - \int_{\mathcal{R}_1} P_0(\boldsymbol{r})d\boldsymbol{r}\right] \;+\; c_{01}\pi_1 \left[1 - \int_{\mathcal{R}_1} P_1(\boldsymbol{r})d\boldsymbol{r}\right] + \\ & c_{10}\pi_0 \int_{\mathcal{R}_1} P_0(\boldsymbol{r})d\boldsymbol{r} \;+\; c_{11}\pi_1 \int_{\mathcal{R}_1} P_1(\boldsymbol{r})d\boldsymbol{r}. \end{aligned} \tag{3.11}$$

Combining terms, this can be written as a single integral;

$$\bar{c}(d,\pi) = c_{00}\pi_0 + c_{01}\pi_1 + \int_{\mathcal{R}_1} \left[\pi_0(c_{10} - c_{00})P_0(\boldsymbol{r}) - \pi_1(c_{01} - c_{11})P_1(\boldsymbol{r})\right] d\boldsymbol{r}. \tag{3.12}$$

The expected cost in (3.12) consists of a constant term, and an integral over the region $\mathcal{R}_1$. If the integrand is positive, then the integral will increase the expected cost. However, when the integrand is negative, the integral reduces the average cost. It is clear that the minimum cost is obtained by choosing $\mathcal{R}_1$ such that the integrand is always negative, or

$$\pi_1(c_{01} - c_{11})P_1(\boldsymbol{r}) \geq \pi_0(c_{10} - c_{00})P_0(\boldsymbol{r}) \tag{3.13}$$

Therefore, the region $\mathcal{R}_1$ in the vector space spanned by $\boldsymbol{r}(\cdot)$ that minimizes the expected cost satisfies the condition,

$$\boxed{\mathcal{R}_1; \qquad \frac{\pi_1 P_1(\boldsymbol{r})}{\pi_0 P_0(\boldsymbol{r})} \geq \frac{c_{10} - c_{00}}{c_{01} - c_{11}}} \tag{3.14}$$

This decision rule is known as the Bayes criterion, and the resulting minimum cost is the Bayes risk.

Bit-Error-Rate in a Binary Communication System In a communication system there is no "cost" in making a correct decision:

$$c_{00} = c_{11} = 0, \tag{3.15}$$

and there is an equal "cost" of making a wrong decision. If this cost is arbitrarily chosen to be unity such that

$$c_{01} = c_{10} = 1, \tag{3.16}$$

then the Bayes risk is just the probability of error, and the decision rule is

$$\mathcal{R}_1; \qquad \pi_1 P_1(\boldsymbol{r}) \geq \pi_0 P_0(\boldsymbol{r}), \tag{3.17}$$

and in the usual case where $\pi_0 = \pi_1 = 1/2$, then

$$\boxed{\begin{array}{ll} \mathcal{R}_1; & P_1(\boldsymbol{r}) \geq P_0(\boldsymbol{r}) \\ \mathcal{R}_0; & P_0(\boldsymbol{r}) \geq P_1(\boldsymbol{r}) \end{array}} \tag{3.18}$$

The probability of error is then from (3.12) found by integrating the conditional pdfs over the optimal decision region.

$$\boxed{P_e = \frac{1}{2}\left[1 - \int_{\mathcal{R}_1} [P_1(\boldsymbol{r}) - P_0(\boldsymbol{r})]\, d\boldsymbol{r}\right]} \tag{3.19}$$

At the boundary where $P_0(\boldsymbol{r}) = P_1(\boldsymbol{r})$, the decision function $d(\boldsymbol{r})$ can be set arbitrarily to either 1 or 0 without affecting the probability of error. However, systems considerations may make one of these choices more desirable than the other.

3.3 PROPERTIES OF GAUSSIAN RANDOM VARIABLES

In the previous section, the criteria for determining optimal decision regions for known, independent binary signals was established. The error performance of the system can also be determined by evaluating the integral in (3.19). However, this general criteria is not very illustrative, and performance evaluation involves performing a multidimensional integration, which is no fun. Greater insight into the problem, and simplifications in the analysis can be obtained by making some assumptions about the statistics of the noise. The standard assumption made is that the noise is additive-white-Gaussian-noise (AWGN). Gaussian distributed random variables have many nice properties that facilitate analysis. In this section we will review a few of these properties.

3.3.1 One-Dimensional Gaussian Random Variables

A Gaussian, or "normal" random variable $x(\cdot)$ has a probability density function given by the familiar expression

$$\text{pdf}_{x(\cdot)}(x) = p(x) = \frac{1}{\sqrt{2\pi}\sigma} e^{-\frac{1}{2}\left(\frac{x-\mu}{\sigma}\right)^2}, \tag{3.20}$$

where μ is the mean of the random variable $x(\cdot)$, and σ^2 is the variance. Two very useful properties of Gaussian random variables are that,

- Any linear combination of Gaussian random variables is also Gaussian.
- The probability density function of a Gaussian random variable is completely defined by a knowledge of only the mean and variance.

The Complementary Error Function We will often be interested in the probability that $x(\cdot)$ is within a certain interval $[a, b]$, given by the integral

$$\Pr[a \leq x(\cdot) \leq b] = \frac{1}{\sqrt{2\pi}\sigma} \int_a^b e^{-\frac{1}{2}\left(\frac{x-\mu}{\sigma}\right)^2} dx. \tag{3.21}$$

Since there is no known closed form of the integral in (3.21), we can make use of the normalized Gaussian distribution. A zero-mean and unit-variance Gaussian random variable $z(\cdot)$ can be defined as

$$z(\cdot) \triangleq \frac{x(\cdot) - \mu}{\sigma}, \tag{3.22}$$

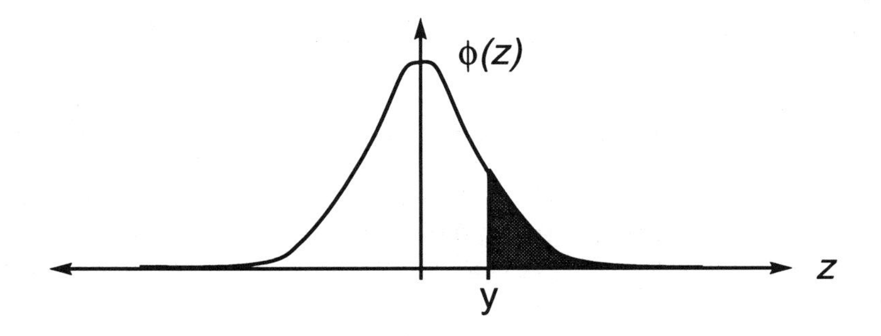

Figure 3.7 Illustration of the complementary error function integral $\Phi(y)$.

and the normalized Gaussian probability density function is given by

$$\text{pdf}_{z(\cdot)}(z) = \phi(z) = \frac{1}{\sqrt{2\pi}} e^{-\frac{1}{2}z^2}. \tag{3.23}$$

We can define a complementary error function $\Phi(y)$ as the probability that $z(\cdot)$ is larger than y. This probability is given by the integral of the tail of the Gaussian pdf from y to infinity. Therefore

$$\Phi(y) \triangleq \int_y^\infty \phi(z)dz \tag{3.24}$$

This integral is illustrated in Fig. 3.7, where it can be seen that $\Phi(-\infty) = 1$, $\Phi(0) = 1/2$, and $\Phi(\infty) = 0$. Using this plot, or a table of values of $\Phi(y)$, the probability for any general Gaussian random variable $x(\cdot)$ can be obtained. Clearly due to symmetry $\Phi(y) + \Phi(-y) = 1$. Now we can express the integral in (3.21) in terms of Φ.

$$\Pr[a \leq x(\cdot) \leq b] = \Phi\left(\frac{a-\mu}{\sigma}\right) - \Phi\left(\frac{b-\mu}{\sigma}\right) \tag{3.25}$$

The logarithm of $\Phi(y)$ is plotted in Fig.3.8a for values of $y \in [1, 8]$. It can be seen from this plot that the probability of a Gaussian random variable being 6 standard deviations away from the mean is 10^{-9}. For values of $y > 3$, $\Phi(y)$ can be approximated by

$$\Phi(y) \simeq \frac{\phi(y)}{y} = \frac{1}{y\sqrt{2\pi}} e^{-\frac{1}{2}y^2} \qquad \text{for} \quad y > 3. \tag{3.26}$$

Using this approximation, a closed form expression for the error probability as a function of the signal-to-noise-ratio (SNR) parameter y is given by

$$-\log\left[\Phi(y)\right] \overset{y \geq 3}{\simeq} \frac{y^2}{2\ln(10)} + \log(y) + \frac{1}{2}\log(2\pi), \tag{3.27}$$

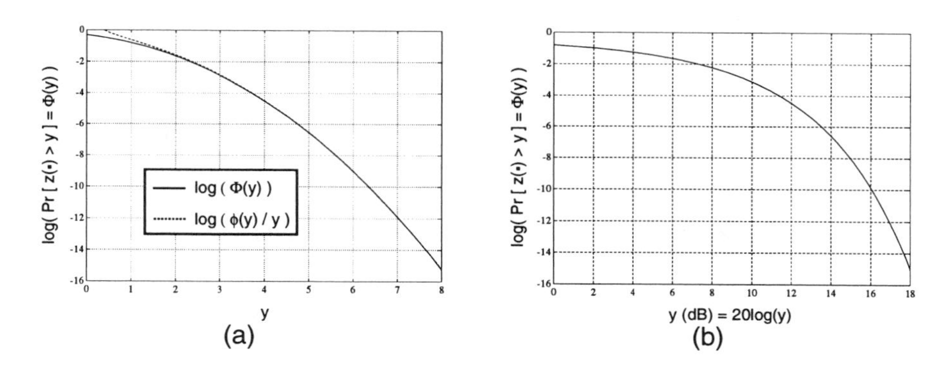

Figure 3.8 Complementary error function: (a) actual and approximate values vs. y, (b) actual values vs. $20\log(y)$.

$\Pr[z(\cdot) > y]$	y	$20\log(y)$
10^{-3}	3.115	9.869
10^{-6}	4.762	13.556
10^{-9}	6.002	15.556
10^{-12}	7.037	16.948
10^{-15}	7.943	18.000

Table 3.1 Values of y required to achieve a given probability.

and putting in numbers

$$\boxed{-\log\left[\Phi(y)\right] \overset{y\geq 3}{\simeq} \frac{y^2}{4.6} + \log(y) + 0.4} \tag{3.28}$$

We will typically be interested in finding a probability of error expressed in terms of $\Phi(y)$, where y is a ratio of a signal to noise, normally given in decibels. Therefore, Fig. 3.8b shows the error function plotted against $20\log(y)$. Values of y required to achieve a given error probability are given in table 3.1.

3.3.2 Multidimensional Gaussian Random Variables

In section 3.2 a random signal $r(t,\cdot)$ was represented by a random vector $\boldsymbol{r}(\cdot)$, and a multidimensional pdf was defined over the vector space spanned by $\boldsymbol{r}(\cdot)$. In this section, the concepts of a one-dimensional Gaussian random variable will be extended to the multidimensional case, so that the results can be applied directly to the problem at hand. The simplest random vector is one composed of N independent random variables.

$$\boldsymbol{x}(\cdot) = [x_1(\cdot), x_2(\cdot), x_3(\cdot), \ldots x_N(\cdot)]^T \tag{3.29}$$

Since the individual components of the random vector are independent, the pdf of the vector is simply equal to the product of the pdfs of each random component,

$$\text{pdf}_{\boldsymbol{x}(\cdot)}(\boldsymbol{x}) = p(\boldsymbol{x}) = \prod_{i=1}^{N} p(x_i), \tag{3.30}$$

and if the the components are Gaussian random variables, then

$$p(\boldsymbol{x}) = \prod_{i=1}^{N} \frac{1}{\sqrt{2\pi}\sigma_i} e^{-\frac{1}{2}\left(\frac{x_i-\mu_i}{\sigma_i}\right)^2}, \tag{3.31}$$

or equivalently

$$p(\boldsymbol{x}) = \frac{1}{(2\pi)^{N/2}} \left(\prod_{i=1}^{N} \frac{1}{\sigma_i}\right) \exp\left[-\frac{1}{2}\sum_{i=1}^{N}\left(\frac{x_i - \mu_i}{\sigma_i}\right)^2\right]. \tag{3.32}$$

(3.32) can be written compactly by making use of vector notation, and the covariance matrix

$$\boldsymbol{R} \triangleq E[\boldsymbol{x}(\cdot)\boldsymbol{x}(\cdot)^*]. \tag{3.33}$$

For a random vector $\boldsymbol{x}(\cdot)$ with independent components, the covariance matrix will be diagonal, with the diagonal entries equal to the variances of each component.

$$\boldsymbol{R} = \text{diag}[\sigma_1^2, \sigma_2^2, \sigma_3^2, \ldots, \sigma_N^2] \tag{3.34}$$

The determinant of $\boldsymbol{R}$ is just the product of the variances for each component. Therefore,

$$\frac{1}{(\det \boldsymbol{R})^{1/2}} = \prod_{i=1}^{N}\left(\frac{1}{\sigma_i}\right), \tag{3.35}$$

Since $\boldsymbol{R}$ is a diagonal matrix, the inverse of $\boldsymbol{R}$ is also diagonal with entries of $1/\sigma_i^2$. Using the inverse covariance matrix, the argument of the exponential can be written as

$$(\boldsymbol{x} - \boldsymbol{\mu})^T \boldsymbol{R}^{-1}(\boldsymbol{x} - \boldsymbol{\mu}) = \sum_{i=1}^{N}\left(\frac{x_i - \mu_i}{\sigma_i}\right)^2. \tag{3.36}$$

Notice that the previous expression is actually a "dot-product", or "inner-product" of two vectors. In the following sections the notation $\langle \boldsymbol{x}, \boldsymbol{y} \rangle$ will be used to represent "dot-products" making this operation more explicit, where

$$\langle \boldsymbol{x}, \boldsymbol{y} \rangle \triangleq \boldsymbol{x}^T \boldsymbol{y} = \boldsymbol{x} \cdot \boldsymbol{y} \tag{3.37}$$

Using vector notation the multidimensional probability density function for a Gaussian random vector is given by

$$\boxed{\text{pdf}_{\boldsymbol{x}(\cdot)}(\boldsymbol{x}) = p(\boldsymbol{x}) = \frac{1}{(2\pi)^{N/2}(\det \boldsymbol{R})^{1/2}} \exp\left[-\frac{1}{2}\langle(\boldsymbol{x}-\boldsymbol{\mu}), \boldsymbol{R}^{-1}(\boldsymbol{x}-\boldsymbol{\mu})\rangle\right]} \tag{3.38}$$

In the previous discussion we assumed that the components of the random vector $\boldsymbol{x}(\cdot)$ were mutually uncorrelated to arrive at the above result. Although we have considered only a special case it can be shown that (3.38) is the general form of the pdf for a multidimensional Gaussian random vector when the components are correlated [3, p. 172], [2, ch. 2]. In the general case the covariance matrix $\boldsymbol{R}$ will no longer be diagonal.

Average of a Random Vector A statistic that will be of primary interest to us is the average of a random vector. If we consider a vector $\boldsymbol{z}(\cdot)$ comprising N zero-mean, unit-variance, independent random variables, then from (3.38)

$$\text{pdf}_{\boldsymbol{z}(\cdot)}(\boldsymbol{z}) = p(\boldsymbol{z}) = \frac{1}{(2\pi)^{N/2}} e^{-\frac{1}{2}\langle \boldsymbol{z}, \boldsymbol{z} \rangle}. \tag{3.39}$$

We can define a new random variable as the average of the components of $\boldsymbol{z}(\cdot)$ as

$$a_v(\cdot) = \frac{1}{N} \sum_{i=1}^{N} z_i(\cdot). \tag{3.40}$$

Since the pdfs of all z_is are known, in principle we can find the pdf of $a_v(\cdot)$. This is a complicated procedure at best. However, since all z_is are Gaussian random variables, any linear combination of these is also Gaussian. Therefore, $a_v(\cdot)$ is Gaussian, and we need only find the mean and variance of $a_v(\cdot)$ to completely determine the pdf. The mean of $a_v(\cdot)$ is clearly zero;

$$E[a_v(\cdot)] = \frac{1}{N} \sum_{i=1}^{N} E[z_i(\cdot)] = 0, \tag{3.41}$$

and since the elements of $z(\cdot)$ are independent, the variance is found simply by

$$E[a_v^2(\cdot)] = \frac{1}{N^2} \left[\sum_{i=1}^{N} E[z_i(\cdot)]\right]^2 = \frac{1}{N^2} \sum_{i=1}^{N} E[z_i^2(\cdot)] = \frac{1}{N}. \tag{3.42}$$

Therefore the pdf of the average is

$$\text{pdf}_{a_v(\cdot)}(a_v) = p(a_v) = \frac{1}{\sqrt{2\pi N}} e^{-\frac{1}{2}a_v^2/N} \tag{3.43}$$

This analysis illustrates the power of assuming the random disturbances to be Gaussian. The pdf of any linear combination of Gaussian random variable can be easily found when the mean and variance are known. The mean of the random process is just the dc value, and the variance is the average ac power. We saw in chapter 2 that the variance can be determined by integrating the ac power spectral density over frequency. Obtaining the pdf from the dc value and the average ac power of a Gaussian random process is a useful technique that will be widely used in the following sections. Before moving on, however, we should note that the standard deviation of the average of N unit variance independent random variables from (3.42) is $\sigma_{av} = 1/\sqrt{N}$. As the number of independent samples increases, the variance approaches zero inversely with the square-root of the number of samples. Therefore, if we have a signal of unit value added to noise with unit variance, then it will be difficult to detect the signal in this noise. If we took 100 samples of the signal-plus-noise and averaged it, then the signal would still have a unit average, but the noise standard deviation is now 1/10. If we wanted to reduce the noise standard deviation to 1/1000, or approximately a 10 bit resolution, then we would need to average 1000^2 or one-million samples.

3.4 OPTIMAL DECISION RULE FOR ADDITIVE-WHITE-GAUSSIAN-NOISE

In section (3.2), the general criteria for determining the optimal decision region for the case of known, independent, binary pulses was established. The optimal decision region from (3.18) is for equal a priori probabilities

$$\mathcal{R}_1; \qquad P_1(\boldsymbol{r}) \geq P_0(\boldsymbol{r}) \tag{3.44}$$

and the probability of error from (3.19) is the following multidimensional integral

$$P_e = \frac{1}{2}\left[1 - \int_{\mathcal{R}_1} [P_1(\boldsymbol{r}) - P_0(\boldsymbol{r})]\, d\boldsymbol{r}\right] \tag{3.45}$$

When the noise is assumed Gaussian, the expression for the optimum decision rule, and evaluation of the probability of error is greatly simplified. The received signal $\boldsymbol{r}(\cdot)$, under the two hypotheses are from (3.5)

$$\begin{aligned} \boldsymbol{r}(\cdot)|H_0 &= \boldsymbol{s}_0 + \boldsymbol{n}(\cdot) \\ \boldsymbol{r}(\cdot)|H_1 &= \boldsymbol{s}_1 + \boldsymbol{n}(\cdot). \end{aligned} \tag{3.46}$$

If we assume that the noise is zero-mean, Gaussian noise, then the conditional means of $\boldsymbol{r}(\cdot)$ will be $\boldsymbol{s}_0$, and $\boldsymbol{s}_1$ respectively, and the conditional pdfs in the time interval $[0, T]$ can be written explicitly as

$$P_0(\boldsymbol{r}) = \frac{1}{(2\pi)^{N/2}(\det \boldsymbol{R}_n)^{1/2}} \exp\left[-\frac{1}{2}\left\langle(\boldsymbol{r}-\boldsymbol{s}_0), \boldsymbol{R}_n^{-1}(\boldsymbol{r}-\boldsymbol{s}_0)\right\rangle\right] \quad (3.47)$$

when $\boldsymbol{s}_0$ is sent, and

$$P_1(\boldsymbol{r}) = \frac{1}{(2\pi)^{N/2}(\det \boldsymbol{R}_n)^{1/2}} \exp\left[-\frac{1}{2}\left\langle(\boldsymbol{r}-\boldsymbol{s}_1), \boldsymbol{R}_n^{-1}(\boldsymbol{r}-\boldsymbol{s}_1)\right\rangle\right] \quad (3.48)$$

when $\boldsymbol{s}_1$ is sent. The optimal decision region $\mathcal{R}_1$ can be written as a simple likelihood ratio

$$\mathcal{R}_1; \qquad \frac{P_1(\boldsymbol{r})}{P_0(\boldsymbol{r})} \geq 1, \quad (3.49)$$

which is always greater than zero. Alternatively, since the logarithm is a monotonicly increasing function for positive arguments, we can take the log of both sides;

$$\mathcal{R}_1; \qquad \ln\left[\frac{P_1(\boldsymbol{r})}{P_0(\boldsymbol{r})}\right] \geq \ln(1) = 0. \quad (3.50)$$

Therefore, the optimal decision region for additive Gaussian noise is

$$\mathcal{R}_1; \qquad -\frac{1}{2}\left\langle(\boldsymbol{r}-\boldsymbol{s}_1), \boldsymbol{R}_n^{-1}(\boldsymbol{r}-\boldsymbol{s}_1)\right\rangle \geq -\frac{1}{2}\left\langle(\boldsymbol{r}-\boldsymbol{s}_0), \boldsymbol{R}_n^{-1}(\boldsymbol{r}-\boldsymbol{s}_0)\right\rangle. \quad (3.51)$$

If in addition to being Gaussian, the noise is assumed to be uncorrelated and stationary, then

$$\boldsymbol{R}_n^{-1} = \frac{1}{\sigma_n^2}\boldsymbol{I} \quad (3.52)$$

The decision rule for additive white Gaussian noise (AWGN) is then

$$\mathcal{R}_1; \qquad \left\langle(\boldsymbol{r}-\boldsymbol{s}_1), (\boldsymbol{r}-\boldsymbol{s}_1)\right\rangle \leq \left\langle(\boldsymbol{r}-\boldsymbol{s}_0), (\boldsymbol{r}-\boldsymbol{s}_0)\right\rangle. \quad (3.53)$$

Using the following notation for the norm of a vector

$$||\boldsymbol{r}||^2 \triangleq \langle \boldsymbol{r}, \boldsymbol{r}\rangle, \quad (3.54)$$

then the decision rule can be expanded in the form

$$\mathcal{R}_1; \qquad ||\boldsymbol{r}||^2 - 2\langle \boldsymbol{r}, \boldsymbol{s}_1\rangle + ||\boldsymbol{s}_1||^2 \leq ||\boldsymbol{r}||^2 - 2\langle \boldsymbol{r}, \boldsymbol{s}_0\rangle + ||\boldsymbol{s}_0||^2. \quad (3.55)$$

Therefore, the final form of the optimal decision rule for AWGN is

$$\boxed{\mathcal{R}_1; \qquad \langle \boldsymbol{r}, \boldsymbol{s}_1\rangle - \langle \boldsymbol{r}, \boldsymbol{s}_0\rangle \geq \frac{||\boldsymbol{s}_1||^2 - ||\boldsymbol{s}_0||^2}{2}} \quad (3.56)$$

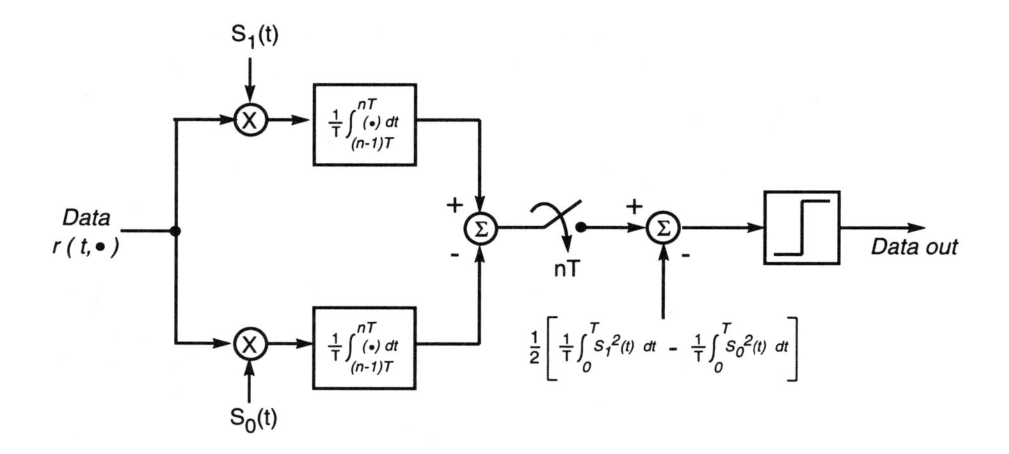

Figure 3.9 Optimal receiver for Binary signals in AWGN assuming that the time of arrival is known, and that the transmitted signals $s_0(t)$ and $s_1(t)$ are precisely known at the receiver.

Interpretation of Decision Rule as a Correlation Receiver In the limit as the number of samples of the random received vector approaches infinity, the "dot-product" becomes an integral

$$\lim_{N\to\infty} \frac{1}{N}\langle \boldsymbol{r}, \boldsymbol{s}_1 \rangle = \frac{1}{T}\int_0^T r(t)s_1(t)dt \tag{3.57}$$

The resulting receiver that implements this decision rule is shown in Fig. 3.9. The receiver has stored at the destination a copy of both pulses $s_0(t)$ and $s_1(t)$. The arrival time of the pulses is assumed to be precisely known. Over each time interval of length T, the receiver correlates the random observed signal with the two possible signals. If $s_0(t)$ and $s_1(t)$ have a difference in energy, then this bias is subtracted out, and the remaining signal is compared to zero. If the result is positive then $s_1(t)$ is chosen, and $s_0(t)$ is chosen if the result is negative. The optimal receiver for binary pulses is even simpler, when it is assumed that the data is balanced around zero, so that $s_1(t) = -s_0(t) = s(t)$. In this case the optimal decision rule is

$$\boxed{\mathcal{R}_1; \qquad \langle \boldsymbol{r}, \boldsymbol{s} \rangle \geq 0.} \tag{3.58}$$

For balanced symmetric signals with equal a priori probabilities the optimal receiver is shown in Fig. 3.10.

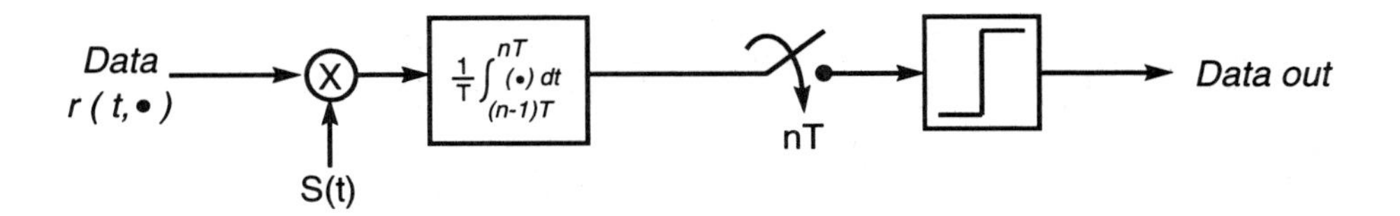

Figure 3.10 Optimal receiver for Binary signals in AWGN where $s_1(t) = -s_0(t) = s(t)$ is known, and the time of arrival is also known.

3.4.1 Optimal Matched filter for AWGN

Often it is more convenient, both for implementation and for analysis, to replace the optimal correlator with a matched filter, whose output at discrete sample $t = nT$ is the same as the output of the correlator. Describing the correlation as matched filters will enable us to evaluate the performance of the receiver for non-white noise, and for receivers with clock jitter. The correlation operation can be replaced by an equivalent matched filter. The operation of a weighted integration is accomplished by the shape of the impulse response of the filter as it is convolved with the data signal. For rectangular data, the correlation is equivalent to filtering the data with a rectangular impulse response filter. In general, if the data pulse has the shape $s(t)$, then the operation

$$\frac{1}{T}\int_0^T s(t)r(t,\cdot)dt \tag{3.59a}$$

with a change of variables, is equivalent to the following integral

$$\frac{1}{T}\int_0^T s(T-\tau)r(T-\tau,\cdot)d\tau. \tag{3.59b}$$

This integral is equivalent to

$$\frac{1}{T}\int_0^T s(T-\tau)r(t-\tau,\cdot)d\tau \tag{3.60}$$

only at the time $t = T$. We could write (3.60) as

$$\int_0^T h_s(\tau)r(t-\tau,\cdot)d\tau \qquad \text{where} \quad h_s(\tau) = \frac{1}{T}s(T-\tau) \tag{3.61}$$

Finally, since the pulse shape $s(t)$ is defined to be zero outside the interval $[0, T]$, we

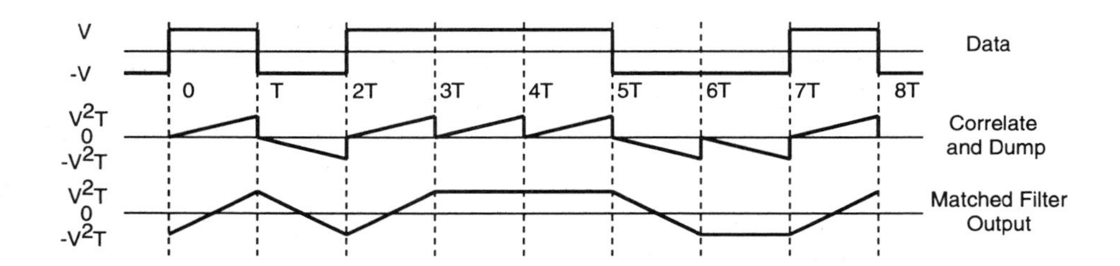

Figure 3.11 Illustration of the equivalence of a correlation receiver and a matched filter receiver. The Correlate and dump signal is equal to the matched filter signal at integer multiples of the bit-period T.

can write (3.61) as a convolution,

$$\int_{-\infty}^{\infty} h_s(\tau)r(t-\tau,\cdot)d\tau = h_s(t) * r(t,\cdot). \tag{3.62}$$

The function $h_s(t)$ can now be thought of as the impulse response of a linear filter that is matched to the data signal $s(t)$. Fig. 3.11 illustrates the equivalence of a correlation receiver and a matched filter receiver when the two are sampled at the same time intervals (intervals of nT in this case).

3.4.2 Comparison with Qualitative Analysis

The receiver of Fig. 3.10 is intuitively satisfying because it corresponds to the same operation that we described qualitatively in section 3.1. If the transmitted data pulses are rectangular NRZ, then the optimal receiver simply takes the average of the data over the bit interval. The signal adds coherently to the average, while the noise average tends to zero. It was shown in (3.43) that the average of identically distributed zero-mean Gaussian random variables can be made arbitrarily close to zero, if enough samples are taken. Therefore, we find a fundamental tradeoff between SNR and bandwidth. If the SNR is high, then the pulse period T can be made small. However, if the SNR is low, then T must be increased so as to average out the noise over a longer interval to achieve the same probability of error.

It is important at this point to reflect on the assumptions that have been made in deriving the optimal receiver. These assumptions are that:

- The noise is assumed to be additive and Gaussian.

- The noise is white $\Rightarrow$ the noise power in a bandwidth of B Hz is $N_0 B$ for all B. (Adjacent samples of the noise process are uncorrelated, no matter how close the two samples are taken in time, which implies that the autocorrelation function of the noise is an impulse function of weight $N_0/2$).
- The transmitted symbol is known precisely at the receiver, (Any distortion due to the channel has been equalized).
- There is no intersymbol-interference (ISI), (Any dispersion of adjacent signals has also been equalized).
- All data pulses are uncorrelated, (A knowledge of the past data pulses gives no information about the current data pulse).
- The time of arrival of the signals is known, (A clock exists at the receiver that is in perfect phase synchronization with the arrival of the data).

If any of the above assumptions are violated, the correlation receiver is no longer optimal. For practical broadband fiber-optic receivers the following conditions will hold.

- Since the clock at the receiver has to be recovered from the random data itself, there will always be jitter in the recovered clock, and the arrival time of the data will not be known precisely.
- The noise power spectral density is non-white. The dominant noise source will be due to the preamplifier. Since the photodiode has a large capacitance $\sim$ (0.5pF), creating a dominant pole with the preamplifier's input impedance, negative feedback is required to broadband the amplifier. The PSD will be shown in chapter 7 to have a "zero" due to the dominant pole at the amplifier input, causing the PSD to increase with the frequency squared.

Even when the assumptions needed for optimality are violated, we often still use the correlation receiver because it corresponds to our intuition about how the noisy signal should be processed and is proven to provide the best performance under nominal conditions. In section 3.9, the shape of the correlation pulse will be altered to provide better performance in the presence of clock-jitter, and non-white noise. Before considering these extensions, we will evaluate the performance of the correlation receiver under ideal conditions.

3.5 PERFORMANCE EVALUATION OF THE CORRELATION RECEIVER IN AWGN

We are interested in evaluating the performance of a correlation receiver shown in Fig. 3.10 in AWGN. For every time interval of length T, the received signal is multiplied by a copy of the original signal, and integrated. A decision is made at times $(n+1)T$ based on the polarity of the result of the test statistic $p_n(\cdot)$, where

$$p_n(\cdot) = \frac{1}{T}\int_{nT}^{(n+1)T} s(t-nT)r(t,\cdot)dt. \tag{3.63}$$

If the random variable $p_n(\cdot)$ is positive, then it is assumed that $s(t)$ was the transmitted signal, and if $p_n(t)$ is negative, then it is decided that $-s(t)$ was sent. There are two different errors that can occur: the receiver could choose $-s(t)$ when $s(t)$ was sent, or it can choose $s(t)$ when $-s(t)$ is the actual transmitted signal. The purpose of this analysis is to determine the probability of these errors. In the first case, $s(t)$ is the actual transmitted signal. Therefore, in the interval for $n = 0$, the receiver signal is given by

$$r(t,\cdot)|H_1 = s(t) + n(t,\cdot), \tag{3.64}$$

and the random test statistic, obtained by performing a correlation with a noiseless copy of the signal, is then

$$p_0(\cdot)|H_1 = \frac{1}{T}\int_0^T s^2(t)dt + \frac{1}{T}\int_0^T s(t)n(t,\cdot)dt \tag{3.65}$$

We recognize the first integral as the average power in the zeroth bit P_0 of the noiseless received signal. Therefore, the test statistic is reduced to

$$p_0(\cdot)|H_1 = P_0 + \frac{1}{T}\int_0^T s(t)n(t,\cdot)dt. \tag{3.66}$$

Since the noise is assumed to have zero mean, the expected value of $p_0(\cdot)|H_1$ is simply

$$E[p_0(\cdot)|H_1] = P_0 + \frac{1}{T}\int_0^T s(t)E[n(t,\cdot)]dt = P_0. \tag{3.67}$$

The variance of the test statistic is given by

$$\sigma_0^2|H_1 = \frac{1}{T^2}E\left[\int_0^T s(t)n(t,\cdot)dt \int_0^T s(t)n(t,\cdot)dt\right]. \tag{3.68}$$

Now we can make use of the assumption that the noise is white with a constant power spectral density of $N_0/2$. The ensemble autocorrelation function of a stationary random process was shown in (2.195) to be the inverse fourier transform of the PSD. Therefore the autocorrelation function is simply an impulse of magnitude $N_0/2$;

$$R_n(\tau) = \mathcal{F}^{-1}\left\{\frac{N_0}{2}\right\} = \frac{N_0}{2}\delta(\tau). \tag{3.69}$$

Now the variance of the test statistic can be found as follows

$$\sigma_0^2|H_1 = \frac{1}{T^2}\int_{t_1=0}^{T}\int_{t_2=0}^{T} E[n(t_1,\cdot)n(t_2,\cdot)]s(t_1)s(t_2)dt_1dt_2 \tag{3.70a}$$

$$= \frac{1}{T^2}\frac{N_0}{2}\int_{t_1=0}^{T}\int_{t_2=0}^{T} \delta(t_1-t_2)s(t_1)s(t_2)dt_1dt_2 \tag{3.70b}$$

$$= \frac{1}{T^2}\frac{N_0}{2}\int_{t_1=0}^{T} s^2(t_1)dt_1 \tag{3.70c}$$

$$= \frac{N_0}{2T}P_0. \tag{3.70d}$$

Since the noise is assumed Gaussian, $p_0(\cdot)|H_1$, which is a linear combination of the noise, is also Gaussian. Therefore, we can write the pdf since we know the mean and variance.

$$\mathrm{pdf}_{p_0(\cdot)|H_1}(p) = \frac{1}{\sqrt{2\pi\frac{N_0}{2T}P_0}}\exp -\frac{1}{2}\left[\frac{(p-P_0)^2}{\frac{N_0}{2T}P_0}\right] \tag{3.71}$$

Repeating the procedure for the case when the transmitted signal is actually $-s(t)$, the results will be the same except that the mean is now $-P_0$. Therefore the conditional pdf is

$$\mathrm{pdf}_{p_0(\cdot)|H_0}(p) = \frac{1}{\sqrt{2\pi\frac{N_0}{2T}P_0}}\exp -\frac{1}{2}\left[\frac{(p+P_0)^2}{\frac{N_0}{2T}P_0}\right]. \tag{3.72}$$

The total probability of error is the sum of the probabilities of each type of error;

$$P_e = \pi_1 \Pr[p_0(\cdot)|H_1 \leq 0] + \pi_0 \Pr[p_0(\cdot)|H_0 > 0]. \tag{3.73}$$

This is a general result, but we recall that the correlation receivers is only optimal for $\pi_0 = \pi_1 = 1/2$. If this is not the case a correlation receiver can still be used, but the decision threshold will not be zero, but will be biased in favor of the more likely signal. The error probability is illustrated graphically in Fig. 3.12. This figure shows that the two-conditional pdfs are identical Gaussians centered at P_0 and $-P_0$. The error probability is the weight in the part of the tails of these Gaussian pdfs that cross the origin. The error probability can be expressed in terms of the complementary error

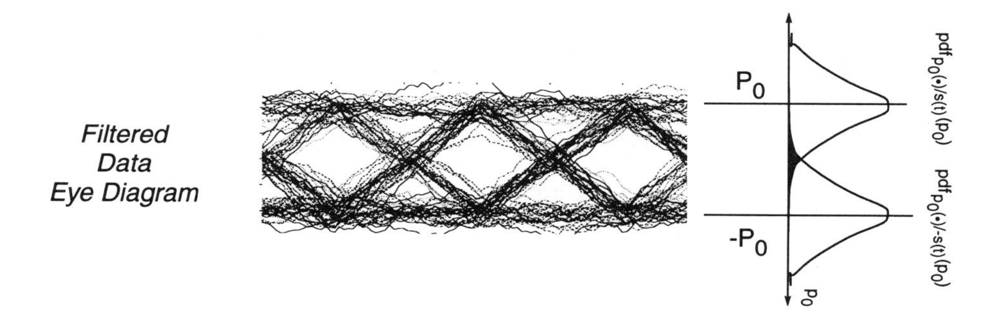

Figure 3.12 Eye-diagram of rectangular NRZ data after passing through a matched filter with the conditional probability density function of the test statistics shown to the right.

function.

$$\Pr[p_0|H_1 \le 0] = \Phi\left[\frac{P_0}{\left(\frac{N_0}{2T}P_0\right)^{1/2}}\right] = \Phi\left[\left(\frac{P_0}{N_0/2T}\right)^{1/2}\right] \tag{3.74}$$

Due to symmetry $\Pr[p_0|H_0 > 0]$ is given by the same expression. Therefore, the error probability is

$$P_e = (\pi_1 + \pi_0)\Phi\left[\left(\frac{P_0}{N_0/2T}\right)^{1/2}\right], \tag{3.75}$$

or simply

$$\boxed{P_e = \Phi\left[\left(\frac{P_0}{N_0/2T}\right)^{1/2}\right].} \tag{3.76}$$

Signal-to-Noise Ratio The parameter $P_0/(N_0/2T)$ is an important quantity. This is the ratio of the average power in the zeroth bit to the noise power in a bandwidth of $1/2T$. Since the bit-rate B_T is equal to $1/T$, the noise power is equivalent to passing the white-noise through an ideal lowpass filter with a bandwidth of half the bit-rate $B_T/2$. This signal-to-noise power ratio (SNR) can also be written in terms of energies as follows,

$$\boxed{\text{SNR}_0 = \frac{P_0 T}{N_0/2} = \frac{E_0}{N_0/2}.} \tag{3.77}$$

Therefore, the error probability is determined by the complementary error function of the square-root of the SNR, where the SNR is the total energy in the bit E_0 divided by the constant white-noise power spectral density $N_0/2$ which also, obviously, has units of energy. The complementary error function $\Phi(y)$ was plotted in Fig. 3.8, and from table 3.1 we find that for $P_e = 10^{-9}$ then

$$\text{SNR}_0 = \frac{P_0}{N_0/2T} = \frac{E_0}{N_0/2} = 6^2 = 15.566 \quad \text{dB}. \tag{3.78}$$

Therefore, to achieve $P_e = 10^{-9}$ the rms signal amplitude should be 6 times larger than the rms noise. At this error probability, a fiber-optic communication system operating at 10-Gb/s will experience 10 errors-per-second on average. It is useful to have a closed form expression for P_e in terms of the SNR. It was shown in (3.28) that for an SNR $> 9 = 9.542$ dB, that P_e can be approximated by

$$\boxed{-\log\left[P_e\right] \overset{\text{SNR}>9}{\simeq} \frac{\text{SNR}}{4.6} + \frac{1}{2}\log(\text{SNR}) + 0.4} \tag{3.79}$$

In following sections, the concept of maximizing the SNR of a test statistic will be used to evaluate the performance of correlation receivers in the presence of non-white noise, and clock jitter. First however, we will consider a fundamental limitation on the maximum achievable SNR.

3.6 QUANTUM LIMIT IN OPTICAL COMMUNICATION SYSTEMS

Before proceeding further to consider the effect of clock jitter and colored noise on receiver performance, we might rightfully ask whether there is a fundamental limit on the performance of an optical receiver, and if so, what is that limit? Due to the quantum nature of light, the energy delivered to a receiver occurs in discrete packets called photons. The arrival of these photons is random, so that the fundamental nature of the signal itself exhibits noise called quantum noise [4, ch. 7, p. 279], [5, sec. 8.4]. In this section we will evaluate the "quantum limit" of a fiber-optic receiver, which is the minimum number of photons that must be incident on a photo-detector, on average, in order to achieve a given error probability.

3.6.1 Approximate Quantum Limit Using AWGN Assumption

First we will consider a non-physical approximation to an optical receiver, and evaluate the quantum limit based on this model. Later we will determine the actual quantum limit based on a consideration of arrival statistics of photons. In this example we will assume that we have bipolar light, so that when we want to transmit a one we send a light pulse, and when we want to transmit a zero we send the negative of this light pulse. We also assume that we have a photo-detector that can produce a bipolar current output with no bias current (I for a "one" and $-I$ for "zero"). We will further assume that the photo-detector is 100% efficient, so that each photon produces an electron-hole pair. Therefore, if n photons are incident on a photo-detector in a time interval T, then the current will be $I = nq/T$, where q is the charge on an electron. Since we would like to get a reasonable estimate of the quantum limit based on concepts that are familiar to circuit designers, we will also assume that the quantum noise has the same statistics as shot-noise. Therefore the two-sided noise PSD is equal to qI. In the anti-case, negative photons are sent giving rise to a current of $-I$, which also has a noise power spectral density of qI.

Approximate Quantum Limit Derived from Shot-Noise Assumption For this symmetric example, the energy delivered to a 1Ω resistor is the same for every bit, and is given by

$$E_B = I^2 T. \tag{3.80}$$

The SNR is given by the ratio of the average bit energy to the white-noise PSD;

$$\text{SNR} = \frac{I^2 T}{qI} = \frac{I}{q/T} \tag{3.81}$$

Therefore we have the interesting interpretation that the SNR is the ratio of the average current produced in one bit period, to the current produced by one single charge in the same time interval. Substituting for I we obtain the SNR in terms of the average number of photons incident on the detector in a time T.

$$\text{SNR} = \frac{\overline{n}q/T}{q/T} = \overline{n}, \tag{3.82}$$

or the SNR is just the average number of photons per bit period. We know from table 3.1 that for $P_e = 10^{-9}$ we require an SNR of approximately 36. Therefore,

$$\overline{n} \simeq 36 \qquad \text{photons/bit for } P_e = 10^{-9} \tag{3.83}$$

Improved Estimate for On-Off Modulation We can get a more realistic approximation to the actual quantum limit if we realize that the light pulse will be on only half of the time. Consider an optical system which transmits a light pulse for a *one* and no light pulse for a *zero*. We will further assume that there is no noise when a *zero* is sent. Therefore the error probability P_0 of detecting a *one* when a *zero* is sent is zero ($P_0 = 0$). Therefore the total error probability is given by

$$P_e = \frac{1}{2}P_0 + \frac{1}{2}P_1 = \frac{1}{2}P_1, \tag{3.84}$$

and

$$P_1 = 2P_e = 2 \times 10^{-9}. \tag{3.85}$$

The average number of photons needed to achieve this higher probability will be slightly less than 36. We find from (3.79) that the average number of photons needed per *one* symbol $\overline{n_1} = 34.64$. Therefore the number of photons per bit $\overline{n}$ is given by

$$\overline{n} = \frac{1}{2}\overline{n_0} + \frac{1}{2}\overline{n_1} = \frac{1}{2}\overline{n_1} = 17.32 \simeq 18. \tag{3.86}$$

The interpretation of this result is that the arrival of photons is a random event. If we want to reduce the chances of getting a negative result in the bit interval to once out of every one-billion bits, then we have to make sure that on average we have 18 photons arriving per bit interval. If we let the probability of error rise to one-in-a-million ($P_e = 10^{-6}$), then the SNR needed is 22.7, which requires 11.35 photons per bit on average.

We have used the familiar concept of shot-noise to approximate the quantum limit as $\overline{n_1} \simeq 36$ and $\overline{n} \simeq 18$. However, we know that the noise can not be negative (either we detect a photon or we do not, but there are no negative photons). Therefore if only one photon is detected, the result will be positive and we will interpret this as a *one*, and no amount of noise will turn this positive result negative. Therefore we should be able to reduce the average number of photons needed to obtain $P_e = 10^{-9}$ to less than 18. To find the actual quantum limit we must consider the arrival statistics of photons at the receiver more carefully.

3.6.2 Actual Quantum Limit from Poisson Distribution

In the previous section, the results for AWGN were applied to a fictitious example to get an estimate of the quantum limit in terms of parameters that are familiar to a circuit designer. In this section, we will determine the actual quantum limit based on the random arrival statistics of photons. To determine the absolute minimum power needed, we will assume that we have the capability to detect a single photon. Furthermore, we'll assume that we have a light source with a 100% extinction ratio (when the light is off, it is *really* off). The received signal will consist of either a light pulse, or no light pulse. When there is no light pulse, there is absolutely no way we can detect a photon (this assumes that there is no ISI). Therefore the conditional probability density function when no pulse is sent is

$$P_0(\boldsymbol{r}) = \delta(\boldsymbol{r}), \tag{3.87}$$

which means that the received signal is identically zero with probability one. We can express the optimal decision rule as

$$\begin{aligned} &\mathcal{R}_0; \qquad \boldsymbol{r} = 0 \\ &\mathcal{R}_1; \qquad \boldsymbol{r} \neq 0. \end{aligned} \tag{3.88}$$

That is, we choose $s_1(t)$ when the received signal is anything but zero. The total error probability for this system is

$$P_e = \frac{1}{2}\Pr[d(\boldsymbol{r}) = 1 | H_0] + \frac{1}{2}\Pr[d(\boldsymbol{r}) = 0 | H_1], \tag{3.89}$$

and since there is no noise when there is no light, there is no chance of making an error when no light is transmitted. The only chance of making an error is when we don't detect any photons when we were supposed to. In other words, we turn on the light source, but because of the random nature of photon emission, no photons are emitted in a time T, even when the light source is on. The total error probability is just due to the later situation, and is given by

$$P_e = \frac{1}{2}\Pr[d(\boldsymbol{r}) = 0 | H_1]. \tag{3.90}$$

To find the probability of this event, we must consider the arrival statistics of the photons themselves.

Derivation of Error Probability If we look at very short time intervals Δt, we will assume that the probability of the arrival of one photon in this time interval is proportional to Δt,

$$P_1(\Delta t) = a\Delta t, \tag{3.91}$$

where the significance of the proportionality constant a will be demonstrated later. Since the time interval is short, either one photon arrives or it doesn't, but the time interval is too short to allow more than one arrival. Therefore,

$$\begin{aligned} P_0(\Delta t) + P_1(\Delta t) &= 1 \\ P_0(\Delta t) &= 1 - a\Delta t \end{aligned} \tag{3.92}$$

We are interested in finding the probability that no arrivals occurred in a time interval of length T. We can consider an interval of length $T + \Delta t$, and we further assume that the arrival of a photon in the time Δt is independent of the arrival of a photon in the previous time interval T. The probability of no emission in a time period of $T + \Delta t$ is then given by

$$P_0(T + \Delta t) = P_0(T)P_0(\Delta t) = P_0(T)[1 - a\Delta t], \tag{3.93}$$

and writing this as a difference equation we get

$$\frac{P_0(T + \Delta t) - P_0(T)}{\Delta t} = -aP_0(T). \tag{3.94}$$

In the limit as Δt goes to zero, the difference equation becomes a differential equation

$$\frac{dP_0(T)}{dT} = -aP_0(T), \tag{3.95}$$

with the solution of

$$P_0(T) = e^{-aT}, \tag{3.96}$$

where we have made use of the boundary condition

$$P_0(0) = \lim_{\Delta t \to 0} P_0(\Delta t) = 1. \tag{3.97}$$

Relationship Between Parameter** a **and Observable Statistics (3.96) gives the desired result, but is expressed in terms of the parameter a. In order to determine a relationship for this parameter in terms of observable statistics it is necessary to continue the derivation to determine $P_n(T)$, the probability of obtaining precisely n photons in a given time interval T. Following Davenport and Root [6, ch. 7, pp. 115–118], we will consider the probability of observing n photons in a time interval of length $T + \Delta t$. We can restrict Δt to be so small that no more than one photon can arrive in this time; therefore only two possibilities exist: either one photon is emitted, or none are. Since we have assumed that emissions at any time t are independent of emissions in the past, for small Δt the probability of observing n photons in an interval of length $T + \Delta t$ is simply given by

$$P_n(T + \Delta t) = P_n(T)P_0(\Delta t) + P_{n-1}(T)P_1(\Delta t). \tag{3.98}$$

Recalling that $P_1(\Delta t) = a\Delta t$, and $P_0(\Delta t) = 1 - a\Delta t$, it follows that

$$\frac{P_n(T + \Delta t) - P_n(T)}{\Delta t} + aP_n(T) = aP_{n-1}(T). \tag{3.99}$$

In the limit as $\Delta t \to 0$, we obtain a differential recursion equation

$$\frac{dP_n(T)}{dT} + aP_n(T) = aP_{n-1}(T), \tag{3.100}$$

which has a solution given by[3]

$$P_n(T) = ae^{-aT} \int_0^T P_{n-1}(\tau)e^{a\tau}\,d\tau, \tag{3.101}$$

[3] Davenport and Root reference Richard Courant, "Differential and Integral Calculus," I. rev. ed., 1937; II, 1936, Interscience Publishers, New York.

where we have utilized the boundary condition $P_n(0) = 0$. For the case of $n = 1$ we can make use of the result $P_0(T) = e^{-aT}$ from (3.96) to obtain

$$P_1(T) = ae^{-aT} \int_0^T e^{-a\tau} e^{a\tau} d\tau = (aT)e^{-aT}. \tag{3.102}$$

For the case of $n = 2$, we can substitute the previous result to obtain

$$P_2(T) = ae^{-aT} \int_0^T (a\tau)e^{-a\tau} e^{a\tau} d\tau = \frac{(aT)^2 e^{-aT}}{2}. \tag{3.103}$$

It is not difficult to see the pattern that emerges from the recursion and therefore determine the probability for any arbitrary n as

$$P_n(T) = \frac{(aT)^n e^{-aT}}{n!} \tag{3.104}$$

Using (3.104) we can find the expected number of arrivals in time T as

$$\overline{n_1} = \sum_{n=0}^{\infty} nP_n(T) = \sum_{n=0}^{\infty} \frac{n(aT)^n e^{-aT}}{n!} \tag{3.105}$$

This sum can be evaluated explicitly as follows. First consider the Taylor series expansion for e^{aT}.

$$\sum_{n=0}^{\infty} \frac{(aT)^n}{n!} = e^{aT} \tag{3.106a}$$

Taking the derivative with respect to both sides gives

$$\sum_{n=0}^{\infty} \frac{n(aT)^{n-1}}{n!} = e^{aT}, \tag{3.106b}$$

and multiplying by (aT) gives a series expansion for $(aT)e^{aT}$:

$$\sum_{n=0}^{\infty} \frac{n(aT)^n}{n!} = (aT)e^{aT} \tag{3.106c}$$

Finally, multiplying by e^{-aT} puts this in the form of (3.105).

$$\overline{n_1} = \sum_{n=0}^{\infty} \frac{n(aT)^n e^{-aT}}{n!} = aT \tag{3.106d}$$

Now we can see the significance of the parameter a and substitute $aT = \overline{n_1}$ into (3.96) to obtain

$$P_0(T) = e^{-\overline{n_1}}, \tag{3.107}$$

which is the probability of not getting any photons in a time T, when on average we get $\overline{n_1}$.

Quantum Limit in Terms of Number of Photons per Bit Using the above results, the desired probability $P_0(T)$ can now be expressed in terms of $\overline{n_1}$, which is an observable quantity. The total error probability for a fiber-optic receiver operating at the quantum limit is given by

$$P_e = \frac{1}{2}P_0(T) = \frac{1}{2}e^{-\overline{n_1}}. \tag{3.108}$$

We have now arrived at the desired result that, due to quantum noise, we require on average $\overline{n_1}$ photons per *one* symbol to insure an error probability of P_e, where

$$\boxed{\overline{n_1} = -\ln(2P_e)} \tag{3.109}$$

Since there are no photons transmitted for a *zero* symbol, the average number of photons per bit $\overline{n}$ is given by

$$\overline{n} = \frac{1}{2}\left(\overline{n_0} + \overline{n_1}\right) = \frac{1}{2}\overline{n_1}, \tag{3.110}$$

therefore

$$\boxed{\overline{n} = -\frac{\ln(2P_e)}{2}} \tag{3.111}$$

The quantum limits $\overline{n_1}$ and $\overline{n}$ are given in table 3.2, where we see that, approximately, an additional 7 photons per *one* symbol are required on average to reduce P_e by 3 decades.

Plots of $P_n(T)$ are given in Figs. 3.13(a) and (b) for $\overline{n_1} = 10$ and 20 respectively. It can also be shown using the same method as outlined in (3.106) that the variance of the Poisson distribution is also equal to $\overline{n_1}$. Therefore the standard deviation is equal to the square-root of the average number of photons, and the average SNR is equal to $1/\sqrt{\overline{n_1}}$, which is a familiar result for independent random variables.

Quantum Limit in Terms of Optical Power We can relate the quantum limit to optical power for a given bit interval T. Continuing with our example of a 10 Gb/s system, the bit-interval T is equal to 100 ps. The energy in a photon is given by

$$e_{\text{ph}} = \frac{hc}{\lambda}, \tag{3.112}$$

where h is Planck's constant, and c is the speed of light. Substituting these constants the photon energy is

$$e_{\text{ph}} = \frac{198.6 \times 10^{-12}(\text{nJ} \cdot \mu\text{m})}{\lambda}. \tag{3.113}$$

P_e	$\overline{n_1}$	$\overline{n}$
10^{-3}	6.2	3.1
10^{-6}	13.1	6.6
10^{-9}	20.0	10.0
10^{-12}	26.9	13.5
10^{-15}	33.8	16.9

Table 3.2 Quantum limit in terms of average photons per *one* symbol $\overline{n_1}$ and average photons per bit $\overline{n}$ incident on the photo-detector in one bit period T to insure a given error probability P_e.

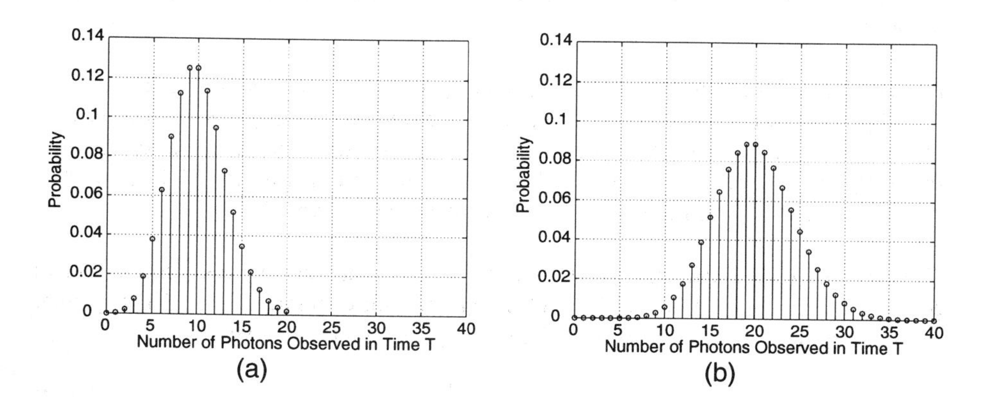

Figure 3.13 Possion distribution for $\overline{n_1}$ equal: (a) 10, (b) 20.

The photon power is the energy divided by the interval T;

$$p_{\text{ph}} = \frac{e_{\text{ph}}}{T} = \frac{1.986(\text{nW} \cdot \mu\text{m})}{\lambda} \quad \text{for} \quad T = 100\text{ps}. \tag{3.114}$$

Since the light pulse is only on half of the time, the average optical power is

$$p_{\text{av}} = \frac{1}{2}(0) + \frac{1}{2}(\overline{n_1} p_{\text{ph}}) = \overline{n} p_{\text{ph}}, \tag{3.115}$$

or substituting (3.109), we can write the average power at the quantum limit in terms of the desired error probability, for a bit-rate of 10-Gb/s such that,

$$p_{\text{av}} = -\log(2P_e)\left(\frac{2.29(\text{nW} \cdot \mu\text{m})}{\lambda}\right). \tag{3.116}$$

The following equation expresses the result for an arbitrary bit-rate B_T,

$$p_{\text{av}} = \frac{-\ln(2P_e)}{2}\left[\frac{hc}{\lambda}\right] B_T, \tag{3.117}$$

and substituting for the numerical constants we obtain the general expression for the quantum limit in an optical system using on-off modulation and no coding.

$$\boxed{p_{\text{av}} = -\log(2P_e)\left(\frac{0.229(\text{nW} \cdot \mu\text{m})}{\lambda}\right)\left[\frac{B_T}{\text{1-Gb/s}}\right]} \tag{3.118}$$

Table 3.3 gives the quantum limit in terms of the minimum average optical power required to achieve various error probabilities for different wavelengths of light at a data rate of 10-Gb/s. We can see that, at best, we need an optical power of about -48 dBm for $P_e = 10^{-9}$. This analysis gives us a theoretical limit on the minimum received power. However practical implementation problems will limit the sensitivity of the receiver such that many more photons above the quantum limit will be needed for accurate communication.

3.6.3 Practical Limitations on the Minimum Number of Photons Required

Although the quantum-limit tells us the absolute minimum average number of photons needed to obtain a given P_e, the actual number will be much higher in practice. The reason is that all the electronic circuitry used to detect the optical signal is also governed by the same quantum statistics, thus adding to the noise level. Even worse is that the

P_e	$\overline{n_1}$	$p_{av}(\lambda = 0.80\mu m)$ dBm	$p_{av}(\lambda = 1.30\mu m)$ dBm	$p_{av}(\lambda = 1.55\mu m)$ dBm
10^{-3}	6.2	−51.1	−53.2	−54.0
10^{-6}	13.1	−47.9	−50.0	−50.8
10^{-9}	20.0	−46.0	−48.1	−48.9
10^{-12}	26.9	−44.7	−46.9	−47.6
10^{-15}	33.8	−43.8	−45.9	−46.6

Table 3.3 Quantum limit for various wavelengths of light in terms of average optical power incident on the photo-detector for a 10 Gb/s optical receiver (T=100 ps) to insure a given error probability.

magnitude of the electronic noise sources of the receiver circuitry will be several orders of magnitude larger than the quantum noise. This is best illustrated with an example.

Example of Practical Limitations In a 10 Gb/s communication system fabricated with integrated circuits, with typical parasitic capacitances at the input of 0.5 pF, a voltage swing of 400 mV in 100 ps requires a current of 2 mA. The number of charges needed to produce a current of 2 mA in 100 ps is

$$\overline{N} = \frac{2\text{mA} \cdot 100\text{ps}}{1.6 \times 10^{-19}\text{C}} = 1.25 \times 10^6 \tag{3.119}$$

The chances of trying to resolve an incremental 20 charges ($\overline{n_1}$ for $P_e = 10^{-9}$), compared to the one-and-a-quarter million charges due to the bias current, is pretty slim. This implies a quantum resolution of

$$\text{RES}_Q = \frac{1.25 \times 10^6}{20} = 62.5 \times 10^3 \sim [16 - \text{bits}]. \tag{3.120}$$

16-bits is a very high resolution; typically current due to the photodetector will be far below the noise floor of the preamplifier. If we consider the shot noise current alone at the output of the optimal correlator, the rms current in a bandwidth of $B_T/2$ is

$$i_{rms} = \sqrt{\frac{Iq}{T}} = \sqrt{2\text{mA} \cdot 1.6\text{nA}} = 1.79\mu\text{A}. \tag{3.121}$$

We can express this in terms of the average number of charges $\overline{N}$

$$i_{rms} = \sqrt{\frac{Iq}{T}} = \sqrt{\overline{N}\left(\frac{q}{T}\right)^2} = \sqrt{\overline{N}}\left(\frac{q}{T}\right) \tag{3.122}$$

Therefore the number of charges giving rise to the rms current in a time of 100 ps is just [4]

$$n_{\text{rms}} = \frac{1.79\mu\text{A}}{100\text{ps}} = \sqrt{N} = 1118. \tag{3.123}$$

Comparing the bias current to the rms noise current, the circuit's maximum resolution, due to shot-noise alone is limited to

$$\text{RES}_S = \frac{I_{\text{bias}}}{I_{\text{rms}}} = \frac{1.25 \times 10^6}{1118} = \frac{N}{\sqrt{N}} = \sqrt{N} = 1118 \sim [10 - \text{bits}]. \tag{3.124}$$

If the input signal power level is at the quantum limit, then the receiver will have to resolve 16 bits in a system that is inherently limited to 10 bits of resolution! This can not be easily accomplished, therefore the signal level will have to be increased. For a system that is dominated by the shot-noise of a single transistor running with a 2 mA bias current the noise-level will be constant and will be determined by the bias current. The signal peak-to-peak value will be determined by $\overline{n_{Q1}}$, which is the is the quantum limit in photons per *one* symbol. This signal current will appear as a deviation of $\pm\overline{n_{Q1}}/2 = \pm\overline{n_Q}$ from the bias current. The SNR for a signal with power at the quantum limit for $P_e = 10^{-9}$ is therefore

$$\sqrt{\text{SNR}} = \frac{I_{\text{sig}}}{I_{\text{rms}}} = \left(\frac{20/2}{1118}\right) \simeq \left(\frac{1}{112}\right) = \frac{\overline{n_{Q1}}/2}{\sqrt{N}}, \tag{3.125}$$

So we need to increase the signal level by approximately a factor of 112, just to get the SNR to unity. Since we know that $P_e \leq 10^{-9}$ requires an SNR$\geq 6^2$, we need to increase the signal level approximately (6×112) times, or 671 times above the quantum limit level in a practical system to reach this performance objective. Therefore, the number of photons-per *one* symbol required in this practical receiver is

$$\overline{n_1} = 6\overline{n_{Q1}}\left[\frac{\sqrt{N}}{\overline{n_{Q1}}/2}\right] = 12\sqrt{N} \tag{3.126a}$$

$$\overline{n_1} \simeq 20 \times 671 = 13,416 \qquad \text{Photons per } \textit{one} \text{ symbol}, \tag{3.126b}$$

The average number of photons-per-bit is just half of $\overline{n_1}$ or

$$\overline{n} = \frac{\overline{n_1}}{2} = 6\sqrt{N} = 6708, \tag{3.127}$$

[4]Notice that averaging shot-noise gives us the same result of averaging identically distributed random variables. If the average number of charges passing a barrier in a time T is N, then the rms deviation of the average of this number is $\sqrt{N}$. Therefore, the SNR is just $\sqrt{N}$. This gives us a very simple way to determine the amount of filtering required to get a given deviation in the dc current. If we require an SNR of 1000, then we need 10^6 charges passing. For a 1.6 mA current in 100 ps we get 10^6 charges, so averaging over this time period gives a result with an rms deviation from the mean of one part in 1000. If we want to increase the resolution by 2 bits, or a factor of 4, the time interval must be increased by a factor of 16 to 1.6 ns.

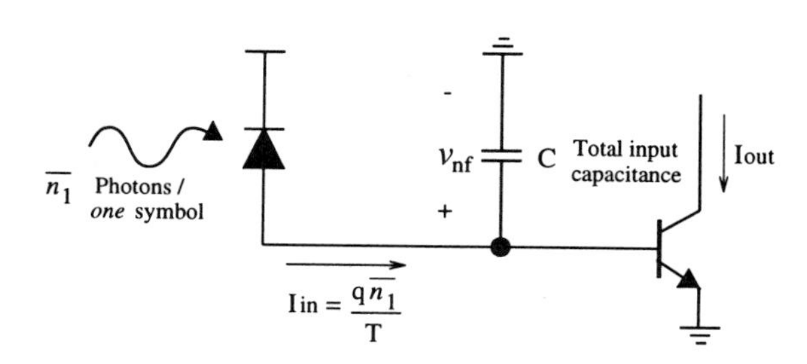

Figure 3.14 Simplified diagram of a fiber-optic receiver for approximate sensitivity calculations.

which implies an incident optical power of -19.8 dBm for $\lambda = 1.3\mu$m. We can now estimate the current required at the output of the photo-detector. Assuming each photon produces an electron-hole pair, the current produced is

$$\overline{I} = \frac{q\overline{n_1}}{T}, \tag{3.128}$$

where T=100-ps for a 10-Gb/s system. Therefore $\overline{n_{Q1}} = 20$ corresponds to a peak current of 32-nA. For a practical system with $\overline{n_1} = 13,416$, the peak current required is approximately 21.5-μA. This is 671 times or (28.3-dB of optical power) larger than the quantum limited current of 32-nA. The average current is 10.7-μA, which is half the peak current and 6 times the shot-noise rms current of 1.79-μA given in (3.121).

Receiver Sensitivity in Terms of Distance Between Repeaters From our previous discussion of limitations in integrated receivers, we know that the number of photons required in a practical systems needs to be increased by a factor of approximately 671, or 28 dB over and above the quantum limit. From table 3.3 the quantum limited power is approximately -48 dBm for a 10-Gb/s system at a wavelength of 1.3-μm. Therefore, we would expect a practical receiver to require approximately -20 dBm, or 10 μW of optical power, producing an average output current of about 11-μA at the photodetector. For a low-loss optical fiber (0.15-dB/km), and an optical source capable of launching 1-mW (0 dBm) of optical power at the transmitter, a received power greater than -20 dBm implies that the maximum repeater spacing is (20/0.15) or 133 kilometers.

Receiver Sensitivity in Terms of Circuit Parameters Thus far we have not talked about the actual preamplifier circuit needed to detect the arrival of photons. Nevertheless we can continue with this approximate analysis to obtain a good indication of how a real circuit might behave. The circuit we have been considering implicitly is shown in Fig. 3.14. The number of charges $\overline{N}$ needed to charge the capacitor C to the noise-floor voltage v_{nf} is found from

$$I = \frac{q\overline{N}}{T} = \frac{Cv_{nf}}{T}, \tag{3.129}$$

from which we obtain

$$\overline{N} = \frac{Cv_{nf}}{q} \tag{3.130}$$

We saw that the number of photons needed per bit to achieve $P_e = 10^{-9}$ is given by

$$\overline{n} = 6\sqrt{\overline{N}} = 6\left[\sqrt{\frac{Cv_{nf}}{q}}\right]. \tag{3.131}$$

The average optical power is then

$$p_{\text{av}} = \overline{n}p_{\text{ph}} = \frac{\overline{n}e_{\text{ph}}}{T} = \frac{\overline{n}hc}{\lambda T}, \tag{3.132}$$

and writing p_{av} in terms of circuit parameters we obtain

$$p_{\text{av}} = \frac{6hc}{\lambda T}\sqrt{\frac{Cv_{nf}}{q}}. \tag{3.133}$$

Hence for a given wavelength λ,

$$\boxed{p_{\text{av}} \propto B_T\sqrt{Cv_{nf}}.} \tag{3.134}$$

The minimum optical power needed to achieve a given BER is proportional to the bit-rate $B_T = 1/T$ and to the square-root of the input capacitance and the noise-floor voltage. The reasons for this relationship are clear. The term involving C and v_{nf} determine the number of photons needed per bit. The square-root dependence occurs due to independent random events having a standard deviation proportional to the square-root of the observations. The number of photons per bit is independent of B_T, but as B_T increases, more photons are required per second, thus increasing the optical power linearly. The preceding analysis is just a first-order estimate that gives us a feeling for how circuit parameters will effect the receiver sensitivity. The actual values of the optical power needed will depend on the quantum efficiency of the photo-detector, and the noise PSD of the preamplifier. We will see that circuit parameters will effect the noise-floor voltage v_{nf}, and this voltage will also increase with frequency. These issues will be discussed in greater detail in chapter 7.

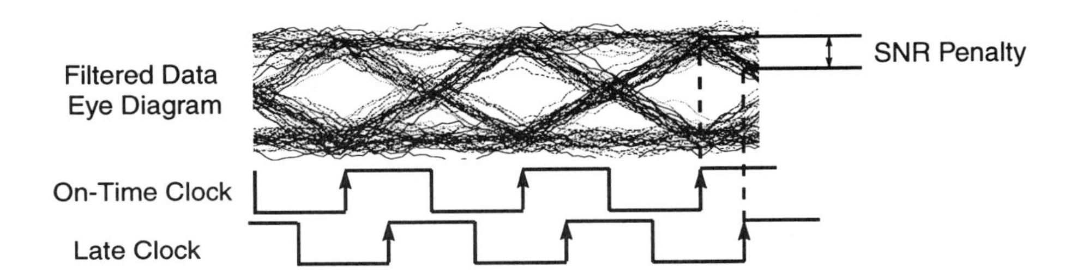

Figure 3.15 Diagram showing the error incurred in the output of a correlation receiver when an error in the clock phase is present.

3.7 CORRELATION RECEIVER PERFORMANCE IN THE PRESENCE OF CLOCK-JITTER

Until now we have only analyzed receivers with no error in the estimation of the arrival time of each symbol, however, as mentioned previously, there will always be jitter in the recovered clock, and a well designed receiver must be robust against such jitter. It is beyond the scope of this work to derive the form of optimal receivers in the presence of clock jitter, and non-white, or colored-noise. Our approach will be to constrict ourselves to a correlation receiver, and we will consider different windowing functions that maximize the SNR at specified sampling intervals.

Windowing Functions to Reduce Susceptibility to Clock Jitter When there is uncertainty in the arrival time of the received data pulses, errors in the correlator, or matched filter output will occur. The errors will be due to the overlap of the correlation pulse with the adjacent data bits as illustrated in Fig. 3.15. If the data is rectangular, then the reduction in the signal power will be linearly proportional to the clock offset when adjacent bits are different, but there will be no penalty when adjacent bits are identical. This error can be reduced by using a windowing function that reduces the contributions of errors at the edges, by concentrating the majority of the pulse energy in the center of the time interval. However, the maximum obtainable SNR will be reduced. In this section we will analyze the performance degradation of a receiver in the presence of clock-jitter, and consider different windowing functions that can reduce the receivers susceptibility to timing errors.

SNR Degradation in Rectangular Pulse Correlation Receivers For rectangular NRZ data there will be no degradation in performance unless a bit transition occurs. This situation is illustrated in Fig. 3.16. For the case of the fourth bit, the SNR is the same as if there were no timing error. For the first bit, however, the correlation pulse overlaps a portion of the second bit. Since the two bits are of opposite polarity, the contribution

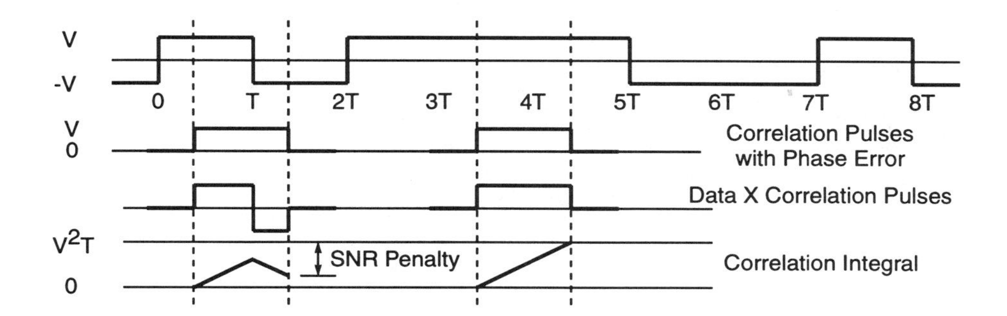

Figure 3.16 Illustration of SNR degradation due to timing errors for rectangular NRZ data.

from the overlap subtracts directly from the SNR. The sample statistic of a correlation receiver is then

$$\begin{aligned} p_0(\cdot)|H_1 = & \frac{1}{T}\int_{\Delta t}^{T} s(t)s(t-\Delta t)dt + \\ & \frac{1}{T}\int_{T}^{T+\Delta t} -s(t)s(t-\Delta t)dt + \\ & \frac{1}{T}\int_{\Delta t}^{T+\Delta t} n(t,\cdot)s(t-\Delta t)dt \end{aligned} \tag{3.135}$$

The mean of this statistic is a maximum at $\Delta t = 0$, and falls off linearly to zero when $|\Delta t| = 0.5$. Therefore,

$$\mu = P_0\left(1 - 2\frac{|\Delta t|}{T}\right) \qquad \text{for} \quad |\Delta t| \leq 0.5. \tag{3.136}$$

The variance of the statistic is independent of timing errors, and from (3.70) is

$$\sigma^2 = \frac{N_0}{2T}P, \tag{3.137}$$

where $P_0 = P$, because the power in each bit is identical for rectangular data, whether or not a transition occurs. The SNR for this case is given by

$$\boxed{\text{SNRrec} = \mu^2/\sigma^2 = \frac{P}{N_0/2T}\left(1 - 2\frac{|\Delta t|}{T}\right)^2} \tag{3.138}$$

This reduction in SNR is illustrated in Fig. 3.17. This is similar to Fig. 3.12, but now the mean of the Gaussians are moved closer to the origin, while the variance remains constant.

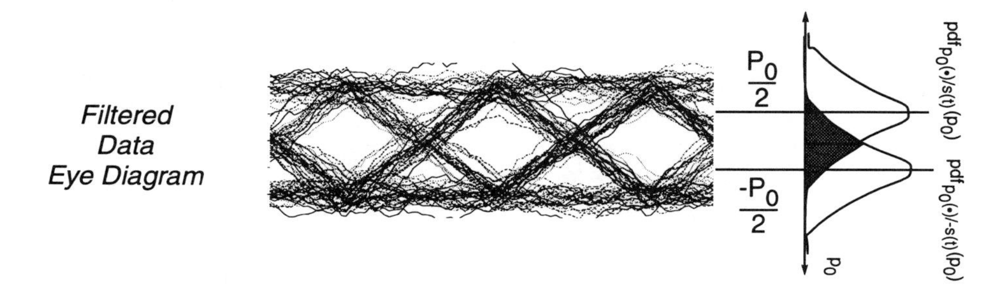

Figure 3.17 Eye-diagram of rectangular NRZ data after passing through a matched filter with the conditional probability density function of the test statistics shown to the right: (a) for the case of a timing error of magnitude $|\Delta t| = 1/4$.

The total probability of error, for a correlation receiver in white noise, with a timing error of $\Delta t \in [-1/2, 1/2]$, is

$$\begin{aligned} P_e =& \Phi\left[\left(\frac{P}{N_0/2T}\right)^{1/2}\right] \times \Pr[\text{no transition}] + \\ & \Phi\left[\left(\frac{P}{N_0/2T}\left(1 - 2|\Delta t|/T\right)^2\right)^{1/2}\right] \times \Pr[\text{transition}]. \end{aligned} \tag{3.139}$$

Since there is a 50% chance that a transition will occur between adjacent bits, the error probability is

$$P_e = \frac{1}{2}\Phi\left[\left(\frac{P}{N_0/2T}\right)^{1/2}\right] + \frac{1}{2}\Phi\left[\left(\frac{P}{N_0/2T}\left(1 - 2|\Delta t|/T\right)^2\right)^{1/2}\right] \tag{3.140}$$

for $|\Delta t| \leq 1/2$.

For the case of an offset of half a bit ($|\Delta t| = 1/2$), and a very large SNR, the receiver will almost always be correct when no transition occurs, but during a transition the SNR will drop to zero, and the receiver can only guess at the actual bit value, and will be correct only half of the time. The error probability for this case is

$$P_e = \frac{1}{2}\Phi(\infty) + \frac{1}{2}\Phi(0) = 1/4. \tag{3.141}$$

The probability of error for a nominal SNR of 6.002^2 is plotted in Fig. 3.18 as a function of the timing error. It can be seen that the error is 10^{-9} at $\Delta t = 0$, and increases to approximately 1/4 at ($|\Delta t| = 1/2$).

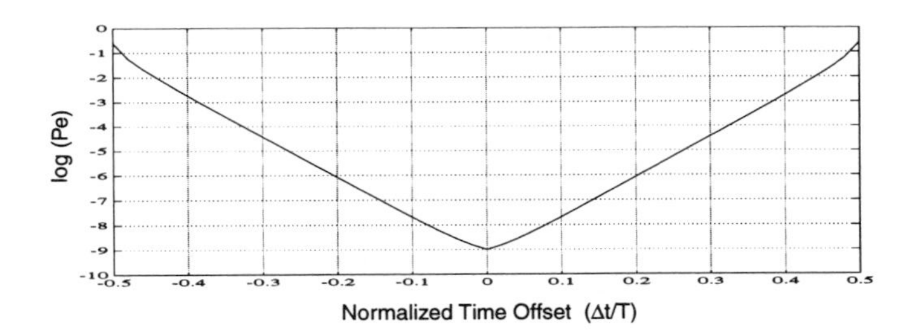

Figure 3.18 The probability of error for a correlation receiver in white noise with a nominal SNR of 6.002^2 as a function of the timing error.

The reduction in SNR due to a timing error results from the overlap of the correlation pulse with adjacent bits. It has been alluded to earlier that by windowing the correlation pulse to minimize edge effects, and concentrating most of the pulse energy in the center of the bit-interval, the receiver can be made less sensitive to timing errors. We will find that the curve in Fig. 3.18 can be flattened, at the expense of increasing the minimum attainable error probability.

3.7.1 Simple Windowing Functions for Reduced Edge Effects

Half-Cosine Window A simple windowing function that is practical to implement, and has the desired characteristics, is a half-cosine pulse, as illustrate in Fig. 3.19. The correlation pulse can be written as

$$c_h(t) = \sqrt{P}\frac{\pi}{2}\sin\left(\frac{\pi t}{T}\right)\text{rect}\left(\frac{t-T/2}{T}\right) \tag{3.142}$$

For no transitions, the pulse has been normalized to give a mean of P, such that

$$\mu = \frac{P}{T}\int_0^T \frac{\pi}{2}\sin\left(\frac{\pi t}{T}\right)dt = P. \tag{3.143}$$

For the interesting case when a transition occurs, the mean of the test statistic is

$$\mu = \frac{P}{T}\int_{\Delta t}^{T}\frac{\pi}{2}\sin\left(\frac{\pi(t-\Delta t)}{T}\right)dt - \int_{T}^{T+\Delta t}\frac{\pi}{2}\sin\left(\frac{\pi(t-\Delta t)}{T}\right)dt \tag{3.144}$$

Shifting the time axis and evaluating the integral we obtain

$$\mu = \frac{P}{2}\left[-\cos\left(\frac{\pi t}{T}\right)\Bigg|_0^{T-\Delta t} + \cos\left(\frac{\pi t}{T}\right)\Bigg|_T^{T-\Delta t}\right], \tag{3.145a}$$

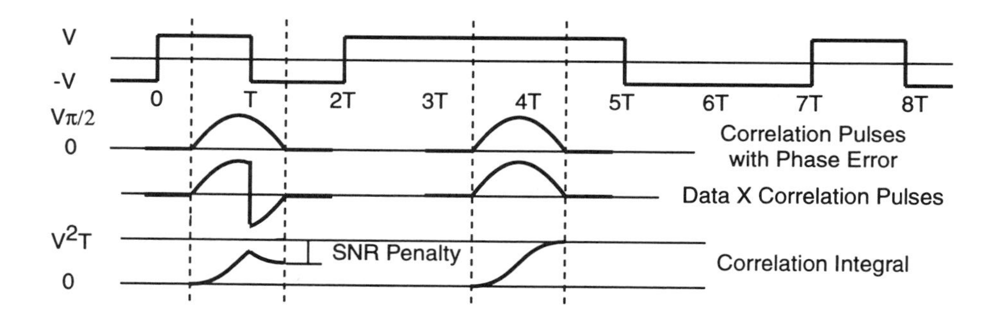

Figure 3.19 Illustration of SNR degradation due to timing errors for rectangular NRZ data with a half-cosine correlation pulse.

or

$$\mu = -P\cos\left(\frac{\pi(T-\Delta t)}{T}\right) = P\cos\left(\frac{\pi\Delta t}{T}\right) \tag{3.145b}$$

The noise variance is independent of the mean of the test statistic and is given by

$$\sigma^2 = \frac{P}{T^2}E\left[\int_0^T n(t,\cdot)\frac{\pi}{2}\sin\left(\frac{\pi t}{T}\right)dt\right]^2 \tag{3.146}$$

For white noise the integral reduces to

$$\sigma^2 = \frac{PN_0}{2T}\left(\frac{\pi}{2}\right)^2\frac{1}{T}\int_0^T \sin^2\left(\frac{\pi t}{T}\right)dt = \frac{PN_0}{2T}\left(\frac{\pi}{2}\right)^2\frac{1}{2} \tag{3.147}$$

The SNR for a transition in adjacent bits is then given by the ratio of the mean-squared, to the variance.

$$\boxed{\mathrm{SNR}_{\mathrm{hcos}}(|\Delta t|/T) = \frac{E_B}{N_0/2}\frac{8}{\pi^2}\cos^2\left(\frac{\pi\Delta t}{T}\right)} \tag{3.148}$$

The resulting probability of error for the half-cosine windowing function is then given by

$$P_e = \frac{1}{2}\Phi\left[\left(\frac{E_B}{N_0/2}\frac{8}{\pi^2}\right)^{1/2}\right] + \frac{1}{2}\Phi\left[\left(\frac{E_B}{N_0/2}\frac{8}{\pi^2}\cos^2\left(\frac{\pi\Delta t}{T}\right)\right)^{1/2}\right] \tag{3.149}$$

Raised-Cosine Window A raised cosine has a more gradual role off at the edges than a half-cosine, so we would expect the raised cosine pulse to have even less susceptibility to timing errors. A raised cosine pulse that is centered around the origin can be written as

$$c_r(t + T/2) = \sqrt{P}\left[1 + \cos\left(\frac{2\pi t}{T}\right)\right] \mathrm{rect}(t/T). \tag{3.150}$$

The mean of the correlation test statistic, when there is no transition in adjacent bits, is just the integral of the pulse itself, multiplied by a constant rectangular pulse of magnitude $\sqrt{P}$. Therefore,

$$\mu = \frac{P}{T}\int_{-T/2}^{T/2} dt + \frac{P}{T}\int_{-T/2}^{T/2} \cos\left(\frac{2\pi t}{T}\right) dt = P. \tag{3.151}$$

When there is a transition between adjacent bits, the correlation pulse will overlap a positive, and a negative bit. Therefore, the mean of the correlation output is

$$\mu = \frac{P}{T}\left[\int_{-T/2}^{T/2-\Delta t} 1 + \cos\left(\frac{2\pi t}{T}\right) dt\right] - \frac{P}{T}\left[\int_{T/2-\Delta t}^{T/2} 1 + \cos\left(\frac{2\pi t}{T}\right) dt\right], \tag{3.152}$$

evaluating this expression the mean is therefore

$$\mu = P\left[1 - \frac{2|\Delta t|}{T} + \frac{1}{\pi}\sin\left(\frac{2\pi|\Delta t|}{T}\right)\right] \quad \text{for} \quad |\Delta t| \leq 1/2. \tag{3.153}$$

The noise variance will be the same independent of the timing error, and is given by

$$\sigma^2 = \frac{P}{T^2}E\left[\int_{-T/2}^{T/2} n(t,\cdot)\left(1 + \cos\left(\frac{2\pi t}{T}\right)\right)\right]^2 \tag{3.154}$$

For white noise with $R_n(\tau) = (N_0/2)\delta(\tau)$, the noise variance is

$$\sigma^2 = \frac{PN_0}{2T}\left[\frac{1}{T}\int_{-T/2}^{T/2} 1 + 2\cos\left(\frac{2\pi t}{T}\right) + \cos^2\left(\frac{2\pi t}{T}\right) dt\right] \tag{3.155}$$

$$\sigma^2 = \frac{PN_0}{2T}\cdot\frac{3}{2} \tag{3.156}$$

The signal-to-noise ratio is then given by

$$\boxed{\mathrm{SNRrcos}(|\Delta t|/T) = \left(\frac{E_B}{N_0/2}\right)\frac{2}{3}\left[1 - 2\frac{|\Delta t|}{T} + \frac{1}{\pi}\sin\left(2\pi\frac{|\Delta t|}{T}\right)\right]^2,} \tag{3.157}$$

and the total error probability is

$$P_e(|\Delta t|/T) = \frac{1}{2}\Phi\left[\sqrt{\mathrm{SNRrcos}(0)}\right] + \frac{1}{2}\Phi\left[\sqrt{\mathrm{SNRrcos}(|\Delta t|/T)}\right] \tag{3.158}$$

Correlation Pulse	$\lvert\Delta t\rvert/T$	degrees
half-cosine	5.7%	20.5°
raised-cosine	9.6%	34.6°

Table 3.4 Time offsets when SNR degradation equals SNR loss with a rectangular correlation pulse.

3.7.2 Comparison of Simple Windowing Functions with a Rectangular Pulse

We can now make some comparisons and observations about the performance of the correlation receiver in white noise with clock-jitter and systematic phase-offsets. Defining an SNR degradation factor such that

$$\Gamma = \frac{\text{SNR}}{\left(\frac{E_B}{N_0/2}\right)} \tag{3.159}$$

Then from (3.138), (3.148), and (3.157) the SNR degradation factors for rectangular, half-cosine, and raised-cosine correlation pulses are respectively

$$\boxed{\begin{aligned} \Gamma_{\text{rec}} &= \left[1 - \frac{2|\Delta t|}{T}\right]^2 \\ \Gamma_{\text{hcos}} &= \frac{8}{\pi^2}\left[\cos\left(\frac{\pi|\Delta t|}{T}\right)\right]^2 \\ \Gamma_{\text{rcos}} &= \frac{2}{3}\left[1 - \frac{2|\Delta t|}{T} + \frac{1}{\pi}\sin\left(\frac{2\pi|\Delta t|}{T}\right)\right]^2 \end{aligned}} \tag{3.160}$$

These SNR degradations are plotted in Fig. 3.20. It can be seen that the rectangular correlation pulse achieves the maximum SNR with no clock phase offset. However, the SNR falls off quickly when a timing error occurs. The half-cosine pulse has a lower peak SNR, but its reduction is more gradual than for a rectangular pulse, and for time offsets larger than about 5.7% of the bit interval, the SNR is higher than for a rectangular correlation pulse. The time offsets where the SNR degradation crosses the rectangular degradation are given in table 3.4. Using a raised-cosine pulse can further flatten the SNR curve, but due to the more severe penalty in the peak SNR, this pulse has little advantage over a half-cosine pulse at large time offsets.

It was shown in table 3.1 that an SNR of 15.556 dB is required to achieve $P_e = 10^{-9}$, and that a 2 dB loss in SNR increases P_e by 3 orders of magnitude to $P_e = 10^{-6}$. If

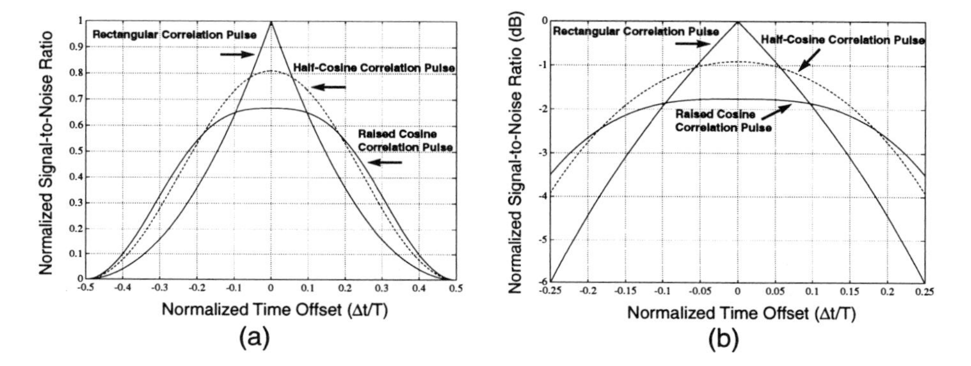

Figure 3.20 Degradation in SNR for a correlation receiver in white noise using a: rectangular correlation pulse, a half-cosine correlation pulse, and a raised cosine correlation pulse. (a) linear scale, (b) decibels.

Correlation Pulse	$\|\Delta t\|/T$ at 2dB loss	$\|\Delta t\|/T$ at 3dB loss
rectangular	10.3%	14.6%
half-cosine	15.6%	21.2%
raised-cosine	12.9%	22.4%

Table 3.5 Time offsets for 2dB and 3dB SNR degradation.

we require our receiver to maximize the time offset that can be accommodated, and still maintain better than 2 dB loss, then we see from table 3.5 that a half-cosine pulse extends the allowable time offset from 10.3% when a rectangular pulse is used, to 15.6%.

3.7.3 Practical Limitations on Timing Estimation

Although $10.3\% = 37.1°$, which is the point where the SNR for a rectangular correlation pulse drops by 2-dB, seems to be a large offset, at a data rate of 10 Gb/s, this corresponds to a time offset of only 10.3 picoseconds! We can compare this time offset with the delay-time of a single differential pair with resistive loads, constructed of transistors with an $f_{max} = 50$ GHz. It will be shown in chapter 8 that the delay through this circuit is on the order of 20–40 ps. Therefore, it is essential to match all the critical delay paths in the system. A 40 ps time-offset will have devastating effects of the error probability, and will render the receiver useless. Even when care is taken to match all delay paths, random delay mismatches, and inevitable mismatches in signal lines in a planar IC process can easily contribute 5ps–10ps offsets. To avoid degrading the system performance in the presence of clock jitter and systematic time offsets, the technique that will be used in the design of the receiver is both to design the physical delay paths in the circuit so that the best matching is obtained, and to adopt a system approach that has low sensitivity to phase-errors, such as using a half-cosine windowing function.

3.8 OPTIMUM CORRELATION RECEIVERS IN COLORED NOISE

When the noise is colored, the common-sense best strategy for optimal detection is to bias the spectrum of the correlation pulse in favor of where the signal power is the strongest, and the noise power is the weakest. If the noise PSD increases with the square of frequency, then using a correlation pulse, or matched filter, that provides good high-frequency attenuation, is desirable. The resulting receiver can be derived from the optimal correlation receiver in AWGN, by using windowing functions to change the correlation pulse in a manner that provides better high-frequency attenuation. Fig. 3.21(a) shows a colored noise spectrum processed by a filter matched to a rectangular pulse, while Fig. 3.21(b) shows the same noise spectrum filtered by a half-cosine impulse response filter. It can be seen that windowing the rectangular correlation pulse with the half-cosine pulse provides desirable high-frequency attenuation.

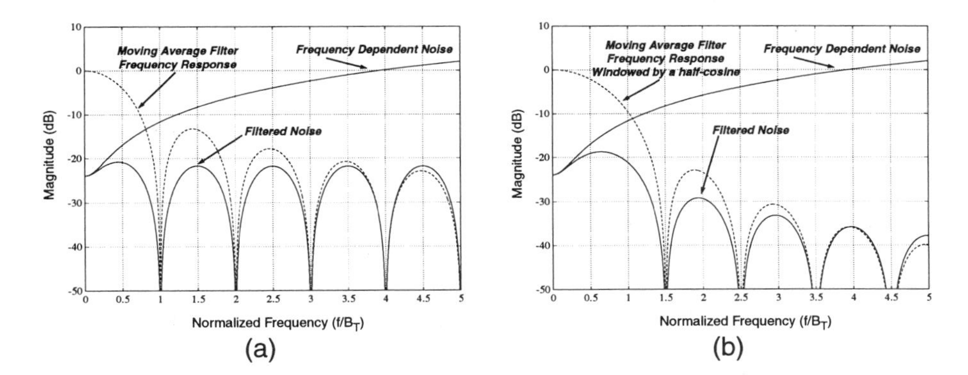

Figure 3.21 Colored noise PSD filtered by: (a) a rectangular impulse response filter, (b) a half-cosine impulse response filter.

3.8.1 Condition for Maximizing SNR of the Test Statistic

We saw earlier in (3.61) the optimal correlation receiver in AWGN can be written as

$$\int_0^T h_s(\tau)r(T-\tau,\cdot)d\tau \tag{3.161}$$

so that if we have a matched filter output of the form

$$p_n(t,\cdot) = \int_{nT}^{(n+1)T} h_s(\tau)r(t-\tau,\cdot)d\tau \tag{3.162}$$

then the samples of the signal $p_n(t,\cdot)$ at values of $(n+1)T$ are equivalent to the optimal test statistics for a correlation receiver in AWGN. If we now are operating in non-white noise, we wish to find the shape of the windowing function that will maximize the signal-to-noise ratio of the test statistic. It can be shown [1, Ch.6, p. 173] that the windowing function $h_0(\tau)$ that maximizes the SNR for a noise process with an autocorrelation function $R_n(\tau)$ satisfies the condition

$$\int_0^T h_0(\tau)R_n(\tau)d\tau = s(T-t) \quad \text{for} \quad 0 \le t \le T, \tag{3.163}$$

3.8.2 Matched Filter Approximation to Optimal Receiver in Colored Noise

Since the integral in (3.163) is only over $[0, T]$ instead of $[-\infty, \infty]$, then $h_0(t)$ can not be considered to be an impulse response of a matched filter. Notice if $h_0(t)$ extends beyond a bit period, then the filtering operation will overlap adjacent bits and cause intersymbol interference (ISI), unless additional care is taken to insure that $h_0(t)$ is orthogonal to shifted data bits at specified sampling points. We can however, gain additional insight into the the shape of $h_0(t)$ if we make the approximation that $h_0(t) * R_n(t)$ is negligible outside the interval $[0, T]$. In this case we can replace the integral in (3.163) with a convolution;

$$\int_0^T h_0(\tau)R_n(\tau)d\tau \simeq \int_{-\infty}^{\infty} h_0(\tau)R_n(\tau)d\tau. \tag{3.164}$$

Therefore,

$$h_0(t) * R_n(t) = s(T-t) \quad \text{for} \quad 0 \leq t \leq T. \tag{3.165}$$

Under this approximation, $h_0(t)$ can now be considered as the impulse response of a matched filter. Taking the Fourier transform of both sides of (3.165) gives,

$$H_0(j2\pi f)P_n(f) = F_s^*(j2\pi f)e^{-j2\pi fT}, \tag{3.166}$$

where $P_n(f)$ is the power spectral density of the noise. Therefore the magnitude response of the filter is given by

$$\boxed{|H_0(j2\pi f)| = \frac{|F_s(j2\pi f)|}{P_n(f)}} \tag{3.167}$$

This result corresponds to the common-sense approach of making the frequency response of the matched filter large where the SNR is high, and weak where the SNR is low. The warping of the frequency spectrum of the matched filter is illustrated in Fig. 3.22 for rectangular NRZ data. The signal spectrum is a sinc function. The PSD of the noise is shown with a break frequency, where the noise begins to rise in proportion of the square of the frequency. The resulting spectrum of the matched filter that maximizes the SNR at sample intervals of T is then shown in Fig. 3.22b. After taking the inverse FFT of the optimal spectrum, we obtain the impulse response $h_0(t)$ as is shown in Fig. 3.23. By windowing this impulse response so that it goes to zero outside the interval $[0, T]$, we can obtain a correlation pulse that improves the SNR of the test statistic, and does not introduce any ISI.

Comparison With Optimal Correlator in White Noise In the previous sections we were dealing with white noise with a constant PSD of $N_0/2$. In this case $|H_0(j2\pi f)| \propto$

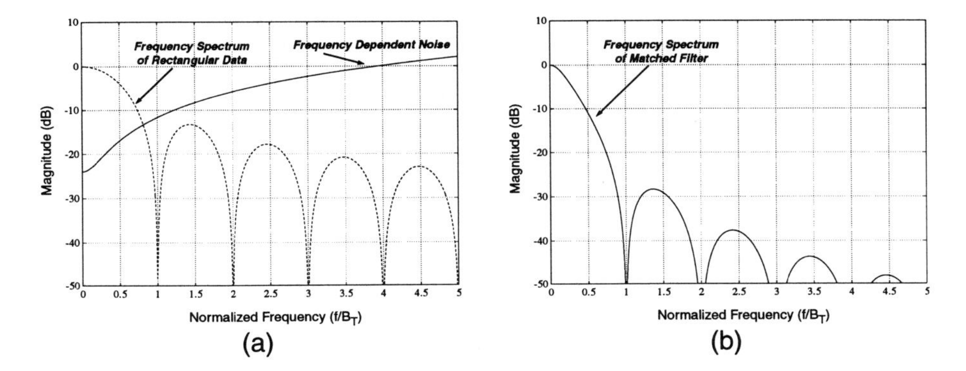

Figure 3.22 Illustration of optimal matched filter frequency response in colored noise: (a) magnitude of rectangular NRZ pulses and colored noise PSD, (b) magnitude response of matched filter.

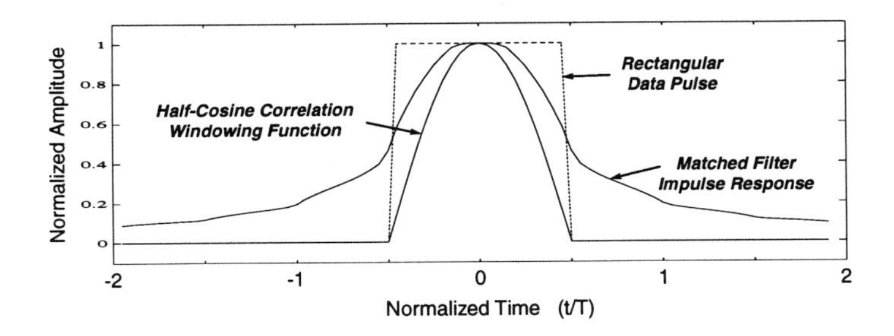

Figure 3.23 Impulse response of a matched filter in colored noise that increase as a function of frequency.

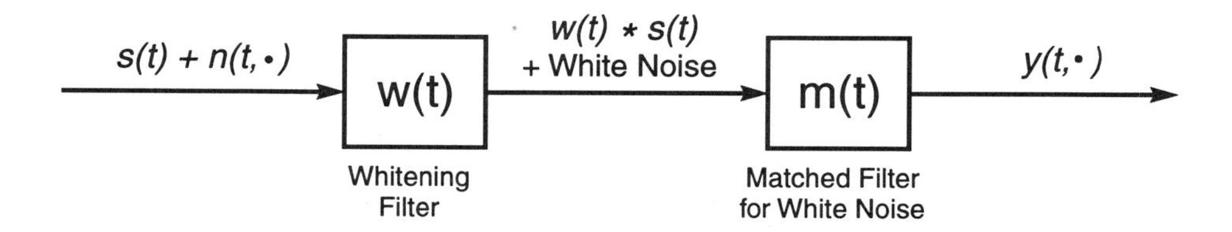

Figure 3.24 Block diagram of a matched filter in colored noise represented as a whitening filter, and a matched filter in white noise.

$|F_s(j2\pi f)|$, and we can show that the optimal filter impulse response for white noise is

$$h_O(t) \propto s(T-t) \qquad \text{for white noise.} \tag{3.168}$$

This is equivalent to the optimal matched filter given in (3.61). Since $s(t)$ is zero outside the interval $[0, T]$, our assumption that the integral in (3.163) could be replaced by a convolution is valid. The fact that $h_O(t)$ is confined to the interval $[0, T]$ for white noise results from $R_n(t)$ being an impulse function so that the spread in time of the convolution integral is no greater than the integration limits. Conversely, the higher the correlation in the noise, or the larger the spread of $R_n(t)$ compared to $s(t)$, the less valid is our assumption made in (3.165).

3.8.3 Whitening Filter

The optimal matched filter in colored noise can be understood more intuitively by splitting the filter into two parts as shown in Fig. 3.24. The first filter whitens the noise producing a constant spectral density at the output. Therefore, the PSD at the output is given by

$$|W(j2\pi f)|^2 P_n(f) = 1, \tag{3.169}$$

and the magnitude of the whitening filters frequency response must satisfy

$$|W(j2\pi f)| = \frac{1}{P_n(f)^{1/2}}. \tag{3.170}$$

Now the second filter is just the matched filter in white-noise for a signal $w(t) * s(t)$, which is the original signal warped by the prewhitening filter. We know that the impulse response of the optimal matched filter in white noise is given by

$$m(T-t) = w(t) * s(t). \tag{3.171}$$

The magnitude response of the second filter is easily found by taking the Fourier transform;

$$|M(j2\pi f)| = |W(j2\pi f)||F_s(j2\pi f)| \tag{3.172}$$

Since we know the magnitude response of the whitening filter, then substituting gives

$$|M(j2\pi f)| = \frac{|F_s(j2\pi f)|}{P_n(f)^{1/2}} \tag{3.173}$$

The overall transfer function of the two filters is then given by the product of the individual transfer functions, so that

$$|H_0(j2\pi f)| = |W(j2\pi f)||M(j2\pi f)| = \frac{|F_s(j2\pi f)|}{P_n(f)}, \tag{3.174}$$

which is the same as that obtained in (3.167).

It is still important to keep in mind that it has been assumed that $h(t)$ is only non-zero for $t \in [0, T]$, When this is not the case, (ISI) will be introduced, and the conditions under which this receiver was assumed optimum will be violated. Nevertheless, this discussion illustrates how the correlation pulse windowing functions can be altered to improve the performance in colored noise. In the following section the performance of a correlation receiver will be evaluated for various windowing functions in one particular type of colored noise. The results will be compared to see the improvement gained over using a correlation receiver that was optimized under the assumption that the noise was white.

3.9 CORRELATION RECEIVER PERFORMANCE IN COLORED NOISE

In this section we will consider a correlation receiver operating in colored noise of one particular form, and we will analyze the receiver's performance when a rectangular correlation pulse is used. For the initial analysis we will assume a simple form of the colored noise spectrum and we'll make some general observations based on the results. Later, we'll make the colored noise spectrum more complicated, and find the SNR by numerical integration. The initial noise PSD will be assumed to have the form as shown below

$$P_n(f) = \frac{1}{2}\left[N_0 + N_0\left(\frac{f}{f_0}\right)\text{rect}(f/2F)\right]$$

where;

$N_0/2$ is the white noise PSD

f_0 is the corner frequency (3.175)

F is the band-limiting frequency

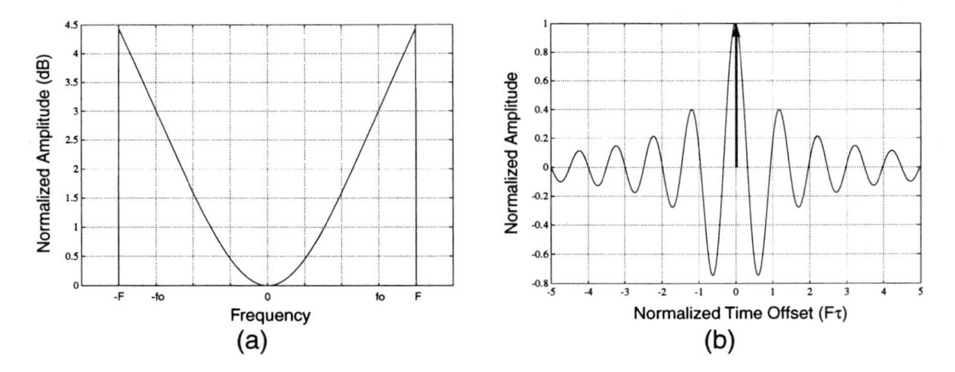

Figure 3.25 Colored Noise: (a) frequency spectrum, (b) autocorrelation function.

This noise spectrum is shown in Fig. 3.25a. It will be shown in chapter 7 that this noise spectrum is a reasonable approximation of the output spectrum of a low-noise preamplifier for a fiber-optic receiver.

3.9.1 Time Domain SNR calculations

We can find the SNR directly in the time domain when we know the functional form of the autocorrelation function of the noise. The autocorrelation can be found by taking the inverse Fourier transform of the PSD in (3.175). Therefore,

$$R_n(\tau) = \frac{N_0}{2}\delta(\tau) + \frac{N_0}{2}\mathcal{F}^{-1}\left\{\left(\frac{f}{f_0}\right)^2 \text{rect}(f/2F)\right\}. \tag{3.176}$$

Realizing that a sinc function in the time domain transforms to a rectangle in the frequency domain;

$$\text{sinc}(2Ft) \leftrightarrow \frac{1}{2F}\text{rect}(f/2F), \tag{3.177}$$

and that taking the derivative in time corresponds to multiplying the frequency domain function by $(j2\pi f)$, then we know that

$$\frac{d^2}{dt^2}\text{sinc}(2Ft) \leftrightarrow \frac{1}{2F}(j2\pi f)^2\text{rect}(f/2F) = -\frac{(2\pi f)^2}{2F}\text{rect}(f/2F), \tag{3.178}$$

and the desired autocorrelation function has the form

$$R_n(t) = \frac{N_0}{2}\left[\delta(t) - \frac{F}{2(\pi f_0)^2}\frac{d^2}{dt^2}\text{sinc}(2Ft)\right]. \tag{3.179}$$

Evaluating, the first derivative gives

$$\frac{d}{dt}\text{sinc}(2Ft) = \frac{1}{t}\left[\cos(2\pi Ft) - \text{sinc}(2Ft)\right], \tag{3.180}$$

and the second derivative is

$$\frac{d^2}{dt^2}\text{sinc}(2Ft) = -\frac{2}{t^2}\left[\cos(2\pi Ft) - \text{sinc}(2Ft)\left[1 - \frac{1}{2}(2\pi Ft)^2\right]\right]. \tag{3.181}$$

Therefore the autocorrelation function of the colored noise is given by

$$\boxed{R_n(\tau) = \frac{N_0}{2}\left[\delta(\tau) + \frac{F}{(\pi f_0 \tau)^2}\left(\cos(2\pi F\tau) - \text{sinc}(2F\tau)\left[1 - \frac{1}{2}(2\pi F\tau)^2\right]\right)\right].} \tag{3.182}$$

A plot of this autocorrelation function is shown in Fig. 3.25b. The cutoff frequency F controls the spread of $R_n(\tau)$, and the corner frequency f_0 controls the amplitude. The ringing in $R_n(\tau)$ is due to Gibbs phenomenon; when the frequency spectrum has an abrupt cutoff, the time domain response will always exhibit ringing.

Evaluation of the SNR for a Rectangular Correlation Pulse We can now use the explicit form of $R_n(\tau)$ given in (3.182) to find the SNR of the sample statistic of a correlation receiver in colored noise. For a correlation pulse $c(t)$, the variance of the test statistic is given by

$$\sigma^2 = \frac{1}{T^2}E\left[\int_{-T/2}^{T/2} n(t,\cdot)c(t)dt\right]^2. \tag{3.183}$$

Writing this as a double integral we obtain

$$\sigma^2 = \frac{1}{T^2}\int_{t_1=-T/2}^{T/2}\int_{t_2=-T/2}^{T/2} R_n(t_1 - t_2)c(t_1)c(t_2)dt_1 dt_2, \tag{3.184}$$

and letting $\tau \triangleq t_1 - t_2$, then

$$\sigma^2 = \frac{1}{T^2}\int_{t_1=-T/2}^{T/2} c(t_1)\int_{\tau=t_1-T/2}^{t_1+T/2} R_n(\tau)c(t_1 - \tau)d\tau dt_1. \tag{3.185}$$

For a rectangular pulse $c(t_1 - \tau) = \text{rect}[(t_1 - \tau)/T]$ is unity between the limits of integration. Therefore, the noise variance for a rectangular pulse is given by

$$\sigma^2 = \frac{P}{T^2}\int_{t_1=-T/2}^{T/2}\int_{\tau=t_1-T/2}^{t_1+T/2} R_n(\tau)d\tau \tag{3.186}$$

This integration can be expedited when we realize that $R_n(\tau)$ was originally expressed as an impulse plus a second derivative. Therefore the variance can be written as

$$\sigma^2 = \frac{PN_0}{2T}\left[1 - \frac{1}{2T}\int_{-T/2}^{T/2} \frac{F}{(\pi f_0)^2}\frac{d}{d\tau}\text{sinc}(2F\tau)\Big|_{t_1-T/2}^{t_1+T/2} dt_1\right], \quad (3.187)$$

and carrying out the second integration we get

$$\sigma^2 = \frac{PN_0}{2T}\left[1 - \frac{FT}{2(\pi f_0 T)^2}\left[\text{sinc}\left(2F\left(t_1 + T/2\right)\right) - \text{sinc}\left(2F\left(t_1 - T/2\right)\right)\Big|_{t_1=-T/2}^{t_1=T/2}\right]\right]. \quad (3.188)$$

We finally arrive at the expression of the colored-noise variance using a rectangular correlation pulse;

$$\sigma^2 = \frac{PN_0}{2T}\left[1 + \frac{FT}{(\pi f_0 T)^2}\left[1 - \text{sinc}(2FT)\right]\right]. \quad (3.189)$$

Finally, since we know that the mean of the test statistic is P, the SNR which is equal to μ^2/σ^2 is

$$\boxed{\text{SNR} = \frac{E_B}{N_0/2}\left[1 + \frac{FT}{(\pi f_0 T)^2}\left[1 - \text{sinc}(2FT)\right]\right]^{-1}.} \quad (3.190)$$

We can make some useful observation about the SNR given in (3.190) by realizing that the bandwidth limiting parameter F will be close to the data rate $B_T = 1/T$. Therefore we can define a parameter α with a value in the vicinity of unity as

$$\alpha \triangleq FT, \quad \Longrightarrow \quad F = \alpha B_T. \quad (3.191)$$

Therefore the SNR can be written as

$$\text{SNR} = \frac{E_B}{N_0/2}\left[1 + \left(\frac{B_T}{f_0}\right)^2 \frac{\alpha}{\pi^2}\left[1 - \text{sinc}(2\alpha)\right]\right]^{-1}. \quad (3.192)$$

This SNR is plotted in Fig. 3.26(a), as a function of the normalized corner frequency f_0/B_T, for various values of α. For the case of $\alpha = 1$, the SNR is simply

$$\text{SNR} = \frac{E_B}{N_0/2}\left[1 + \left(\frac{B_T}{\pi f_0}\right)^2\right]^{-1} \quad (3.193)$$

Typical colored noise PSDs for $\alpha = 1$ are shown in Fig. 3.26b for various corner frequencies. From (3.193) we can see that the SNR is reduced by 3 dB for a corner frequency of $f_0 = B_T/\pi$, and the SNR is reduced by 6 dB when B_T is $\pi\sqrt{3} = 5.44$ times the corner frequency.

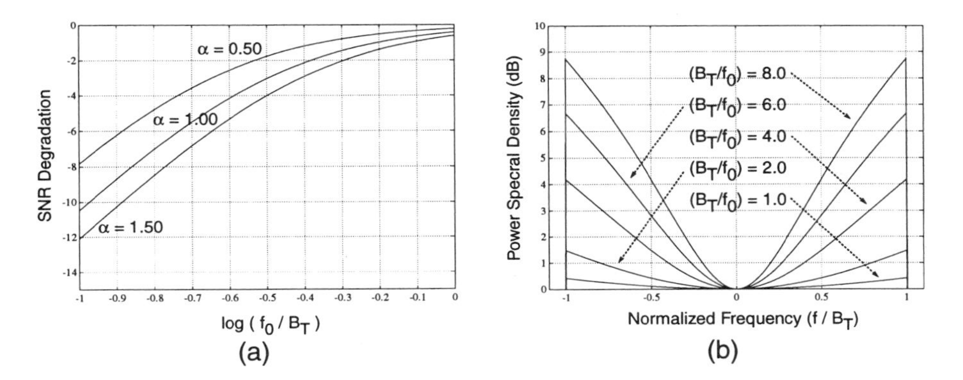

Figure 3.26 (a) SNR reduction for a correlation receiver in colored noise a a function of the corner frequency for various values of α; (b) Power spectral densities of a simple type of colored noise for various corner frequencies and $\alpha = 1$.

3.9.2 Frequency Domain SNR calculations

In chapter 2 we saw that the variance of a random process can be obtained by integrating the PSD in the frequency domain. The correlation receiver is equivalent to a matched filter with a rectangular impulse response sampled at specified intervals. The matched filter impulse response is of the form

$$m(t) = c(T - t), \tag{3.194}$$

where c(t) is an arbitrary correlation pulse. For a rectangular pulse

$$c(t) = \frac{\sqrt{P}}{T}\text{rect}(t/T). \tag{3.195}$$

The magnitude response of the matched filter is therefore,

$$|M(j2\pi f)|^2 = P\text{sinc}^2(fT) \tag{3.196}$$

At the output of the matched filter the PSD of the noise is

$$P_m(f) = P_n(f)|M(j2\pi f)|^2, \tag{3.197}$$

and the variance of the test statistic is therefore just the integral of $P_m(f)$ over all frequencies.

$$\sigma^2 = \frac{PN_0}{2}\int_{-\infty}^{\infty}\left[1 + \left(\frac{f}{f_0}\right)^2 \text{rect}(f/2F)\right] \text{sinc}^2(fT)df \tag{3.198}$$

This can be evaluated easily recalling that the integral of $\text{sinc}^2(fT)$ function is just $1/T$. Therefore,

$$\sigma^2 = \frac{PN_0}{2T} + \frac{PN_0}{2} \cdot \frac{1}{(\pi f_0 T)^2} \int_{-F}^{F} \sin^2(\pi f T) df, \tag{3.199}$$

from which we obtain

$$\sigma^2 = \frac{PN_0}{2T} \left[1 + \frac{FT}{(\pi f_0 T)^2} [1 - \text{sinc}(2FT)] \right]. \tag{3.200}$$

This result agrees with (3.190), obtained from a time domain approach. However, doing the analysis in the frequency domain is not only simpler, but it provides a much better intuitive approach on how one can go about altering the frequency response of the correlation pulse to obtain better performance. We will normally forego the calculation of noise variances in the time domain for the windowing functions, and skip directly to the frequency domain.

3.9.3 Constrained Optimization in Colored Noise

We saw in the previous analysis, that the actual shape of the optimal correlation pulse depends on the parameters of the colored noise spectrum. If we were using a correlate-and-dump receiver in our high-speed fiber-optic system, we could continue with this type of analysis to find the shape of the correlation window that maximizes the SNR of the test statistic. However, the shape of the correlation function will be sensitive to the placement of the noise peak in relationship to the nulls in the matched filter spectrum. These nulls are a result of the impulse response of the matched filter being non-zero only in the interval $[0, T]$. In a real, high-speed system, these nulls won't exist, so that continuing an optimization in this manner is rather pointless.

In a high-speed system, we can only approximate a matched filter. Typically we will use a simple one- or two-pole, approximation, and we will take advantage of the parasitics of the transistors themselves to do our noise-reduction filtering. We can therefore perform a constrained optimization for such a system, by varying a few parameters of the preamplifier and postamplifier to alter pole locations. We will use the frequency domain techniques described above to find the best SNR under the constraints of the system. We will find that the resulting SNR will be only slightly less than what we could obtain with an ideal matched filter. These and other issues will be investigated in more detail in in chapter 7, where we will consider the actual circuit implementation of the low-noise preamplifier, and determine the precise functional form of its colored-noise spectrum.

3.10 SUMMARY

In this chapter we have addressed the problem of deriving an optimal receiver in the presence of both non-white noise, and phase-jitter. Although several books on communication theory cover this topic adequately, (Whalen's popular book is an excellent example [1]), we have specifically discussed the application of this theory to the design of high-speed IC receivers. The types of questions that we considered were as follows.

- What is the optimal receiver in the presence of additive white gaussian noise, and what is its performance?
- How can a correlation receiver be modified to reduce its sensitivity to phase-jitter and systematic timing offsets?
- What is the quantum limit of a receiver, and how does this affect the minimum optical power that must be received?
- How do practical considerations affect the minimum receiver power, and how does this translate to the maximum distance that optical repeaters can be spaced?
- How can a correlation receiver be approximated by a matched filter, and what is the penalty of using the parasitic bandwidth limitations of the preamplifier and postamplifier for noise filtering, as opposed to an optimal matched filter?
- How can a correlation receiver be modified to produce the best signal-to-noise ratio in the type of non-white noise that can be expected in fiber-optic receivers?

This chapter has provided the theoretical background to answer the above questions. In a practical systems we can only approximate an optimal receiver, but the theory gives us a benchmark for performance characterization, and gives a guide to the design and optimization of the essential building blocks of the receiver.

REFERENCES

[1] Anthony D. Whalen. *Detection of Signals in Noise*. Academic Press, New York, 1971.

[2] Richard E. Mortensen. *Random Signals and Systems*. John Wiley & Sons, New York, 1987.

[3] Frederic de Coulon. *Signal Theory and Processing*. Artech House, Inc., Dedham MA, 1986. Translation of *Theorie et Traitement des Signaux*, originally published in French as volume VI of the *Traité d'Electricité* by The Presses Polytechniques Romandes, Lausanne, Switzerland. ©1984.

[4] Gerd Keiser. *Optical Fiber Communications*. McGraw-Hill, Inc., New York, second edition, 1991.

[5] Paul E. Green, Jr. *Fiber Optic Networks*. Prentice-Hall, Inc., Englewood Cliffs, New Jersey, 1993.

[6] Wilbur B. Davenport, Jr. and William L. Root. *An Introduction to the Theory of Random Signals and Noise*. IEEE Press, New York, 1987. IEEE PRESS edition of a book published by McGraw Hill Book Company in 1958 under the same title.

4

CLOCK RECOVERY IN BROADBAND COMMUNICATION SYSTEMS

When random data is transmitted over a channel, in the form of a sequence of symbols belonging to a given alphabet, a receiver designed to interpret these signals must perform two separate tasks. The primary task is to decide which of the signals from the original alphabet was transmitted. But the receiver can not do this until it first performs the equally important task of estimating the time of arrival of the data symbols. Both tasks are complicated by the presence of additive noise, nonlinear distortion, and dispersions that cause intersymbol interference. In addition, for a full-duplex system, echoes of a response signal being transmitted in the opposite direction add to the difficulty in detecting the received pulses. In the previous chapter, techniques for determining the optimal signal processing operations required to minimize the probability of error in a binary decision circuit were presented. In this chapter, circuits for deriving the necessary clock signal from random data will be discussed. Since in any high-efficiency signaling scheme, the clock signal is completely suppressed, and has to be *recovered*, or *extracted* from the data itself by nonlinear operations, the process of estimating the time of arrival of random data is often referred to as clock-recovery, or clock-extraction, and we will use these terms interchangeably.

Nyquist Limited Signals and Narrowband Modulation Schemes

Approaches for recovering a clock from a random data signal vary depending upon the modulation scheme used. For communication over a bandlimited channel, the pulses of each individual symbol can extend far beyond the bit interval (e.g. 100–200 symbol periods). This causes a great deal of intersymbol interference (ISI) . There is generally a smaller opening in the data-eye, where the ISI goes to zero in any given symbol period, especially for multilevel symbol pulses. A typical eye-diagram for a 16-quadrature-amplitude-modulated (16 QAM) communication channel is shown in

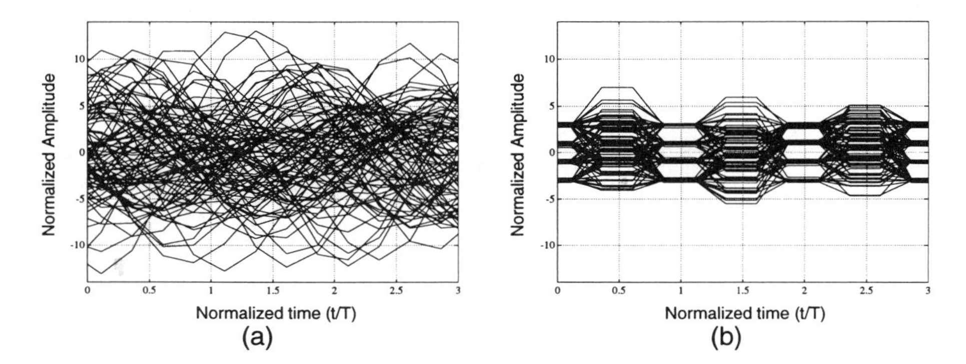

Figure 4.1 Eye diagrams of one quadrature component of a 16 QAM communication system over, copper wire with a 3 dB bandwidth of 4 kHz, and with a signal rate of 400 kbaud/s = 1.6 Mbit/s.

Fig. 4.1. [1] Only after careful channel equalization to compensate for the distortions in the transmission can the data be properly detected. And in the case of a full duplex system, the transmitted signal, and its echoes, must also be separated from the received signal. Recent efforts to increase data rates over twisted pairs of copper wire have shown impressive results. The High Bit-Rate Digital Subscriber Line standard (HDSL) provides for full-duplex communication over two twisted-pairs at bit-rates of 800-kb/s over each pair, for a total bit-rate of 1.6-Mb/s; while the Asymmetric Digital Subscriber Line standard (ADSL) provides for one-way communication on a single twisted pair at 1.6-Mb/s. For a typical phone-line twisted-pair transmission channel, the 3 dB bandwidth is about 4 KHz. In an ADSL system, a pair of quadrature pulses, with 4 levels of amplitude modulation each (16 QAM), centered at a frequency of 300-KHz, are sent on the transmission line at a symbol rate of 400 kbaud (100 times the 3 dB bandwidth). By the time the 2.5 V-peak-signal reaches its destination, the amplitude is approximately 2 mV (-62 dB). Aside from the attenuation, channel bandwidth limitations cause severe smearing of adjacent symbols. Techniques for recovering a clock in these circumstances are usually based on optimal stochastic estimation theory. Often the baud rate is slow enough to afford a significant amount of signal processing. For example, all-digital systems are proposed, that implement sophisticated algorithms for channel equalization and clock recovery. Since the opening in the post-equalized data-eye is still narrow even after equalization, tight controls on the phase jitter of the recovered clock are required. For purposes of implementation, multi-phase clocks can be generated easily at a low baud rate, and the *best* phase can be chosen from among them. Also, digitally controlled variable frequency oscillators, or a direct-digital frequency synthesizer (DDFS) , can be utilized.

[1] Communication channel simulations and eye diagrams provided by Babak Daneshrad.

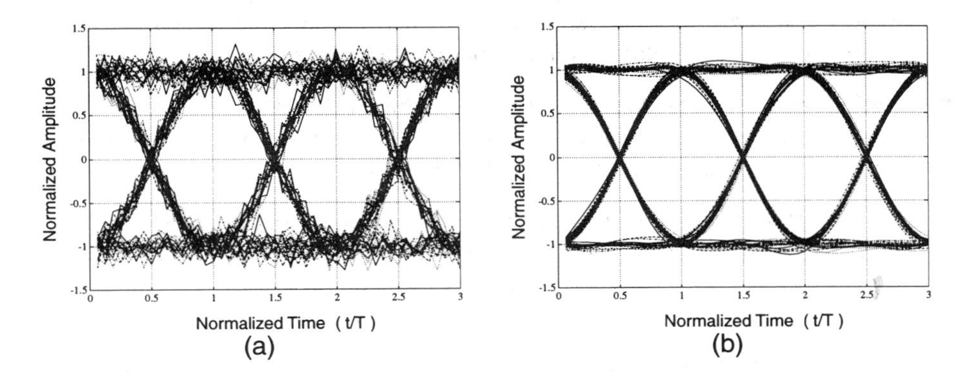

Figure 4.2 Eye diagram of broadband NRZ binary data with additive noise: (a) before filtering, (b) after filtering.

Broadband Signaling Schemes

In contrast to bandlimited channels, broadband communication systems, such as optical fiber networks, have primary data-rate limitations due to receiver and transmitter electronics, and are not restricted by channel characteristics. ISI is generally minimal, and the data-eye opening is wide. A typical eye diagram for a broadband binary NRZ system is shown in Fig. 4.2. In a broadband receiver, the clock recovery operation is the most difficult to perform, and is often the limiting factor on the speed of the overall system. Techniques for extracting a clock have almost exclusively centered on spectral-line techniques, where a clock-tone component is generated from the data by a nonlinear operation; the resulting tone, plus random, data-dependent, noise, is passed through a bandpass filter producing a periodic clock waveform. Alternatively, the data-derived signal can be input to the phase detector of a phase-lock loop (PLL); the filtered phase-error is used to synchronize a tunable oscillator to the data rate. Although the mathematics of determining optimal estimates of arrival times of a random pulse are the same for narrowband and broadband systems, many simplifying techniques exist for broadband systems, where ISI is ignored. These simplifications are not applicable to narrowband systems. For example, edge detection is a technique used in broadband systems to generate a spectral-line at the bit-rate, but can not be applied directly to a signal with large ISI, since no clear edges exist. Clock recovery circuits in broadband systems normally perform operations on data over only one bit-period to arrive at an immediate estimate of the phase error, whereas in narrowband systems, several symbol periods must be observed so that ISI contributions of adjacent symbols can be removed. For the remainder of this chapter, we will concentrate only on techniques that are suitable for implementation in a *broadband* system. Key features of broadband clock recovery circuits are as follows:

- **Speed**. Since the clock recovery circuit limits the maximum obtainable data rate, we will be primarily interested in the speed of the these circuits.
- **Self-Correction**. The data rates are so high that even a 10 ps systematic timing error can reduce the SNR by 2 dB. The delay of the decision circuit must be accounted for in the final estimate of the optimal clock phase.
- **Phase-Jitter**. Aside from the systematic phase-error, the random phase-jitter can also substantially reduce the effective SNR. This jitter is data-dependent, and can lead to errors whenever specific data patterns are encountered.

In the following section we will qualitatively discuss different techniques for extracting a clock signal from random NRZ data. This will help in developing our intuition about such systems. Later, we will describe the problem mathematically, and compare our intuition with various systems which derive clocks using both, spectral-line techniques, and maximum a posteriori (MAP) estimates.

4.1 QUALITATIVE ANALYSIS OF CLOCK RECOVERY SCHEMES

An NRZ signaling scheme is often used to conserve bandwidth in a baseband communication system. Since the data does not return to zero in one bit period, the maximum fundamental frequency in the data is half of the data rate, and occurs when the data is alternating ones and zeros. A typical waveform of an NRZ data signal is shown in Fig. 4.3(a), and the PSD of this data signal is shown in Fig. 4.3(b). The 3 dB bandwidth, required of a lowpass filter to pass 80% of the data signal power, is about $0.80B_T$, as shown in Fig 4.4. Therefore, a 10 Gb/s system can, in principle, be implemented with circuits limited to a bandwidth of approximately 8 GHz, with a penalty in maximum SNR of 20%, or approximately 1 dB, by having suppressed the high frequency edges. We would like to extract a clock signal directly from the random data. However, from Fig. 4.3(b) we see there is a spectral-null at the bit-rate. The reason for this spectral-null was discussed in detail in chapter 2. From the eye-diagram of Fig. 4.2 we notice a definite timing structure embedded in the data, despite its random nature. When the data does not change values, the signal stays either high or low, and there is no way to obtain any timing information from a constant signal. However, the cross-overs in the eye-diagram occur at integer multiples of the bit period T. Therefore, in an NRZ data signal timing information is only contained in the transitions between different bits, and we can extract a clock by synchronizing a periodic signal with these data transitions. This procedure can be illustrated more clearly with an example.

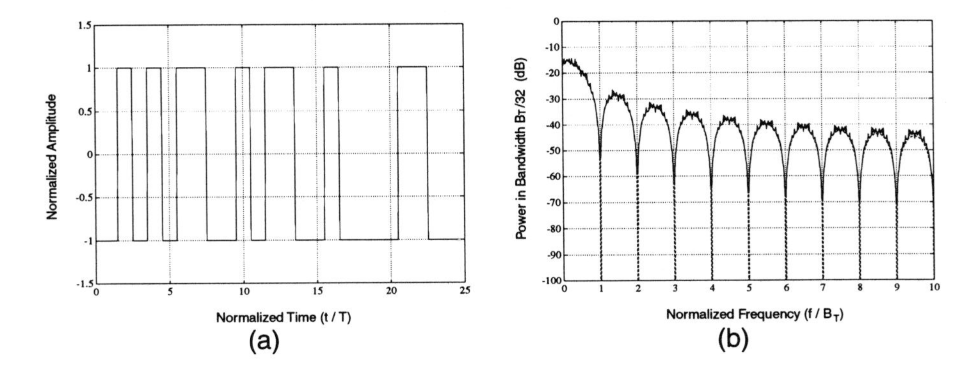

Figure 4.3 Random NRZ data: (a) typical time domain sample, (b) power spectral density.

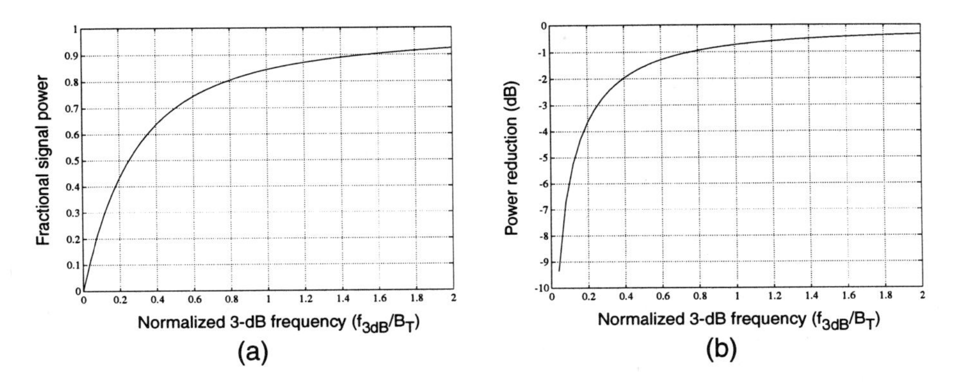

Figure 4.4 Cumulative power in rectangular NRZ data after passing through a lowpass filter with a 3-dB frequency of f_{3dB}: (a) linear scale, (b) decibels.

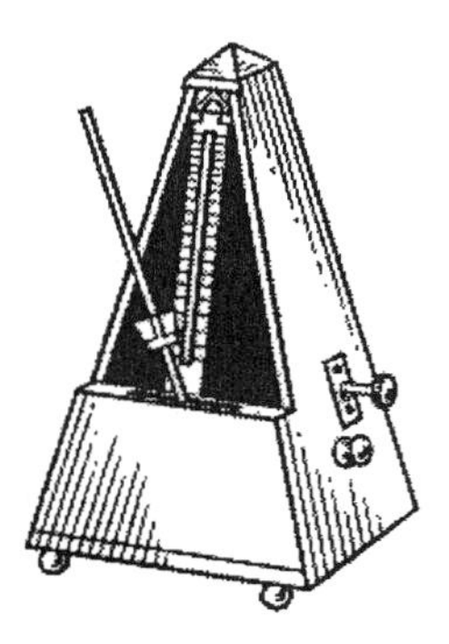

Figure 4.5 Metronome, as an analogy of a variable frequency oscillator used to recover a clock from random data.

4.1.1 Traffic Light Analogy

We could imagine ourselves trying to recover a clock from random data manually. Imagine sitting on a park bench in Munich, just after having purchased a metronome, like the one shown in Fig. 4.5, for our piano at home. While waiting for our train, we decide to pass the time by synchronizing the metronome with the traffic-light across the street. Our goal is to find the lowest fundamental clock period used to control the traffic lights. As we are watching, we see long periods where the light stays either red or green. When the light is constant on one color, we have no idea as to the timing information controlling the traffic signal. Suddenly, the light switches to yellow, and we start our pendulum swinging; we want to try to get the pendulum to return by the time the light turns red. If the pendulum doesn't get there in time, we speed it up by sliding the weight down on the pendulum; if there was more than one cycle of the pendulum during one yellow light, then we slow the pendulum by moving the weight higher. Over several cycles of the traffic light we get the pendulum swinging so that it has exactly one cycle on every yellow light, and has an integer, but not necessarily equal, number of cycles when the light is red or green. We will notice that the pendulum will need a slight adjustment every now-and-then because there will be drift in both the metronome, and the traffic-light timing; therefore feedback is required to keep the two clocks synchronized. Adjustments are made by measuring the position of the pendulum whenever a change occurs in the colors of light being transmitted. This system, albeit operating at a very low data rate, is a model of a wavelength-shift-keyed (WSK) optical communication system, where different wavelengths (or colors) of light are transmitted across the same channel. In this case there are three colors transmitted, each with a distinct interpretation. In our analogy we used a PLL to extract the clock from the data by applying feedback to adjust a variable frequency oscillator in accordance with phase-error estimates obtained by looking at transitions in the data.

Instead of the metronome we could have used a *slinky*[2] with a weight on the end. We can vary the natural frequency of this harmonic oscillator by holding the *slinky* in different places, thus altering the effective spring-constant. We will try to match the self-resonance of the spring and mass system to the clock rate of the traffic lights. This is analogous to pre-tuning a bandpass filter to the bit-rate of a communication system. Each time that we notice the traffic light changing colors we give the slinky a push downward. When the light stays constant, the slinky keeps oscillating, but the amplitude gets smaller due to dissipation in the spring. Then the light changes and we give the slinky another push to keep in going. This example illustrates clearly how dissipation (finite Q) in the resonator leads to random amplitude modulation in the clock signal.

From the above analogies we see there is no mystery in extracting a clock from a system such as this. We have just outlined how the clock can be recovered from random data using either a PLL, or a BPF. Our challenge will be to design a circuit that will do this clock extraction automatically and considerably faster.

4.2 INTERMITTENT PHASE-READJUSTING APPROACHES TO CLOCK RECOVERY

An approach to clock recovery, that can be understood simply from a qualitative point of view, is intermittent adjustment of the phase of a local oscillator, in jumps, at discrete time increments, so as to synchronize it with the data. This approach can tolerate slight frequency errors in the local clock at the receiver. Perhaps *clock-recovery* is a slight misnomer, and *phase-recovery* is a more appropriate term. When recovering a clock, we are interested in frequency-recovery, and phase-recovery, and both are explicitly implied. However, some systems have there own local clocks at the receiver, that are not synchronized in frequency with the data-rate. An effective sampling rate, equal to the data-rate, can be achieved by restarting the clock phase in the center of the data-eye before a cycle-slip occurs. This method is illustrated conceptually in Fig. 4.6.

Readjusting the phase at discrete time intervals is analogous to the synchronization method used for wall-clocks in public schools, that many may remember. The clock in each room was allowed to run freely; slightly before the end of the hour, each clock

[2] *Slinky* is the brand name of a toy that is merely a long, loose spring with a small spring-constant. Despite its simplicity the *slinky* is a wonderful, wonderful toy, that's fun for a girl or a boy.

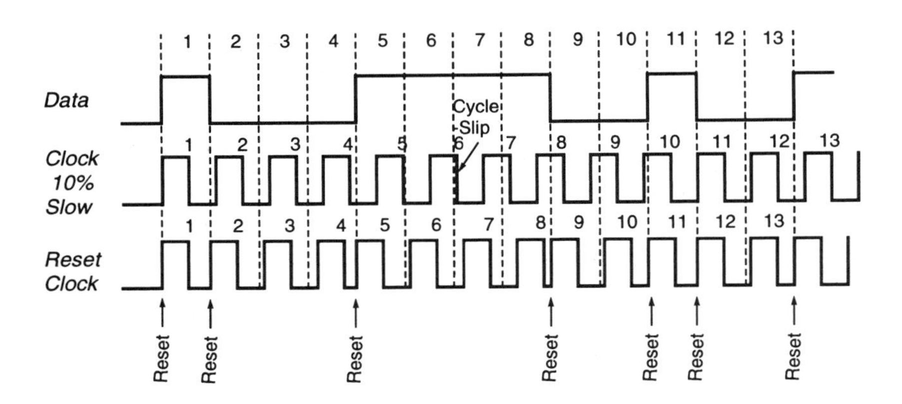

Figure 4.6 Conceptual block diagram of discrete phase-readjusting method of timing recovery.

was sped-up, forcing the second-hand to a held position on "12," until the master clock simultaneously released all of clocks. For the remainder of the next hour, each clock again ran open-loop. As long as the individual clock frequency errors, relative to the master clock, were small, timing throughout the building remained within acceptable levels of synchronization.

Synchronization by this method uses feedback only at discrete, times, and phase adjustments are made in discrete jumps. This is not a particularly good approach for low SNR systems, or ones with tight controls on the allowable phase-jitter. However, there are systems operating with very high SNRs over short transmission distances, such as local area networks (LANs), where the primary goal is to make the receiver circuitry simple since there is plenty of signal power to spare. We can think of this method as "living" with an error, but correcting it every chance that we get. Naturally we make corrections every time that a transition in the data occurs. One of the problems with this approach is that phase errors accumulate when there are no transitions. If the frequency of the local clock differs from the bit-rate by 1%, then in 50 transitionless data bits, the clock-phase will be sampling at the data cross-overs instead of the maximum data value, and communication through the network will cease. Even after 10 bits in a row without a transition, we will have a 10% phase error which reduces the equivalent SNR by 2 dB. For an optical fiber with a loss of 0.15 dB/Km, this corresponds to a 13 Km reduction in the maximum repeater spacing.

Since phase errors accumulate when no transitions occur, the maximum obtainable phase-error can be limited by using coding to force a data transition every few bits. For a system using Manchester coded data, there are guaranteed transitions in each

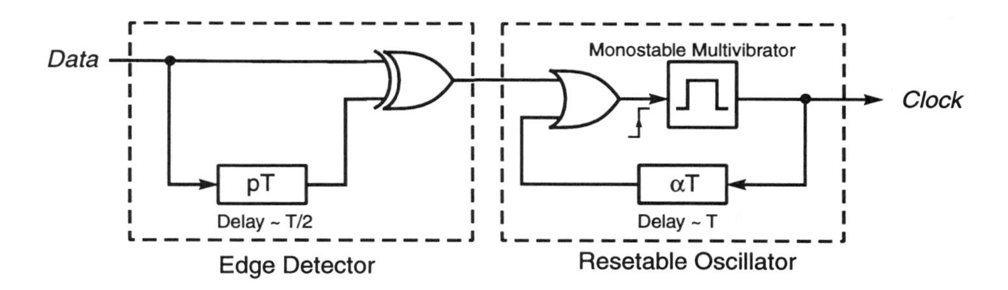

Figure 4.7 Simple clock recovery circuit that uses the edges of NRZ data to retrigger a multivibrator.

bit period. As a result, the phase can be constantly corrected, and there will never be more than one period of error accumulation. The FDDI (Fiber Distributed Data Interface) standard for local area networks (LANs) provides for communication over multi-mode fibers using light-emitting-diode (LED) sources at a symbol rate of 125 MHz, and a bit-rate of 100 Mb/s. The reason for the discrepancy between the bit-rate and baud-rate is that a 4b/5b code is used to code every 4 bits into a block-code 5-bits long. The code is designed in such a way as to produce at least one transition every 5 bits. Therefore, in our previous example of a clock frequency error of 1%, the maximum phase error accumulates over 4 bits without a transition and is equal to 4%. We saw in chapter 3 that a 4% timing error reduces the effective SNR by approximately 1 dB. Although, this phase readjusting technique generates significant phase-errors, it may be a penalty worth paying when instantaneous frequency acquisition is required. Unlike narrowband filters, or PLLs which act like heavy flywheels, and take a long time to start spinning, the retriggered multivibrator scheme generates a clock after the first data transition. This property can be extremely important in various types of communication systems, other than long-haul fiber-optic trunk-line. Two specific examples of intermittent phase-readjusting clock-recovery circuits will now be discussed.

4.2.1 Retriggering a Multi-Vibrator

A simple technique used to recovery a clock from NRZ data is to use the data edges to retrigger a multi-vibrator. One such circuit is described by Witte and Moustakes [1], and is illustrated in Fig. 4.7. The first circuit block generates a positive pulse of width pT, where $p \simeq 1/2$, whenever a transition in the data occurs. The pulses are used to reset the free-running oscillator, constructed by using feedback around a monostable multivibrator. The delay in the feedback loop αT is set as close to T as possible to achieve a frequency of oscillation equal to the bit-rate. However, there will be

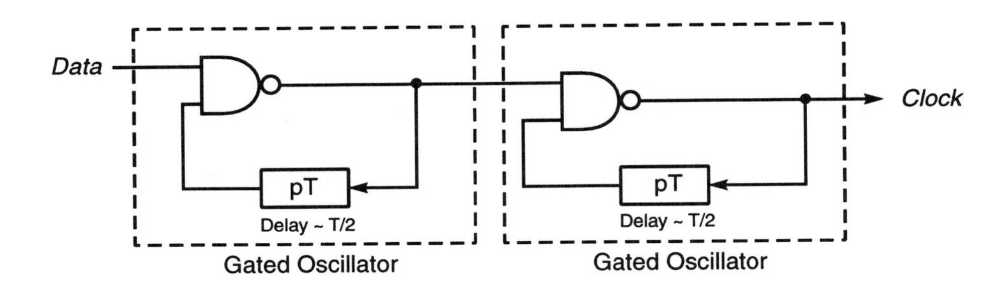

Figure 4.8 Clock recovery circuit using two gated oscillator.

inevitable errors and α will differ from unity, causing phase-error accumulation when no data transitions occur.

An even simpler implementation of the circuit of Witte and Moustakes was reported by Eng *et al.* [2]. This circuit is illustrated in Fig. 4.8, and consists of two gated oscillators. When the gating signal is high, the oscillator is free-running. When the gating signal is low, the output of the oscillator is held high. The effect of cascading two such oscillators is that the second oscillator can operate in one of three conditions. It is free-running whenever the data input is low, is reset to the data-transition whenever the data changes from low to high, and is reset to the transitions of the first free-running oscillator when the data stays high. The net result is that the clock-phase only gets realigned on a positive data transition, and synchronization information contained in the negative transitions are ignored.

4.2.2 Choosing One Phase of a Multi-Phase Clock

An implementation of a simple clock recovery circuit that derives its active decision clock from among only two clock phases was recently reported by Yamanaka *et al.* [3] for a 2-Gb/s system. A block diagram of this clock recovery circuit is shown in Fig. 4.9. Digital logic is used to control a multiplexer that selects the clock phase closest to the center of the data-eye. Since only two clock phases exist, there will be a severe SNR penalty due to errors in timing. However, for the designed purpose of this chip-set, namely high-speed interconnect of VLSI modules, the SNR degradation is not a primary concern. A similar circuit that chooses the best of two clock phases was also reported by Bagheri *et al.* [4, 5]. This circuit used AlGaAs/GaAs HBTs, and functioned at a bit-rate of 6.1-Gb/s.

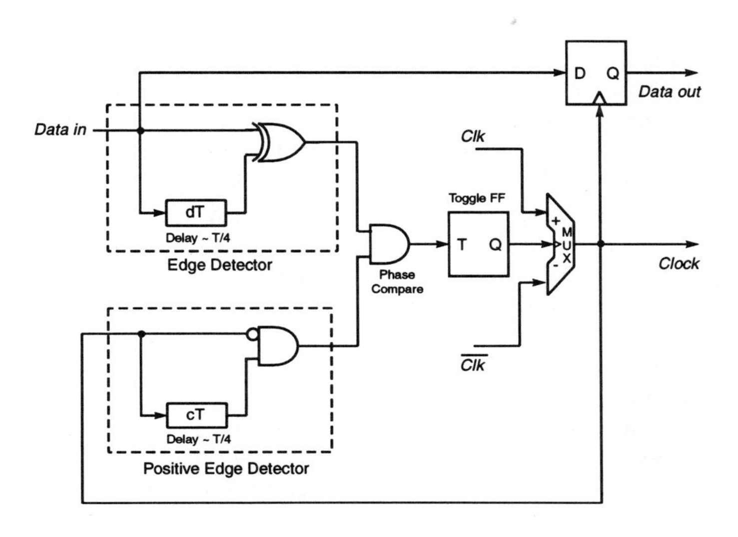

Figure 4.9 Clock recovery circuit that chooses the best among two clock phases.

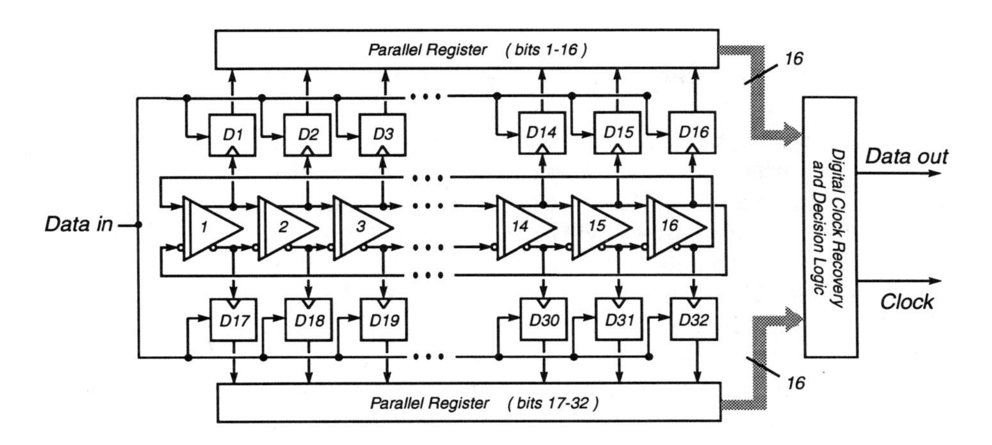

Figure 4.10 Clock recovery circuit based on a hybrid analog/digital approach of choosing the best clock among 32 separate phases.

An implementation of a multi-phase clock recovery scheme, with 32 separate clock phases, was described by Kim *et al.* in [6], and is explained in more detail in Kim's Ph.D. thesis [7]. A block diagram of this circuit is shown in Fig. 4.10. The local oscillator is a 16-stage, fully-differential, tapped delay line. Using both outputs of each differential delay cell, provides 32 separate clock phases, equally spaced across the bit-interval. There is one decision circuit for every clock phase, and the resulting decisions are clocked into parallel registers. A final decision as to the data value can then be made by the digital logic circuit. The digital logic is also used to determine which of the 32 clock phases is closest to the middle of the data-eye. An obvious disadvantage of this circuit is its complexity. However, with complexity comes added flexibility. Various clock phase-shifting schemes can be implemented. This is important in a long chain of repeaters when jitter accumulation may necessitate a large discrete jump in the clock phase. Also, various decision circuit voting algorithms can be implemented easily by programming the digital logic block.

This circuit runs open-loop in the sense that the recovered clock is not locked to the bit-rate. However, in Kim's implementation, a PLL is used to lock the tapped delay-line oscillator to a crystal reference clock. This keeps the center frequency of the multi-phase clock close to the data rate, avoiding frequent cycle-slips. This circuit was designed to operate at a bit-rate of 30-Mb/s, and is therefore not directly applicable to high-speed systems. However, we will see similar looking approaches for high-speed implantations that use multi-phase clocks for interleaving parallel decision circuits, such as Pottbäcker's approach to be described in chapter 5 (see Fig. 5.15).

4.2.3 Clock Recovery Using Line-Coding

Clock extraction can be simplified if a coding scheme is used to impose a predictable structure in the data signal. For example, coding can be used to install a framing pulse at periodic intervals. A low frequency clock can be locked to these framing pulses during start up. Synchronization will be maintained during operation provided that the PLL frequency doesn't drift far enough in one frame, so as to confuse a data-pulse with a framing-pulse. This technique was adopted for a 1.5-Gb/s computer data interface chipset, designed at Hewlett-Packard by Walker *et al.* [8, 9]. An earlier 5-Gb/s circuit employing the same approach was described by Bentland *et al.* [10]. Although line coding can simplify the clock recovery circuitry, with minimal bandwidth overhead, we will assume for the remainder of this book that no such coding is used.

4.3 EDGE DETECTION

Synchronizability in Relation to Transition Density We have seen that the timing information for random NRZ data is contained in the transitions between different symbols. If the SNR is high enough for the receiver to distinguish between different pulses, then it should also be able to estimate the time at which the data-pulses change value. The accuracy with which the receiver can estimate the data transition time, will, to a large extent, determine the probability of error for the overall receiver. Since timing information is contained in the data transitions, the more transitions available to observe in a given time interval, the smaller will be the rms error in the estimated arrival time. We have seen that using NRZ data signaling format reduces the required bandwidth by a factor of 2 over RZ signaling. However, what we gain in bandwidth, we sacrifice in synchronizability. Moeneclaey [11] has shown that the lower bound on the variance of a minimum mean-square estimate (MMSE) of the data arrival time is inversely proportional to the average number of transitions $\overline{N_T}$ per bit-period. Moeneclaey gives an expression for the Cramér-Rao lower bound on the timing error for a signal in additive white Gaussian noise as

$$\sigma_T \geq \frac{T}{2}\left[\frac{1}{\sqrt{Q_{eq}}}\frac{1}{\sqrt{\mathrm{SNR}}}\frac{1}{\sqrt{2\overline{N_T}}}\right], \tag{4.1}$$

where the SNR is the ratio of the average energy-per-bit, divided by the two-sided white noise PSD value,

$$\mathrm{SNR} = \frac{E_B}{N_0/2}. \tag{4.2}$$

There are three contributions to the lower bound as seen from (4.1). First, is the bandwidth of the filtering used, which is represented by the parameter Q_{eq}. Second, is the ratio of the bit-energy to the additive white noise spectral density. And third, is the average number of data transitions. The significance of the first two factors are clear. We can understand the significance of the parameter $\overline{N_T}$ if we recall from chapter 3 that the standard deviation of the average of several independent observations was reduced by the square-root of the number of observations. Since we can only make timing measurements when a transition occurs, then the number of observations possible in a given time period is proportional to $\overline{N_T}$. Table 4.1 gives values of $\overline{N_T}$, and the 80% power bandwidth for the binary signaling formats discussed in section 2.3.3.

Importance of Edge-Detection in Clock Recovery Schemes In a binary communication system, changes in the data manifest themselves as either rising, or falling edges in the data signal. Therefore, it's not surprising that edge-detection of the data will play an important roll in clock-recovery circuits. The PSD of edge-detected data was studied extensively in chapter 2. We saw that simply detecting an edge by differentiating the signal is not, by itself, sufficient to generate a spectral-line at the bit-rate. Since the data is random, the polarity of the edge pulses will also be random. To generate a strong clock component, the random phase reversals have to be removed. This can be easily accomplished with either a squaring circuit, or a rectifier. A typical sample of rectangular edges detected from an NRZ data sequence is shown in Fig. 4.11. This signal can be separated into the sum of a deterministic, periodic waveform, with a fundamental frequency at the data rate, and a random, zero-mean, data-dependent

PCM Signaling Format	$\overline{N_T}$	B_{80}
NRZ	$1/2$	$0.50B_T$
RZ	1	$1.00B_T$
Manchester	$3/2$	$1.25B_T$
Miller	1	$0.66B_T$

Table 4.1 Average number of transitions per bit-period $\overline{N_T}$, and the 80% bandwidth for various binary PCM signaling formats.

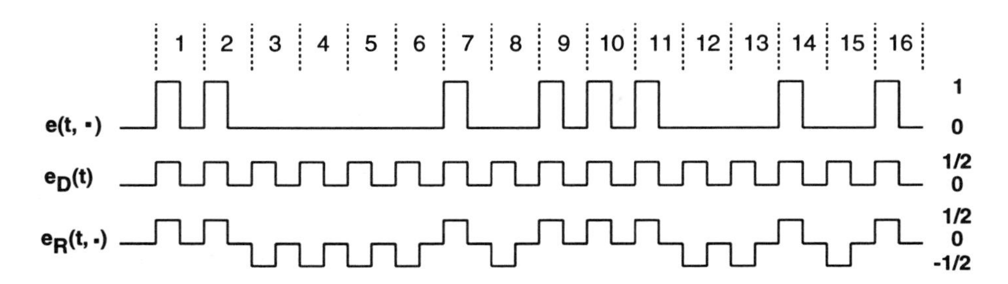

Figure 4.11 Detection of transitions in random NRZ data, and its decomposition into the sum of a deterministic and random part.

signal. The deterministic part gives rise to the clock tone, and its harmonics, while the random part generates amplitude modulation and phase-jitter.

The important feature of an edge detection circuit for NRZ data is that it produces a pulse, always in the same direction, whenever a transition in the data occurs. The shape of this pulse will determine the harmonic content of clock signal, and the functional form of the continuous noise spectrum. There are several circuits that can be used to generate these pulses. Five of them will be illustrated here in block diagram form.

4.3.1 Delay and EXOR

A common technique for detecting the edges of rectangular data is to exclusive-or (EXOR) the data with a delayed version of the same signal, as illustrated in Fig. 4.12. It can be seen from the timing diagram in Fig. 4.12, that the circuit will generate a rectangular pulse of width pT whenever a transition in the data stream occurs. We

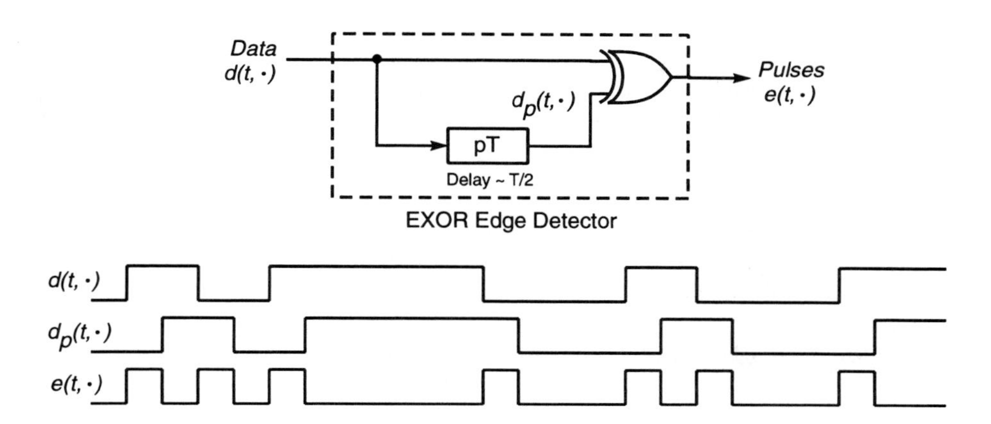

Figure 4.12 Edge detection circuit using an EXOR gate.

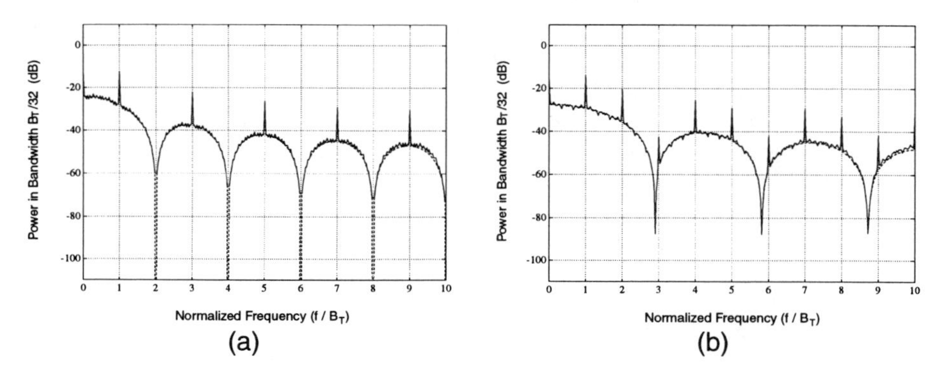

Figure 4.13 Simulated and calculated power in harmonics of an edge-detected NRZ data signal for: (a) $p = 0.5$, (b) $p = 0.3438$.

saw in chapter 2 that for $p = 0.5$, all even harmonics of the bit-rate are nulled, and the power in the fundamental clock tone is maximized. The power spectral densities of the edge-detected signals were given in Fig. 2.13 for $p = 0.5$, and $p = 0.3438$, and are repeated here in Fig. 4.13 for convenience. The functional form of the PSD was derived in (2.135), and is given by

$$P_e(f) = \left[\frac{p}{2}\text{sinc}(fpT)\right]^2 \left[T + \sum_{M=-\infty}^{\infty} \delta\left(f - \frac{M}{T}\right)\right]. \tag{4.3}$$

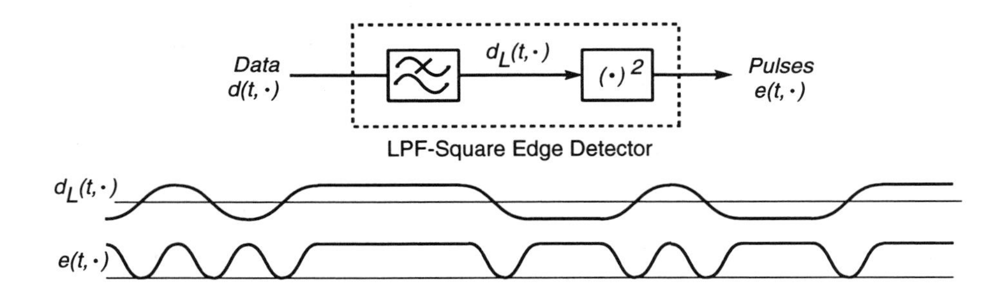

Figure 4.14 Edge detection circuit using a lowpass filter followed by a squaring circuit.

4.3.2 Lowpass Filter and Square.

Another example of an edge-detection circuit is shown in Fig. 4.14. The data is first lowpass filtered so that the transitions are smeared over a greater percentage of the bit-period. After squaring, the new signal has a constant dc value when there are no data transitions, and has negative pulses whenever there is a change in the data. It was shown in chapter 2 that for raised cosine kernel pulses, all harmonics of the clock signal are nulled. The dc component can also be nulled with an appropriate level shift. The zero-mean edge-detected signal, and the power-spectral density were shown in Fig. 2.28, and are repeated here in Fig. 4.15. The functional form of the PSD was also derived in (2.169), and was shown to be of the form

$$P_e(f) = \frac{1}{4}\left[\sum_{m=-1}^{1}\left(\frac{1}{2}\right)^{|m|}\text{sinc}\left(\frac{f - mB_T}{B_T}\right)\right]^2\left[T + \sum_{M=-\infty}^{\infty}\delta(f - MB_T)\right] \tag{4.4}$$

for the special case of raised cosine pulses.

4.3.3 Lowpass Filter, Highpass Filter, and Square

Another technique that will give results similar to the previous circuit is shown in Fig. 4.16. In this example the data is also lowpass filtered initially to smear the edges over the bit-period. The changes in the edges are detected by a highpass filter, often implemented as a differentiator [12]. The random phase reversals of the detected edges are removed by squaring the signal. For data with sinusoidal transitions, the edge-detected pulses after differentiation will be raised cosines.

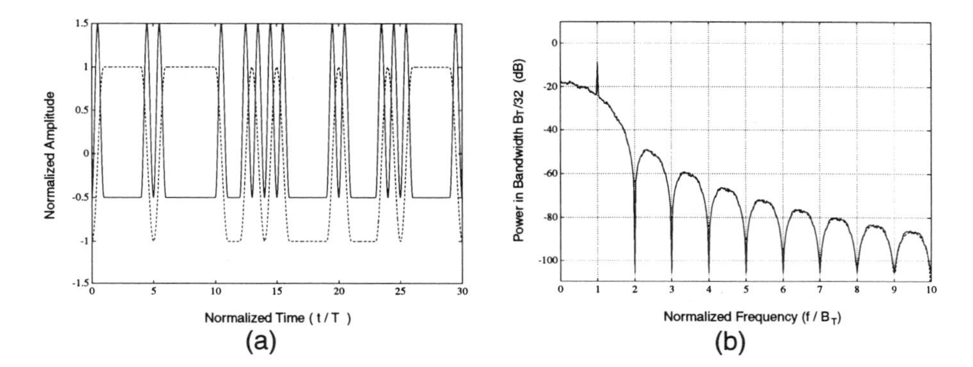

Figure 4.15 Transition detected pulses using a raised cosine kernel function: (a) the zero-mean pulse stream in the time domain and the original NRZ random data, (b) calculated and simulated normalized power in a bandwidth of $B_T/32$.

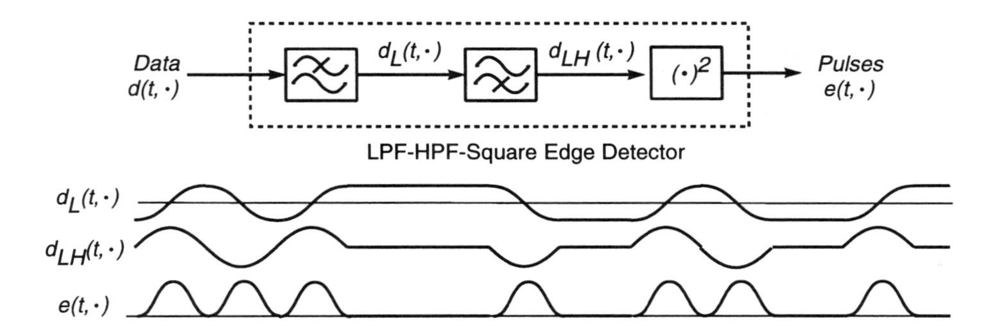

Figure 4.16 Edge detection circuit using LPF followed by an HPF and a squaring circuit.

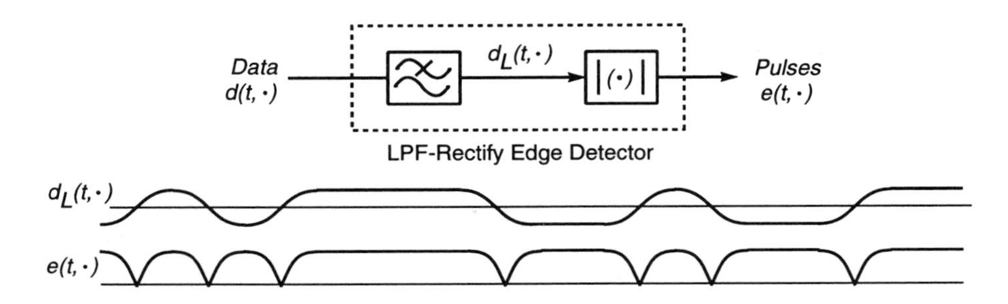

Figure 4.17 Edge detection circuit using a lowpass filter followed by a rectifier.

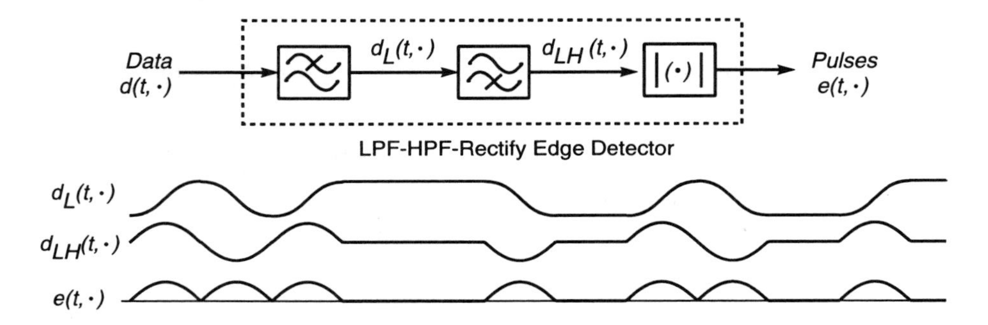

Figure 4.18 Edge detection circuit using LPF followed by an HPF and a rectifier.

4.3.4 Lowpass Filter and Rectify.

The circuit of Fig. 4.14 could have been implemented with a rectifier instead of a squaring circuit. This modification is shown in Fig. 4.17. Implementing the phase reversal circuit as a rectifier has advantages for broadband operation, because diodes can perform this operation at high-speeds.

4.3.5 Lowpass Filter, Highpass Filter, and Rectify

Likewise, in the circuit of Fig. 4.16, the squaring operation can be replaced by rectification. The resulting circuit, and the edge-detected pulses, are illustrated in Fig. 4.18. Unlike the smooth pulses generated by squaring circuits, the abrupt nonlinear rectification creates harmonics at the bit-rate much the same as a rectangular edge-detected pulse. Typical power-spectral-densities for the circuits of Fig. 4.17 and Fig. 4.18 are given in Figs. 4.19(a) and (b) respectively.

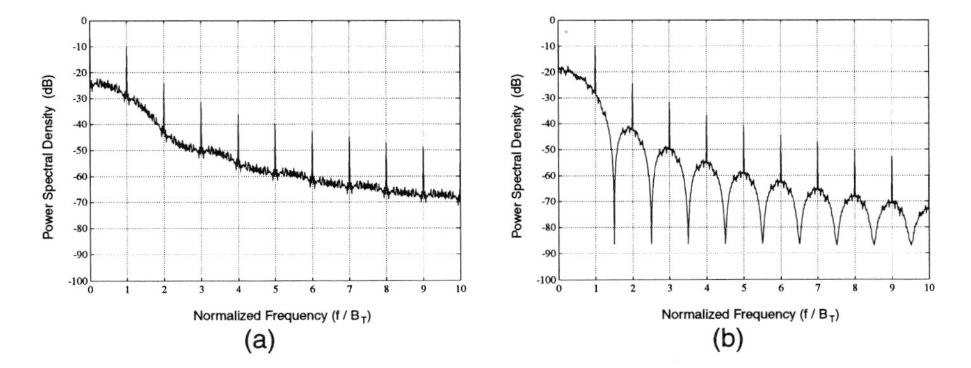

Figure 4.19 Typical power spectral densities for edge-detected pulse obtain from: (a) Lowpass filtering and rectifying, (b) Lowpass filtering, highpass filtering, and rectifying.

4.3.6 Alternative Representations of Identical Circuits

Any particular edge-detection circuit can be derived using several different, and seemingly distinct, approaches. At high-speeds there are no clean signal edges; signals thought to be *digital* are in effect *analog* signals. When it comes to implementing a clock-recovery architecture as an interconnection of transistors, most techniques will look rather similar. For example, we saw that for *digital* signals, an EXOR gate, together with a time delay, can be used to detect edges. For an *analog* signal, squaring the data, in conjunction with highpass filtering, is also a viable technique. Fig. 4.20 shows how a high-pass filter, and a squaring circuit, used for edge-detection, can be thought of as a delay-and-EXOR circuit. First consider the block diagram representation of this circuit shown in Fig. 4.20(a). In the analog domain, delaying a signal and subtracting it from itself performs a high-pass function, as in Fig. 4.20(b). Therefore, the same circuit could be used either as a delay, or as part of an HPF. In Fig. 4.20(c) the squaring operation is shown conceptually as the sum of three multiplications. If we assume that the data is rectangular, then the squared data, and the squared delayed-data, will be dc signals with equal value. Fig. 4.20(d) shows the equivalent circuit for rectangular data where the new edge-detected signal $\hat{e}(t, \cdot)$ is a level shifted version of the signal $e(t, \cdot)$. This delay and multiply edge-detection circuit was described by Millicker and Standley [13, 14]. However, it is essentially equivalent to a delay and EXOR. In the transistor level design, an EXNOR gate can have the same circuit topology as a balanced multiplier. Whether one calls the circuit a multiplier, or an EXNOR gate, is a matter of interpretation, and the level of signals one is using. Fig. 4.20(e) shows the analog multiplier represented as an EXNOR gate with one inverting input, which is logically equivalent to an EXOR gate, Fig. 4.20(f). Therefore, we have illustrated how one clock recovery circuit can be thought of as an extension of the other. The

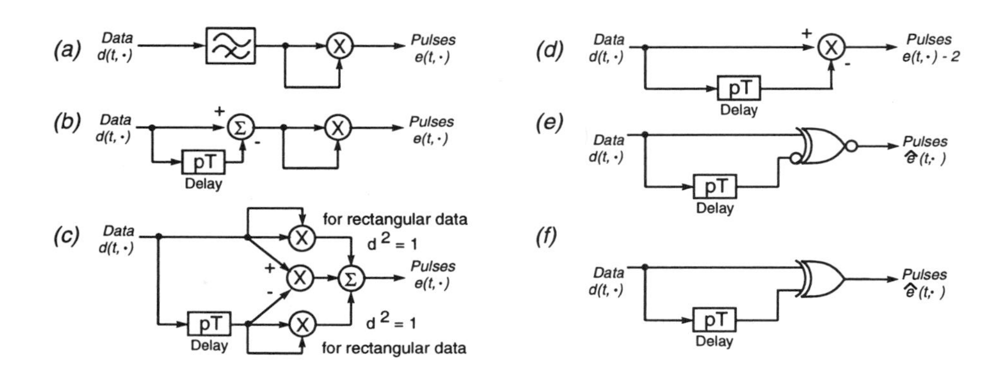

Figure 4.20 Edge-detection circuit using an HPF and squaring circuit showing relationship to a delay-and-EXOR circuit.

authors have found it very useful to look at a given clock recovery scheme from as many points of view as possible. This not only leads to a better understanding of the signal processing being performed, but also to circuit embellishments that improve performance.

4.4 SPECTRAL LINE TECHNIQUES

In the previous section we saw that several different nonlinearities can be used to extract a tone component from random NRZ data. The operation of edge-detection creates a spectral-line at the bit-rate, and techniques of clock recovery employing this method are often referred to as spectral-line clock extraction circuits. The clock can be recovered using either a bandpass resonator or a PLL. In the case of a resonator tuned to the bit-rate, the edge-detected data is used as an input signal to keep the resonator ringing in response to the edge-detected signal. In the case of a PLL, a variable frequency oscillator is adjusted by feeding back a comparison of the clock-phase with the phase of the edge pulses. There are advantages and disadvantages of each method, which will now be discussed.

4.4.1 Clock Recovery using High-Q Bandpass Filters

Once we have the edge detected signal, we now want to separate the pure tone at the bit rate from the random data-dependent variations. One method is to filter out the unwanted signal with a bandpass filter tuned to the clock bit-rate. The BPF is

a resonant circuit that will ring in response to an input pulse. The signal $e(t, \cdot)$ is a random stream of identical pulses at integer multiples of the bit-period. A pulse will be present when there was a transition in the data, and no pulse will be present when the data does not change states. Clearly this signal can be used to keep a resonator ringing at the bit-rate, provided that the pulse repetition rate is within the bandwidth of the BPF. Since there will be missing pulses whenever no data transition occurs, the ringing will tend to die away during long periods of missing pulses due to dissipation in the resonator. This dissipation will cause both amplitude, and phase modulation in the extracted tone. This effect can easily be seen in the time domain. For a simple second-order BPF with a transfer function of the form

$$H(s) = \frac{2\zeta\omega_n s}{s^2 + 2\zeta\omega_n s + \omega_n^2} \tag{4.5}$$

there is a zero at the origin, and two complex poles, as shown in the pole-zero plot of Fig. 4.21(a). The dissipation of the filter is the real-part of the complex poles $-\zeta\omega_n$, where ζ is the damping ratio, and ω_n is the undamped natural frequency. We saw in (2.217) and (2.227) that random amplitude and phase modulations were related to the equivalent selectivity of the filter by

$$\sigma\text{am} = \frac{1}{\sqrt{Q_{eq}}} \tag{4.6}$$

for pure amplitude modulation, and

$$\sigma_\phi = \frac{1}{\sqrt{Q_{eq}}} \tag{4.7}$$

for pure phase modulation. We can relate the selectivity to the dissipation in this simple filter as the inverse of the integral of the normalized frequency response.

$$\begin{aligned} \frac{1}{Q_{eq}} &= \frac{1}{\omega_n |H(j\omega_n)|^2} \int_0^\infty |H(j\omega)|^2 d\omega \\ &= \int_0^\infty \frac{4\zeta^2\hat{\omega}^2}{1 + (4\zeta^2 - 2)\hat{\omega}^2 + \hat{\omega}^4} d\hat{\omega} \\ &= \pi\zeta \end{aligned} \tag{4.8}$$

We can also define a selectivity Q_{3dB} that is the ratio of the 3-dB bandwidth to the center frequency. For a second-order bandpass this can be shown to be

$$Q_{3dB} = \frac{1}{2\zeta} \tag{4.9}$$

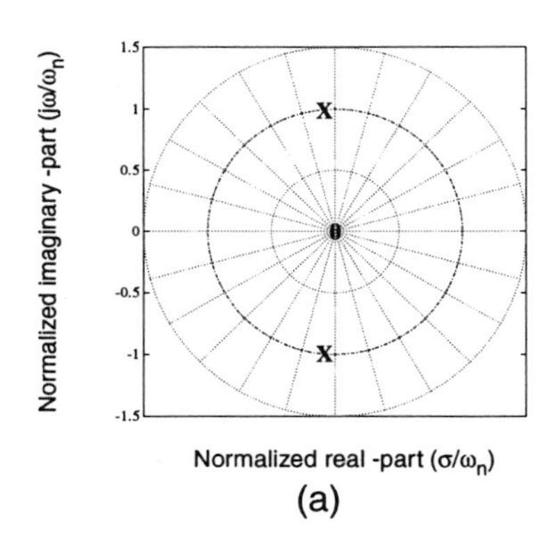

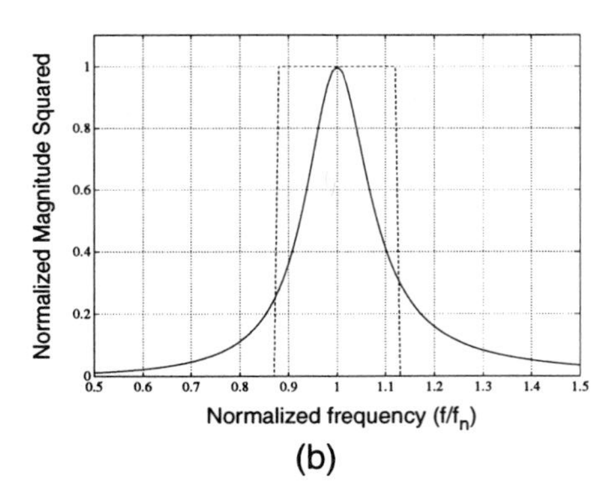

Figure 4.21 Second-order bandpass filter, (a) pole-zero diagram, (b) frequency response of real filter and equivalent ideal filter for $Q_{eq} = 4$.

Therefore, the relationship between these two selectivities is

$$Q_{eq} = \frac{1}{\pi\zeta} = \frac{2}{\pi} Q_{3dB} \tag{4.10}$$

The frequency response of this second-order bandpass filter is shown in Fig. 4.21(b) together with an ideal BPF of normalized bandwidth $1/Q_{eq}$. In this plot $Q_{eq} = 4$, which correspond to a damping ratio of $\zeta = 1/4\pi$. Taking the inverse Laplace transform we know that the impulse response of the filter has a decaying envelope of the form

$$\widehat{\text{env}}(t) = e^{-\zeta\omega_n t} \tag{4.11}$$

For a filter that is tuned to the bit-rate, then $\omega_n = 2\pi B_T$, and if we normalize time by the bit-period such that

$$n_t \triangleq \frac{t}{T}, \tag{4.12}$$

then the decaying envelope of the impulse response of the filter is

$$\begin{aligned} \text{env}(n_t) &= \exp(-\zeta 2\pi n_t) \\ &= \exp\left(\frac{-2n_t}{Q_{eq}}\right) \end{aligned} \tag{4.13}$$

This can be written in terms of a normalized time constant n_τ, where n_τ is the number of bit-periods before the envelope decreases to a value of $1/e = 0.37$;

$$\text{env}(n_t) = \exp\left(\frac{-n_t}{n_\tau}\right), \tag{4.14}$$

and the normalized time constant is given by

$$n_\tau = \frac{Q_{eq}}{2} = \frac{Q_{3dB}}{\pi}. \tag{4.15}$$

The decay in the power envelope is proportional to the voltage squared, and is simply

$$\text{Penv}(n_t) = \exp\left(\frac{-n_t}{n_\tau/2}\right), \tag{4.16}$$

which is the same as the result given in [15] with $Q = Q_{3dB}$.

Physical Interpretation of Quality Factor Q (4.15) can be written in a form that adds physical insight [16, ch. 10, p. 297]. The envelope of the stored energy in the system will have the same functional form as the envelope of the dissipated power. For example, we could consider the signal of interest to be the voltage across a capacitor, in which case the energy stored on this capacitor is $E = 1/2(CV^2)$, and the envelope of the stored energy can be written as

$$\text{Eenv}(n_t) = E_0 \exp\left(\frac{-n_t}{n_\tau/2}\right), \tag{4.17}$$

where E_0 is the initial stored energy at time t_0. Differentiating both sides gives

$$\frac{d\text{Eenv}(n_t)}{dn_t} = \frac{-E_0}{n_\tau/2} \exp\left(\frac{-n_t}{n_\tau/2}\right) = \frac{-1}{n_\tau/2}\text{Eenv}(n_t), \tag{4.18}$$

which is a first-order differential equation relating the rate of energy dissipation to the total energy stored, from which we observe that

$$n_\tau = 2\left[\frac{\text{Eenv}(n_t)}{-d\text{Eenv}(n_t)/dn_t}\right]. \tag{4.19}$$

Therefore the normalized time constant n_τ is twice the ratio of the stored-energy to the energy-lost-per-cycle; substituting for values of Q we obtain the following physical interpretation for the filter's quality factor.

$$\begin{aligned} Q_{eq} &= 4 \left[\frac{\text{stored energy for the } n^{th} \text{ cycle}}{\text{energy lost in the } n^{th} \text{ cycle}}\right] \\ Q_{3dB} &= 2\pi \left[\frac{\text{stored energy for the } n^{th} \text{ cycle}}{\text{energy lost in the } n^{th} \text{ cycle}}\right] \end{aligned} \tag{4.20}$$

Since Q is a constant, the fractional energy-lost-per-cycle is constant and equal to $4/Q_{eq}$. Hence for a bandpass filter with $Q_{eq} = 10$, the resonator will lose 40% of its stored energy per cycle if no input is applied.

Deviations in Clock Signal Envelope in Terms of Q Due to energy dissipation in the resonator, the voltage envelope is reduced by 86.5%, and the power envelope is reduced by 98% in Q_{eq} clock periods. Statistical analysis shows that the rms envelope deviation for a linear phase filter is $1/\sqrt{Q_{eq}}$. For example, a 3σ variation in the clock envelope of within 50% requires $Q_{eq} \geq 36$. However, this result is derived from an ensemble average, and there will be time intervals when the deviation in the clock envelope is significantly worse. The above time domain analysis gives us another means to estimate the selectivity of a BPF needed to meet desired specifications. If we have a requirement that the clock envelope can not drop below 50% of the nominal for N_B consecutive bits without a transition, then we required

$$e^{-N_B/n_\tau} \geq 1/2 \quad \Longrightarrow \quad n_\tau \geq \frac{N_B}{\ln(2)}, \tag{4.21}$$

therefore,

$$Q_{eq} \geq \frac{2N_B}{\ln(2)} = 2.89 N_B \simeq 3N_B. \tag{4.22}$$

So as a rule of thumb for an arbitrary BPF, the number of consecutive bits without a transition that can be tolerated before the clock amplitude is cut in half is

$$\boxed{N_B \simeq \frac{Q_{eq}}{3},} \tag{4.23}$$

and $N_B \simeq Q_{eq}/6$ before the clock power is halved. If all bits are independent and equally likely, then the probability that a sequence of N_B bits will not have a transition is

$$P_{NB} = \Pr[\text{no transition}] = 2^{-(N_B - 1)} \tag{4.24}$$

Therefore, for a given probability, the sequence length is given by

$$N_B = 1 + \frac{-\log(P_{NB})}{\log(2)}, \tag{4.25}$$

and the required filter selectivity is therefore

$$\boxed{Q_{eq} = 2.89 - 9.6\log(P_{NB}).} \tag{4.26}$$

Therefore, a probability of less than 10^{-9} that the clock amplitude will fall below 50% of the nominal value requires $Q_{eq} \geq 90$. For the same probability that the clock power falls below 50%, Q_{eq} must be greater than 180. For a 10-Gb/s data signal, the probability of an event of duration 100 ps happening once in ten years of operation is 3.17×10^{-19}. This corresponds to a transitionless string of bits of length $N_B = 62$. The resulting Q_{eq} values needed are 180 for 50% envelope reduction, and 360 for 50%

power reduction; the Q_{3dB} values are 283 and 566 respectively. It should be stressed that all of this analysis is approximate because it is assumed that the clock signal was at the nominal value when the string of no transitions started. In reality, shorter strings of data with no transitions will cause the same envelope reduction, provided that the shorter strings occur in rapid succession. This analysis, however, does provide useful information about Q_{eq} and its relationship between the transition density of data and the amplitude modulation; (4.26) supplements the information derived previously that the rms amplitude and phase modulations are approximately equal to $1/\sqrt{Q_{eq}}$. These results together provide the fundamental guidelines for determining the maximum selectivity of a BPF required to meet a given specification.

4.4.2 Clock Recovery Using Surface-Acoustic-Wave Filters

To reduce the random amplitude and phase modulations and improve the accuracy of the data arrival-time estimate, a very high selectivity filter is required. For a 1% rms envelope deviation, which implies a peak-to-peak deviation of approximately 6σ, or 6% in the clock envelope, we require a filter Q of 100^2, or $10,000$. From the analysis of the previous section we see that for $Q_{eq} = 10,000$ the impulse response of the filter rings approximately 3333 clock cycles before reducing in amplitude by 50%, and will ring $20,000$ cycles before reducing by 98%. Special design considerations are required to achieve such a low dissipation and narrow bandwidth in a bandpass filter. Lumped element bandpass filters, for example, can achieve Q values in the hundreds, and mechanical and crystal filters can achieve Q values on the order of 1000. One clock recovery method that has been very popular in recent years involves the use of surface-acoustic-wave (SAW) filter. SAW techniques have made possible stable resonators with very high Q values; practical filters achieve Q's in excess of 50,000 [17, p. 887].

Brief Overview of Transversal SAW Filters SAW filters are constructed using transducers on a piezoelectric material, usually quartz, that converts electrical energy to acoustic waves and back again. Both resonator-type, and transversal filters are realizable. We will briefly describe the operation of a transversal SAW filter, such as the one shown in Fig. 4.22. The filter operation is analogous to a finite-impulse-response (FIR) filter. The electrical input signal transduces an acoustic wave that propagates in the direction of the output transducer, where it is reconverted to an electrical signal. An illustrative analogy describes a transversal SAW filter as "beating on one side of drum, and picking up the vibrations on the other side.[3]" During each period of the input signal, a new acoustic wave is launched. If the input signal frequency is such that the new wave constructively interferes with the old ones, then a large acoustic wave

[3] Analogy given by Dr. Scott Willingham

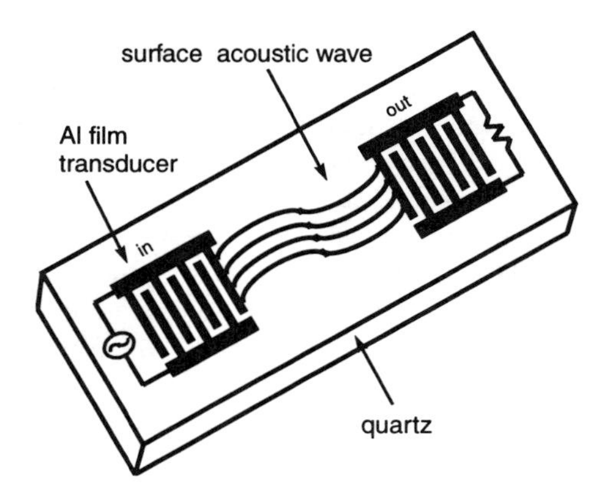

Figure 4.22 Conceptual diagram of a transversal SAW filter.

builds up, and is detected by the output transducer. If the input signal is not at the proper frequency, then the waves interferes destructively, and no signal is transmitted to the output.

The transducer typically consists of several hundred metal fingers. As each acoustic wave travels in space, it interacts with new waves launched by the input signal at different finger locations. The final acoustic wave transmitted to the output transducer is, therefore, a convolution of the input signal, with a sinusoidal, time-limited, acoustic wave. The impulse response of the filter is essentially the portion of a surface-acoustic wave that overlaps the transducer, and is therefore of finite duration. If the finger spacing and the velocity of the SAW are uniform, then the impulse response is symmetric, and the filter will have a linear-phase, or a constant group-delay. The frequency-domain magnitude response will be a narrowband filter, where the center frequency depends on the finger spacing, and the propagation speed of the SAW. The more fingers in the transducer, the longer the convolution pulse, and therefore, the narrower the bandwidth. The Q value as given in [15] is

$$Q = \frac{\pi N_T}{2}, \tag{4.27}$$

where N_T is the number of transducer fingers. For a typical value of $N_T = 500$, Q is approximately 800. Problems with SAW filters are that they are generally very lossy, and the sub-micron finger spacing required for high-speed operation limits their applicability to 3–5 GHz.

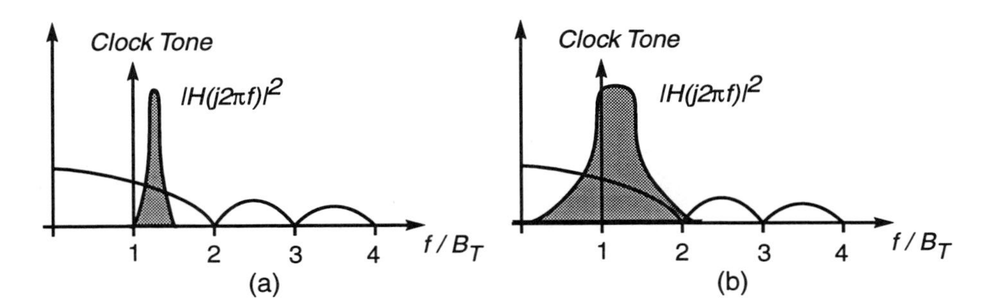

Figure 4.23 Frequency instability of a narrowband BPF: (a) When Q is too large, the clock tone is out of the filter bandwidth. (b) When Q is relaxed, the tone lies within the passband.

Frequency Stability and Detuning Limitations on Maximum Q

Center-frequency stability places limitations on the maximum Q value of a filter for clock recovery circuits. In order to filter as much of the unwanted noise and random data dependent modulation as possible, the filter should *zoom-in* very close in frequency to the clock tone. However, if the filter *zooms-in* too closely, it runs the risk of missing the clock tone itself. This situation is illustrated in Fig. 4.23(a), where the clock tone falls outside of the filter passband. When the BPF is not tuned to the bit-rate, then we say that the filter is *detuned*. Factors that contribute to detuning are:

- Fluctuations in the data rate due to frequency instabilities in the clock at the transmitter.
- Limited accuracy in which the filter can be manufactured and tuned after manufacturing
- Drift in the filter's center frequency with temperature.
- Drift in the filter's center frequency due to aging.

When all of these detuning factors are taken into consideration they impose a limitation on the maximum Q value needed to insure that the clock tone lies within the passband of the filter for worst-case center frequency deviations. Fig. 4.23(b) illustrates the situation when Q is reduced. The clock tone now lies within the passband, but not necessarily at the center frequency. The penalty incurred by increasing the filter bandwidth is of course an increase in noise. If we define qualitative measures for the performance degradation due to filter detuning, versus increased noise, then we can derive an optimum Q value, or at least, a range of Q values that simultaneously satisfy both the detuning and noise requirements.

Quantitative Q Limits The actual limitations on Q will depend on the choice of filter and the accuracy with which the center frequency can be maintained. We can illustrate the procedure for deriving the allowable range of Q values for a simple example. If we return to the second-order bandpass filter example we recall that the transfer function is given by

$$H(j\hat{\omega}) = \frac{2\zeta j\hat{\omega}}{(1-\hat{\omega}^2) + 2\zeta j\hat{\omega}}, \tag{4.28}$$

where $\hat{\omega} = \omega/\omega_n$. The phase is therefore given by

$$\theta = \angle 2\zeta\hat{\omega} + j(1-\hat{\omega}^2) = \tan^{-1}\left[\frac{1-\hat{\omega}^2}{2\zeta\hat{\omega}}\right]. \tag{4.29}$$

This expression is more enlightening when we write the frequency in terms of the deviation from the center frequency $\Delta\hat{\omega} = \hat{\omega} - 1$. Recalling that $Q_{3dB} = 1/2\zeta$ we obtain

$$\theta = \tan^{-1}\left[-Q_{3dB}\Delta\hat{\omega}\frac{2+\Delta\hat{\omega}}{1+\Delta\hat{\omega}}\right]. \tag{4.30}$$

We can take this normalization one step further, and express $\Delta\hat{\omega}$ in terms of the one-sided 3-dB bandwidth. Therefore,

$$\Delta\omega_Q = 2Q_{3dB}\Delta\hat{\omega} \tag{4.31}$$

so that $|\Delta\omega_Q| = 1$ at approximately the points of 3-dB attenuation, as illustrated in Fig. 4.24. If the BPF were symmetric about the center frequency, then $|\Delta\omega_Q|$ would equal unity at exactly the 3-dB attenuation frequencies. The phase response is then given by

$$\theta = \tan^{-1}\left[-\frac{\Delta\omega_Q}{2}\left(\frac{2Q_{3dB} + \Delta\omega_Q/2}{Q_{3dB} + \Delta\omega_Q/2}\right)\right], \tag{4.32}$$

and for large values of Q_{3dB}, the result simplifies to

$$\theta \simeq \tan^{-1}\left[-\Delta\omega_Q\right]. \tag{4.33}$$

Furthermore, since we will be interested in small phase deviations, where the tangent function is approximately linear,

$$\theta \simeq -\Delta\omega_Q = -2Q_{3dB}\left(\frac{f-f_n}{f_n}\right). \tag{4.34}$$

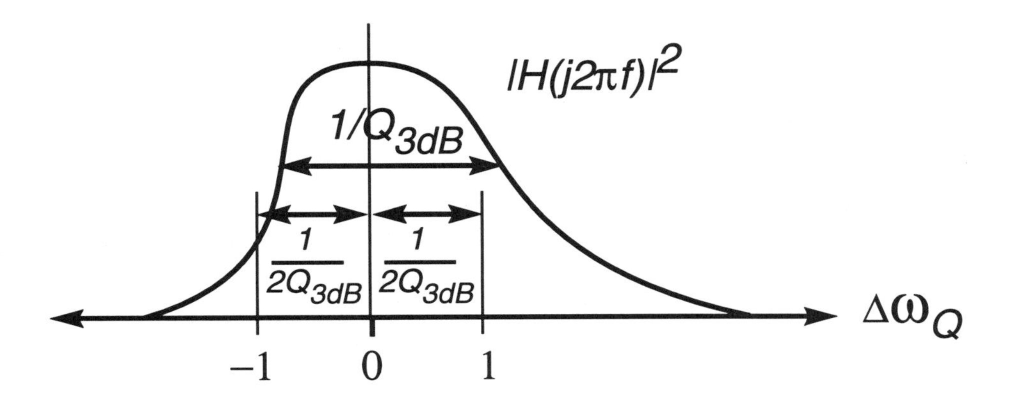

Figure 4.24 Illustration of frequency normalization in a bandpass filter.

Relating Q to Maximum Phase Deviation To find a limit on the maximum value of Q, we need to determine the maximum allowable deviation in the phase due to detuning of the filter. We saw in table 3.5 and Fig. 3.20 that a steady-state phase error in the clock signal reduced the effective SNR of the test statistic. In particular for a rectangular correlation pulse, a 10% error in the clock-phase caused a 2-dB drop in the SNR, which increases the error probability from 10^{-9} to 10^{-6}. If we allow for a phase deviation budget of 10% in our design, then we might arbitrarily allocate 2.5% of the phase deviation to the filter detuning. From (4.34) we can find the maximum frequency deviation that will produce a 2.5% phase deviation;

$$\theta\text{max} = 2\pi\left(\frac{2.5}{100}\right) \geq \left|2Q_{3dB}\left(\frac{\Delta f}{f_n}\right)\right|. \tag{4.35}$$

This imposes an upper limit on Q_{3dB} of

$$Q_{3dB} \leq \frac{\theta\text{max}}{2}\left(\frac{f_n}{|\Delta f|}\right) \tag{4.36}$$

We can further express the filter detuning in terms of the deviations caused in the center frequency of the filter. If each contribution to filter detuning is Δf_i, then in the worst-case

$$|\Delta f| = \sum_i |\Delta f_i|. \tag{4.37}$$

Since the filter is nominally tuned to the bit-rate ($f_n = B_T$), the detuning upper limit on Q_{3dB} is given by

$$Q_{3dB} \leq \frac{\theta\text{max}}{2}\left(\frac{B_T}{\sum_i |\Delta f_i|}\right), \tag{4.38}$$

and in terms of the equivalent noise power selectivity

$$Q_{eq} \leq \frac{\theta_{\max}}{\pi}\left(\frac{B_T}{\sum_i |\Delta f_i|}\right), \tag{4.39}$$

To obtain a lower limit on Q_{eq} we recall from chapter 2 that when all of the energy in the amplitude modulation of the clock is converted to phase-noise, the rms phase deviation is given by

$$\sigma_\theta = 1/\sqrt{Q_{eq}}. \tag{4.40}$$

If we assume that the peak deviation in the phase-noise is approximately $3\sigma_\theta$ in each direction, then we obtain the rough lower limit

$$\theta_{\max} \geq 3\sigma_\theta = 3/\sqrt{Q_{eq}} \tag{4.41}$$

$$Q_{eq} \geq 9/\theta^2_{\max}. \tag{4.42}$$

Putting the two limits together we finally obtain the desired relationship

$$\boxed{\frac{9}{\theta^2_{\max}} \leq Q_{eq} \leq \frac{\theta_{\max}}{\pi}\left(\frac{B_T}{\sum_i |\Delta f_i|}\right).} \tag{4.43}$$

For our numerical example of $\theta_{\max} = 2.5\% = 0.157\,\text{rad} = 9^\circ$ we find that

$$365 \leq Q_{eq} \leq 5\%\left(\frac{B_T}{|\Delta f|}\right). \tag{4.44}$$

In order to simultaneously satisfy both requirements then

$$\left(\frac{|\Delta f|}{B_T}\right) \leq \frac{5\%}{365} = 137 \quad \text{ppm}. \tag{4.45}$$

Therefore, the worst-case detuning of the BPF can not exceed 137 ppm. If the bit-rate is 2-Gb/s, this requirement imposes a total worst case drift in the BPF center frequency of 274-KHz. This detuning allocation is within typical specifications of commercially available filters as reported in [18].

Summary of Clock Recovery Using SAW Filters

Although the above analysis is only approximate, it does illustrate the trade-offs that must be made between noise suppression and center-frequency stability in choosing a Q value for the BPF. Typical Q values for such systems are on the order of 1000. Extensive analytical and experimental studies of SAW filter for use in undersea long-haul fiber-optic systems were undertaken in the mid 1980s at Bell Labs. The results

Clock Recovery Using SAW Filters	
Advantages	Disadvantages
• Simple to implement • Proven Reliability • No problems such as frequency acquisition and cycle-slipping common in phase-locked loops • Variations due to aging and temperature changes are manageable	• Fixed center frequency doesn't track the data rate • Noise bandwidth is fixed • Limiter circuit is required to eliminate amplitude modulation, which generates additional phase-noise • Phase adjustment is required, open-loop adjustment doesn't track variations in operating conditions • Maximum frequency limited to about 3–5 GHz • Maximum Q limited by detuning requirements • SAW Filter not compatible with IC process. I/O buffers add excess phase-shifts that must be cancelled

Table 4.2 Advantages and Disadvantages of using SAW filters for clock recovery in broadband communication systems.

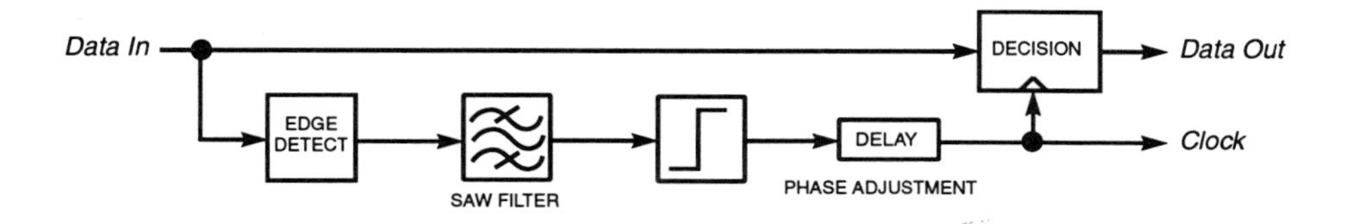

Figure 4.25 Block diagram of a clock recovery circuit using a SAW filter.

are summarized in two papers by Rosenberg *et al.* [15, 18]. SAW filters have proven their reliability in practical systems, and as a result have been used extensively in clock recovery circuit for multi-gigabit-per-second fiber-optic systems [15, 18, 19, 20, 21, 22, 23, 24]. However, there are several disadvantages of using a SAW filter, (namely, the filter is not compatible with standard IC processes, and must be external to the signal processing electronics). The advantages and disadvantages of using SAW filters are listed in table 4.2.

A block diagram of a clock recovery system using a SAW filter is shown in Fig. 4.25. Since the filter is external to the chip, input- and output-buffers, with indeterminate phase delays, are required. The output-buffer couples the edge-detected signal, from the chip, into the external filter, while the input-buffer is needed to couple the filtered clock-signal back into the chip. Compensation must be made for the phase-shifts caused by these interface circuits by adding an adjustable phase shifter. This adjustment must be manually tuned during an evaluation phase, where it will be set to a nominal value that minimizes the error during test. However, this phase adjustment is a one-time adjustment and can not compensate for variations in the bit-rate, or environmental changes once the filter is in operation. Fig. 4.25 also shows a limiter that is needed to remove the amplitude modulation in the recovered clock. The limiter contributes excess phase-noise to the clock by two distinct methods. One is the nonlinear phase-shift variations as a function of frequency, which is a characteristic of any causal infinite-impulse-response circuit. The other is the conversion of amplitude modulation into phase-noise by a nonlinear, amplitude-dependent, phase delay of the buffer, which is often the dominant phase-jitter contribution. This is a characteristic of any semiconductor device, where the parasitic capacitances are voltage dependent. Therefore, delays will vary with the power level of the input signal. A technique for reducing both types of these nonlinear delays is described by Nakamura *et al.* [25].

A further severe limitation on the use of SAW filters is that the maximum center frequency is currently limited to about 3 GHz, with 5 GHz projected as the maximum [17]. When a SAW filter can not be used, other techniques such as lumped-element, microstrip-line, resonant-cavity, or dielectric resonators [23, 26, 13, 27] ($Q \simeq 1000$) can be substituted for bit-rates up to 20 Gb/s. However, limitations in the maximum center frequency can be circumvented by using mixers, or frequency dividers, to heterodyne the clock-tone to a lower frequency, where the filtering can be done by a SAW filter. A system that uses this approach to mix the clock frequency down by a factor of 2 is described by Wang *et al.* [19, 20]. In the extreme case the signal could be mixed all the way down to dc, and the noise filtering can be done in the baseband. However, this doesn't solve the detuning problem. Mismatches between the local oscillator and bit-rate still need to be accounted for. However, if the local oscillator can be made to track the bit-rate, then the detuning restriction can be eliminated. One such tracking system is known as a phase-lock loop (PLL). Clock recovery using PLLs is the subject of the following section.

4.4.3 Using PLLs to Synchronize a VCO to the Data Rate

Most of the disadvantages of using SAW filters, or other fixed frequency bandpass filters, for clock recovery can be overcome by using a PLL. This comes at the expense

Clock Recovery Using Phase-Locked Loops	
Advantages	**Disadvantages**
• Can achieve arbitrarily high Q, and therefore a narrow noise bandwidth • Clock tracks the bit-rate, eliminating detuning safeguards • Clock has no amplitude modulation eliminating the need for a limiter amplifier • Can be used to implement clock recovery systems based on optimal stochastic estimation • With appropriately designed phase detectors can be self-adjusting to compensate for the phase-errors due to other circuits in the system	• Requires frequency acquisition aids • Complex circuit design • Nonlinear frequency acquisition and cycle-slipping limit performance

Table 4.3 Advantages and Disadvantages of using PLLs for clock recovery in broadband communication systems.

of increased design complexity. In addition to overcoming several of the disadvantages of BPFs, PLLs are directly applicable to clock extraction using optimal stochastic estimation techniques, to be described in section 4.5, whereas fixed filters would require a feedback loop to be added for controlling an electronically tunable delay in response to an error signal. The advantages and disadvantages of using a PLL for clock recovery are given in table 4.3. Since the loop tracks the input bit-rate, detuning constraints are eliminated and the effective Q of the PLL can be arbitrarily large. Ultimately, limitations on the effective Q, which is controlled by the closed-loop noise bandwidth of the PLL, will be set by nonlinear transient behavior constraints, such as frequency acquisition, and frequency tracking. There are, however, analogies to detuning that place limits on the maximum possible noise suppression. PLLs can only naturally acquire frequency errors on the order-of-magnitude of the closed-loop bandwidth. Therefore, if we depended on natural acquisition of the PLL alone, we would be faced with the same detuning limitations discussed in the previous section. However, we rarely depend on natural acquisition, and supplement the process with a frequency acquisition aid of one kind or another to be discussed further in chapter 5.

A block diagram of a spectral-line clock recovery technique using a PLL is shown in Fig. 4.26. Since a PLL can be fabricated on the same chip as signal processing

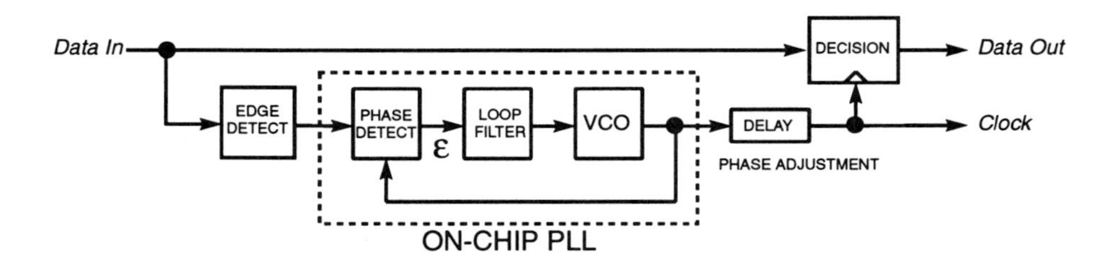

Figure 4.26 Block diagram of a spectral-line clock recovery circuit using a PLL.

circuitry, the need for interface circuits, to bring signals on and off chip, and their associated phase delays, are eliminated, thereby substantially reducing the phase-lag in the lower-arm of the circuit. However, this doesn't eliminate the need for the phase adjustment altogether. There are still residual differences in the delays of signal propagation in the data path and the clock path. Even in the decision circuit itself, it is typical to find unequal delays in the data, and clock paths. The result is that for very high data-rates, phase adjustments are ultimately required to center the clock edge in the data-eye. Although, elimination of interface circuits reduces the magnitude of the phase adjustment, we are still faced with the same problem that we had when using a BPF for clock recovery, (namely, the open-loop phase-adjustment will not track variations in the bit-rate due to temperature, or aging). We then have two options in the design: we can perform open-loop phase compensation to account for the worst-case detuning effects in the design, or we can design a special phase detector that measures all of the excess phase errors, which can be zeroed using the negative feedback of the PLL. Techniques for implementing the former approach are the topic of this section. The later, self-adjusting systems, will be discussed in section 4.6.

PLL as a Bandpass Filter A PLL can is some respects be considered as an adaptable BPF where the center frequency is automatically tuned to the bit-rate. If we look at the operation in the frequency domain we see that the phase-detector functions as a mixer to heterodyne the edge-detected input signal down to the baseband. This is illustrated in Fig. 4.27(a). When the loop is in lock, the clock signal of the VCO is in quadrature with the spectral line tone of the edge-detected signal. There will be no resulting dc component since the two signal are orthogonal. The action of the PLL tracks the phase of the edge-detected signal and mixes the signal energy, from a band of frequencies around the clock rate, down to dc where it can be suppressed by the loop filter. The mixer has the effect of *zooming-in* directly on the interesting part of the edge-detected signal spectrum. Since the PLL is automatically tuned, the loop filter bandwidth doesn't have to be made large to account for various detuning factors. Therefore, the loop filter can be be made narrowband, and excess noise is not added by processing the signal in guard-band frequencies that contain only noise with no information. The

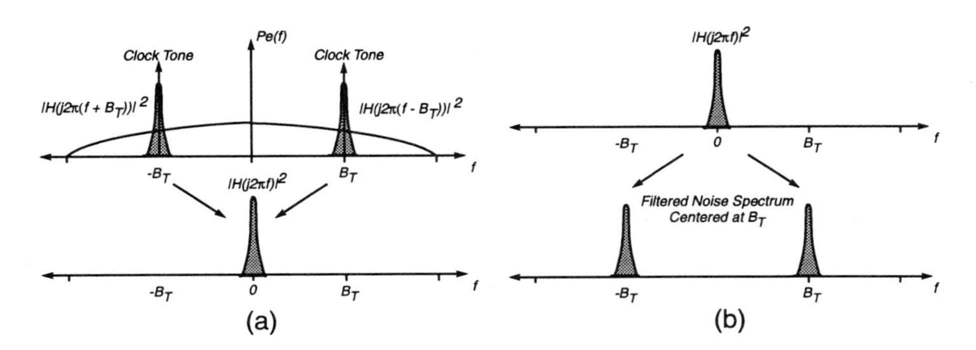

Figure 4.27 Illustration of a PLL converting: (a) passband energy to baseband energy, (b) baseband energy back to passband energy.

tuning of the PLL is accomplished by filtering the phase-error signal and using the filtered signal to adjust a variable frequency oscillator. This baseband tuning signal frequency modulates the VCO, and therefore shifts the spectrum of energy spectrum to that of an FM signal center around B_T. This operation is illustrated in Fig. 4.27(b).

Extremely high Q values are possible using a PLL without requiring a high-quality resonator, although in many respects, since a low-phase-noise clock requires a low-phase-noise VCO, we have just passed the problem of designing a good resonator from the filter designer to the VCO designer. However, when the majority of the phase-noise in the recovered clock is due to random modulations in the data, or due to additive noise, as is typically the case for recovery of a clock from random data, the bandwidth of the noise-suppression filter is the critical parameter in determining the phase-noise in the recovered clock, and the added jitter of the free-running oscillator is of secondary importance. Therefore we can use a somewhat noisy VCO with a low-Q resonance together with a narrowband loop filter to achieve the same jitter performance of a SAW filter with a high-Q resonance. Since the PLL is free from the detuning constraints that limited the maximum Q in a bandpass filter, we can easily achieve an effective Q of one million. If we design a PLL with a lag-lead loop filter such that the closed-loop transfer function is second-order with a damping ratio of $\zeta = 1/\sqrt{2}$, and a natural frequency of f_n=5-KHz, then for a clock tone at 10-GHz, the effective Q is approximately

$$Q_{\text{PLL}} = \frac{10 \text{ GHz}}{2 \cdot 5 \text{ kHz}} = 10^6. \tag{4.46}$$

This effective Q can be interpreted by realizing that the PLL averages the phase-error over several cycles; in this case it takes approximately one-million clock-cycles before the loop filter can accumulate a large enough signal on the VCO control line to respond to the error in phase. We can think of a PLL as a flywheel that is spinning at a rate

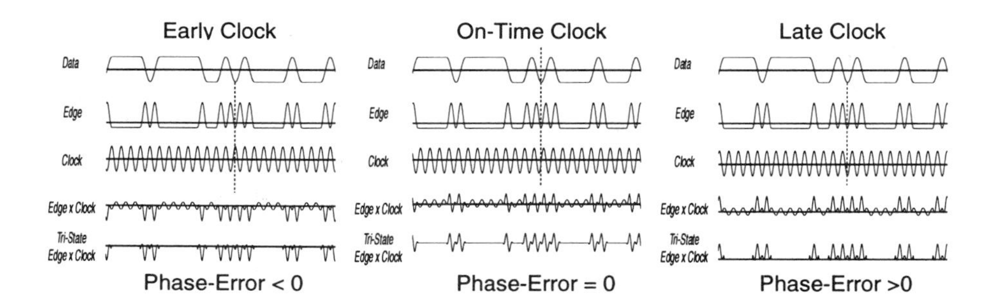

Figure 4.28 Phase detection of edge-detected pulses in a direct implementation of a spectral-line clock recovery system using a PLL.

close to the data rate. The flywheel has a timing mark on it. Input data signal acts like a strobe light that flashes every time that a data transition is detected, revealing the current phase-error of the timing mark. Feedback is used to align the timing mark to the desired position. Increasing the time constant of the loop filter is analogous to increasing the mass of the flywheel. A narrowband loop acts like a very heavy flywheel that takes a lot of energy to alter its momentum. Whereas in the case of a BPF we saw that the effective Q was determined by how many cycles the resonator could ring, in a PLL the Q is determined by how many clock cycles it takes for the VCO to respond to a phase error.

Direct Implementation of Spectral-Line PLL Clock Extractors

A balanced multiplier and a lowpass filter are typically used for phase detection in PLLs. The phase detection process for random data is best illustrated in the time domain. We will assume that an edge-detection scheme has been used that generates raised cosine pulses. Timing diagrams for early, on-time, and late clocks are shown in Fig. 4.28. During data transitions, the circuit acts as a traditional phase detector. The dc output of the phase detector goes to zero when the two signals are in quadrature, is a maximum when they are in-phase, and is a minimum when they are 180° out of phase. When there is no data transition, we have already reasoned that there is no phase information. The phase detector, therefore, contributes nothing to the average phase error signal. When no transition occurs the edge-detected signal is steady at some dc value. Multiplying by the recovered clock produces a pure ac signal that is suppressed by an ideal lowpass filter. However, the ripple is not suppressed completely, and residual ripple leads to excess clock phase jitter. One technique for reducing this jitter is to use a tri-state phase detector that switches to a zero-state when no transitions occur. It can be seen from Fig. 4.28, that the ripple in the tri-state phase detector is significantly reduced as compared to a standard phase detector.

Data Density Dependence and Pattern-Dependent Jitter Non-ideal effects will cause degradations in performance. We have already mentioned that the transmission of high-frequency ripple through the lowpass filter will modulate the VCO, resulting in increased phase-jitter. In addition, noises in the circuit will modulate the phase-error around zero, and constant adjustments have to be made by the negative feedback of the PLL to maintain average synchronization. Since contributions to the phase-error only occur during a data transition, the phase error magnitude is dependent on the transition density of the data. Therefore, the dynamic behavior of the loop will vary significantly for dense, and sparse transitions, leading to data pattern-dependent jitter in the recovered clock (Certain data patterns will contribute much more jitter than others. As a result the receiver is more likely to make an error when these patterns are transmitted.). Pattern-dependent jitter is always present in a direct implementation. However, this problem can be avoided by using alternative phase-detection methods. In section 4.6 we will present a technique that is similar to direct implementations, but uses a special phase detector circuit, which is insensitive to data-density, thereby significantly reducing pattern-dependent jitter. For now we will briefly review three different clock recovery circuits that are direct implementations of spectral-line techniques using a PLL.

The Circuit of Cordell et al. (Bell Labs 1979)

A direct implementation of a spectral-line clock recovery using a PLL was designed at Bell Labs in 1979, and is described by Cordell *et al.* [12]. The circuit operates at a data rate of only 50-Mb/s, however, the circuit was fabricated in a 300-MHz bipolar process. Therefore, the transistor-speed-to-bit-rate ratio, $f\text{max}/B_T \simeq 6$, is favorable. Modern transistors are 100 times faster, so that the techniques described by Cordell are applicable to 5-Gb/s systems using technologies available in 1992. A block diagram of the circuit used by Cordell is given in Fig. 4.29. The edge detection is performed using a lowpass filter, differentiate, and rectify technique. The differentiation is performed using a differential pair with capacitive emitter coupling, and the rectification is done simply by tapping the emitters of an emitter-coupled pair. Cordell uses a tri-state phase detector that turns off when no data transition occurs. As we saw in Fig. 4.28, this prevents the double frequency ripple from coupling to the VCO and increasing the phase jitter when the data is constant.

Cordell gives a very clear and concise overview of clock recovery in broadband systems. Helpful timing diagrams are given as well as practical bipolar transistor-level circuit realizations of critical functional building blocks. A frequency discriminator was used

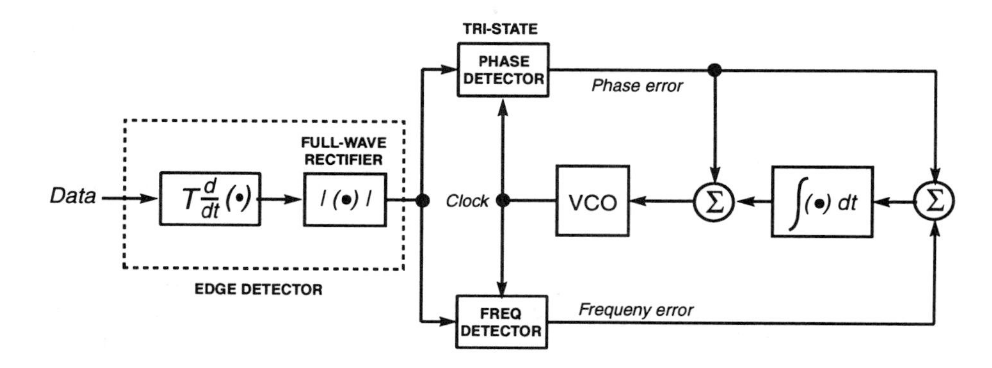

Figure 4.29 Block diagram of clock recovery circuit used by Cordell *et al.*

to aid in PLL frequency acquisition. The frequency detector was based on a circuit described earlier by Bellisio [28], which was a quantized version of a quadricorrelator first introduced in 1954 by Richman [29] in his classic paper on phase synchronization accuracy in color television. The quadricorrelator and other frequency detectors will be discussed in chapter 5.

The Circuit of Ransijn and O'Connor (AT&T 1991)

The circuit of Ransijn and O'Connor confirms that the technique of Cordell *et al.* can be used to implement multi-gigabit-per-second systems using modern technologies. Ransijn and O'Connor use AlGaAs heterojunction FETs to operate at data rates of 4-Gb/s with transistor f_ts of 26-GHz ($f_t/B_T = 6.5$). This represented the state-of-the-art in PLL based clock recovery circuits in 1991. And it demonstrated that monolithic PLL clock recovery circuit were approaching the speeds of 10-Gb/s hybrid circuits using dielectric resonator bandpass filters [26, 27]. A block diagram of the clock recovery and data retiming circuit is shown in Fig. 4.30. The data is first passed through a limiter. The edges of the data are detected using a delay and EXOR circuit. The phase and frequency of these edge pulses are detected using a quadricorrelator. The resulting clock phase depends on the half-bit delay of the edge-detection circuit as shown in Fig. 4.31. A tunable shorted strip-line is used to generate the delay. The optimum clock phase is determined by adjusting this delay. The delay is adjusted in both directions until the BER increases above a certain threshold. The final delay is then set in the center of this interval. Although this may, nominally, not be at the optimal sampling point in terms of maximizing the SNR, it does provide good immunity to parasitic effects. Since the decision circuit and phase detector are fabricated using similar circuits, their respective delays will track to a first order. Furthermore, as long

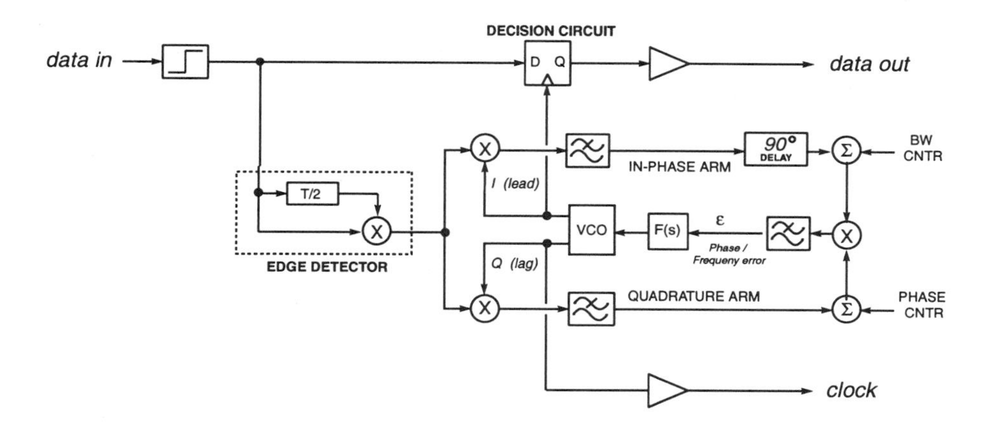

Figure 4.30 Block diagram of the clock recovery and data retiming circuit of Ransijn and O'Connor.

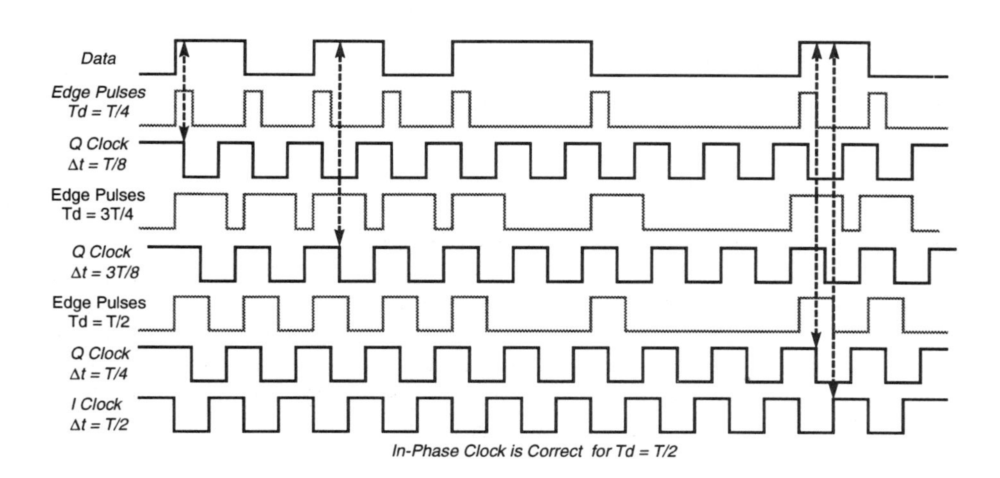

Figure 4.31 Timing diagram showing the dependence of final clock phase on the delay time t_d.

as t_d is stable, the clock phase will be relatively fixed at the proper sampling point over a broad range of operating conditions.

Ransijn and O'Connor give several helpful details concerning testing, and photographs of high-speed hybrid circuits required for system integration are given. They also share the belief with these authors, that the primary challenge of high-speed receiver design is in minimizing parasitic effects that can render an otherwise good design useless. This idea is probably best stated by Ransijn and O'Connor as follows:

> "Although parameters such as input ambiguity, clock (phase), and attainable bit rate are prime objectives, the real challenges in a circuit such as this, with its various types of signals, are in finding ways to route the high-speed signals and bypass the bias signals without introducing crosstalk interference that could easily result in reduced sensitivity, or worse, injection locking of the PLL. The physical layout of the chip as well as its environment are as important as the electrical design."

When operating at a bit rate of 2.5-Gb/s, the 3-dB closed loop bandwidth of the PLL is 1.2-MHz, which corresponds to $Q \simeq 1000$. The measured rms clock jitter was $2°$, which is approximately equal to the simple estimate derived in chapter 2 $(1/\sqrt{Q})180/\pi = 1.8°$. The reported frequency acquisition time is approximately 4-ms. Ransijn and O'Connor surmised that the fundamental limitation in the maximum bit rate is due to the decision circuit. We will now present methods for overcoming speed limitations in the decision circuit, by using bit interleaving.

Interleaving for Reduced Bandwidth Requirements

Direct implementations result in straightforward circuit design, but are rather wasteful of precious bandwidth. If we were to implement the circuit of Fig. 4.26 directly, it must pass the clock tone at a rate of B_T. To pass 80% of the clock power requires a circuit with a 3-dB bandwidth close to $2B_T$, which is more bandwidth than we may care to sacrifice. We must keep in mind that our goal is to cram as much data through transistors with limited speed as possible. For NRZ data, 80% of the signal power can be passed by a lowpass filter with a 3-dB bandwidth of $0.8B_T$. The frequency content of the data establishes a fundamental limitation on the speed of the circuitry required. Since the speed of the electronics is the bottleneck in system throughput, we don't want to impose a more restrictive limit, due to our own sloppy circuit design than is absolutely necessary. One might ask how we can reduce the bandwidth requirement when we need a clock at a rate of B_T? The answer is that we need a *clock* at a rate B_T, but we don't necessarily need a *signal* with a bandwidth of B_T. Fig. 4.32

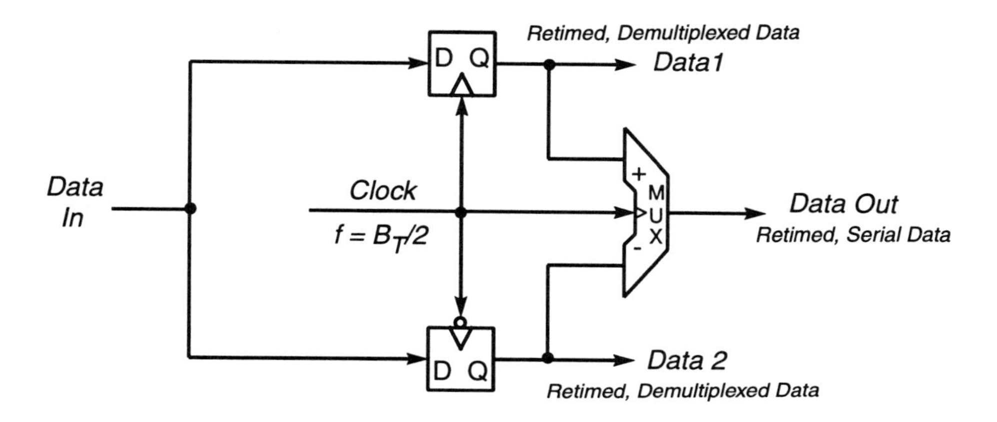

Figure 4.32 Block diagram of a clock recovery and decision circuit using two-level interleaving and a clock frequency of $B_T/2$.

illustrates how a signal with a fundamental frequency of $B_T/2$ can be used in a two-level interleaved system to provide clocking at a rate of B_T. Two identical decision circuits are used. One is triggered on the positive edge of the clock, and the other is triggered on the negative edge. The retimed data can be multiplexed back to the original data rate, or the decision circuit interleaving can function as the first level of demultiplexing of the data. The maximum required speed of the decision circuit is cut in half, as is the maximum clock rate.

Potential Problems with Interleaving One should always be suspicious of claims about increased throughput; in reality there will always be second-order effects to counteract the proposed gains. One potential problem is that the half-rate clock may not have a 50% duty-cycle. If this is the case, the sampling-instant will appear to have jitter, and this jitter will be pattern-dependent. Another limitation is the setup-time of the interleaved flip-flops. Looking at Fig. 4.32 we see that the flip-flops are clocked at half the data-rate, however, the input to each flip-flop is still the high-speed data. Such a flip-flop must be fast in order to *grab* the data as it goes by, because no matter how slowly the flip-flop is clocked, the setup time remains short (one bit interval). It is still an open questions as to how much speed improvement one gains in using a flip-flop as a decision circuit in an interleaved receiver. Ideally the gain in throughput from using bit interleaving will be somewhere N, where N is the number of stages of interleaving, but in practice that gain will be somewhere between 1 and N. We will discuss this matter in a slightly different context in section 4.6.3, and in chapter 5 we will present

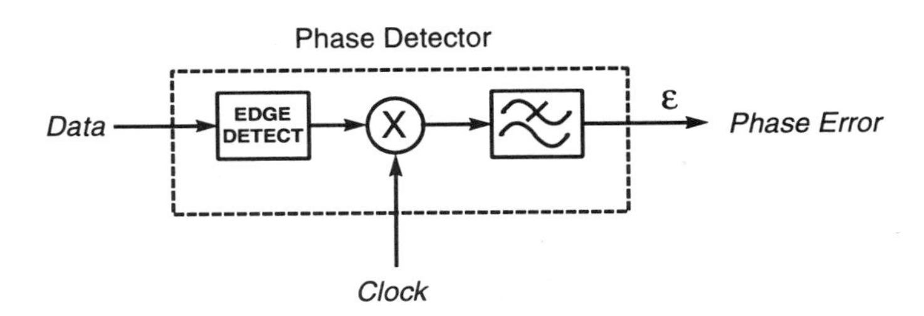

Figure 4.33 Conceptual diagram combining the function of edge-detection and phase-detection into one circuit.

an alternative approach to using a flip-flop as a decision circuit which may circumvent these problems.

Clock Recovery Circuit of Enam and Abidi (UCLA 1992)

Considering the block diagram of Fig. 4.26, we see that the essential control signal, required to adjust the VCO, is the phase-error. Since the input signal and the recovered clock will be very close in frequency, the phase-error signal will be a slowly varying baseband signal. If we can combine the functions of the edge-detector and the phase-detector, as illustrated conceptually in Fig. 4.33, into one circuit that produces a slowly varying phase-error output, without producing an intermediate signal at a frequency of B_T, then no internal circuits are required with a bandwidth of $2B_T$. This is the goal of an ideal bit-interleaved approach by insuring that no circuit nodes within the clock-recovery or decision circuits place limitations on the maximum obtainable data rate that the circuit can process.

A realization of a bit-interleaved approach was reported by Enam and Abidi [30, 31]. The circuit, as illustrated in Fig. 4.34, uses two-levels of interleaving. The VCO produces an in-phase and a quadrature clock at a frequency of half the bit-rate. The function of the circuit can be understood simply as a spectral-line PLL clock extractor. Edge-detection is performed by squaring the data, which has been pre-conditioned by a lowpass filter so that the data transitions are smeared across one bit-interval. A second multiplier acts as a frequency doubler by mixing the in-phase and quadrature signals from the VCO. Therefore the input to the third frequency detection multiplier is an edge-detected signal containing a spectral-line at B_T, and a clock signal at a frequency close to B_T. Multiplying these signals, and then lowpass filtering the product, produces the desired phase error. Thus far, nothing has been gained in terms of reduced bandwidth requirements, because the circuit, as described, still requires high-speed internal signals. However, the benefit of Enam and Abidi's implementation

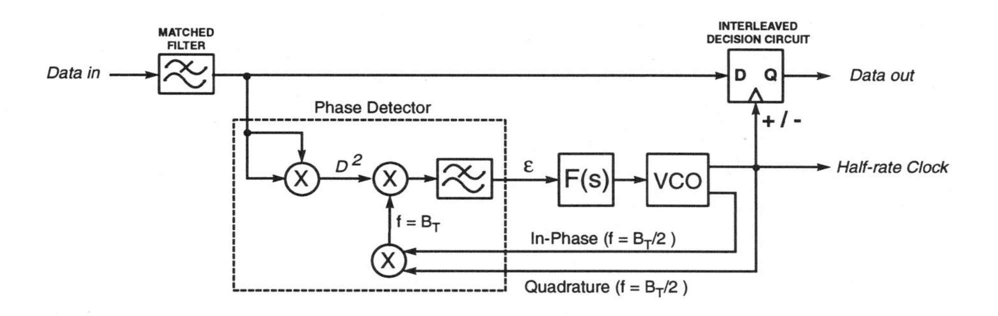

Figure 4.34 Block diagram of a clock recovery circuit using two-levels of interleaving and a clock at a rate of $B_T/2$

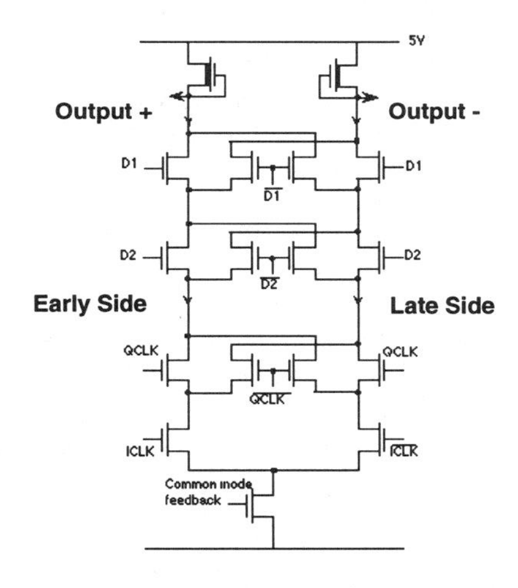

Figure 4.35 Enam's phase detector combining edge-detection, frequency doubling and phase-detection in one circuit that produces a low-frequency output.

is in the clever design of the phase-detector shown in dashed lines in Fig. 4.34. This circuit is a quadruple-stacked multiplier, as shown in Fig. 4.35.

Although we have developed an understanding of the operation of this circuit in the frequency domain, in terms of spectral-line techniques, we also could have derived the same function in the time domain. The quadruple-stacked phase detector can be thought of as an early-late clock recovery circuit. We will develop the early-late concept more thoroughly in section 4.5. The phase detector can be viewed as ideal switches that steer the tail current either, down the early-side, or down the late-side.

The difference between the quiescent current in each leg is integrated by the parasitic load capacitance at the output nodes. The bottom two levels of the phase detector switch the current under control of the clock signals only. The end result is that the tail current will be sourced down the early-side for the first half of each timing interval, and down the late-side for the last half. The top two levels of the phase detector redirect the current depending on the data polarity. The result of this current switching will now be described for the various cases of data transitions.

data high When the data stays high during a timing interval, the current will be sourced straight down the early-leg for the first half, and straight down the late-leg during the last half. The average current in each leg will be half the tail current, so there is no net current diverted to the parasitic integration capacitors, and the resulting phase error obtained by subtracting the early signal from the late signal will be zero.

data low The condition is similar when the data stays low. However, now the current will follow a zig-zag path to the negative power supply. But the final result will be that no phase-error signal is accumulated.

on-time transition The interesting case occurs when a data transition falls within the timing interval. Since the data has been pre-filtered, the transition is smeared across the bit interval. For this discussion we will assume that the transition is symmetric. If a transition occurs so that the zero crossing falls precisely in the middle of the timing interval, then during the first half of the interval, most of the current will flow down the early-side, with some residual current being steered to the late-side. During the next half cycle, the reverse will be true, and since the transition in the data is symmetric there will be no net difference between the early and late outputs.

early transition We can now see what will happen if the transition occurs early. Now, too much of the current that was supposed to flow on the early side gets passed to the late side. Therefore, the early output will be higher than the late output. If we subtract the late signal from the early signal we get a negative result that can be used to slow down the VCO.

late transition Clearly when the pulse is late in the timing interval, the late output will be greater than the early output and the positive difference will speed up the VCO.

In the following section we will derive the operation that a receiver must perform in order to produce a maximum a posteriori (MAP) estimate. We will see that the early-late technique implemented by the phase detector of Enam and Abidi is a limiting case of a MAP clock extractor.

4.5 MAXIMUM A POSTERIORI (MAP) SYMBOL SYNCHRONIZATION

In this section, the problem of estimating the random arrival time, or epoch, of a random data-bearing signal will be posed in mathematical terms. As has been the trend in this book, several of the intermediate steps of the derivation will be shown explicitly. These authors believe this approach makes the treatment more readable for the intended audience of circuit designers, who may not be as familiar with probability theory as the systems engineer. We will find that the basic operation required of the optimal receiver is to perform multiple correlations of the received data signal with stored replicas of the original data pulses, each replica pulse being shifted in time, to varying degrees, relative to the received data signal. This is similar to template matching, where the time offset corresponding to the template producing the highest correlation is declared the maximum a-posteriori (MAP) epoch estimate. The mathematical derivation to come, could well have been placed in an appendix. However, it has been included here for continuity. The reader wishing to skip the mathematical details and get straight to the results can proceed to section 4.5.2 on page 219.

4.5.1 Mathematical Derivation of MAP Clock Extractors

The analysis to follow is a summary of that given by Lindsey and Simon [32, ch. 9]. Similar analysis can be found for maximum-likelihood (ML) symbol synchronization in the book by Stiffler [33, ch. 7]. The reader is also referred to a discussion on minimum-mean-square estimation (MMSE) of arrival time, in the book by Lee and Messerschmitt [34, ch. 15]. To make the problem tractable, we must necessary impose certain conditions on the system. The primary assumptions made are the following:

- A clock exists at the receiver of exactly the same frequency as the bit-rate — only the phase of the clock is unknown.
- There exists an exact replica of the transmitted pulse at the receiver. This assumes either no distortion due to the channel, or that all distortion has been equalized.
- The interfering noise is assumed additive, white, and Gaussian (AWGN).
- The parameters of the observed data signal do not change during the observation interval (time invariance).

A Word or Two About Notation It is appropriate at this point to say a few words about notation. Often the same variable name is used to represent several different things. For example, in the literature, y can be: all possible outcomes of a random experiment, one possible outcome, or a dummy variable of integration. Such notation has been a great source of confusion to these authors. We will use the notation $y(\cdot)$ to represent the ensemble random variable. $y(\xi)$ is the result of the random experiment ξ. And y is a dummy variable that spans the space of all possible outcomes of the experiment. Therefore, a random signal may be represented as $\boldsymbol{y}(\cdot)$, and the pdf of the random variable is $p(\boldsymbol{y})$. But don't make the mistake of thinking that $\boldsymbol{y}(\cdot)$ and $\boldsymbol{y}$ are the same thing. $\boldsymbol{y}$ is just a dummy variable that we integrate over to find probabilities. We could give $\boldsymbol{y}$ any name, but we give it a name similar to $\boldsymbol{y}(\cdot)$ so we are reminded of which random variable we're dealing with.

Statement of the Problem

The problem can be stated as follows. The bit-interval is known to be T seconds long. However, the arrival time of the bits $t_\varepsilon(\cdot)$ is unknown, and can take on any value in the interval $[0, T]$. After having observed the signal over the specified time period, we want to evaluate the probability, at the receiver, that the actual phase error is equal to t_ε, given that we have observed one particular sample function of $\boldsymbol{y}(\cdot)$, namely $\boldsymbol{y}(\xi)$. We will choose for our timing estimate the value $\hat{t_\varepsilon}$ that maximizes this probability. Stated mathematically,

$$\hat{t_\varepsilon}; \qquad \max^{t_\varepsilon} \left[\Pr(t_\varepsilon | \boldsymbol{y})\Big|_{\boldsymbol{y}(\xi)} \right]. \tag{4.47}$$

Since $t_\varepsilon(\cdot)$ can take on a continuum of values, the probability of any particular value t_ε is zero. Therefore, instead of maximizing the probability, the optimal estimate is the value of t_ε that maximizes the a posteriori probability density function (pdf):

$$\hat{t_\varepsilon}; \qquad \max^{t_\varepsilon} \left[p(t_\varepsilon | \boldsymbol{y})\Big|_{\boldsymbol{y}(\xi)} \right] \tag{4.48}$$

Naturally, the parameters of the received signal will change over time, and the phase estimate will have to be periodically updated. We can restrict our attention to time intervals of length $[0, KT]$, over which the signal parameters are assumed to remain constant. We observe a data signal $y(t, \cdot)$ for $(K+1)$ bits. For every time interval we choose $\hat{t_\varepsilon}$, such that $p(t_\varepsilon | \boldsymbol{y})$ is maximized. We see already that this receiver requires storage of $(K+1)$ bits. Upon arrival of the last bit in this sequence, the receiver must go back in time to make decisions about the polarity of the previous data. This *analog* data storage is not practical; indeed, distortion-free storage is not even possible. In reality, the clock phase will be continuously adjusted, and the next bit will be clocked with a phase derived from the previous $(K+1)$ bits, thus eliminating the storage requirements. The rest of this section is devoted to finding an explicit expression

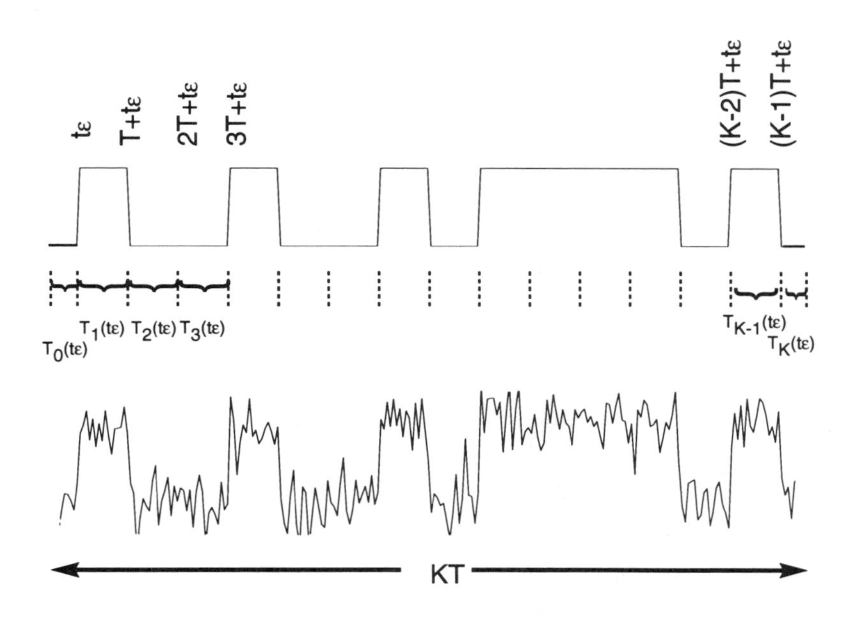

Figure 4.36 Sample data waveform with random phase $t_\varepsilon(\cdot)$ both with and without additive noise.

for $p(t_\varepsilon|\boldsymbol{y})$ as a function of t_ε, under a given set of assumptions. Maximizing this expression with respect to t_ε will reveal the mathematical operations that the receiver must perform to derive a MAP arrival-time estimate.

Towards an Explicit Expression for the a Posteriori PDF

We will restrict our attention here to rectangular signals. Over a time interval $[0, KT]$, the data can be represented as

$$y(t,\cdot) = \sum_{k=0}^{K} r_k(\cdot) p_T[t-(k-1)T-t_\varepsilon(\cdot)] + n(t,\cdot) \tag{4.49}$$

There are three independent random variables in the above expression. The first $r_k(\cdot)$ is due the data polarity, the second $t_\varepsilon(\cdot)$ is the data phase, and the third $n(t,\cdot)$ is the random additive noise, assumed to be zero-mean, white, and Gaussian. This data signal is shown in Fig. 4.36, both with, and without, additive noise. Referring to Fig. 4.36, we can make the following definitions. The time interval corresponding to the kth subinterval is given by

$$T_k(t_\varepsilon); \qquad t \in [(k-1)T + t_\varepsilon, kT + t_\varepsilon]. \tag{4.50}$$

Notice that this time interval definition is a function of the dummy variable t_ε that we will vary in our optimization procedure. But $T_k(t_\varepsilon)$ is independent of the true time offset $t_\varepsilon(\xi)$. Since the time is restricted to be within $[0, KT]$, we see that the 0th and the Kth subintervals are truncated to t_ε and $T - t_\varepsilon$ respectively. Although this condition will cause some of the simplifying assumptions made later to be violated, as long as K is sufficiently large, these edge effects will be insignificant.

Vector Representation of Signals We have already expressed the signal $y(t, \cdot)$ as a vector. Now we will justify this more rigorously. Without loss of generality, we can sample the data signal using M samples-per-bit. The number of sample can later be made to approach infinity. Since we are dealing with white noise, we have a problem, in that the variance of the noise sample is infinite, because the bandwidth is also infinite. Therefore, we need some method of limiting the bandwidth, and letting the bandwidth approach infinity together with the number of samples-per-bit. We will now describe two conceptual methods of bandwidth limitation.

Bandwidth Limitations for Sampled White Noise If we assume that we have non-ideal sampling, so that the sampler produces the average of the signal over the sampling interval $\Delta t = T/M$, then the resulting noise will be averaged, and the variance will be finite. As M gets large, the sampler becomes closer to an ideal impulse sampler. The sampling interval Δt is assumed so small that the signal doesn't changes in this interval. Therefore replacing the signal value with the average doesn't affect the result. The autocorrelation function for the white noise is given by

$$r_n(\tau) = \frac{N_0}{2}\delta(\tau) \tag{4.51}$$

and the variance of the average noise in the interval Δt is

$$\sigma_n^2 = E\left[\frac{1}{\Delta t}\int_0^{\Delta t}\frac{N_0}{2}\delta(\tau)d\tau\right]^2 = \frac{N_0}{2\Delta t} = \frac{N_0}{2}\frac{M}{T} \tag{4.52}$$

Each noise sample in all sampling windows are uncorrelated.

We could also consider a method whereby the receiver is preceded by an ideal lowpass filter of bandwidth $B = M/2T$. The frequency response of the ideal filter is given by

$$H(f) = \text{rect}(fM/T). \tag{4.53}$$

The variance of the noise passing through this filter is just

$$\sigma_n^2 = \frac{N_0}{2}\frac{M}{T}. \tag{4.54}$$

The autocorrelation function is given by

$$r_n(\tau) = \frac{M}{T}\text{sinc}\left(\frac{\tau}{T/M}\right). \tag{4.55}$$

The sinc function has nulls at multiples of $T/M = \Delta t$. Therefore noise samples separated in time by Δt are uncorrelated.

Whether we assume that the system is preceded by an ideal lowpass filter, or the samples arise from a non-impulse sampler, we will end up with the same result when M is allowed to grow arbitrarily large. In both cases the noise samples are uncorrelated, and the variance of each sample is finite and given by $\sigma_n^2 = N_0 M/2T$. We can now use vector representations for the signals and noise without worrying about problems when the noise variance becomes infinite. We can order the samples in row vectors. Therefore the received data signal can be expressed as

$$\begin{aligned} y(t,\cdot) &= \sum_{k=0}^{K} r_k(\cdot) p_T[t-(k-1)T - t_\varepsilon(\cdot)] + n(t,\cdot) \\ \boldsymbol{y}(\cdot) &= \sum_{k=0}^{K} [r_k(\cdot)\boldsymbol{p_T}[k, t_\varepsilon(\cdot)] + \boldsymbol{n}(\cdot)], \end{aligned} \tag{4.56}$$

where

$$p_T[t] = \begin{cases} V_0 & \text{for } t \in [0,T] \\ 0 & \text{elsewhere.} \end{cases} \tag{4.57}$$

We can define each shifted version of the original data pulse as

$$\begin{aligned} p_k[t_\varepsilon(\cdot)] &= p_T[t-(k-1)T - t_\varepsilon(\cdot)] \\ \boldsymbol{p_k}[t_\varepsilon(\cdot)] &= \boldsymbol{p_T}[k, t_\varepsilon(\cdot)]. \end{aligned} \tag{4.58}$$

Relationships Between Various Conditional Probabilities

We want to maximize the a posteriori pdf $p(t_\varepsilon|\boldsymbol{y})$. This task can be made simpler by expressing $p(t_\varepsilon|\boldsymbol{y})$ in terms of a priori pdfs, and likelihoods. From Bayes' rule [35, p. 30] the joint pdf can be expressed as

$$p(t_\varepsilon, \boldsymbol{y}) = p(t_\varepsilon|\boldsymbol{y})p(\boldsymbol{y}) = p(\boldsymbol{y}|t_\varepsilon)p(t_\varepsilon), \tag{4.59}$$

therefore

$$p(t_\varepsilon|\boldsymbol{y}) = \frac{p(t_\varepsilon)}{p(\boldsymbol{y})}p(\boldsymbol{y}|t_\varepsilon). \tag{4.60}$$

The conditional pdf $p(\boldsymbol{y}|t_\varepsilon)$ is the likelihood of observing $\boldsymbol{y}$ when in fact the timing error is t_ε. However, $\boldsymbol{y}$ is also a function of the random data $r_k(\cdot)$. Letting $\boldsymbol{r}(\cdot)$ be a row vector of the $(K+1)$ data values such that[4]

$$\boldsymbol{r}(\cdot) = [r_0(\cdot), r_1(\cdot), r_2(\cdot), \ldots, r_K(\cdot)], \tag{4.61}$$

then the desired likelihood can be obtained by averaging over all possible data sequences $\boldsymbol{r}$. We would like to represent $p(\boldsymbol{y}|t_\varepsilon)$ in terms of $p(\boldsymbol{y}|t_\varepsilon, \boldsymbol{r})$, for which we can find an explicit representation. This can be accomplished with further applications of Bayes' rule. We begin by writing the likelihood expression in terms of the joint pdf;

$$p(\boldsymbol{y}|t_\varepsilon) = \frac{p(\boldsymbol{y}, t_\varepsilon)}{p(t_\varepsilon)}. \tag{4.62}$$

Further, the joint pdf can be extracted by integrating $p(\boldsymbol{y}, t_\varepsilon, \boldsymbol{r})$ over all possible data sequences $\boldsymbol{r}$;

$$p(\boldsymbol{y}, t_\varepsilon) = \int_{\boldsymbol{r}} p(\boldsymbol{y}, t_\varepsilon, \boldsymbol{r}) d\boldsymbol{r}. \tag{4.63}$$

Therefore, the likelihood is

$$p(\boldsymbol{y}|t_\varepsilon) = \frac{p(\boldsymbol{y}, t_\varepsilon)}{p(t_\varepsilon)} = \int_{\boldsymbol{r}} \frac{p(\boldsymbol{y}, t_\varepsilon, \boldsymbol{r})}{p(t_\varepsilon)} d\boldsymbol{r}, \tag{4.64}$$

and the desired a posteriori pdf is given by

$$p(t_\varepsilon|\boldsymbol{y}) = \frac{p(t_\varepsilon)}{p(\boldsymbol{y})} \int_{\boldsymbol{r}} \frac{p(\boldsymbol{y}, t_\varepsilon, \boldsymbol{r})}{p(t_\varepsilon)} d\boldsymbol{r}. \tag{4.65}$$

Continuing, the combined joint pdf can be expressed in terms of the the double conditional density

$$p(\boldsymbol{y}, t_\varepsilon, \boldsymbol{r}) = p(\boldsymbol{y}|t_\varepsilon, \boldsymbol{r}) p(t_\varepsilon, \boldsymbol{r}). \tag{4.66}$$

Since the data is independent of the phase error, then

$$p(t_\varepsilon, \boldsymbol{r}) = p(t_\varepsilon) p(\boldsymbol{r}). \tag{4.67}$$

Finally, substituting (4.66) and (4.67) into (4.65), we get the first simplification of the a posteriori conditional pdf

$$\boxed{p(t_\varepsilon|\boldsymbol{y}) = \frac{p(t_\varepsilon)}{p(\boldsymbol{y})} \int_{\boldsymbol{r}} p(\boldsymbol{y}|t_\varepsilon, \boldsymbol{r}) p(\boldsymbol{r}) d\boldsymbol{r}.} \tag{4.68}$$

[4] Notice that this vector is different from the ones defined previously. The signal is represented by a KM dimensional vector obtained by taking M samples per bit. The vector $r(\cdot)$ is a $(K+1)$ dimensional vector that has only one value per bit.

Assumptions Concerning the Data and the Additive Noise Further simplifications can be made by making the reasonable assumption that all data symbols are uncorrelated. This implies

$$p(\boldsymbol{r}) = \prod_{k=0}^{K} p(r_k). \tag{4.69}$$

In addition, when each symbol is equally probable, the random variable r_k, which determines the data polarity, has a pdf that consists of two impulse functions of magnitude 1/2 at +1, and -1;

$$p(r_k) = \frac{1}{2}\delta(r_k \pm 1). \tag{4.70}$$

Since the initial noise is assumed to be white, each bandlimited noise sample is uncorrelated. With both the data, and additive noise being uncorrelated in different sampling windows, the received signal pdf can be separated into the product of pdfs over each of the $(K + 1)$ bit intervals.

$$p(\boldsymbol{y}|t_\varepsilon, \boldsymbol{r}) = \prod_{k=0}^{K} p(\boldsymbol{y}_k|t_\varepsilon, r_k). \tag{4.71}$$

This condition implies that any sample in one bit interval, provides no information about the data value, or the noise value in any other bit interval. Therefore, the integral in (4.68) can be expressed as the integral of the product of the pdfs for each bit interval;

$$\int_{\boldsymbol{r}} p(\boldsymbol{y}|t_\varepsilon, \boldsymbol{r})p(\boldsymbol{r})d\boldsymbol{r} = \int_{\boldsymbol{r}} \prod_{k=0}^{K} p(\boldsymbol{y}_k|t_\varepsilon, r_k)p(r_k)d\boldsymbol{r}. \tag{4.72}$$

This integral of products can be grouped as the product of integrals, so that

$$p(t_\varepsilon|\boldsymbol{y}) = \frac{p(t_\varepsilon)}{p(\boldsymbol{y})} \prod_{k=0}^{K} \int_{r_k} p(\boldsymbol{y}_k|t_\varepsilon, r_k)p(r_k)dr_k. \tag{4.73}$$

Substituting $p(r_k)$, as given in (4.70), into the above expression we obtain the second simplification of the a posteriori pdf.

$$p(t_\varepsilon|\boldsymbol{y}) = \frac{p(t_\varepsilon)}{p(\boldsymbol{y})} \prod_{k=0}^{K} \frac{1}{2}\left[p(\boldsymbol{y}_k|t_\varepsilon, 1) + p(\boldsymbol{y}_k|t_\varepsilon, -1)\right]. \tag{4.74}$$

Assumptions Concerning the Time Offset and Received Data To obtain the final form of the expression that we desire, we make the reasonable assumption that the random arrival time t_ε is uniformly distributed over the interval $[0, T]$. This implies that the the pdf $p(t_\varepsilon)$ is a constant, independent of t_ε. We also assume hat the statistics of signal $\boldsymbol{y}(\cdot)$, over the entire observable time interval $[0, KT]$, are independent of t_ε. This will be true for $K >> 1$, where the edge effects mentioned earlier become insignificant. Therefore, both $p(t_\varepsilon)$ and $p(\boldsymbol{y})$ are independent of t_ε, so that maximizing $p(t_\varepsilon|\boldsymbol{y})$ with respect to t_ε is equivalent to maximizing

$$\prod_{k=0}^{K} \left[p(\boldsymbol{y}_k|t_\varepsilon, 1) + p(\boldsymbol{y}_k|t_\varepsilon, -1)\right]. \tag{4.75}$$

We can see that maximizing the above expression is nothing more than maximizing the likelihood. Hence, for the assumptions made, the MAP estimate is equivalent to the ML estimate. We now have the condition that the MAP epoch estimate $\hat{t_\varepsilon}$ is given by

$$\boxed{\hat{t_\varepsilon}; \qquad \max^{t_\varepsilon} \prod_{k=0}^{K} \left[p(\boldsymbol{y}_k|t_\varepsilon, 1) + p(\boldsymbol{y}_k|t_\varepsilon, -1)\right] \Big|_{\boldsymbol{y}_k(\xi)}.} \tag{4.76}$$

MAP Estimator in Gaussian Noise

If we now assume that the noise is zero-mean and Gaussian, we can find an explicit expression for the pdfs given in (4.76). The data in each bit interval can be expressed as

$$\begin{aligned} \boldsymbol{y}_k(\cdot)|(t_\varepsilon, \ \ 1) &= \ \ \boldsymbol{p}_k[t_\varepsilon(\cdot)] + \boldsymbol{n}_k(\cdot) \\ \boldsymbol{y}_k(\cdot)|(t_\varepsilon, -1) &= -\boldsymbol{p}_k[t_\varepsilon(\cdot)] + \boldsymbol{n}_k(\cdot) \end{aligned} \tag{4.77}$$

From section 3.3.2 we know that the multi-dimensional Gaussian pdf can be written as

$$p(\boldsymbol{y}_k|t_\varepsilon, r_k) = \frac{1}{(2\pi)^{M/2}(\det \boldsymbol{R}_n)^{1/2}} \exp\left[-\frac{1}{2}\langle(\boldsymbol{y}_k - r_k\boldsymbol{p}_k(t_\varepsilon)), \boldsymbol{R}_n^{-1}(\boldsymbol{y}_k - r_k\boldsymbol{p}_k(t_\varepsilon))\rangle\right] \tag{4.78}$$

Since the noise is uncorrelated, the covariance matrix is diagonal;

$$\boldsymbol{R}_n = \sigma_n^2 \boldsymbol{I}, \qquad \boldsymbol{R}_n^{-1} = \frac{1}{\sigma_n^2}\boldsymbol{I}. \tag{4.79}$$

And due to the bandwidth limitations of (4.52) and (4.54) that we have conceptually imposed, the variance is equal to

$$\sigma_n^2 = \frac{N_0}{2}\frac{M}{T}. \tag{4.80}$$

The pdf can then be simplified to

$$p(\boldsymbol{y}_k|t_\varepsilon, r_k) = \frac{1}{(2\pi\sigma_n^2)^{M/2}} \exp\left[-\frac{1}{2\sigma_n^2}||\boldsymbol{y}_k - r_k\boldsymbol{p}_k(t_\varepsilon)||^2\right]. \tag{4.81}$$

The sum of the two conditional densities,

$$S(\boldsymbol{y}_k, t_\varepsilon) = p(\boldsymbol{y}_k|t_\varepsilon, 1) + p(\boldsymbol{y}_k|t_\varepsilon, -1), \tag{4.82}$$

is therefore given by

$$S(\boldsymbol{y}_k, t_\varepsilon) = \frac{1}{(2\pi\sigma_n^2)^{M/2}}\left[\exp\left[-\frac{1}{2\sigma_n^2}||\boldsymbol{y}_k - \boldsymbol{p}_k(t_\varepsilon)||^2\right] + \exp\left[-\frac{1}{2\sigma_n^2}||\boldsymbol{y}_k + \boldsymbol{p}_k(t_\varepsilon)||^2\right]\right] \tag{4.83}$$

Expanding the arguments of the exponentials we obtain

$$S(\boldsymbol{y}_k, t_\varepsilon) = \frac{C}{2}\left[\exp\left[\frac{1}{\sigma_n^2}\langle\boldsymbol{y}_k, \boldsymbol{p}_k(t_\varepsilon)\rangle\right] + \exp\left[-\frac{1}{\sigma_n^2}\langle\boldsymbol{y}_k, \boldsymbol{p}_k(t_\varepsilon)\rangle\right]\right], \tag{4.84}$$

where

$$\frac{C}{2} = \frac{1}{(2\pi\sigma_n^2)^{M/2}} \exp -\frac{1}{2\sigma_n^2}\left[||\boldsymbol{y}_k||^2 + ||\boldsymbol{p}_k(t_\varepsilon)||^2\right]. \tag{4.85}$$

We see that $||\boldsymbol{p}_k(t_\varepsilon)||^2 = P_0$ is the transmitted power in a bit interval T. Since we have defined the time intervals $T_k(t_\varepsilon)$ such that exactly one bit is contained in that interval, then t_ε only translates the received data and has no effect on the power. Indeed for rectangular data, it wouldn't matter if a transition had fallen in the middle of a bit interval; the average-power-per-bit would still be constant. $||\boldsymbol{y}_k||^2 = P_y$ is the received average power per bit interval. This is only a function of the dummy variable $\boldsymbol{y}_k$ and is independent of the the time offset t_ε. Further realizing that

$$\cosh(x) = \frac{e^x + e^{-x}}{2}, \tag{4.86}$$

then the sum $S(\boldsymbol{y}_k, t_\varepsilon)$ in (4.84) is given by

$$C \cosh \frac{1}{\sigma_n^2}\langle\boldsymbol{y}_k, \boldsymbol{p}_k(t_\varepsilon)\rangle \tag{4.87}$$

The arrival time estimate that maximizes the a posteriori probability is therefore

$$\hat{t_\varepsilon}; \qquad \max^{t_\varepsilon} \prod_{k=0}^{K} \cosh \frac{1}{\sigma_n^2}\langle\boldsymbol{y}_k, \boldsymbol{p}_k(t_\varepsilon)\rangle\Big|_{\boldsymbol{y}_k(\xi)}. \tag{4.88}$$

The hyperbolic cosine function is always positive. Since the natural logarithm is a monotonic function for positive values, then maximizing η over positive values is equivalent to maximizing $\ln(\eta)$. By taking the natural logarithm we obtain the desired function that the receiver must maximize in order to estimate the arrival time;

$$\hat{t_\varepsilon}; \qquad \max^{t_\varepsilon} \sum_{k=0}^{K} \ln \left[\cosh \langle \frac{\boldsymbol{y}_k, \boldsymbol{p}_k(t_\varepsilon)}{\sigma_n^2} \rangle \right] \Big|_{\boldsymbol{y}_k(\xi)}. \tag{4.89}$$

In the following section we will see how this result can be extended to the continuous time case when M approaches infinity.

Extension to Continuous time

The dot-product operation in (4.89) can be written as

$$\frac{1}{\sigma_n^2} \langle \boldsymbol{y}_k(\xi), \boldsymbol{p}_k(t_\varepsilon) \rangle = \sum_{m=1}^{M} \frac{y_{km}(\xi) p_{km}(t_\varepsilon)}{N_0/2} \frac{T}{M}. \tag{4.90}$$

In the limit as $M \to \infty$ the dot-product becomes an integral;

$$I_k(y, t_\varepsilon) = \frac{1}{N_0/2} \int_{T_k(t_\varepsilon)} y(t, \xi) p_k(t_\varepsilon) dt. \tag{4.91}$$

In the case of rectangular pulses the integral simplifies to

$$I_k(y, t_\varepsilon) = \frac{V_0}{N_0/2} \int_{(k-1)T+t_\varepsilon}^{kT+t_\varepsilon} y(t, \xi) dt. \tag{4.92}$$

We can define a normalized received signal $z(t, \cdot)$ such that

$$z(t, \cdot) = \frac{y(t, \cdot)}{V_0}. \tag{4.93}$$

The correlation integral can then be written in terms of the normalized signal

$$I_k(z, t_\varepsilon) = \frac{V_0^2 T}{N_0/2} \left[\frac{1}{T} \int_{(k-1)T+t_\varepsilon}^{kT+t_\varepsilon} z(t, \xi) dt \right]. \tag{4.94}$$

We recognize $V_0^2 T$ as the energy per bit E_0. Recalling that the signal-to-noise ratio from chapter 3 is given by

$$\text{SNR} = \frac{E_0}{N_0/2}, \tag{4.95}$$

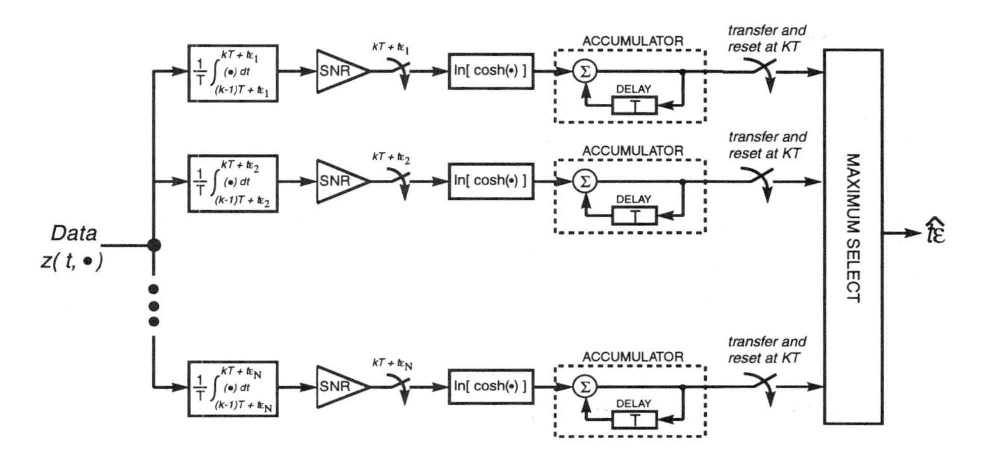

Figure 4.37 Open-loop implementation of a MAP timing estimate circuit.

we arrive at the expression that must be maximized to obtain a MAP timing estimate,

$$\hat{t_\varepsilon}; \qquad \max_{t_\varepsilon} \sum_{k=0}^{K} \ln \left[\cosh \left(\text{SNR} \left[\frac{1}{T} \int_{(k-1)T+t_\varepsilon}^{kT+t_\varepsilon} z(t,\xi)dt \right] \right) \right] \tag{4.96}$$

In the following sections, block diagrams of receivers will be presented that implement the search algorithm given in (4.96).

4.5.2 Open Loop Correlator

An open-loop system that approximates the operation outlined by (4.96) is illustrated in Fig. 4.37. In this implementation, only a discrete number of correlators are used. The number of correlations needed depends on the desired accuracy of the estimate. This circuit is not practical because it requires perfect frequency synchronization at the receiver, several parallel correlations, and a multi-phase clock. Nevertheless, it is instructive to consider the operation of the open-loop estimator.

In the absence of noise, and at the optimal sampling phase, the integral

$$\frac{1}{T} \int_{(k-1)T+t_\varepsilon}^{kT+t_\varepsilon} z(t,\xi)dt, \tag{4.97}$$

will be equal to unity. When a transition occurs in the bit-interval the integral will be linearly proportional to the timing error, dropping to a value of zero for an error

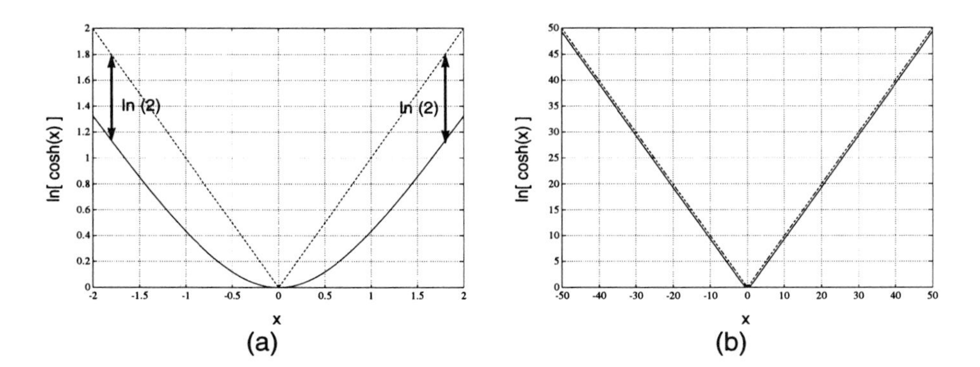

Figure 4.38 Plots of the weighting function $\ln(\cosh(x))$ for: (a) small values of x, (b) large values of (x).

of $|T/2|$. In the presence of additive noise however, this integral will fluctuate. We multiply the integral by the SNR and take a weighted average over K bits to get an indication of the degree to which our timing estimate is accurate. The weighting function $\ln[\cosh(x)]$ is plotted in Fig. 4.38. For those readers familiar with bipolar transitor circuits, the function $\ln[\cosh(x)]$ is perhaps better understood by looking at a circuit which produces it. The circuit of Fig. 4.39 shows two identical differential pairs biased with the same current. One is driven with a differential input voltage and the other provides a reference voltage. The difference between the voltages at the coupled emitters of the two pairs is given by

$$\frac{\Delta V}{V_T} = \ln\left[\cosh\left(\frac{V_d}{2V_T}\right)\right]. \tag{4.98}$$

From Fig. 4.38(a) we can see that for low SNR the weighting function is approximately equivalent to squaring the signal, while at high SNR, the weighting function amounts to rectification. Therefore at high SNRs all correlations, large and small, are given approximately equal weighting. At low SNR however, the signal is close to the noise floor, and a large part of the correlation output is due to noise. Therefore, a large correlation event is biased much more heavily than a small one. For example, a single correlation value of 10 is given much more weight than 10 separate correlations of unit value. However, the nonlinear weighting is only relevant for SNR less than unity. We saw in chapter 3 that in order to achieve an error probability less than 10^{-9}, an SNR greater than 36 is required. For broadband fiber-optic communications systems the SNR will be large, so that the weighting function can, for all practical purposes be replaced by a rectifier. For our circuit analogy this means $V_d >> V_T$.

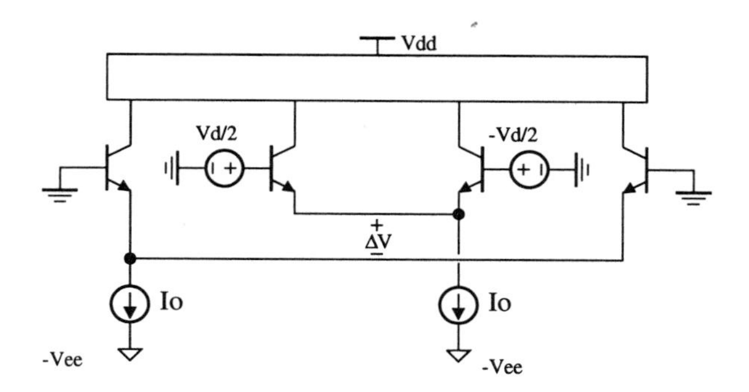

Figure 4.39 Circuit used to produce the logarithm of a hyperbolic cosine

Although it has been stated that this open-loop system is not practical it can be used when the desired accuracy of the timing estimate is not crucial. In this case an intermittent phase readjusting approach could be used as outlined in section 4.2. Only a small number of discrete clock phases need be used, and slight frequency errors in the clock at the receiver can be tolerated. In the following sections we will show practical closed-loop clock recovery circuits based on the MAP estimate.

4.5.3 Closed-Loop Stochastic Gradient Based Clock Extractors

The correlation function $\Lambda[z(t,\xi), t_\varepsilon]$ given in (4.96) is obtained by taking a weighted average of the following correlations performed on each bit

$$I_k[z(t,\xi), t_\varepsilon] = \frac{\text{SNR}}{T} \int_{T_k(t_\varepsilon)} z(t,\xi)\overline{p_k}(t_\varepsilon)dt. \tag{4.99}$$

The pulse $\overline{p_k}(t_\varepsilon)$ is normalized by the average energy such that

$$\overline{p_k}(t_\varepsilon) = \frac{p_k[t_\varepsilon]}{V_0} = \frac{p_T[t-(k-1)T-t_\varepsilon]}{V_0} = \overline{p_T}[t-(k-1)T-t_\varepsilon], \tag{4.100}$$

where V_O is defined for a general pulse as

$$V_0 \triangleq \left[\frac{1}{T}\int_0^T p_T^2(t)dt\right]^{1/2} \tag{4.101}$$

The optimal open-loop correlator sums all integral outputs I_k for $(K+1)$ bits through the weighting function $[\ln\cosh(I_k)]$, which is an even function and thus removes random phase reversals due to the data polarity. We can plot the output of the correlator

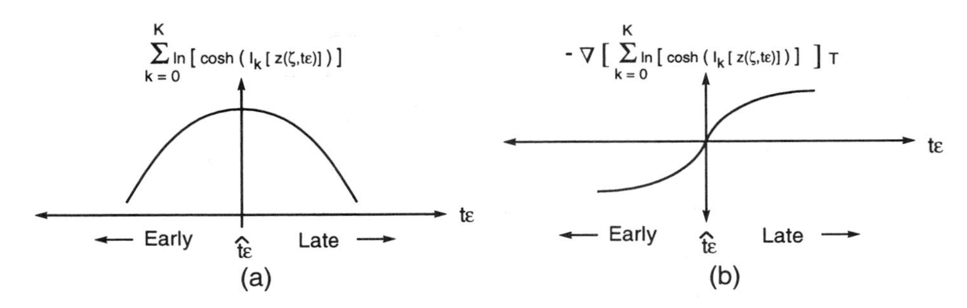

Figure 4.40 (a) Correlation output as a function of t_ε. (b) Negative gradient of correlation function.

as a function of the phase estimate t_ε as shown in Fig. 4.40(a). The optimal phase estimate $\hat{t_\varepsilon}$ is shown at the point where $\Lambda[z(t,\xi), t_\varepsilon]$ is a maximum. The negative gradient of the correlation function is shown in Fig. 4.40(b). We can see that the gradient goes to zero at the optimal estimate, is negative for an early clock, and is positive for a late clock. Instead of building a receiver to find the correlation function, we can design one that finds the gradient of the correlation function, and use the gradient in a closed loop system to synchronize the clock to the optimal phase. A closed-loop system is desirable because the clock at the transmitter and the channel characteristics drift over time, and need to be continuously tracked. To reveal the operations that such a receiver must perform, we will find an explicit expression for the gradient by differentiating $\Lambda[z(t,\xi), t_\varepsilon]$ with respect to t_ε.

Explicit Expression for the Gradient

The correlation function is given by

$$\Lambda[z(t,\xi), t_\varepsilon] = \sum_{k=0}^{K} \ln\left[\cosh\left(I_k[z(t,\xi), t_\varepsilon]\right)\right]. \tag{4.102}$$

The negative gradient by the chain rule is then

$$-\frac{\partial \Lambda[z(t,\xi),t_\varepsilon]T}{\partial t_\varepsilon} = -T\sum_{k=0}^{K}\frac{1}{\cosh(I_k)}\frac{\partial \cosh(I_k)}{\partial t_\varepsilon}$$
$$= -T\sum_{k=0}^{K}\frac{\sinh(I_k)}{\cosh(I_k)}\frac{\partial I_k}{\partial t_\varepsilon} \tag{4.103}$$
$$= -T\sum_{k=0}^{K}\tanh(I_k)\frac{\partial I_k}{\partial t_\varepsilon}.$$

We now have the problem of evaluating

$$\frac{\partial I_k[z(t,\xi),t_\varepsilon]}{\partial t_\varepsilon} = \frac{\partial}{\partial t_\varepsilon}\left[\frac{\text{SNR}}{T}\int_{(k-1)T+t_\varepsilon}^{kT+t_\varepsilon} z(t,\xi)\overline{p_{\overline{k}}}(t_\varepsilon)dt\right], \tag{4.104}$$

where the variable t_ε appears both in the integrand, and in the limits. Derivatives of this type can be evaluated using an extension of *Leibniz's rule* [36, p. 360], stating that if

$$F(t_\varepsilon) = \int_{\alpha(t_\varepsilon)}^{\beta(t_\varepsilon)} f[t,t_\varepsilon]dt, \tag{4.105a}$$

then

$$\frac{\partial F(t_\varepsilon)}{\partial t_\varepsilon} = \int_{\alpha(t_\varepsilon)}^{\beta(t_\varepsilon)}\frac{\partial f}{\partial t_\varepsilon}dt + \frac{\partial \beta(t_\varepsilon)}{\partial t_\varepsilon}f[\beta(t_\varepsilon),t_\varepsilon] - \frac{\partial \alpha(t_\varepsilon)}{\partial t_\varepsilon}f[\alpha(t_\varepsilon),t_\varepsilon]. \tag{4.105b}$$

Therefore, the derivative in (4.104) is given by

$$\frac{\partial I_k[z(t,\xi),t_\varepsilon]}{\partial t_\varepsilon} = \frac{\text{SNR}}{T}\int_{(k-1)T+t_\varepsilon}^{kT+t_\varepsilon} z(t,\xi)\frac{\partial}{\partial t_\varepsilon}\overline{p_T}[t-(k-1)T-t_\varepsilon]dt +$$
$$\frac{\text{SNR}}{T}\left[z[kT+t_\varepsilon,\xi]\overline{p_T}(T) - z[(k-1)T+t_\varepsilon,\xi]\overline{p_T}(0)\right]. \tag{4.106}$$

Since we can build a circuit to process the data signals in real time, we would like to relate the derivative with respect to t_ε to derivatives with respect to time. The pulse $\overline{p_T}[t-(k-1)T-t_\varepsilon]$ is plotted in Fig. 4.41(a) as a function of time for a fixed value of t_ε, and conversely in Fig. 4.41(b). We see that

$$-\frac{\partial}{\partial t_\varepsilon}\overline{p_T}[t-(k-1)T-t_\varepsilon] = \frac{\partial}{\partial t}\overline{p_T}[t-(k-1)T-t_\varepsilon], \tag{4.107}$$

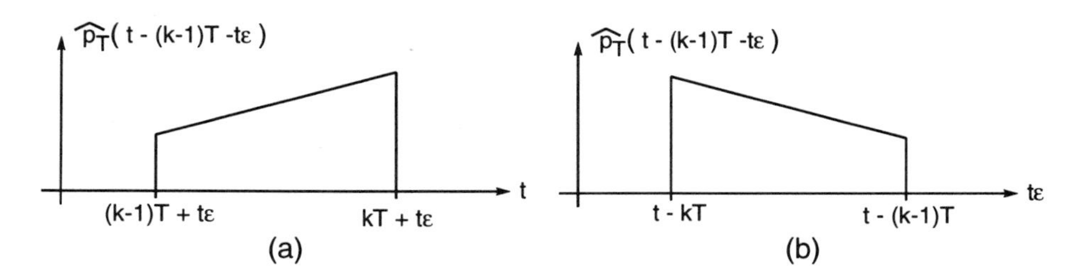

Figure 4.41 Plots of the normalized data pulse: (a) as a function of t for a fixed offset t_ε, (b) as a function of t_ε for a fixed time t.

which is obvious by a substitution of variables. However, illustrating the derivative of the pulse shapes aids in understanding the resulting clock recovery circuit on an intuitive level. For the case of rectangular pulses, $\overline{p_T}$ equals unity in the interval $[0, T]$. Therefore,

$$\begin{aligned} -\frac{\partial}{\partial t_\varepsilon}\overline{p_T}[t-(k-1)T-t_\varepsilon] &= -\delta[t_\varepsilon-(t-kT)]+\delta[t_\varepsilon-(t-(k-1)T)] \\ &= \delta\left[t-((k-1)\,T+t_\varepsilon)\right]-\delta\left[t-(kT+t_\varepsilon)\right]. \end{aligned} \tag{4.108}$$

For a half-cosine pulse of the form

$$\overline{p_T}[t] = \begin{cases} \sqrt{2}\sin\left(\frac{\pi t}{T}\right) & \text{for } t \in [0,T] \\ 0 & \text{elsewhere,} \end{cases} \tag{4.109}$$

The derivative is given by

$$-\frac{\partial}{\partial t_\varepsilon}\overline{p_T}[t-(k-1)T-t_\varepsilon] = \frac{\pi\sqrt{2}}{T}\cos\left[\frac{\pi}{T}\left(t-(k-1)T-t_\varepsilon\right)\right]. \tag{4.110}$$

These derivatives are shown in Fig. 4.42. We can see that the derivative of the pulse makes a transition from positive to negative over the bit interval, with a zero crossing at the center of the bit. The steeper the transition in the data, the more the energy in the derivative signal will be concentrated at the edges. Substituting into (4.103) we can now write the negative gradient of $\Lambda[z(t,\xi), t_\varepsilon]$ in terms of operations with respect to

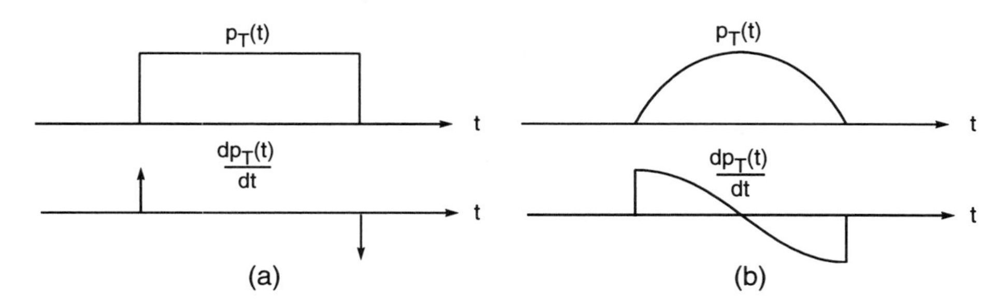

Figure 4.42 Negative gradients of pulse shapes as a function of the offset time t_ε for: (a) a rectangular pulse, (b) a half-cosine pulse.

time. The result is given by

$$\begin{aligned}
&-\frac{\partial\Lambda[z(t,\xi),t_\varepsilon]T}{\partial t_\varepsilon} = \\
&\sum_{k=0}^{K}\tanh\left[\frac{\text{SNR}}{T}\int_{(k-1)T+t_\varepsilon}^{kT+t_\varepsilon} z(t,\xi)\overline{p_T}[t-(k-1)T-t_\varepsilon]dt\right]\times \\
&\left\{\frac{\text{SNR}}{T}\int_{(k-1)T+t_\varepsilon}^{kT+t_\varepsilon} z(t,\xi)\left[T\frac{\partial}{\partial t}\overline{p_T}[t-(k-1)T-t_\varepsilon]\right]dt+\right. \\
&\left.\text{SNR}\quad [z[(k-1)T+t_\varepsilon,\xi]\overline{p_T}(0)-z[kT+t_\varepsilon,\xi]\overline{p_T}(T)]\right\}.
\end{aligned} \tag{4.111}$$

Direct Implementation of Gradient Based Clock Recovery

The signal processing required to produce the gradient given in (4.111) can be understood much more easily in block diagram form, as shown in Fig. 4.43, for the special case of a signal pulse that is equal to zero at the end points 0, and T [37, p. 233] [32, p. 431]. The hyperbolic tangent function can be implemented easily using bipolar transistors as a simple emitter-coupled pair. For large SNR the tanh function approaches a hard-limiter with the transfer characteristic of a signum function. The accumulation of correlation values over $(K+1)$ bits has been replaced by a filter with a transfer function $F(s)$. Since the clock extraction circuit is a negative-feedback system, $F(s)$ must be designed appropriately for loop-stability and the desired dynamic behavior. Effective accumulation of the phase-errors over the $(K+1)$ bit sequence is accomplished by a convolution with the impulse response of the loop filter. Since the statistics of the signal are constantly changing, we prefer to weight recent bit correla-

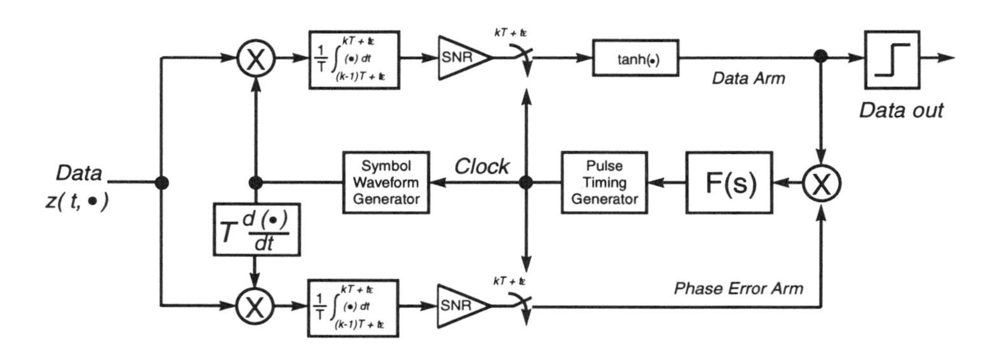

Figure 4.43 Block diagram of a direct implementation of MAP gradient-based closed-loop clock extractor.

tions more heavily than those from the past. Therefore, the impulse response should have a decaying envelope and effectively go to zero beyond $(K + 1)$ bits. Traditional filters have this type of impulse function, and we favor them over a moving average accumulator.

Phase-Detector Characteristic The timing recovery circuit of Fig. 4.43 is reminiscent of a Costas Loop used for carrier recovery in binary phase-shift keyed (BPSK) systems. The operation of the circuit is understood simply if we assume the clock is nearly synchronized with the data. In this case the data-arm signal is the retimed data, and the phase-error-arm produces an estimate of the timing offset. However, the polarity of the phase-error varies randomly with the data. By multiplying the phase-error with the retimed data, the random polarity ambiguity is removed, and an error signal is produced with an average value proportional to the time offset. This error signal is negative for an early clock, and positive for a late clock. We are interested in determining the error function produced at the input of the loop filter $F(s)$ as a function of the timing offset t_ε. The actual phase-error function will be random. Its magnitude will depend on the number of data transitions, and it will have ripple components that will depend on the data pattern. However, we can determine the basic operation of the phase detector be considering maximum data density (alternating *ones* and *zeros*, or a periodic input with a frequency of half the data rate). The inherent nonlinearities of the circuit will warp the phase-error estimate, producing a nonlinear function of the time offset in general. This warping will depend on the data-pulse shape, and will now be illustrated for some special cases.

Half-Cosine Pulses Correlations for an early, on-time, and late clock are shown in Fig. 4.44. For an early clock, the data and the derivative are out of phase, and the correlation is negative — slowing down the clock. When the clock is on time, the

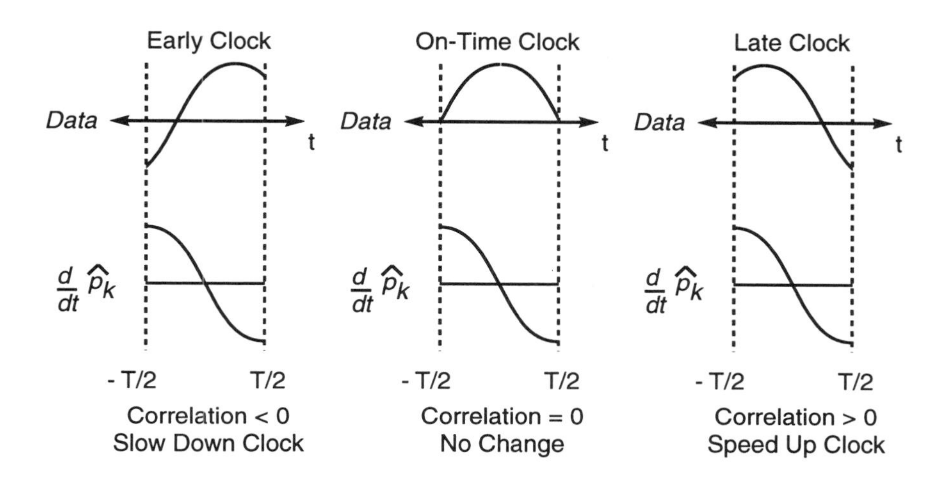

Figure 4.44 Illustration of correlation of a half-cosine pulse with its derivative for and early, on-time, and late clock.

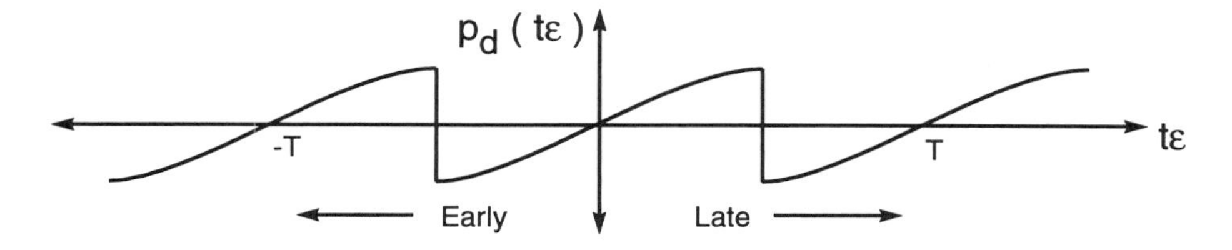

Figure 4.45 Phase error as a function of timing offset for half-cosine data pulses and maximum data density.

data and the derivative are in quadrature, producing an error of zero, so that the clock phase stays fixed. For the case of a late clock, the data and derivative are in-phase. The phase-error produced is positive, and the clock frequency will be increased to compensate for the error.

We have shown results for a positive data pulse. The results will be the same for a negative pulse, because the polarity of the error is determined by the product of the correlations with the sign of the retimed data value. The resulting error signal is plotted in Fig. 4.45 as a function of the actual time offset. We see that the error function is a switched sinusoidal, with stable equilibrium points at multiples of T, and is monotonic over the bit interval $[-T/2, T/2]$.

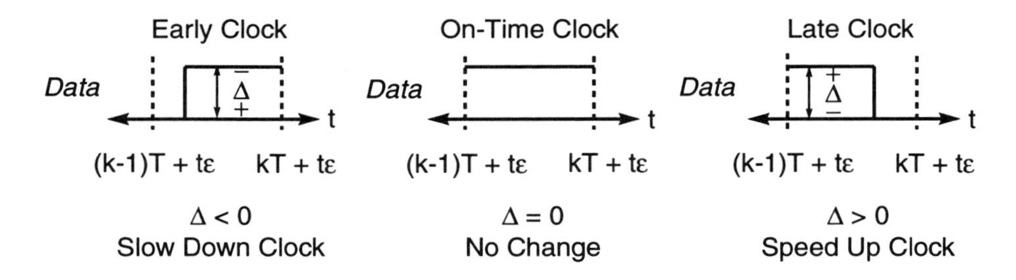

Figure 4.46 Illustration of phase error for rectangular pulses obtained by taking the difference of the value at the start of the timing interval and the data value at the end of the timing interval.

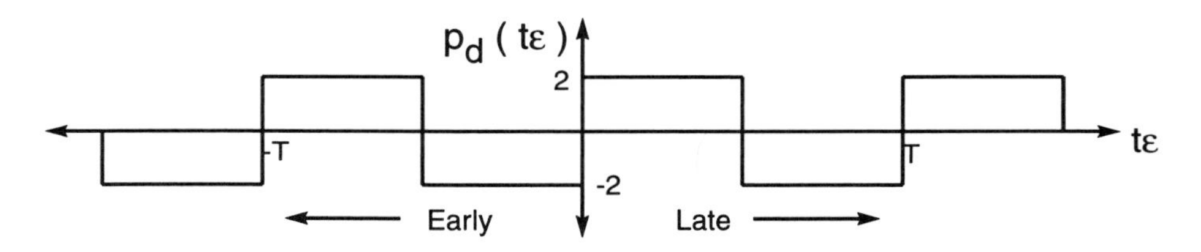

Figure 4.47 Phase error as a function of timing offset for rectangular data pulses and maximum data density.

Rectangular Pulses For the case of rectangular pulses, the data is non-zero at the end points. Substituting the impulse functions for the derivative into (4.111) we arrive at the result

$$-\frac{\partial \Lambda[z(t,\xi),t_\varepsilon]T}{\partial t_\varepsilon} = \sum_{k=0}^{K} \tanh\left[\frac{\text{SNR}}{T}\int_{(k-1)T+t_\varepsilon}^{kT+t_\varepsilon} z(t,\xi)dt\right] \times \\ 2\text{SNR}\big[z[(k-1)T+t_\varepsilon,\xi] - z[kT+t_\varepsilon,\xi]\big]. \tag{4.112}$$

We see that the phase error is obtained by taking the difference of the data at the two end points and multiplying by the retimed data. The difference at the end points is illustrated in Fig. 4.46 for timing errors of magnitude less than $T/2$. It can be seen that the difference is negative for $t_\varepsilon \in [-T, 0]$ and positive for $t_\varepsilon \in [0, T]$. However, the data changes sign at the points $\pm T/2$, and we end up with a square wave phase-error function as shown in Fig. 4.47. This phase error function is undesirable from a stability standpoint because of the steep transition through the equilibrium point. However, this square-wave characteristic is equivalent to quantizing the phase-error to one-bit. Since the closed-loop bandwidth is much less than the clock rate, these one-bit errors will average out to produce a stable equilibrium.

Altering the Pulse Derivative to Enhance the Phase Error Function We can see from the previous examples that the monotonic range of the phase detector is determined by the time over which the energy is spread in the derivative pulse. For half-cosine pulses the derivative had energy over the entire bit interval, and the phase-error was also monotonic over the interval. Often we will be dealing with rectangular pulses, and we would like to increase the monotonic range of the phase error function. This improves both phase tracking and frequency acquisition properties of PLL-based clock recovery circuits. A straightforward method of increasing the monotonic range is to replace the derivative with a finite difference over a time Δt. This will have the effect of spreading the energy in the impulses over a larger portion of the bit interval. Clock recovery circuits based on this approximation of the derivative are discussed in the following section.

4.5.4 Early-Late Clock Recovery Circuits

We can approximate the time derivative of the data pulse by a difference. If we let t be in the center of an interval Δt, then the derivative is approximately given by

$$\frac{\partial}{\partial t}\overline{p_T}(t) \simeq \frac{\overline{p_T}[t+\frac{\Delta t}{2}] - \overline{p_T}[t-\frac{\Delta t}{2}]}{\Delta t}, \tag{4.113a}$$

or

$$\frac{\partial}{\partial t}\overline{p_T}[t-(k-1)T-t_\varepsilon]\Delta t \simeq \underbrace{\overline{p_T}[t-(k-1)T-(t_\varepsilon-\tfrac{\Delta t}{2})]}_{\text{early}} - \underbrace{\overline{p_T}[t-(k-1)T-(t_\varepsilon+\tfrac{\Delta t}{2})]}_{\text{late}} \tag{4.113b}$$

Approximations of this derivative for various values of Δt are illustrated in Fig. 4.48. We also recall that for rectangular pulses, the term

$$z[(k-1)T+t_\varepsilon,\xi]\overline{p_T}(T) - z[kT+t_\varepsilon,\xi]\overline{p_T}(0) = z[(k-1)T+t_\varepsilon,\xi] - z[kT+t_\varepsilon,\xi] \tag{4.114}$$

produces a square-wave phase-error response, as we saw in Fig. 4.47. Since we don't want a square-wave output from the detector, we can ignore this term in favor of a more gradual transition through the equilibrium point. After substituting (4.113b) into (4.111), and ignoring the last term, we arrive at the following approximation for the

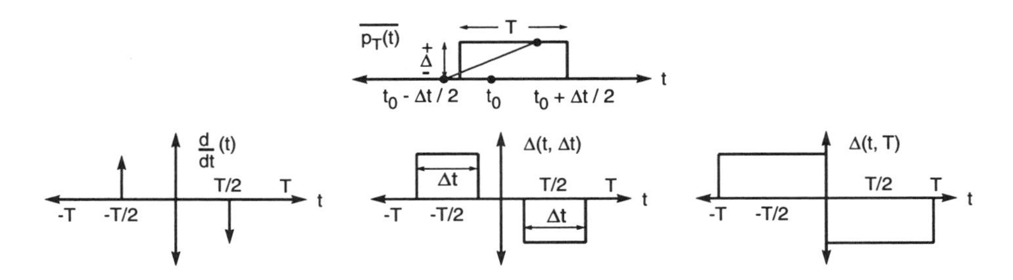

Figure 4.48 Illustration of the approximations to the derivative of a rectangular pulse for various values of Δt.

gradient

$$-\frac{\partial\Lambda[z(t,\xi),t_\varepsilon]T}{\partial t_\varepsilon} \simeq \sum_{k=0}^{K} \tanh \underbrace{\left[\frac{\text{SNR}}{T}\int_{(k-1)T+t_\varepsilon}^{kT+t_\varepsilon} z(t,\xi)dt\right]}_{\text{on-time}} \times$$

$$\frac{T}{\Delta t}\left[\underbrace{\frac{\text{SNR}}{T}\int_{(k-1)T+t_\varepsilon-\Delta t/2}^{kT+t_\varepsilon-\Delta t/2} z(t,\xi)dt}_{\text{early}} - \underbrace{\frac{\text{SNR}}{T}\int_{(k-1)T+t_\varepsilon+\Delta t/2}^{kT+t_\varepsilon+\Delta t/2} z(t,\xi)dt}_{\text{late}}\right] \tag{4.115}$$

A circuit that implements this function is shown in Fig. 4.49, where a hard limiter replaces the tanh function for large SNR. One of the problems with processing signals skewed in time is that they must be "deskewed" before being operated upon. In Fig. 4.49 we show a delay of Δt in the early-arm, and a delay of $\Delta t/2$ in the on-time-arm. This ensures that the signals x, y and d_2 arrive synchronously. Other methods of deskewing, such as sample-and-holding the signals, or using shift-registers in a quantized realization, are also viable techniques.

Self-Adjusting Property of Early-Late Circuits The early-late clock recovery circuit of Fig. 4.49 is an example of a self-adjusting circuit. The self-synchronizing property arises due to the fact that identical circuits are used for both the phase-detector and the decision circuit. Therefore, any parasitic delay in the decision circuit will be accounted for by the phase-detector, and the clock phase will be automatically compensated. More will be said about self-adjusting circuits in section 4.6.

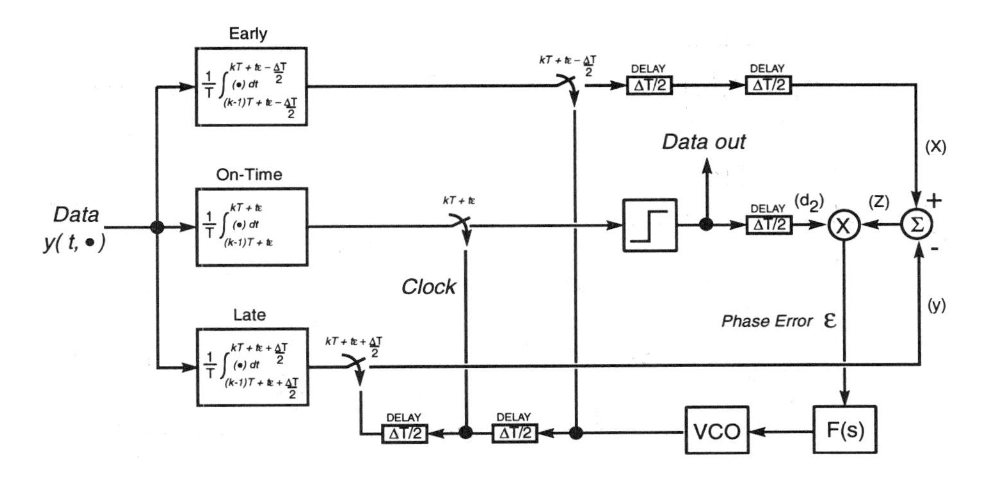

Figure 4.49 Block Diagram of an early-late gate clock recovery circuit.

Early-Late Circuit Using Rectifiers

We can arrive at a slightly different early-late gate structure if we first approximate the $\ln[\cosh(I_k)]$ nonlinearity for large SNR before we differentiate. Recalling that the correlation function is given by

$$\Lambda[z(t, \xi), t_\varepsilon] = \sum_{k=0}^{K} \ln[\cosh(I_k)], \tag{4.116}$$

we realize that for large SNR, the function is approximately equal to the absolute value of the argument;

$$\ln[\cosh(I_k)] = \ln\left[\frac{e^{I_k} + e^{-I_k}}{2}\right] \simeq \ln\left[\frac{e^{|I_k|}}{2}\right] = |I_k| - \ln(2) \simeq |I_k|. \tag{4.117}$$

This property can be seen easily from Fig. 4.38(b). Making this approximation, the correlation function is now simplified to

$$\Lambda[z(t, \xi), t_\varepsilon] \simeq \sum_{k=0}^{K} |I_k(z(t, \xi), t_\varepsilon)|. \tag{4.118}$$

Further, using the difference approximation for the derivative, the negative gradient is of the form

$$-\frac{\partial \Lambda[z(t, \xi), t_\varepsilon]}{\partial t_\varepsilon} \simeq \frac{\Lambda[z(t, \xi), t_\varepsilon - \Delta t/2] - \Lambda[z(t, \xi), t_\varepsilon + \Delta t/2]}{\Delta t}. \tag{4.119}$$

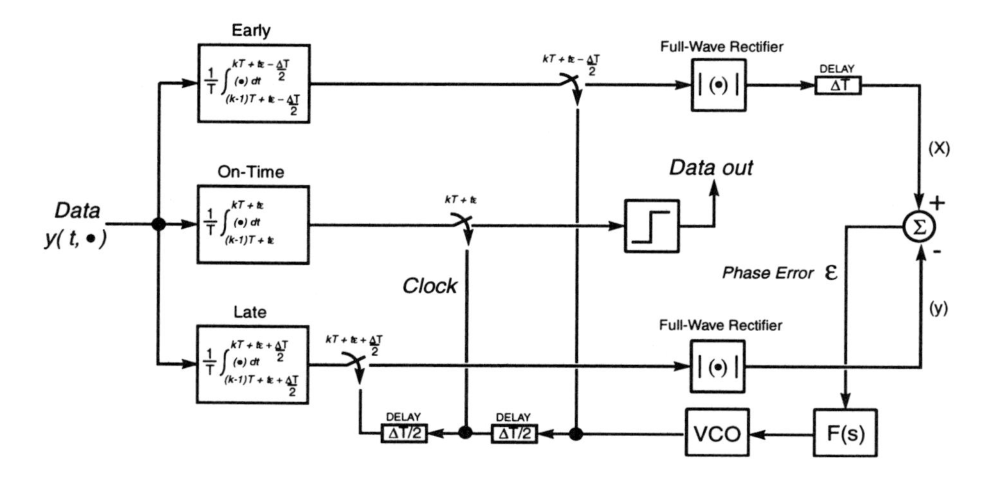

Figure 4.50 Block Diagram of an early-late gate clock recovery circuit using rectifiers in each arm.

Therefore,

$$-\frac{\partial\Lambda[z(t,\xi),t_\varepsilon]T}{\partial t_\varepsilon} \simeq$$

$$\underbrace{\mathrm{SNR}\left(\frac{T}{\Delta t}\right)\left|\frac{1}{T}\int_{(k-1)T+t_\varepsilon-\Delta t/2}^{kT+t_\varepsilon-\Delta t/2} z(t,\xi)\overline{p_T}[t-(k-1)T-(t_\varepsilon-\tfrac{\Delta t}{2})]dt\right|}_{\text{early}} -$$

$$\underbrace{\mathrm{SNR}\left(\frac{T}{\Delta t}\right)\left|\frac{1}{T}\int_{(k-1)T+t_\varepsilon+\Delta t/2}^{kT+t_\varepsilon+\Delta t/2} z(t,\xi)\overline{p_T}[t-(k-1)T-(t_\varepsilon+\tfrac{\Delta t}{2})]dt\right|}_{\text{late}}. \tag{4.120}$$

A block diagram of a circuit that performs this function for rectangular signals is shown in Fig. 4.50. This circuit has been used extensively, and is described throughout the literature [37, p. 235], [34, p. 577]. Performance comparisons of various implantations of this basic structure are given by Lindsey and Simon [32, pp. 458–465].

Comparison of Early-Late Circuits The difference between the early-late circuits of Fig. 4.49 and Fig. 4.50 is the manner in which the phase reversal is implemented; this has an affect on the phase-error characteristic. Consider the timing diagram of Fig. 4.51 for the early-late circuit using multiplication of the error estimate by the retimed data to reverse the phase. The case of an early clock is illustrated. The second

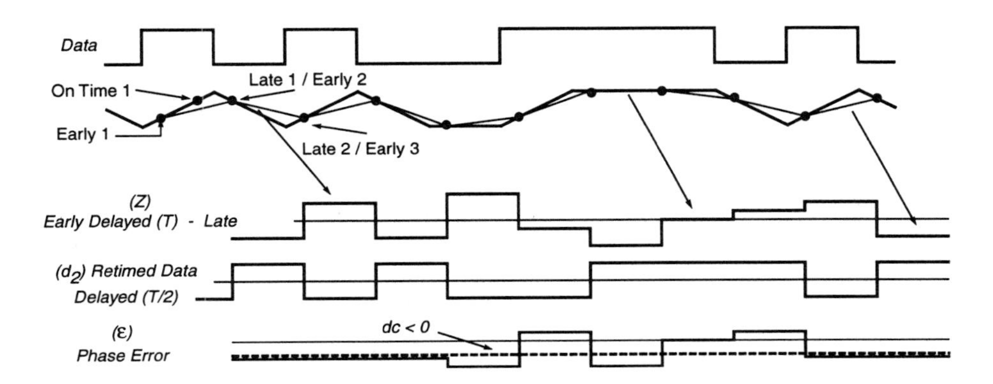

Figure 4.51 Timing diagram for a multiplying early-late circuit for an early clock with $\Delta t = T$.

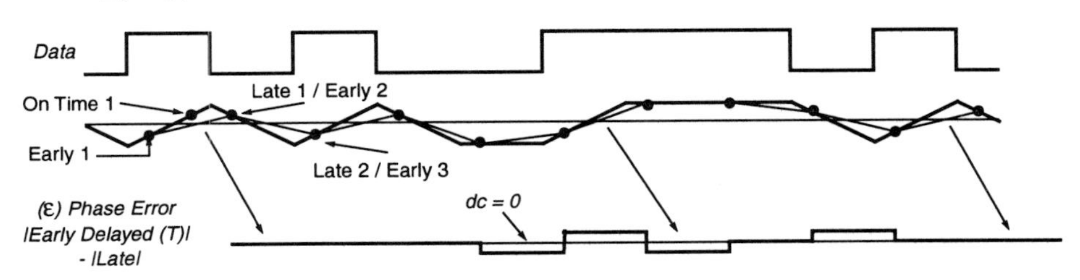

Figure 4.52 Timing diagram for a rectifying early-late gate circuit for an early clock with $\Delta t = T$.

signal is the result of taking a sliding average of the data signal. The waveform Z is the difference of the deskewed early and late correlations, d_2 is the deskewed data, and ϵ is the phase-error signal. It can be seen that the dc value of the error signal is negative, indicating an early clock.

We can compare these results with those obtained for the circuit of Fig. 4.50. Since this circuit takes the absolute value of the early and late correlations before taking the difference, we would expect that there will be a penalty because information contained in the polarity of the individual signal is being thrown away. This penalty is manifested in the reduced amplitude of the phase-detector, and is illustrated by the timing diagram of Fig. 4.52. For the case of $\Delta t = T$, we see that the early and late correlations are symmetric about the zero crossing when a data transition occurs, and therefore have identical absolute values. The dc value of ϵ in this case is zero. Therefore the separation of the early and late clocks, Δt, is restricted when using a rectifier in both arms. The phase-error magnitude increases linearly with Δt, until it reaches a

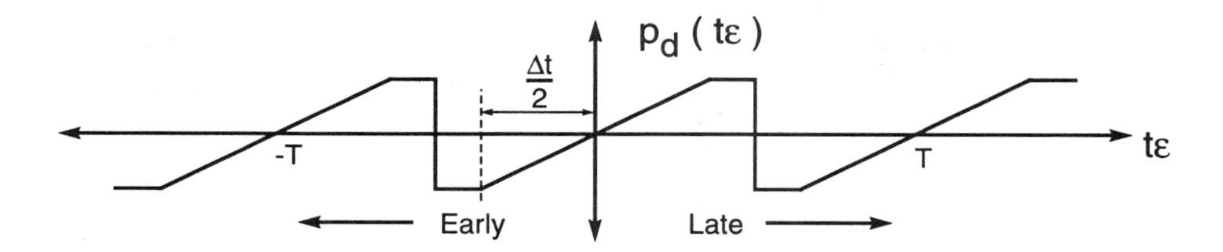

Figure 4.53 Phase error for an early-late gate, with a multiplier for phase reversals, as a function of timing offset for rectangular data pulses and maximum data density.

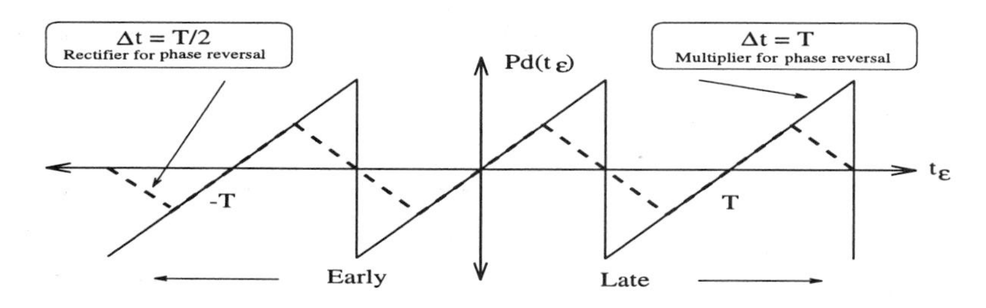

Figure 4.54 Phase detector characteristics for early-late clock recovery circuits using a multiplier (solid-line) and rectifiers (dashed-line).

maximum at $\Delta t = T/2$. This is half of that obtainable with the multiplying early-late circuit. The phase error magnitude decreases linearly with increasing Δt until it reaches a value of zero again at $\Delta t = T$. For an early-late circuit that uses a multiplier for phase polarity reversal, as in Fig. 4.49, the phase detector characteristic is shown plotted in Fig. 4.53. The monotonic range of the phase-detector is Δt. This can be as large as T, but is limited to $T/2$ in a realization using rectifiers in the early and late arms, such as the circuit shown in Fig. 4.50. The phase detector characteristic for the circuits of Fig. 4.49 and Fig. 4.50 are shown in Fig. 4.54 for $\Delta t = T$ and $\Delta t = T/2$ respectively.

Although the amplitude, and maximum monotonic range of the phase detector is reduced when taking the absolute value in each arm, the circuit becomes less sensitive to the deskewing delay. Since taking the average of an early-late difference is a linear operation, identical dc values will result at the loop filter output, whether or not the early arm correlation has been delayed. Therefore, the circuit is functional when the delay Δt, from Fig. 4.50 is removed. However, the deskewing delay is important for the reduction of phase-detector ripple that causes excess clock-phase jitter. Without this delay, a zero-value phase error will be produced by a square-wave, alternating

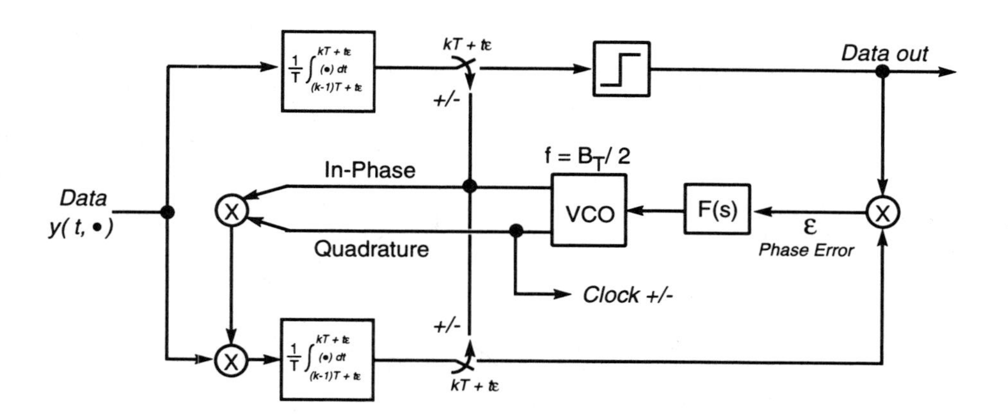

Figure 4.55 Block diagram of a clock recovery circuit using a clock at half the data rate.

between a high and low value. This ripple will not be completely suppressed by the loop filter, and will modulate the VCO, causing jitter.

Special Cases for $\Delta t = T$

Simplified Stochastic Gradient Circuit For the special case of $\Delta t = T$ the phase-detector characteristic in Fig. 4.53 is a sawtooth wave, as can be seen from Fig. 4.54, and the difference approximation of the derivative of the data pulse becomes a square wave, as shown in Fig. 4.48. Realizing that the data-pulse derivative approximation is a square-wave, simplifies its generation. If we have a square-wave VCO operating at half of the bit-rate, then a quadrature shifted clock provides the desired derivative, except the polarity is reversed for alternating bits. This can be corrected if the quadrature-clock is multiplied by the in-phase clock. The resulting signal makes a positive-to-negative transition in the middle of each bit-interval. A block diagram of such a clock recovery circuit is shown in Fig. 4.55. The $\pm$ symbols in the block diagram indicate that the signal is sampled on both the positive, and negative clock transitions. In reality this circuit would be implemented using bit-interleaving with two decision circuit being clocked on alternate phases. Therefore, the integrators are not required to dump instantaneously, and timing constraints can be relaxed.

We notice that the circuit of Fig. 4.55 is virtually identical to the circuit of Enam and Abidi, Fig. 4.34, discussed in the previous section as an example of a spectral-line circuit. The difference between the two circuits is that the polarity reversal in Fig. 4.55 is accomplished using the retimed data. Whereas Enam and Abidi use the data itself, which amounted to squaring the data before it entered the phase-error arm. In Enam

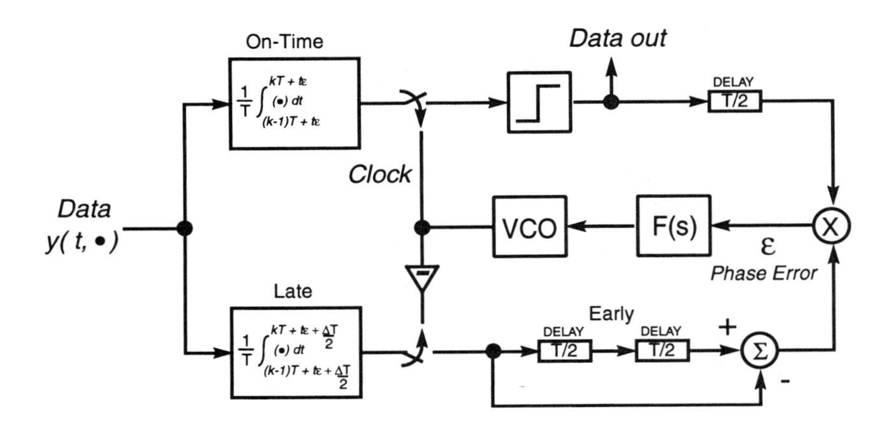

Figure 4.56 Special case of an early-late gate circuit for $\Delta t = T$.

and Abidi's approach, the decision circuit is implemented apart from the clock recovery circuit, and is therefore not self-adjusting.

Simplified Early-Late Gate We recognize that for the special case of $\Delta t = T$ the late sample of the current bit is exactly the same as the early sample of the next bit. Therefore, the early result can be obtained by delaying the late correlator output, thus eliminating the early correlator. A block diagram of this circuit is shown in Fig. 4.56. For a slightly different derivation, the reader is referred to a discussion of this circuit by Stiffler [33, p. 227]. The speed of this circuit is limited by the need to perform the integrate and dump functions. However, by replacing the correlators with a matched filter, we obtain a circuit that is applicable for high-speeds. A block diagram of this circuit is given in Fig. 4.57. In a practical design, the sample and limit function in the on-time arm could be replaced with a single flip-flop. Also if we quantize the phase-error from the late arm to one-bit, then the cross-over sampling switch can also be replaced by a flip-flop. The deskewing time delays, are now operating on quantized data and can be implemented with a shift register. This one-bit phase-error quantization can still maintain accuracy of the phase estimate. Since the bandwidth of the loop filter is much smaller than the data rate, several quantized errors will have to be accumulated before the VCO will respond. Much like a Σ-Δ data converter,[5] the resulting filtered phase-error estimate can be quite accurate. The circuits of Figs. 4.56 and 4.57 are self-adjusting, since the decision element, and the phase-detector are realized using identical circuits. A practical clock recovery circuit implementing a quantized version of the circuit of Fig. 4.57 was first reported by Alexander [38]. This

[5] Also known as a Δ-Σ data converter.

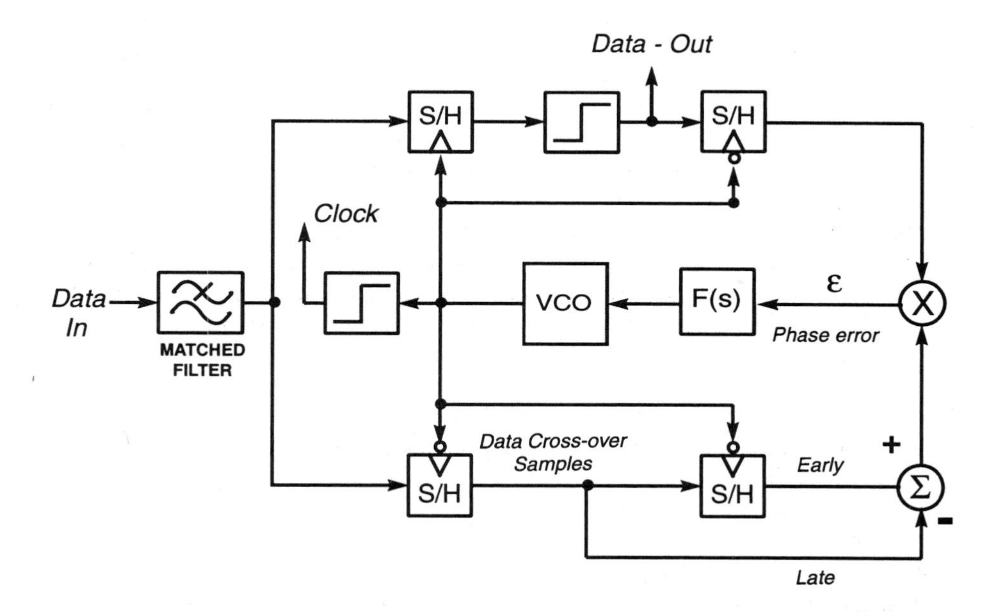

Figure 4.57 Implementation of an early-late clock recovery circuit for high-speed applications using a matched filter.

and other self-adjusting clock extractors designed with familiar circuit building blocks will be presented in the following section.

4.6 PARASITIC-DELAY INSENSITIVE CLOCK RECOVERY SCHEMES

When we refer to the *speed* of a system, we will be speaking in terms of the bit-rate, in comparison with the maximum oscillating frequency of the transistors f_{max}. Since the speed of the electronics in a fiber-optic receiver limits the overall speed of the communication system, we would like to get the symbol-rate as close to f_{max} as possible for a given IC process. Therefore, we will be primarily concerned with high-speed systems, where the bit-rate will be somewhere between $f_{max}/50$ and $f_{max}/4$. Clock recovery circuits using an external BPF will almost always require a phase adjustment, because the delays associated with input and output buffers are substantially larger than delays internal to the integrated circuit. The need for an additional phase adjustment can be eliminated only if the phase-lag due to the additional signal processing in the clock recovery arm is insignificant compared to the bit-period. This will certainly be true at very low data-rates. However, even at moderate bit-rates ($f_{max}/50$), the excess phase shift is significant enough to require compensation.

Clock extraction circuits using a monolithic PLL, at moderate bit-rates, have been able to avoid additional phase compensation, since the steady-state phase offset contribution from parasitic delays is only a few degrees. As bit-rates move closer to f_{max}, the excess phase becomes increasingly problematic and must be compensated in an efficient way, eliminating the need for manual tuning. Such a self-adjusting circuit can automatically center the clock in the data-eye; parasitic delays can be nulled by the feedback loop. Self-adjusting circuits are necessary at high-speeds, where parasitic delays make up a significant portion of the bit interval. To design a self-adjusting circuit, all asymmetries resulting in different delays in the clock-path and the data-path must be taken into account in the phase detector design. Therein lies the challenge of designing a clock recovery circuit for high-speed networks.

4.6.1 Fundamental Requirements of Self-Adjusting Circuits

A block diagram of a self-adjusting clock recovery circuit is shown in Fig. 4.58. The key feature is the inclusion of the decision circuit in the feedback loop. We must keep in mind that the purpose of the clock recovery circuit is to provide a sampling signal at precisely at the moment that the SNR of the test statistic is at a maximum. To obtain

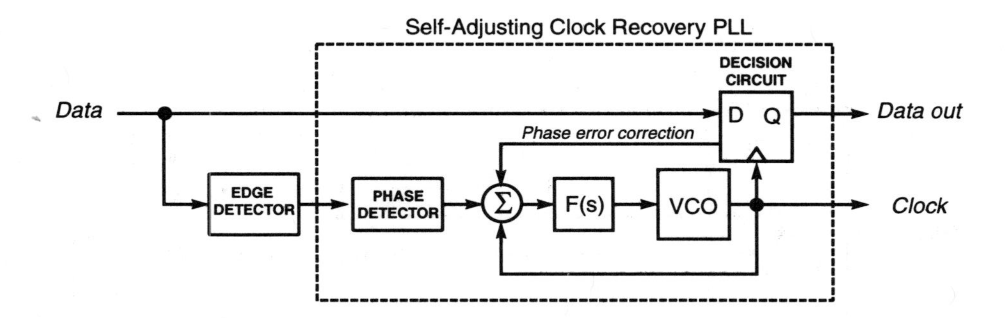

Figure 4.58 Block diagram of a self-adjusting clock extraction circuit.

the desired error signal, a measurement of the clock phase, directly at the decision circuit, is required. Therefore, we need a circuit that will either directly, or indirectly measure the SNR of the test statistic at the decision circuit, producing an error signal that adjusts the VCO phase. Taking the derivative of the absolute value of the test statistic produces a zero output when the SNR is maximum. The value of the output signal gives an indication of both the magnitude and direction of the phase error. We saw in the previous section that this derivative was approximated by a finite difference in an early-late circuit, and the clock phase was optimized assuming,

- the early, on-time, and late decision circuits are matched,
- the peak SNR lies exactly between the early and late samples,
- there is no error in the clock phases (the early, and late clocks are equals spaced around the on-time clock).

Therefore, the measurement of the maximum SNR point is indirect, but the assumptions allowing us to infer a direct measurement are reasonable. In the following sections we will describe several architectures that use the decision circuit as part of the phase detector. The degree to which the maximum SNR is explicitly measured by the phase detector will determine how well the circuit can adjust itself to the optimum sampling phase.

4.6.2 Alexander's Clock Recovery and Data Retiming Circuit

Several clock recovery circuits for moderate bit rates have emerged recently [39, 40, 41, 42, 43, 44, 27], all of which are based on a phase-detector similar in concept to

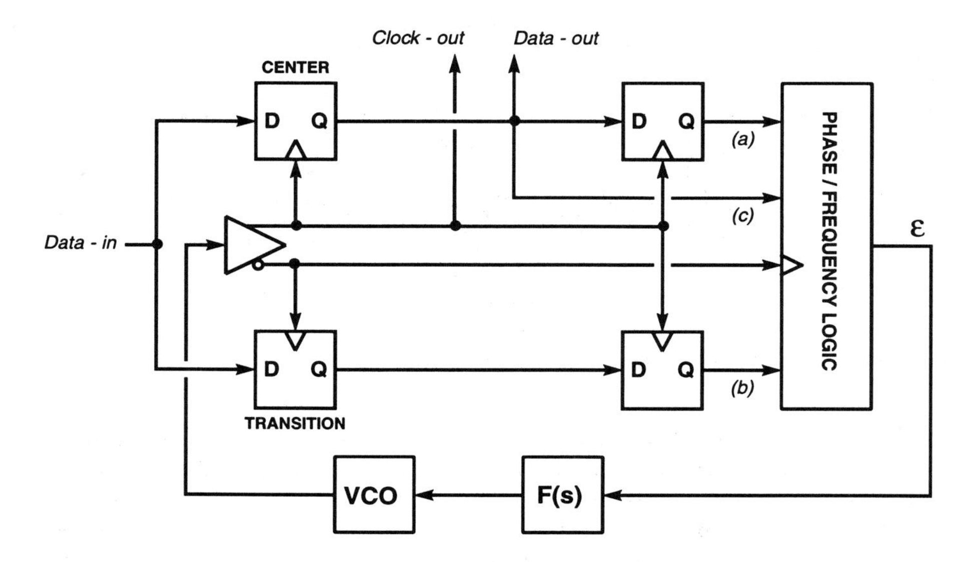

Figure 4.59 Block diagram of Alexander's self-adjusting clock recovery and data retiming circuit.

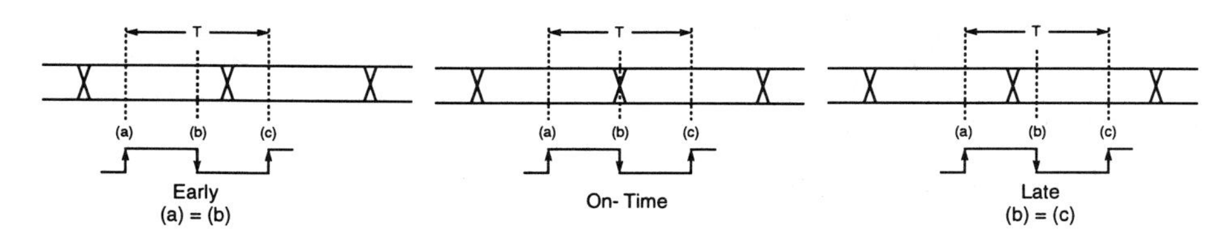

Figure 4.60 Illustration of sampling points (a), (b), and (c) for early, on-time, and late clocks.

circuit described in 1975 by Alexander [38]. This clock recovery circuit is based on a digital approximation to an early-late technique, described in the previous section. The basic circuit is shown in Fig. 4.59. The phase detector is designed so that during any particular clock interval there are three binary samples of the data signal available: (a) is the previous data value, (b) is a sample of the data at the transition, and (c) is the current data value. The ordering of these three samples are illustrated in Fig. 4.60 for early, on-time, and late clocks respectively. The retimed data can be taken from either the (a) or (c), and is usually taken from (a) so as to get an additional squaring of the data pulse by passing through two decision circuits. Based on the binary outcome

Enable	Control	Frequency
$a \oplus c$	$b \oplus c$	
0	0	f_0
0	1	f_0
1	0	$f_0 + \Delta f$
1	1	$f_0 - \Delta f$

Table 4.4 Truth-table enumerating control possibilities for clock recovery using a bang-bang oscillator.

of the samples, we can devise a set of rules used to control the phase of the sampling clock [6].

- If $a = b \neq c$, the clock is early $\Rightarrow$ slow down the clock.
- If $a \neq b = c$, the clock is late $\Rightarrow$ speed up the clock.
- If $a = b = c$, no data transition occurred $\Rightarrow$ do nothing.
- If $a = c \neq b$, shouldn't happen in phase-lock $\Rightarrow$ possible frequency error.

The digital logic block translates the above rules into signals that control the phase of a local oscillator. When the clock is on-time, the center sample (b) will randomly equal (a) or (c), causing the clock to randomly switch between, speeding up, and slowing down. On average, (b) will equal both (a) and (c) half of the time. It is interesting to note that this system, is an early example of a *Fuzzy-Logic* control system.

Discrete Frequency Adjustment Circuit Alexander describes two methods for clock recovery using his phase detector. The first assumes that a VCO exists that can operate at three discrete frequencies: f_0, $f_0 - \Delta f$, and $f_0 + \Delta f$. Such an oscillator is often referred to as a bang-bang oscillator. From the list of control rules, we can see that the truth-table 4.4 provides the necessary control signals for the circuit. A simple circuit to implement this operation is shown in Fig. 4.61. When the enable signal is low, no data transition occurred, and the VCO remains at the center frequency. When a transition does occur, the enable goes high, and the frequency is shifted up or down slightly

[6] Although Alexander states the early-late conditions correctly in the text, he reverses the order in the itemized list, and in the truth-tables.

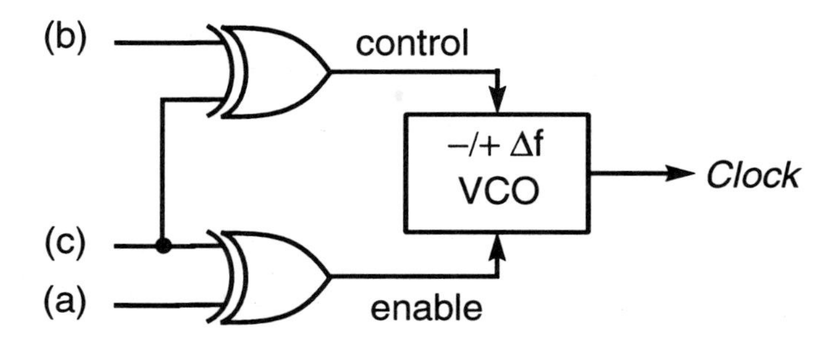

Figure 4.61 Block diagram of a bang-bang VCO used for clock recovery.

depending on the polarity of the control signal. When the control signal is high, the clock was late, so the frequency is increased. A high control signal indicates an early clock, and the VCO frequency is reduced. The signal $a \oplus b$ could also be used for the control signal, in which case the polarity is opposite of the circuit shown in Fig. 4.61.

The Hewlett-Packard 622-Mb/s Circuit A circuit based on Alexander's phase detector was used in a 622-Mb/s clock recovery and data retiming IC designed by Lai and Walker of Hewlett Packard [39]. This circuit used a coarse-tune/fine-tune approach. The average of the phase-error signal is used in a narrowband feedback loop to adjust the nominal center frequency of the VCO, and the fine-tuning of the phase is accomplished using the discrete frequency adjustment. The same phase detector was used in a 1.5-Gb/s system designed by Walker *et al.* [8, 9], which used line-coding to achieve simultaneous frame- and bit-synchronization.

3-Level Phase Error Circuit Alexander also described how his phase detector could be used to produce a 3-level phase error for use in tuning an analog PLL. In this case the desired phase error can be obtained by subtracting the early from the late signal. The truth-table for this situation is given in table 4.5. A circuit that implements this truth table is shown in Fig. 4.62. A lowpass filter is used to average the phase-error over several cycles. The net dc value will give an indication of the phase error, and this filtered signal is used to adjust the VCO. Alternatively these control signal can be used as the inputs of a charge-pump that converts pulse widths into voltage levels by controlling the charging-time of integration currents. Since the phase is adjusted according to the outputs of decision circuits, the clock is automatically adjusted to the proper phase. However, the sampling of the center point (b) varies randomly, so that in steady-state, the phase-error will have a strong ripple component leading to increased clock-jitter. Since this circuit uses concatenated decision circuits, the maximum data rate will be limited by the decision circuit delay. To insure proper circuit operation this

Late $a \oplus b$	Early $b \oplus c$	Phase-Error
0	0	0
0	1	-1
1	0	1
1	1	0

Table 4.5 Truth-table describing the 3-level output phase-detector.

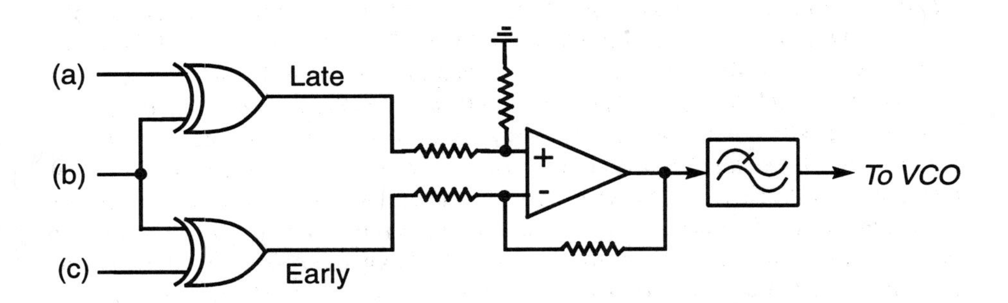

Figure 4.62 Phase detector for 3-level Alexander circuit.

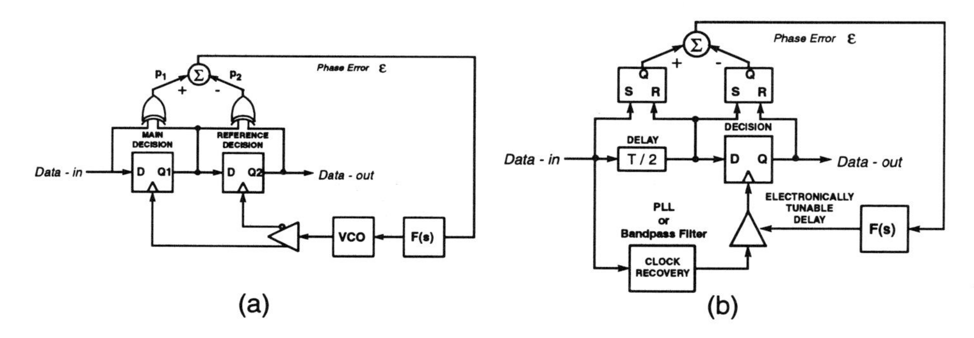

Figure 4.63 Block diagram of self-adjusting phase detector and decision circuit: (a) Hogge's circuit, (b) Whitt's circuit.

delay must not exceed the bit-interval. However, interleaving can be used to increase the throughput when $t_d > T$.

4.6.3 Hogge's Phase Detector and Decision Circuit

A simple self-adjusting phase detection technique, illustrated in Fig. 4.63(a), was reported by Hogge in 1985 [40]. This circuit is very similar to the ones described by Alexander. However, Hogge's circuit does not quantize the phase error, but rather produces a continuous phase measure. A circuit nearly identical to Hogge's was submitted for publication by Whitt six-months after Hogge, but appeared in the literature one month earlier [41]. This circuit is illustrated in Fig. 4.63(b), and was actually described by Hogge [40, Fig. 5] as a variation of his method. The implementations of these two circuits are slightly different, but the basic principle is the same. However, Whitt uses a delay-line, which directly controls the ultimate clock phase. Since this delay-line must be manually tuned, the clock recovery circuit is not self-adjusting. In contrast, Hogge uses a reference decision circuit instead of a delay-line. Ideally, the reference circuit tracks the delays of the decision circuit, and the clock adjusts itself to the proper phase regardless of parasitic delays. We will now explain the operation of Hogge's basic circuit. The delay-line approach of Whitt, although not self-adjusting, will be considered later as a method of overcoming problems in high-speed applications.

In Hogge's basic circuit, a phase-error estimate is obtained by taking the difference of two pulses, both of which are generated whenever a data transition occurs. The width of p_1 is linearly related to the clock phase, and is given by

$$t_1 = T/2 + t_d + t_\phi, \tag{4.121}$$

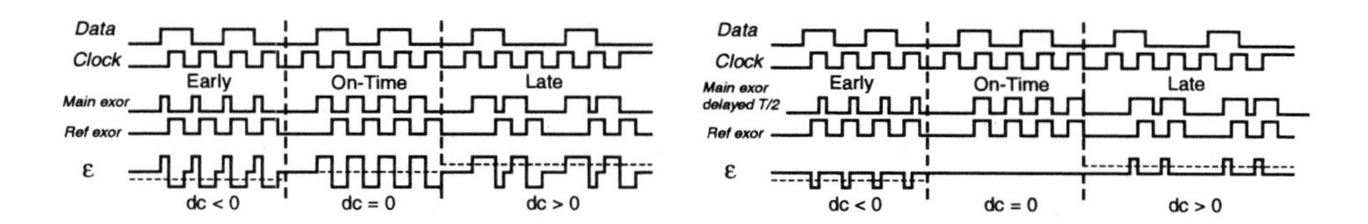

Figure 4.64 Timing diagram of Hogge's phase-detector.

where t_d is the decision circuit delay, and t_ϕ is the timing error in the clock. The second decision circuit and EXOR gates produces a pulse p_2 when a data transition occurs, with a fixed width of $T/2+t_d$. An estimate of t_ϕ is obtained by subtracting p_2 from p_1, which generates a residual pulse of width t_ϕ. By using a reference decision circuit to generate a differential error signal, the common-mode parasitic delays of the decision circuits are cancelled, and the resulting phase error is precompensated for the decision circuit delays. The differential phase-error pulses are lowpass filtered to convert pulse-widths to a dc voltage. A timing diagram for the circuit is given in Fig. 4.64.

In the straightforward implementation we see that the phase-error signal $p_1 - p_2$ produces a dc value that is proportional to the phase error. However, since p_2 is delayed by 90° relative to p_1, the difference of these signals will have a strong ripple component, which causes excess clock phase-jitter. The ripple can be removed by delaying p_1 by 90°, or $T/2$ seconds, before p_2 is subtracted. This delayed signal is represented by $\hat{p_1}$ and the resulting phase error is shown in Fig. 4.64. We see that the dc value is unchanged, but the ripple has been removed. Shin *et al.* [42] describes a circuit that implements Hogge's phase detector, using a delay of $T/2$, and also gives expressions showing the improvement in phase-jitter. The $T/2$ delay used, does not have to be exact. Any errors in this delay will simply result in a residual ripple, causing a second order degradation of the clock-phase jitter. However, the nominal clock phase will not be affected.

Indirect Measure of Maximum SNR The Hogge circuit doesn't measure the point of maximum SNR directly. It only equalizes the time delays between the input and output of two identical circuits. However, even in an idealized situation, this doesn't guarantee that the circuit is sampling at the point of maximum SNR. And worse, in a real circuit, systematic errors exist that will be exaggerated at high speeds. Whenever relying on a cancellation of identical operations, one has to ask what is it about these two, supposedly identical, circuit paths, other than random mismatches, that makes them different. In the case of Hogge's circuit the answer is clear. The first decision circuit is making a decision on the data signal with a clock that is nominally aligned for optimal sampling. The second decision circuit is sampling a retimed data waveform,

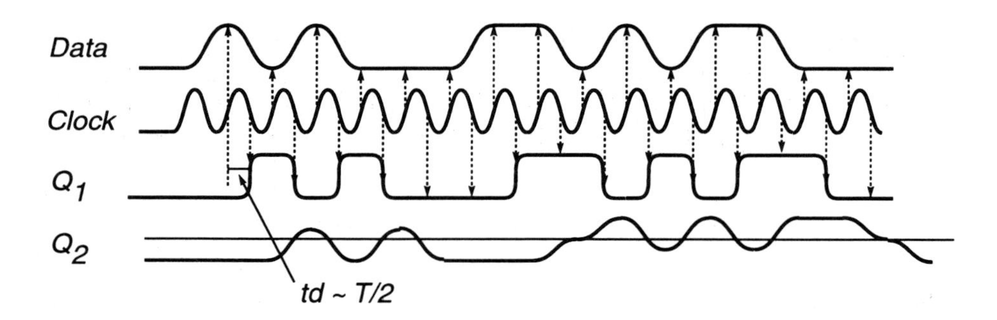

Figure 4.65 Illustrations of asymmetries in Hogge's circuit that reduce its effectiveness at high-speeds.

where the phase error of the sampling depends on the delay of the first decision circuit. In other words, if the first decision circuit is sampling with an optimal clock phase, then the second decision circuit is sampling with a clock timing error of t_d. This asymmetry restricts the use of Hogge's circuit in high-speed applications. This is best illustrated by the sample timing diagram of Fig. 4.65. Assuming that the clock is centered in the data-eye, and the decision circuit delay time is close to half of a bit-period (50-ps for a 10-Gb/s system), the second decision circuit will be sampled very near to the data transition. Since each decision circuit has a finite gain, the rise times of the retimed data will depend on the magnitude of the data at the sample point. Therefore, (Q_2) will not be an exact time shifted replica of (Q_1). Likewise, the retimed signal (Q_1) will not be a replica of the original data signal. The EXOR circuits that generate the phase-error pulses will also have a finite gain. The resulting pulse shapes depend on the shape, magnitude, and rise-times of the data and retimed data signals. The retimed data (Q_2) in Fig. 4.65 is shown with a smaller magnitude than (Q_1) because the second circuit samples near the cross-overs, where the input signal value is weak. Therefore, p_1 will be larger than p_2. The system will interpret this as a late clock, and the phase will be adjusted to make the clock arrive earlier, resulting in a systematic phase-error.

Because of the asymmetries in the circuit, the natural action of the feedback loop will try to adjust the clock phase so that both waveforms (Q_1), and (Q_2) are similar in shape. This will have the effect of forcing the delay of the decision circuit to straddle the center of the data-eye. Therefore, if $t_d = T/2$, the positive clock-edge will occur $T/4$ seconds early and the negative edge will occur $T/4$ seconds late. As the delay time of the decision circuit gets closer to T, the problem gets worse. Obviously if t_d is greater than T, the circuit will not work at all, because the second decision circuits will be sampling the previous bit.

The decision circuit delay t_d also limits the tracking and acquisition range. For a delay free circuit, the phase-detector outputs a correct signal for phase errors in the range $[-T/2, T/2]$, however, t_d subtracts directly from this range. For $t_d = T/2$, the range is reduced to $[-T/4, T/4]$. We can determine a very rough limit on the maximum bit-rate that can be handled by the Hogge circuit. If we assume that the decision circuit can be made with a delay t_d of approximately

$$t_d \simeq \frac{2.5}{f_{max}}, \tag{4.122}$$

and if we only allow a maximum phase error of 10% then t_d can not be greater than 20% of the bit-period;

$$\begin{aligned} t_d &\simeq \frac{2.5}{f_{max}} \leq T/5 \\ B_T &\leq \frac{f_{max}}{10}. \end{aligned} \tag{4.123}$$

Allowing for a safety margin of a factor of two, the maximum data rate that can be handled is approximately limited to

$$B_T \leq \frac{f_{max}}{20}. \tag{4.124}$$

For high-speed bipolar transistors or HBTs with f_{max} ranging from 20–50-GHz, Hogge's circuit is applicable to data rates of approximately 1–2.5-Gb/s. Hogge points out that the delay of the first decision circuit is a problem. In high-speed applications he adds a trimmable delay element to compensate, thereby defeating the purpose of using a self-adjusting circuit in the first place. Once we have given up on the idea of designing a self-adjusting circuit, then the implementation described by Whitt provides a simple method of measuring the phase error, provided that the delay-line has been precalibrated, but we must keep in mind that variations in operating conditions will not be tracked, and large clock-phase deviations can result.

Despite these speed limitations predicted by us, a tunable phase shifter that implements the Hogge phase detector has recently been reported by Wennekers *et al.* for a circuit operating at 10-Gb/s [27] using transistors with f_t = 45-GHz ($f_t/B_T = 4.5$). A high-Q bandpass filter is used to extract the clock whose phase is adjusted using a tunable delay element in a feedback loop. As was predicted in the above analysis, the phase detector was shown to function only over a small phase interval of $\pm 54°$ or $\pm(0.15)T$. It is also not clear from [27] whether the clock-phase at the point of zero phase error is correct or not. Also the circuit relies on a delay of approximately $T/2$ in the data line. This delay directly effects the phase-error signal. But the tuning of this delay was not discussed in the paper. It is believed that either one of two situations occurred which allowed the circuit to function. The first is that the $T/2$ data delay

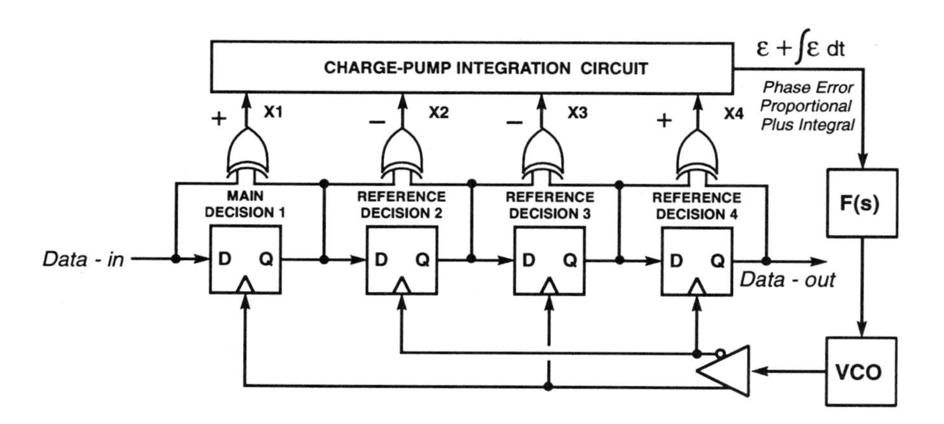

Figure 4.66 Block diagram of a transition-density-independent phase detector.

was adjusted to center the clock in the bit-interval. The second alternative is that the decision circuit, and buffer delays, happen to be at the proper value so as to achieve the desired result. In fact, the phase detector, as drawn in [27], will not function unless these parasitic delays are nominally equal to $T/4$. Although, the circuit of Wennekers *et al.* does demonstrate functionality of a 10-Gb/s circuit utilizing the Hogge phase detector, it is suspected that either external tuning was required, or parasitic delays happened to be of the proper value. The results of Wennekers *et al.* not withstanding, we still believe the Hogge method is not applicable for data rates above $f_{\max}/20$, unless external phase-adjustments are provided.

4.6.4 Analog Devices Transition-Density-Independent Circuit

A problem with all of the phase-detectors that we have discussed thus far, is that the dc value of the phase-error depends on the data transition density. As a result, the phase detector gain, and therefore, the loop gain, are proportional to the data-density. This causes variations in the dynamic response of the PLL, leading to pattern dependent jitter in the recovered clock. Data-density-dependency is an artifact of the phase-error going to zero when no transitions occur. If the phase-error is held in place during periods of no transition, then the phase-detector output will be the same for both dense and sparse data. A phase detector, based on the Hogge circuit, that is data-density-independent was designed at Analog Devices by DeVito *et al.* [43], and utilized in the circuit of Lee and Bulzacchelli [44, 45]. A block diagram of this circuit is shown in Fig. 4.66. The first two decision circuits of Fig. 4.66 are the same as in the Hogge circuit. When a data transition occurs, a pulse p_1 is generated. The width of p_1 is

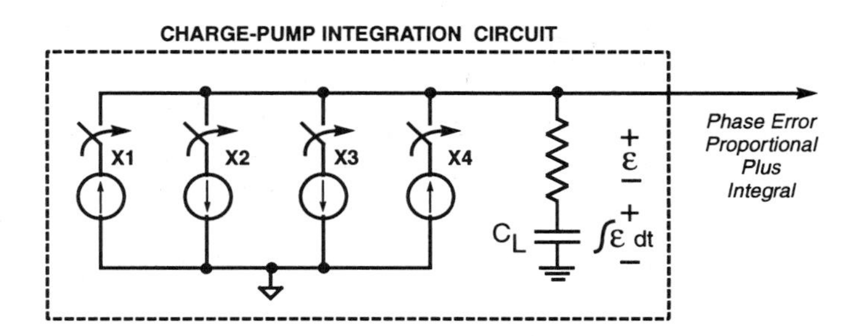

Figure 4.67 Conversion of phase detector pulses to voltages by an up-down-down-up sequence of integrating currents.

linearly related to the clock phase,

$$t_1 = T/2 + t_d + t_\phi. \tag{4.125}$$

The remaining three decision circuits and EXOR gates produce pulses p_2, p_3, and p_4, when a data transition occurs, all of which have a fixed width of $T/2 + t_d$. In Hogge's circuit an estimate of t_ϕ was obtained by subtracting p_2 from p_1, which generates a pulse of width t_ϕ. When no data transitions occur the phase-error pulses are missing, and the resulting dc phase error is modulated by the data density.

The approach adopted by DeVito is to convert the pulse width information directly into a voltage by integration. The pulses are used to control switches as shown in Fig. 4.67 that controls an up-down-down-up sequence of integrating currents on the load capacitor. A timing diagram showing the resulting phase error on the load capacitor for an early, on-time, and late clock is shown in Fig. 4.68. The clever aspect of this design is that the integrated value of the phase-error on the load capacitor in steady-state operation is the same for both dense, and sparse data transitions. Another nice feature is that the integration cycle takes two clock periods to complete. Therefore, when the data is dense, up integrations from one transition will cancel down integrations from the previous transition, and the phase detector will have no ripple for adjacent transitions, reducing the eventual clock-jitter.

Limitations of Devito's Phase Detector Although Devito's circuit solves the problem of data density dependence, it is limited in application to low and moderate bit-rates. We saw that cascading two decision circuits limited the performance of Hogge's circuit at high-speeds. This problem is exacerbated in Devito's circuit because 4 decision circuits are cascaded. Therefore the clock phase error in the last decision circuit will be 3 times worse than in the Hogge circuit. The approximate bit-rate limitation is then

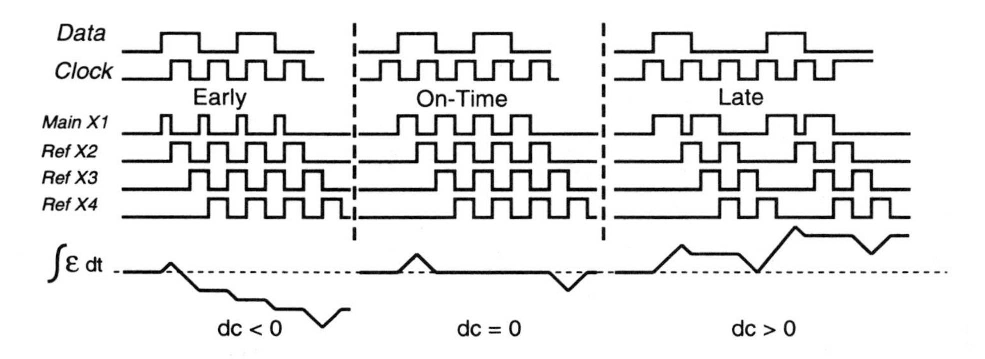

Figure 4.68 Timing diagram for data-density-independent phase detector.

given by

$$B_T \leq \frac{f\text{max}}{60}. \tag{4.126}$$

The designed application of Devito's circuit was for a 52-MHz and a 155-MHz circuit. Using GaAs HBTs the approach may be good for bit-rates of from 300-Mb/s to 1-Gb/s.

Problems with Hogge's and Devito's circuits are that they use serial connections of decision circuits to estimate the phase error. Therefore, the decision circuit delay alters the sampling phase of successive decisions. The serial decision circuit delay is in the *critical path*; as a result, these circuits can not be pipelined. Pipelining, or bit-interleaving is only possible when all sampling is done in parallel. Instead of using resampling with a chain of flip-flops, as in the circuits of Hogge and Devito, appropriate time skewing of the samples can be obtained by using a multi-phase clock. The sampled data can be clocked to deskewing registers for further processing. Since at the front-end, all sampling is performed with matched circuit, all of which are sampling the original data, and not retimed data, the parasitic delays of the sampling circuits will track each other. Variations in parasitics delays due to changes in the environment will be automatically compensated, and won't degrade the accuracy of the final clock-phase estimate.

Based on these considerations, the circuit of Alexander is a prime candidate for high-speed clock recovery, because the two front-end flip-flops, generating the data and phase-control information, are operating in parallel. In the following chapter, practical modifications to the circuit of Alexander will be presented that are applicable for monolithic clock recovery and data retiming at bit-rates near $f\text{max}/4$. In addition, a novel data-density independent phase detector circuit will be presented, which implements all of the desired features efficiently, and is ideally suited to self-adjusting, PLL-based, clock extraction at data rates in excess of 10-Gb/s.

4.7 SUMMARY

A significant amount of material has been covered in this chapter. We have looked at clock recovery circuits in the following ways.

- Qualitative approaches.
- Spectral-line or frequency domain approaches.
- Optimal stochastic estimates or time domain approaches.

Our goal has been to tie all of these approaches together into a unified treatment which clearly exposes the advantages and disadvantages of each circuit. We have utilized the theory developed in chapters 2 and 3 in order to have quantitative methods to measure the *goodness* of competing designs. Comprehensive analysis of a particular circuit have not been presented, rather we have aimed at giving the reader an overall view of clock recovery techniques that are applicable for high-speed broadband receivers. With this information the circuit designer should be able to choose which circuit best meets his design needs. A recapitulation of the main topics of this chapter will now be given in the form of questions that the serious reader ought to be able to answer.

Qualitative Approaches to Clock Recovery

- When a clock is recovered from random data using a BPF, why is there amplitude modulation on this clock, and how is it related to the Q of the filter?
- Intermittent-phase-readjusting clock recovery circuits can begin clocking data immediately after the first data transition is detected. Why can circuits using a PLL or a BPF not do this?
- Are intermittent-phase-readjusting schemes good, or bad for low SNR applications? Are they good, or bad for systems requiring tight phase-jitter tolerance. When would you use this approach? Discuss how such a circuit accumulates phase errors when no data transition occurs.

Spectral-Line Techniques

- Why is it necessary to pre-process the data with a nonlinear edge detection circuit?
- What is the essential requirement of an edge-detection circuit, and why do several different approaches produce the same result?
- How is the selectivity of a bandpass filter related to energy dissipation in the natural response of the resonator?
- What imposes a maximum limitation on Q? What imposes a minimum limitation? What are typical Q values for BPFs in clock recovery circuits?
- A PLL can track the data rate provided that it can first achieve lock. How large of a frequency deviation can be acquired by the PLL, and how stable does the center frequency of the VCO have to be to insure locking occurs?
- Why is a frequency discriminator important for PLL-based clock recovery circuits? What methods other than using a frequency discriminator can be employed to insure that the PLL will achieve lock under worst-case frequency offsets?
- Discuss the advantages and disadvantages of bit-interleaving. If the setup time to a flip-flop is too short, can the flip-flop capture the proper data if the clock is slowed down? How does the setup time limit the effectiveness of bit-interleaving for increasing throughput when a flip-flop is used as a decision circuit?

MAP Estimate Based Clock Recovery

- Show that in AWGN the MAP estimate of data arrival is obtained by correlating each bit with a template of the received data, and finding the maximum sum of these correlations for all observed bits, weighting each correlation through a $\ln(\cosh(x))$ function.
- Discuss how various closed-loop circuits and early-late circuits can be derived by making different approximations to the gradient of the MAP correlation function.
- Why do some early-late circuits have a phase-detector characteristic that is monotonic over the bit interval and some do not? How does a monotonic characteristic improve frequency acquisition and tracking of the PLL?
- What is essential for a clock recovery circuit to be self-adjusting? Why is an early-late circuit self-adjusting?
- Explain Alexander's circuit in terms of a quantized early-late circuit. Is it self-adjusting? What are its limitations in speed of operation?

- What causes the phase error to depend on the data pattern? How can this be reduced, thereby reducing pattern-dependant jitter? How does a tri-state phase detector reduce ripple-induced jitter?

REFERENCES

[1] H.-H. Witte and S. Moustakas. Simple clock extraction circuit using a self sustaining monostable multivibrator output signal. *Electron. Lett.*, 19(21):897–898, October 1983.

[2] Sverre T. Eng, Robert Tell, Torbjorn Andersson, and Bjorn Eng. A 200-Mbit/s ring local computer network using 1.3-μm single-mode optical fibers. *J. Lightwave Technol.*, LT-3(4):820–823, August 1985.

[3] Naoaki Yamanaka, Masaharu Sasaki, Shiro Kikuchi, Thoru Takada, and Masao Idda. A gigabit-rate five-highway GaAs OE-LSI chipset for high-speed optical interconnections between modules or VLSI's. *IEEE J. Select. Areas Commun.*, 9(5):689–697, June 1991.

[4] Mehran Bagheri, Keh-Chung Wang, Mau-Chung F. Chang, Randy B. Nubling, Peter M. Asbeck, and Andy Chen. 11.6 GHz 1:4 demultiplexer with bit-rotation control and 6.1 GHz auto-latching phase-aligner ICs. In *ISSCC Dig. Tech. Papers*, pages 94–95, San Francisco, California, February 1992.

[5] Mehran Bagheri, Keh-Chung Wang, Mau-Chung F. Chang, Randy B. Nubling, Peter M. Asbeck, and Andy Chen. 11.6-GHz 1:4 regenerating demultiplexer with bit-rotation control and 6.1-GHz auto-latching phase-aligner IC's using AlGaAs/GaAs HBT technology. *IEEE J. Solid-State Circuits*, 27(12):1787–1793, December 1992.

[6] Beomsup Kim, David N. Helman, and Paul R. Gray. A 30-MHz hybrid analog/digital clock recovrey circuit in 2-μm CMOS. *IEEE J. Solid-State Circuits*, 25(6):1385–1394, December 1990.

[7] Beomsup Kim. *High Speed Clock Recovery in VLSI using Hybrid Analog/Digital Techniques*. PhD thesis, University of California, Berkeley, June 1990. Memorandum No. UCB/ERL M90/50.

[8] Richard C. Walker, Thomas Hornak, Chu-Sun Yen, Joey Doernberg, and Kent H. Springer. A 1.5 Gb/s link interface chipset for computer data transmission. *IEEE J. Select. Areas Commun.*, SAC-9(5):698–703, June 1991.

[9] Richard C. Walker, Cheryl L. Stout, Jieh-Tsorng Wu, Benny Lai, Chu-Sun Yen, Tom Hornak, and Patrick T. Petruno. A monolithic 622Mb/s clock extraction data retiming circuit. *IEEE J. Solid-State Circuits*, 27(12):1805–1811, December 1992.

[10] B. Bentland, L. A. Bergman, Sverre T. Eng, and Robert Tell. Clock recovery for a 5 Gbit/s fibre-optic system. *Electron. Lett.*, 18(13):547–548, June 1982.

[11] Marc Moeneclaey. Synchronizability of a general class of PCM formats, including NRZ, Manchester, and Miller coding. *IEEE Trans. Communications*, COM-32(9):1020–1024, September 1984.

[12] Robert R. Cordell, J. B. Forney, Charles N. Dunn, and William G. Garrett. A 50 MHz phase- and frequency-locked loop. *IEEE J. Solid-State Circuits*, SC-14(6):1003–1010, December 1979.

[13] Daniel J. Millicker and R. D. Standley. 2 Gbit/s timing recovery circuit using dielectric resonator filter. *Electron. Lett.*, 23(14):738–739, July 1987.

[14] Daniel J. Millicker, R. D. Standley, and Klaus Runge. A delay and multiply timing recovery circuit for lightwave transmission systems using NRZ format. In *OFC'85*, San Diego, California, 1985.

[15] Robert L. Rosenberg, Christodoulos Chamzas, and Daniel A. Fishman. Timing recovery with SAW transversal filters in the regenerators of undersea long-haul fiber transmission systems. *J. Lightwave Technol.*, LT-2(6):917–925, December 1984.

[16] Jan Davidse. *Analog Electronic Circuit Design*. Prentice Hall, New York, 1991.

[17] Inder Bahl and Prakash Bhartia. *Microwave Solid State Circuit Design*. John Wiley & Sons, New York, 1988.

[18] Robert L. Rosenberg, David G. Ross, Patrick R. Trischitta, Daniel A. Fishman, and Carol B. Armitage. Optical fiber repeatered transmission systems utilizing SAW filters. *IEEE Trans. Sonics and Ultrasonics*, 30(3):119–126, May 1983.

[19] Zhigong Wang, Ulrich Langmann, and Berthold Bosch. Mulit-Gb/s silicon bipolar clock recovery IC optical receivers. *IEEE J. Select. Areas Commun.*, SAC-9(5):656–663, June 1991.

[20] Zhigong Wang and Ulrich Langmann. New proposal for multigigabit/s clock recovery IC based on a standard silicon bipolar technology. *Electron. Lett.*, 23(9):454–455, April 1987.

[21] B. Wedding, D. Schlump, E. Schlag, W. Pöhlmann, and B. Franz. 2.24-Gbit/s 151-km optical transmission system using high-speed integrated silicon circuits. *IEEE J. Select. Areas Commun.*, SAC-8(2):227–234, February 1990.

[22] E. Schlag, B. Franz, and W. Pöhlmann. Integrierte Si-bipolar schaltungen für ein optisches übertragungssystem von 2.4 Gbit/s. In *Proc. ITG Fachtagung Mikroelektronik für die Informationstechnik*, pages 221–226, Stuttgart, Germany, October 1989.

[23] Kazuo Hagimoto and Kazuo Aida. Multigigabit-per-second optical baseband transmission system. *J. Lightwave Technol.*, LT-6(11):1678–1685, November 1988.

[24] George E. Andrews, Dennis C. Farley, Stanley H. Kravitz, and Arthur W. Schelling. A 300Mb/s clock recovery and data retiming system. In *ISSCC Dig. Tech. Papers*, pages 188–189, New York, New York, February 1987.

[25] Makoto Nakamura, Yuhki Imai, Eiichi Sano, Yoshiki Yamauchi, and Osaake Nakajima. A limiting amplifier with low phase deviation using an AlGaAs/GaAs HBT. *IEEE J. Solid-State Circuits*, 27(10):1421–1427, October 1992.

[26] Kazuo Hagimoto, Yuuzou Miyagawa, Yutaka Miyamoto, Masanobu Ohhata, Tatsuhito Suzuki, and Hiroyuki Kikuchi. Over 10 Gb/s regenerators using monolithic IC's for lightwave communication systems. *IEEE J. Select. Areas Commun.*, SAC-9(5):673–682, June 1991.

[27] Peter Wennekers, Ulrich Novotny, Axel Huelsmann, Gugrun Kaufel, Klaus Koehler, Brian Raynor, and Joachim Schneider. 10-Gb/s bit-synchronizer circuit with automatic timing alignment by clock phase shifting using quantum-well AlGaAs/GaAs/AlGaAs technology. *IEEE J. Solid-State Circuits*, 27(10):1347–1352, October 1992.

[28] J. A. Bellisio. A new phase-locked timing recovery method for digital regenerators. In *IEEE Int. Conf. Commun.*, pages 10–17–10–20, Philadelphia, Pennsylvania, June 1976.

[29] Donald Richman. Color-carrier reference phase synchronization accuracy in NTSC color television. *Proc. IRE*, 42:106–133, January 1954.

[30] Syed Khursheed Enam and Asad A. Abidi. Mos decision and clock-recovery circuits for Gb/s optical-fiber receivers. In *ISSCC Dig. Tech. Papers*, pages 96–97, San Francisco, California, February 1992.

[31] Syed Khursheed Enam and Asad A. Abidi. NMOS IC's for clock and data regeneration in gigabit-per-second optical-fiber receivers. *IEEE J. Solid-State Circuits*, 27(12):1763–1774, December 1992.

[32] William C. Lindsey and Marvin K. Simon. *Telecommunication Systems Engineering*. Dover Publications Inc., New York, 1991. Dover edition first published in 1991 is an unabridged, unaltered republication of the work first published by Prentice-Hall, Inc., Englewood Cliffs, N.J., 1973 in its "Prentice-Hall Information and System Science Series.".

[33] J. J. Stiffler. *Theory of Synchronous Communications*. Prentice-Hall, Inc., Englewood Cliffs, New Jersey, 1971.

[34] E. Lee and D. Messerschmitt. *Digital Communnication*, chapter 15. Kluwer Academic Publishers, Boston, 1988.

[35] Wilbur B. Davenport, Jr. and William L. Root. *An Introduction to the Theory of Random Signals and Noise*. IEEE Press, New York, 1987. IEEE PRESS edition of a book published by McGraw Hill Book Company in 1958 under the same title.

[36] Peter V. O'Neil. *Advanced Calculus, Pure and Applied*. Macmillan Publishing Co., Inc., New York, 1975.

[37] F. Gardner. *Phaselock Techniques*. Wiley, New York, second edition, 1979.

[38] J. D. H. Alexander. Clock recovery from random binary signals. *Electron. Lett.*, 11(22):541–542, October 1975.

[39] Benny Lai and Richard C. Walker. A monolithic 622Mb/s clock extraction data retiming circuit. In *ISSCC Dig. Tech. Papers*, pages 144–145, San Francisco, California, February 1991.

[40] Charles R. Hogge, Jr. A self correcting clock recovery circuit. *J. Lightwave Technol.*, LT-3(6):1312–1314, December 1985.

[41] S. Whitt. Automatic timing alignment for regenerative repeaters. *Electron. Lett.*, 21(24):1122–1123, November 1985.

[42] D. Shin, M. Park, and M. Lee. Self-correcting clock recovery circuit with improved jitter performance. *Electron. Lett.*, 23(3):110–111, January 1987.

[43] Lawrence DeVito, John Newton, Rosamaria Croughwell, John Bulzacchelli, and Fred Benkley. A 52 MHz and 155 MHz clock-recovery PLL. In *ISSCC Dig. Tech. Papers*, pages 142–143, San Francisco, California, February 1991.

[44] Thomas H. Lee and John F. Bulzacchelli. A 155 MHz clock recovery delay-and-phase-locked loop. In *ISSCC Dig. Tech. Papers*, pages 160–161, San Francisco, California, February 1992.

[45] Thomas H. Lee and John F. Bulzacchelli. A 155-MHz clock recovery delay- and phase-locked loop. *IEEE J. Solid-State Circuits*, 27(12):1736–1746, December 1992.

5

PRACTICAL ARCHITECTURES FOR HIGH-SPEED CLOCK RECOVERY AND DATA RETIMING

In the previous chapter, several techniques for recovering a clock from NRZ data were discussed. There are inherent disadvantages in nearly all of the architectures presented thus far, preventing integrated PLL based, clock recovery circuits from operating above 4-Gb/s [1]. Practical integrated hybrid solutions at 10-Gb/s have been described [2, 3] that use dielectric resonate filters to extract the timing information. However, using external filters for clock extraction has the added drawbacks of requiring precise phase adjustment, and power-hungry I/O buffers. In addition, packaging of the IC chip with the external filter can be problematic.

Thus far we have provided an overview of the underlying theories, and reviewed several circuits that fall short of our goal. In this chapter, practical architectures for clock recovery and data retiming ICs will be presented that are capable of operating at rates exceeding 10-Gb/s. We will illustrate how modifications can be made to some of the circuits discussed in chapter 4, extending their applicability to higher speeds. In section 5.4 a novel structure will be presented, culminating our effort in developing efficient, pipelineable, self-adjusting, data-density-independent structures for high-speed clock extraction. Circuit design techniques required to implement these practical architectures will be deferred to Part II of this book.

5.1 FREQUENCY DETECTION

As we have mentioned previously, any practical clock recovery circuit using a PLL will require some type of frequency acquisition aid. Exceptions are when very stable VCOs, such as crystal oscillators, are used to insure that the frequency error is never larger than the natural acquisition range of the PLL (on the order of the PLLs closed-

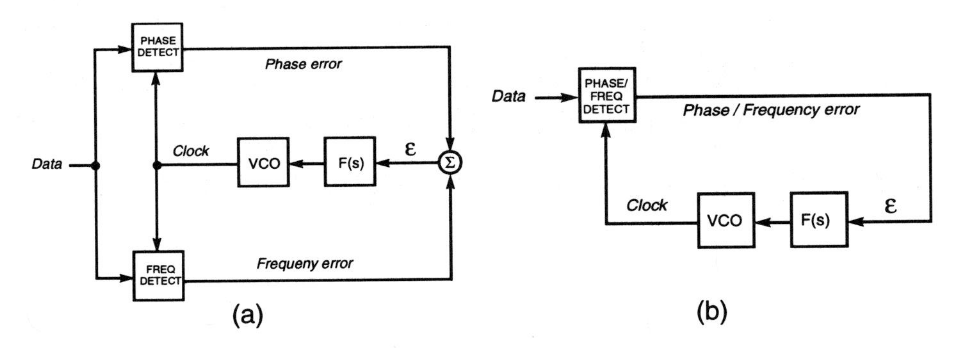

Figure 5.1 Illustrations of clock recovery PLLs using frequency detectors to aid frequency acquisition: (a) circuit summing phase and frequency errors, (b) circuit combines phase and frequency detectors into a single function.

loop bandwidth). Excluding these stable VCOs, and other acquisition aids, such as frequency sweeping, a frequency discriminator is required as an integral part of any phase-detector. Before presenting practical clock recovery and data retiming architectures, we will first briefly present some block diagrams illustrating how a frequency discriminator can be utilized in a PLL-based clock recovery circuit. Then we will present a rotational analogy that is very useful for deriving various frequency detector circuits.

5.1.1 Applications of Frequency Detectors in Clock Recovery Circuits

A straightforward application of a frequency detector to a clock recovery PLL is shown in Fig. 5.1(a). In this application, the error signal ϵ is the sum of a phase-error term, and a frequency error term. One requirement of the frequency detector is that its output go to zero when frequency acquisition has been obtained. Problems with this approach is that ripple from the frequency detector can still exist when the loop is in lock, causing excess phase-jitter. Also the frequency detector output needs to be taken into consideration when optimizing the loops dynamic response. Some systems use a *dead-zone* that breaks the frequency detector (FD) from the loop when the phase-error is within a zone surrounding zero. This prevents the FD from interfering with the phase acquisition process. A second alternative is to use a phase/frequency detector (PFD) as shown in Fig. 5.1(b). Although this may seem a trivial extension, later we will see that, with simple modifications, both phase, and frequency can be detected with the same circuit, thus eliminating duplicate functions.

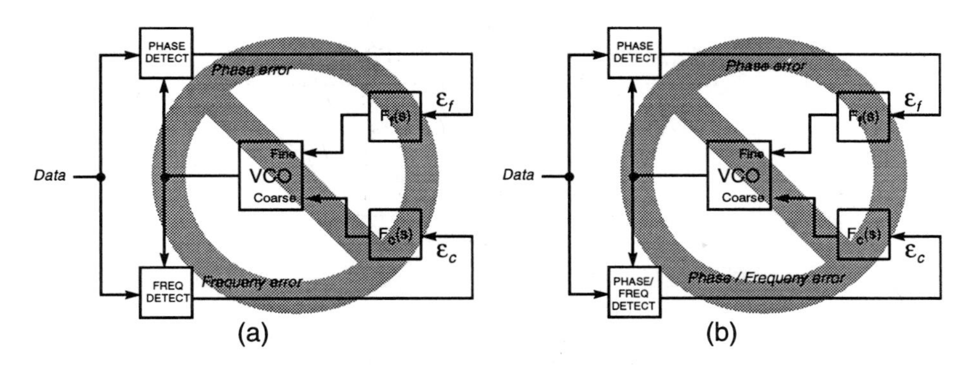

Figure 5.2 Bad ideas for using frequency acquisition aids for a VCO with two coarse and fine tuning inputs.

A Couple of Bad Ideas

In some cases it is desirable to have a VCO with two controls. One is a coarse adjustment used to set the center frequency close to the bit-rate, the other is a fine adjustment that is used to track the input, once the frequency error is within a specified range. A tempting idea, that invites all types of trouble, is shown in Fig. 5.2(a). Since no real integrator can be realized without dissipation, a VCO input signal will have to be continually updated to maintain its value at the proper level. This will require the error signals ϵ_f and ϵ_c to periodically deviate from a zero value. However, any significant deviation from zero in the signal ϵ_c will require a frequency error, and several cycle-slips will have to occur before the tuning signal can be readjusted to the proper value.

Another bad idea is to replace the FD in Fig. 5.2(a) with a PFD, as shown in Fig. 5.2(b). Since a non-zero value can appear at the output of the PFD in the absence of a frequency error, lock can be maintained without intermittent cycle-slips. However, now we have two loops that are fighting each other for control of the VCO phase. Provided that this condition produces a steady-state output, the resulting phase will most likely not be what is desired.

Techniques for Simultaneous Coarse and Fine Tuning

When adjusting two signals simultaneously, there must be sufficient degrees of freedom for a solution to exist. In other words, we can not try to drive the phase of a VCO to two different values simultaneously. A master-slave approach to setting the center frequency is shown in Fig. 5.3. The master-loop is used to acquire the input frequency.

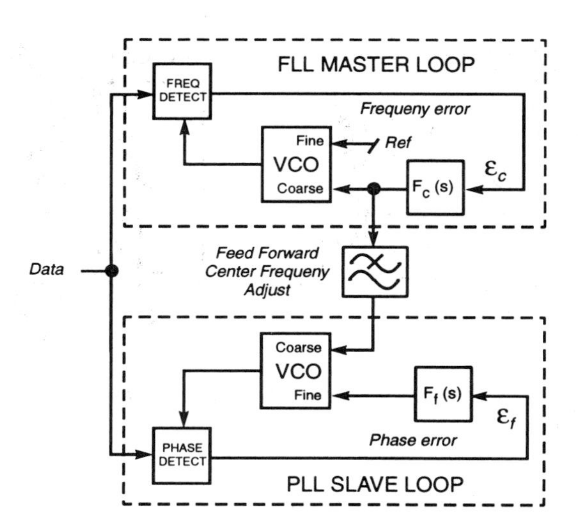

Figure 5.3 Illustration of a master-slave approach for simultaneous coarse and fine adjustment of the VCO controls.

Since the clock from the master loop is not the clock that samples the data, cycle-slipping is allowed in the master clock. It is only in the slave clock where cycle-slipping is forbidden. Therefore, an FD can be used in the master loop, creating a frequency-locked loop (FLL), that will not maintain phase-lock. The tuning signal can be fed forward to the slave loop, which contains a VCO matched to the master VCO. Using a filter in the feedforward path decouples the dynamic response of the two loops. For example, the master filter $F_c(s)$ can be adjusted to meet specific dynamic response requirements. Then using a lowpass filter in the feedforward path, can make the slave-loop appear as if the coarse tuning signal is a dc value. The master-slave approach can be used to reduce the steady-state phase offset, without requiring a high dc gain in the slave loop. The steady-state phase offset in a PLL is proportional to the frequency deviation of the input signal from the center-frequency of the VCO, and inversely proportional to the dc gain. The master-loop will reduce the frequency offset to within the matching accuracy of the VCOs, allowing the slave-loop to operate in the center of the dynamic range, without a high dc gain.

Delay-Locked Loops for Fine-Tuning the Clock Phase An alternative approach for adjusting the clock phase, after frequency and phase acquisition is established is shown in Fig. 5.4. In this circuit a PFD must be used so that the top-loop can maintain phase-lock. The resulting loop is a phase/frequency-locked loop (PFLL). However, the final VCO phase may still need compensation to achieve optimal clocking of the input data stream. This can be achieved by using a delay-locked loop (DLL), where a precise,

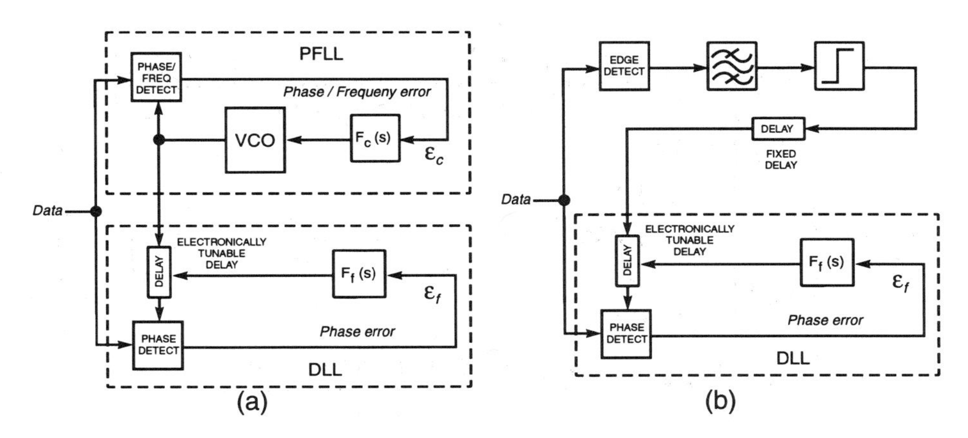

Figure 5.4 Clock recovery schemes using a voltage controlled delay for: (a) a clock extracted using a PLL, (b) a clock extracted using a bandpass filter.

self-adjusting, phase-detector measures the residual phase error, and fine tunes it to zero via a voltage controlled delay (VCD). It is also possible to replace the PFLL in Fig. 5.4(a) with a bandpass filter clock extractor as shown in Fig. 5.4(b). This approach was used by Wennekers *et al.* [3], to achieve 10-Gb/s operation, where the clock was originally extracted with a dielectric resonator filter.

In this section we have illustrated several possible methods for incorporating a frequency detector into the design of clock recovery circuits. Because of the important role FDs play in aiding PLL frequency acquisition, we will now present several implementations of FDs and PFDs.

5.1.2 Quadricorrelator Frequency Detector

We have already seen examples of circuits using a frequency discriminator in chapter 4. Both the circuits of Cordell *et al.* [4], and Ransijn and O'Connor [1] utilized a scheme similar to a quadricorrelator, which was first described, and given its name by Richman [5] in 1957. Richman applied the quadricorrelator to carrier-phase synchronization in color television. In 1976, Bellisio reported on a quantized quadricorrelator for use in clock recovery circuits for NRZ data formats [6]. Before discussing the general requirements of frequency discriminators, it is instructive to look at this quadricorrelator in more detail.

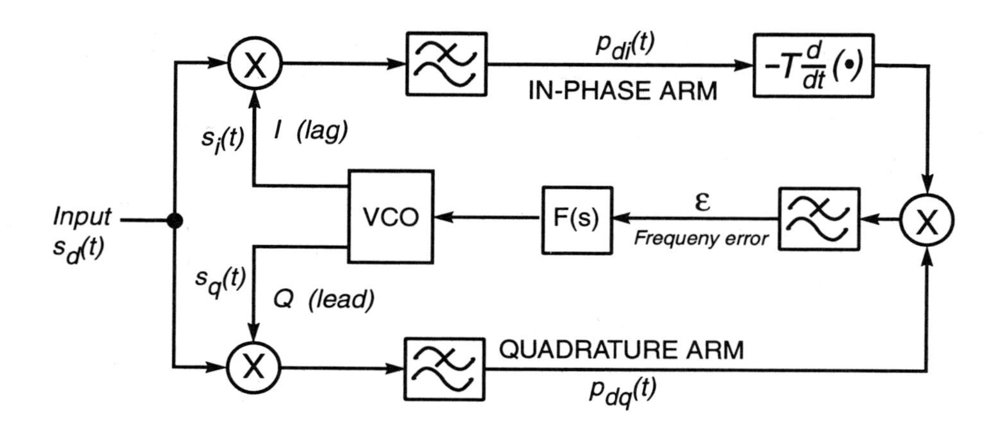

Figure 5.5 Block diagram of a quadricorrelator frequency-error detector.

A block diagram of a quadricorrelator is shown in Fig. 5.5. The circuit consists of two correlators: one is in-phase with the input signal, and the other is in quadrature. The input signals to the mixers have the following form:

$$\begin{aligned} s_d(t) &= \cos(\omega_d + \theta_d), \\ s_i(t) &= \cos(\omega_i + \theta_i), \\ s_q(t) &= -\sin(\omega_i + \theta_i). \end{aligned} \tag{5.1}$$

Modelling the mixers as ideal multipliers, their outputs are given by the sum and difference frequencies. For the in-phase arm,

$$m_i(t) = \frac{1}{2}\cos\big((\omega_i + \omega_d)t + \theta_i + \theta_d\big) + \frac{1}{2}\cos\big((\omega_i - \omega_d)t + \theta_i - \theta_d\big) \tag{5.2a}$$

and for the quadrature arm,

$$m_q(t) = -\frac{1}{2}\sin\big((\omega_i + \omega_d)t + \theta_i + \theta_d\big) - \frac{1}{2}\sin\big((\omega_i - \omega_d)t + \theta_i - \theta_d\big). \tag{5.2b}$$

After lowpass filtering the resulting signals are

$$p_{di}(t) = \frac{1}{2}\cos\big((\omega_i - \omega_d)t + \theta_i - \theta_d\big) \tag{5.3a}$$

and

$$p_{dq}(t) = -\frac{1}{2}\sin\big((\omega_i - \omega_d)t + \theta_i - \theta_d\big). \tag{5.3b}$$

We can define frequency and phase difference quantities such that

$$\begin{aligned} \Delta\omega &= \omega_i - \omega_d \\ \Delta\theta &= \theta_i - \theta_d. \end{aligned} \tag{5.4}$$

Therefore, the lowpass filtered signal for $\Delta\theta = 0$ are given by

$$\begin{array}{lclcl} p_{di}(t) & = & \frac{1}{2}\cos(\Delta\omega t) & = & \frac{1}{2}\cos(|\Delta\omega|t) \\ p_{dq}(t) & = & -\frac{1}{2}\sin(\Delta\omega t) & = & -\frac{1}{2}\text{sgn}(\Delta\omega)\sin(|\Delta\omega|t) \end{array} \tag{5.5}$$

Therefore, we see that the sign of the quadrature correlated signal $p_{dq}(t)$ depends on the sign of the frequency difference, whereas the in-phase correlation $p_{di}(t)$ is an even function of $\Delta\omega$. By taking the negative derivative of $p_{di}(t)$ we can generate a signal with an amplitude that is proportional to the frequency error;

$$-\frac{dp_{di}(t)}{dt}T = -\Delta\omega T\left[-\frac{1}{2}\sin(\Delta\omega t)\right] = \frac{1}{2}|\Delta\omega|T\sin(|\Delta\omega|t). \tag{5.6}$$

This signal is in-phase with the quadrature signal when the frequency error is negative and out-of-phase when the frequency error is positive. multiplying these two signal gives

$$M_{iq}(t) = -\frac{1}{4}\Delta\omega T\sin^2(\Delta\omega t) = -\frac{1}{8}\Delta\omega T + \frac{1}{8}\Delta\omega T\cos(2\Delta\omega t). \tag{5.7}$$

After lowpass filtering to remove the double frequency ripple, we are left with a dc value that is proportional to the frequency error, and opposite in sign;

$$\epsilon = -\frac{\Delta\omega T}{8}. \tag{5.8}$$

Therefore, if the clock is too fast, $\Delta\omega$ is positive, and the frequency detector outputs a negative value that can be used to slow down the clock. With a little thought, the reader will realize that this result is independent of $\Delta\theta$.

5.1.3 Rotating Wheel Analogy

Now that we've seen how a quadricorrelator frequency detector works, we can abstract the notion of frequency detection, and from this abstraction develop ideas that will be useful in alternative schemes. We can visualize the mixing, and lowpass filtering operations, using the analogy of a strobe-light and a rotating wheel with a timing mark. Consider a wheel rotating clockwise at a given rate. This is analogous to a local

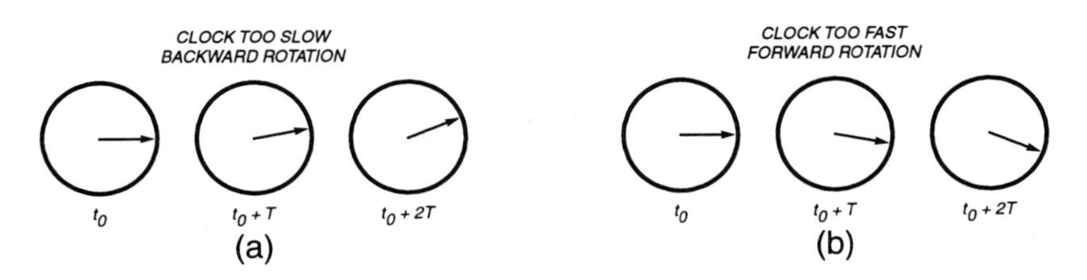

Figure 5.6 Illustration of rotating wheels with timing marks: (a) wheel appears to rotate backwards when clock is slow, (b) wheel appears to rotate forward when clock is fast.

clock with a given angular frequency. The data signal is used to control a strobe-light. Whenever the data makes a transition, the light flashes, revealing the position of the timing mark on the rotating wheel. The value of the clock signal at any point in time is the projection of the timing mark onto the positive x-axis. When the data-rate and clock are in perfect phase-lock, the timing mark will align precisely on the x-axis every-time that the strobe-light flashes; therefore, the mark will appear to be stationary. When a phase-error exists, the timing mark will be offset from the x-axis by the phase-error angle. Fig. 5.6(a) shows the condition when the clock-rate is too slow. In this case the wheel can not make a full revolution in one data interval. As the strobe-light flashes the timing mark appears to be rotating backwards at a rate equal to the difference between the data-rate and the clock frequency. In Fig. 5.6(b) the reverse is true. The clock is too fast, and the timing mark appears to rotate forward at the difference rate.

Phase is One-Dimensional; Rotation is Two-Dimensional

If we look only at the phase-error signal in Fig. 5.6, we see a projection of the timing mark onto the x-axis oscillating back-and-forth at a rate equal to the frequency difference. However, since we have taken a two-dimensional rotational concept, and projected it onto a one-dimensional line, we have no way of knowing the direction of the error; rotation in both directions produces the same shadow on the x-axis. To obtain directional information, we need a second timing mark, preferable one that is orthogonal to the first. An illustration of a rotating wheel with two orthogonal timing marks is shown in Fig. 5.7. For a clockwise rotation we see that the quadrature signal vector has a projection onto the x-axis given by

$$s_q(t) = -\sin(\omega_i t), \tag{5.9a}$$

and the in-phase vector I, has an x-axis projection of

$$s_i(t) = \quad \cos(\omega_i t) \tag{5.9b}$$

It can be seen from Fig. 5.7, that the quadrature signal leads the in-phase signal by $90°$. If the projection of Q onto the x-axis is used as the phase error signal for the PLL,

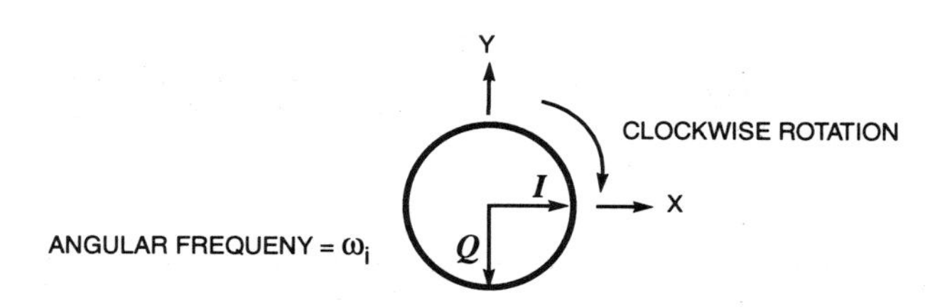

Figure 5.7 Illustration of a wheel rotating clockwise, with two orthogonal timing marks: I is the in-phase signal and lags the quadrature signal Q by 90°.

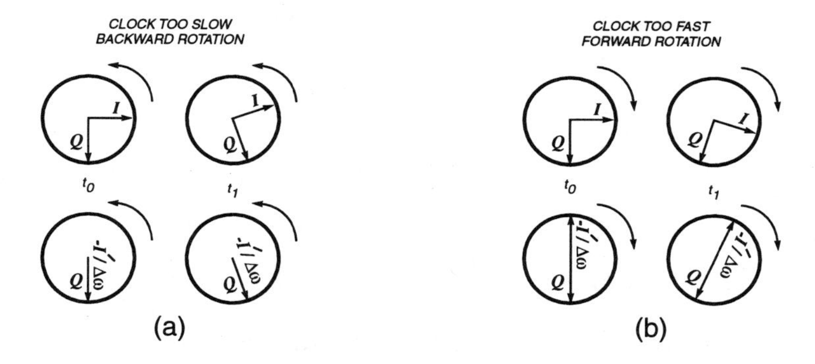

Figure 5.8 Illustration of the operation performed by a quadricorrelator.

then the position of the vectors shown in Fig. 5.7 shows the equilibrium condition, where the PLL will achieve phase-lock. The phase-error signal is zero in this case. If the wheel starts to rotate too fast, then Q will cross the y-axis into the negative x-half-plane, producing a negative error signal that acts to slow the wheel down. This equilibrium condition can be visualized as a marble sitting at the bottom of a cylinder.

Vector Diagram Representation of a Quadricorrelator

Using two orthogonal signal vectors, the apparent direction of rotation of the wheel can now be determined. When the clock is too fast, the beat-note rotation is forward, and Q still leads I by 90°. However, when the clock is too slow, the beat-note rotation reverses, and I now leads Q. We have already seen how a quadricorrelator uses this information to derive a frequency error. This is illustrated using the rotational analogy in Fig. 5.8. Taking the negative derivative of a sinusoidal signal delays the signal by 90°, and scales it by the frequency. In Fig. 5.8(a), we see that the negative derivative of the I vector falls on top of the Q vector when the clock is too slow, producing a positive frequency error proportional to $\Delta\omega$. Whereas in Fig. 5.8(b), delaying I by

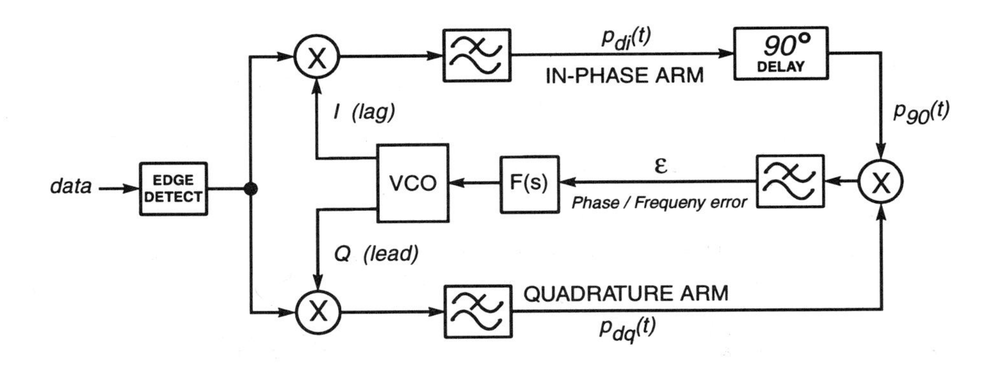

Figure 5.9 Block diagram of a phase-frequency detector.

90° causes it to point in the opposite direction of Q, therefore producing a negative error signal.

A Phase/Frequency Detector

We can use the vector operations of a quadricorrelator to derive a single circuit that produces both a phase, and frequency error. Instead of taking the negative derivative of the I vector, we could simply delay it by 90°. This will keep the magnitude of the resulting vector constant with respect to the frequency error. Therefore, a non-zero result can occur when the frequency error is zero. A circuit that performs this function is illustrated in Fig. 5.9. This phase-frequency detector was used in the clock-recovery circuit of Ransijn and O'Connor [1]. The in-phase and quadrature signals have the same phase relationships as in the quadricorrelator; the phase detector outputs are

$$\begin{aligned} p_{di}(t) &= \cos(\Delta\omega t + \Delta\theta) \\ p_{dq}(t) &= -\sin(\Delta\omega t + \Delta\theta). \end{aligned} \tag{5.10}$$

After delaying the in-phase signal by 90° we obtain

$$p_{90}(t) \propto \operatorname{sgn}(\Delta\omega)\sin(\Delta\omega t + \Delta\theta). \tag{5.11}$$

After multiplying $p_{90}(t)$ with $p_{dq}(t)$ we obtain

$$\begin{aligned} p_{90}(t) \times p_{dq}(t) &\propto -\operatorname{sgn}(\Delta\omega)\sin^2(\Delta\omega t + \Delta\theta) \\ &\propto -\operatorname{sgn}(\Delta\omega) + \cos(2\Delta\omega + 2\Delta\theta). \end{aligned} \tag{5.12}$$

Therefore, when a frequency error exists, the phase-error signal is the sum of a constant dc value, and a double frequency ripple term. If the lowpass filter completely suppresses

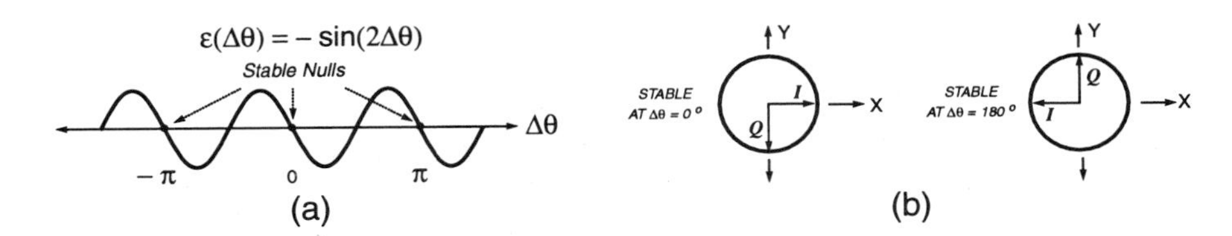

Figure 5.10 (a) Phase detector output vs. $\Delta\theta$ for maximum density data. (b) vector diagram showing the two possible equilibrium points.

the ripple term, then the error signal is given by

$$\epsilon \propto -\text{sgn}(\Delta\omega) \qquad \text{for} \quad \Delta\omega \neq 0. \tag{5.13}$$

When frequency acquisition has been established, $p_{di}(t) = \cos(\Delta\theta)$ will be constant, so delaying it by $90°$ will have no effect. Therefore, the phase error signal is given by

$$\begin{aligned} \epsilon &\propto -\cos(\Delta\theta)\sin(\Delta\theta) \\ &\propto -\sin(2\Delta\theta) \end{aligned} \tag{5.14}$$

It can be seen from Fig. 5.10(a), which plots the phase detector output as a function of $\Delta\theta$, that two stable nulls exist in each bit interval. The presence of two stable nulls can result in ambiguous results. In one case the clock edge will fall in the center of the bit interval as desired. In the other stable point, the clock edge will fall precisely on the data zero-crossings. One simple method of removing the undesired stable null ($\Delta\theta = 180°$ in this example) is to detect when I is negative and add a phase error that is sufficient to force the PLL off of this null. Perhaps a better method is to use a multiplexer to feed both clock phases to the decision circuit. The polarity of the I vector will determine which clock phase gets passed. Another alternative is to use a separate phase detector signal that has only one stable null per cycle, as in the circuit of Cordell.

This is easily obtained by taking the output of a quadricorrelator and summing it with the quadrature arm phase error signal. Before discussing some alternative frequency detectors, we will briefly discuss how nonlinear frequency acquisition and cycle-slipping can be visualized using the rotating vector diagram.

Vector Diagram Representation of Cycle-Slipping

This vector diagram is convenient for visualizing nonlinear frequency acquisition and cycle-slipping. Consider the case where we have a heavy flywheel spinning too fast, as shown in Fig. 5.11. For the case illustrated, the momentum of the wheel is so great, that although the phase-error is negative, forcing the wheel to slow down as the Q vector is in the negative x-half-plane, it can't put the brakes on fast enough to

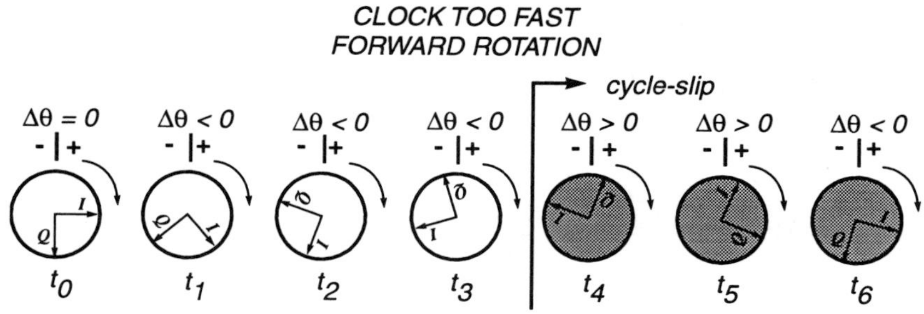

Figure 5.11 Illustration of cycle-slipping and frequency acquisition in a PLL with a sinusoidal phase detector characteristic.

prevent the wheel from rotating another half turn. Once the Q vector crosses back to the positive x-half-plane, then the restoring force on the wheel is positive, which is the wrong direction. As a result of this oscillating behavior, several cycle-slips may occur before a steady-state is reached. We can also reason that if the initial frequency error is large enough, and the wheel is sufficiently heavy (narrow closed-loop bandwidth), then cycle-slips can continue indefinitely, and the loop will never achieve lock.

It may seem slightly counter intuitive, at first glance, that a PLL could ever achieve frequency acquisition once cycle-slipping has begun, because the restoring force will oscillate, causing the wheel to alternately slow down, and then speed up again. However, from Fig. 5.11 we see that when the phase-error is negative, the restoring force acts to reduce the frequency error. Therefore the beat-note frequency will be reduced, and the phase-error Q vector will spend more time in the negative half-plane, than in the positive. The end result is that the wheel was slowed more in one cycle than it was sped up. If this difference accumulates over several cycles, then the PLL will eventually achieve lock.[1] The phase-error signal is plotted in Fig. 5.12(a) and (b) for frequency acquisition of a clock that is too slow, and too fast respectively. In each case the dc

[1] This method of frequency acquisition is reminiscent of a game one of the authors (A.B.) used to play with his brothers. The object of the game was to stop a rotating fan with your finger. The fan had a rubber blade. On the perimeter of the hub (about the diameter of coffee can), was the rubber sleeve of the fan blade, which was about the same thickness as a human finger. When the fan speed was low, stopping the blade with finger pressure was no problem, and produced only a faint smell of burning flesh. As we worked our way to the highest speed, the problem became increasingly difficult. When the speed was set on MAX, we could only leave our finger on the fan sleeve until we felt like it was going to catch on fire. We would then remove it for an instant, and quickly press it hard against the fan sleeve again. Each time we removed our finger, the fan speed increased, but it never got back to full speed; so that the next assault started at slightly more favorable initial conditions (much like taking a brief rest while running up a down escalator). Using this technique we were able to stop the fan on MAX speed. The winner of the game was the one who removed his finger the least amount of times before the fan was stopped. It goes without saying that any audible cries of pain resulted in immediate disqualification.

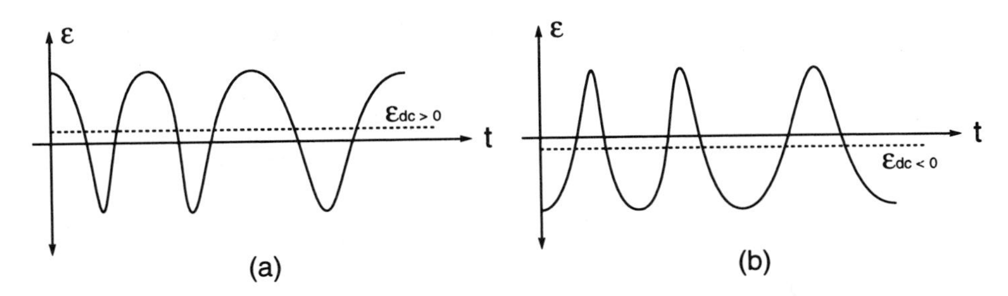

Figure 5.12 Illustration of the output of a phase-detector during closed loop cycle-slip of a PLL: (a) the local oscillator frequency is too slow, (b) the local oscillator frequency is too fast.

value of the phase detector signal is of the proper polarity to reduce the frequency error. If this signal is integrated, and then applied to the VCO control input. The dc value will accumulate and the loop will acquire the frequency. This is illustrated in Fig. 5.12 as the beat-note is shown to reduce in frequency with each cycle-slip. However, the dc value from the phase-detector can be quite small, especially for large frequency errors. In a real circuit, this small dc signal can be indistinguishable from offsets; also, the accumulation of the error signal may not have sufficient strength to overcome dissipation in the integrator. Therefore, in a practical PLL, the frequency acquisition range is similar to the PLLs closed-loop bandwidth (the heavier the flywheel, the smaller the range of frequency offsets that can be acquired).

5.1.4 Frequency Detectors Based on Rotational Analogy

The rotating vector analogy is convenient for understanding the necessary and sufficient conditions for a circuit to produce a frequency error output. If only the error direction is required, then one can develop a small list of rules that will produce this function. Based on these rules, several different, but similar circuits can be derived. In this section we will briefly discuss a few of these options.

Sequential Phase-Frequency Detector Circuits

An example of a circuit based on some heuristic rules is the well-known sequential phase/frequency detector. We realize from our vector analogy that some sort of past history of the signal is required to determine frequency error direction. However, instead of using signals separated by 90°, we could use digital storage elements to hold the quantized signal value at a given time, and use this delayed signal value, together with the current signal information to determine the direction of the frequency error. The sequential phase/frequency detector is such a circuit, and it has been described throughout the literature. In particular, Gardner offers a clear and concise explanation [7, pp. 121–125]. This circuit is commonly used in charge-pump PLLs at low frequencies, and is not applicable for our needs. Since it functions by producing pulses between data and clock transitions, it is not fast enough for high-speed applications, and it is confused by random data with missing transitions. We only mention it here for completeness.

Rotational Phase-Frequency Detector Circuits

Messerschimtt [8] presents a set of heuristic rules, also based on a rotational analogy. The clock-cycle is divided into four quadrants. By monitoring the position of the data transition in these quadrants a frequency error signal can be derived. Messerschimtt shows that a triangular phase-detector function results for periodic square-wave inputs. The triangle wave is monotonic over the range $[-T/2, T/2]$. However, Messerschimtt doesn't give any circuit implementations, and his technique is limited to low-speed applications. We mention it here as another example of a frequency detection scheme derived from a rotational analogy, and we will use a very similar concept in section 5.2.1 when we add frequency detection to Alexander's phase-detector.

A Cycle-Slip-Transition Frequency Detector

We will now consider a practical high-speed frequency detector based on quantizing the sampled outputs of each arm of a quadricorrelator. One very simple method of producing a frequency error is to detect a cycle-slip, and use the polarity of the quadrature signal to determine which direction the cycle-slip occurred. We can illustrate this with the rotating wheel diagram of Fig. 5.13. We can see from Fig. 5.13 that the I vector makes a negative transition at the top of the wheel, while the Q vector is positive, when the clock is too slow. When the clock is too fast, the I vector makes a positive-to-negative transition at the bottom of the wheel, where Q is negative. Therefore, we can derive the following simple rules for producing a frequency error.

- When I makes a negative transition, quantize Q to one-bit and pass it to the output.
- When I makes a positive transition, quantize Q to one-bit and pass the negative of Q to the output.

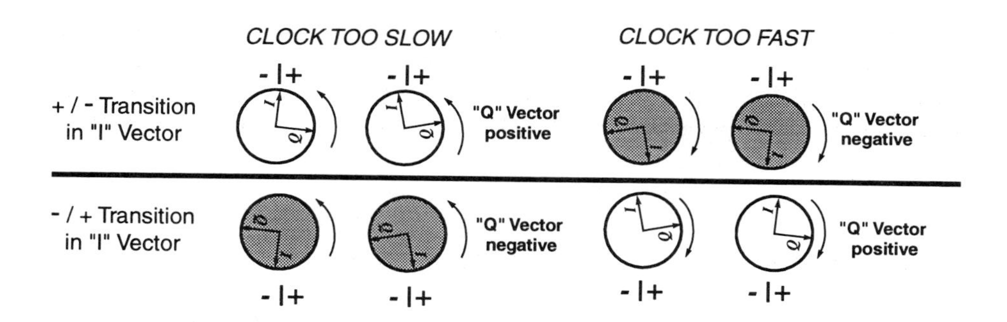

Figure 5.13 Illustration showing the position of the Q vector when a cycle-slip occurs: (a) the clock is too slow, (b) the clock is too fast.

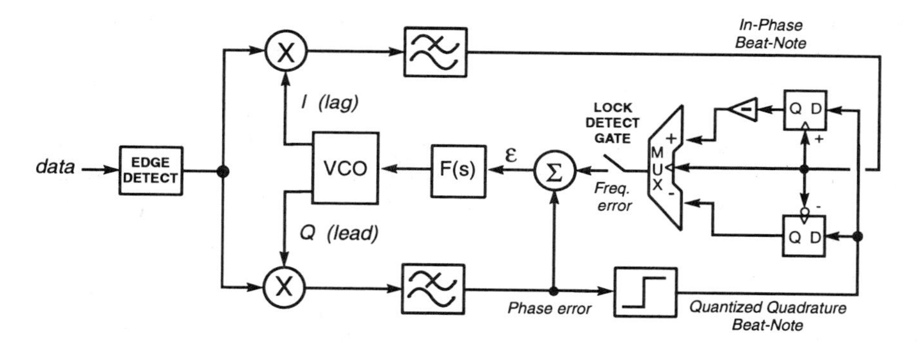

Figure 5.14 Block Diagram of a phase/frequency detector which produces unipolar pulses when a cycle-slip occurs. The pulse polarity depends on the sign of the frequency error.

A phase/frequency detector that implements these rules is shown in Fig. 5.14. In the presence of a frequency error, ϵ will consist of a zero-mean signal due to the phase error, and a dc signal in the opposite direction of the frequency error. We have embellished this circuit with a lock-detector gate. Since the previous frequency error will persist, even after the frequency error has gone to zero, we must detect this condition and force the frequency error signal to zero. One simple method of lock detect gating is to use a tri-state gate that is enabled by a transition in I, and disabled by a Q transition. Therefore the output of the frequency detector will be a series of unipolar pulses with the direction of the pulse determined by the frequency error. The pulses will be activated at a cycle-slip boundary and deactivated at the center of these boundaries. The in-phase signal is used to detect a cycle-slip boundary. Therefore, the quadrature signal can have a time offset relative to the input signal in the range $[-T/4, T/4]$ before a frequency error is detected.

The frequency detectors that we have discussed thus far require an edge-detector preprocessor. It seems reasonable that we could do away with the edge detector and determine the frequency error directly from the data. In the following section a stand-alone phase/frequency detector for NRZ data, which does not require edge detection preprocessing will be described.

5.1.5 Phase/Frequency Detector of Pottbäcker *et al.*

A clever phase/frequency detector (PFD) was recently reported by Pottbäcker *et al.* [9, 10]. This circuit uses the data transitions to sample the clock, and has nearly a one-to-one correspondence with the rotating wheel analogy, where the sampling of the clock signal is analogous to looking at the position of the rotating vector when a strobe-light, controlled by the data transition, flashes. A block diagram of this circuit is shown in Fig. 5.15. Although the circuit is not directly applicable to our needs as a self-adjusting clock recovery and data retiming circuit, it nevertheless implements useful functional building blocks efficiently, and serves as an excellent starting point for our discussion of practical clock extraction circuits. This PFD is fabricated in an advanced silicon bipolar technology, and can operate at a data-rate of 8-Gb/s. Although, the transistor-level design of this PFD is just as important, if not more, than the concept, we will only discuss the architecture of this circuit, and refer the interested reader to [9, 10] for a discussion of the circuit design details.

Since the polarity of the data is random, the sampling is performed on both the positive, and negative data edges. This was the same effect of detecting the data edges, but doesn't preprocess the data to produce an explicit edge-detected signal. When a frequency error occurs, the sampled signals Q_1, and I_1 will be beat-note square-waves at a frequency equal to the magnitude of the frequency error Δf. When $\Delta f > 0$, Q_1 will lead I_1, and the reverse will be true for $\Delta f < 0$. By monitoring the magnitude of I_1, when Q_1 makes a transition, the direction of the frequency error can be determined. When in lock, the in-phase signal I_1 will always by positive. The frequency error detector always outputs a zero value for $I_1 > 0$, so as not to interfere with the normal loop operation in phase-lock. Therefore two events must occur before a frequency error appears. The first is that I_1 must go negative. This occurs at timing errors greater than $|T/4|$ in magnitude. However, this value will not be transferred to the sample-and-hold output until a transition in Q_1 occurs at phase errors of $-T/2$ and $T/2$. Therefore, the circuit has the desirable property that the complete monotonic range of the phase detector is spanned before a cycle-slip is detected.

At high-data rates, this circuit has a fairly narrow pull-in range. It was reported that frequency errors on the order of 100-MHz can be acquired at 8-GHz. This is only

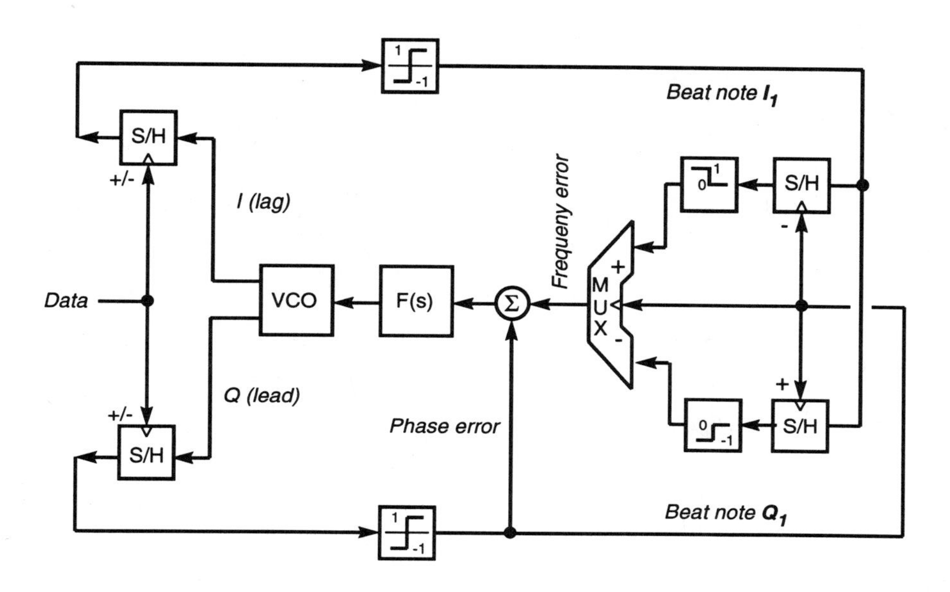

Figure 5.15 Block diagram of the phase frequency detector of Pottbäcker *et al.*.

1.25%, which means that the center frequency of the VCO must be stable to within this accuracy to guarantee acquisition. However, the factors limiting frequency pull-in were not discussed in [9, 10].

Clock Recovery and Data Regenerator IC Pottbäcker utilized this PFD to design a fully-integrated 8-Gb/s clock recovery and data regenerator IC [11]. At the time of this writing Pottbäcker's circuit is the fastest *fully-integrated*, PLL-based fiber-optic receiver.[2] Pottbäcker's circuit generates a clock at the proper frequency, but the steady-state phase is not well controlled and depends on several parasitic delays which depend on processing and temperature. Therefore the resulting clock may be very far away from the proper sampling phase. Pottbäcker addressed this problem by having the VCO generate four separate clock phases; the closest of the four to the optimal sampling phase is used as the active clock. This is an adequate approach in low SNR systems, where the phase of the recovered clock is not required to be precise. However, in long-haul telecommunication applications there can be a significant reduction in sensitivity due to offsets in the sampling phase, as was discussed in section 3.7.

[2] Several fiber-optic receivers have been reported which operate at higher data rates than 8-Gb/s, however these use some type of external filter for clock extraction and are not "fully-integrated."

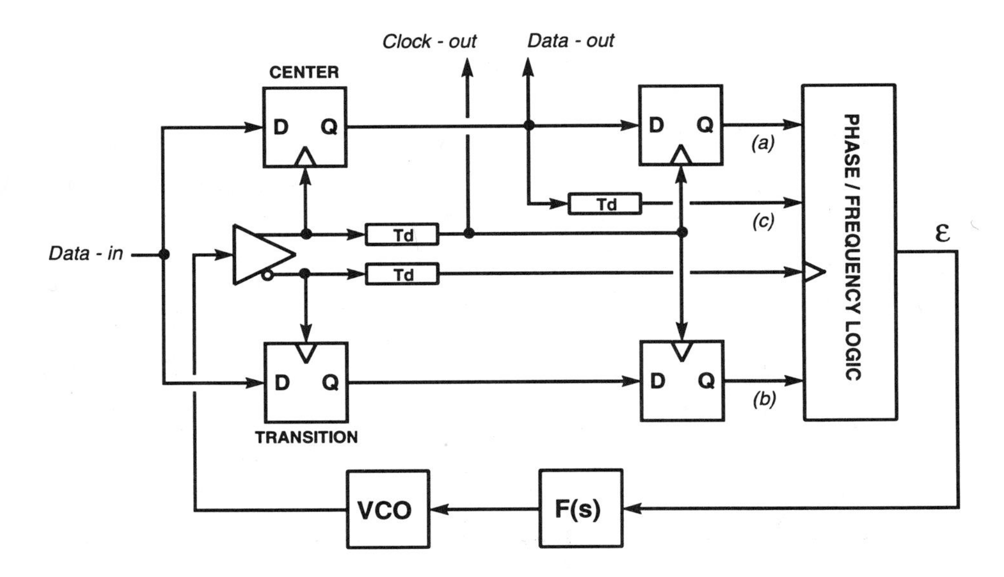

Figure 5.16 Block diagram of Alexander's clock recovery and data retiming circuit with deskewing delays added to account for flip-flop delays.

Despite the success of this PFD and clock recovery IC, it uses the data transitions to sample the clock, which is the opposite of what is required in a self-adjusting approach. Since we ultimately have to use the clock to sample the data in the decision circuit, it behooves us to use this information, if we can, in the phase/frequency detector. The remainder of this chapter describes practical high-speed architectures, using building blocks that function both as phase/frequency detectors and decision circuits. This not only gives us two functions for the price of one, but also leads to circuits that are inherently self-adjusting.

5.2 MODIFIED ALEXANDER CIRCUIT

The first practical clock recovery and data retiming circuit that will be presented is a modification of Alexander's circuit [12] that was shown in Fig. 4.59. The basic operation of the circuit was explained in chapter 4. The modified circuit is illustrated in Fig. 5.16. In normal operation, the flip-flop labelled *center*, samples the data signal in the center of the eye-diagram, and the *transition* flip-flip samples the signal at the data transitions. The binary quantized outputs of these two circuits are clocked into storage locations for further processing. The delay of the decision circuit is denoted as t_d. We have modified the basic structure by adding delays of approximately t_d to skew the clock before resampling the quantized data. This allows the outputs of the *center* and *transition* flip-flops to settle. An identical delay is used to skew the signal (c) so that the samples (a), (b), and (c) arrive at the phase/frequency detection logic block simultaneously.

Notice, that the delays elements do not have to be precise. They only have to be close enough to t_d to insure that the proper value is resampled into the deskewing register. The relative phase information between the data and clock is contained in the output of the *center* and *transition* flip-flops. In a fully-differential circuit, inversion of the clock is accomplished simply by reversing the polarity of the differential signals. Therefore the positive and negative clocks are exactly 180° degrees out of phase, and precise timing is inherent in this structure. When the loop is in lock, the phase separation is $T/2$. It is this time offset that controls the phase-estimate accuracy. Once the samples have been quantized, the timing of the rest of the circuits will not affect the results. As long as the deskewing is sufficient to allow the circuit to work at high data-rates, the phase estimate will be independent of changes in t_d.

5.2.1 A Frequency Detector Based on Sequential Early-Late Decisions

We have stated that the logic block can detect errors in frequency by using the information that it is provided. We have a couple of options for adding frequency detection to the standard Alexander circuit. One technique is to use a start-up sequence. This allows frequency acquisition to occur before the random data is sent, and the frequency error can be detected easily with virtually no additional hardware. The second alternative adds complexity to the receiver design, but does not rely on a start-up sequence to establish frequency acquisition. Both of these techniques will now be briefly described.

Frequency Detection Using a Start-up Sequence

If during a start-up phase we send a sequence of alternating ones and zeros (a periodic waveform at a frequency of $B_T/2$), we will have no trouble detecting a frequency error using the samples that are already available. In fact, we can see that the Alexander circuit is actually the front-end of a quantized quadricorrelator, provided that the input signal is periodic with a period of $2T$. We can therefore use the samples (b) and (c) as the quadrature and in-phase samples of the waveform respectively. This technique was used by Walker *et al.* [13, 14] in a 1.5-Gb/s serial data link.

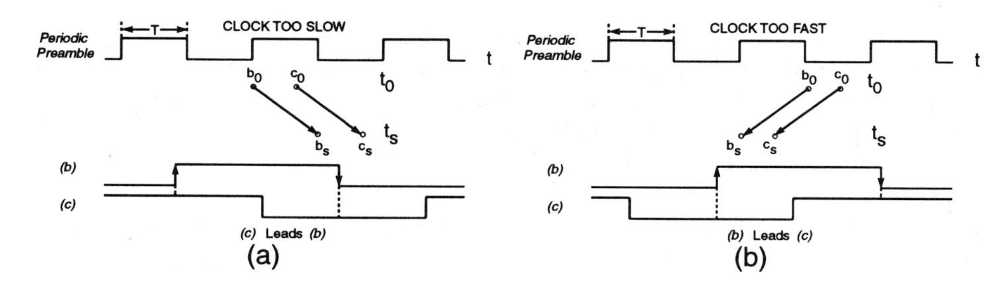

Figure 5.17 Sample of an Alexander circuit for a start-up sequence of a square-wave at a frequency of $B_T/2$. (A) When the clock is too slow, the sampling points move to the right and (c) leads (b). (B) When the clock it too fast, the sampling points move to the left, and (b) leads (c).

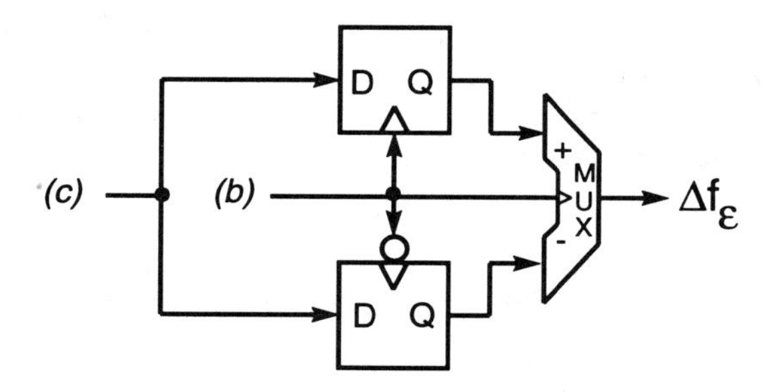

Figure 5.18 Conceptual block diagram of a frequency detector of Alexander's circuit when a start-up sequence is used.

Consider the case illustrated in Fig. 5.17. The frequency error is detected easily with one flip-flop and a multiplexer. When (b) makes a positive transition we pass (c) to the output, and when (b) makes a negative transition, we pass the negative of (c). This is shown conceptually in Fig. 5.18.

Rotational Four-Quadrant Frequency Detector

In many cases, it may be undesirable from a systems standpoint to use a start up sequence to insure frequency acquisition. In these instances, we need to obtain frequency error information from the random data itself. The quadricorrelator of the previous section will be confused by random polarity variation in the data. Therefore, we need to adopt a different approach, and we turn again to the rotational analogy. One alternative is too add two more sampling flip-flops at the front-end. If we offset these samples (x) and (y) by $T/4$ from the original samples, then we can arrive at the sampling

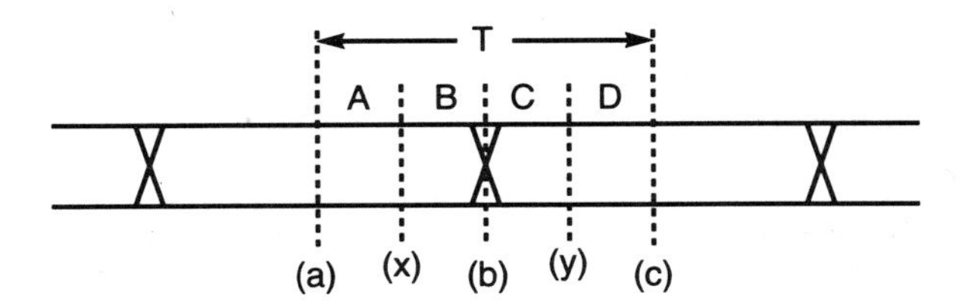

Figure 5.19 Illustration of ordering of samples in a modified Alexander circuit with additional samples (x) and (y) added at an offset of $T/4$ from (a) and (b) respectively.

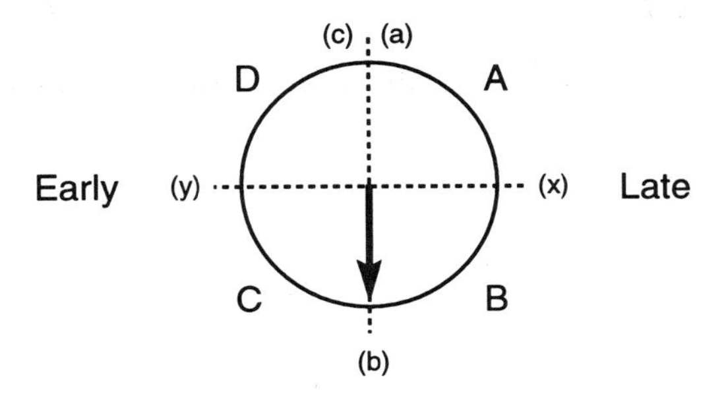

Figure 5.20 Rotational analogy of data transitions

order illustrated in Fig. 5.19. We can also represent four quadrants, [A,B,C,D], as the time intervals between successive samples. In normal operation there will either be no transitions, or one transition between the samples (a) and (c). When the loop is in phase-lock, the transition should fall precisely at the sample (b). We can represent a clock-cycle of length T on a circle, and consider the transition location, as if it were rotating. This circle is shown in Fig. 5.20, which illustrates the locked condition, where the data-transition occurs at sample (b). In normal locked operation this transition will fall either in quadrant B or C. We can devise our frequency detection circuit so as not to interfere with the normal locked condition. Therefore, we can set the frequency error to zero whenever the data-transition is in quadrant B or C. A cycle-slip is detected when the transition crosses into quadrant A or D, at which time, the frequency error signal is activated. This provides a time-offset range of $[-T/4, T/4]$ over which the frequency error is always equal to zero.

The cases of a clock that is too slow, and one that is too fast are shown in Figs. 5.21(a) and (b) respectively. We can now use the direction of the rotation of the transition to derive a frequency error. A conceptual circuit for obtaining this error is shown in Fig. 5.22. The output of the SR flip-flop is a series of positive pulses. The signal is equal to zero in quadrants B or C, and is high in quadrants A or D. This signal is

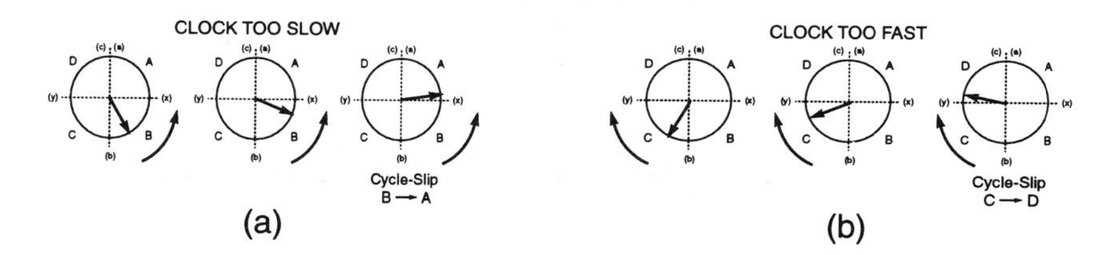

Figure 5.21 Rotational analogy of cycle-slipping: (a) clock is too slow, (b) clock it too fast.

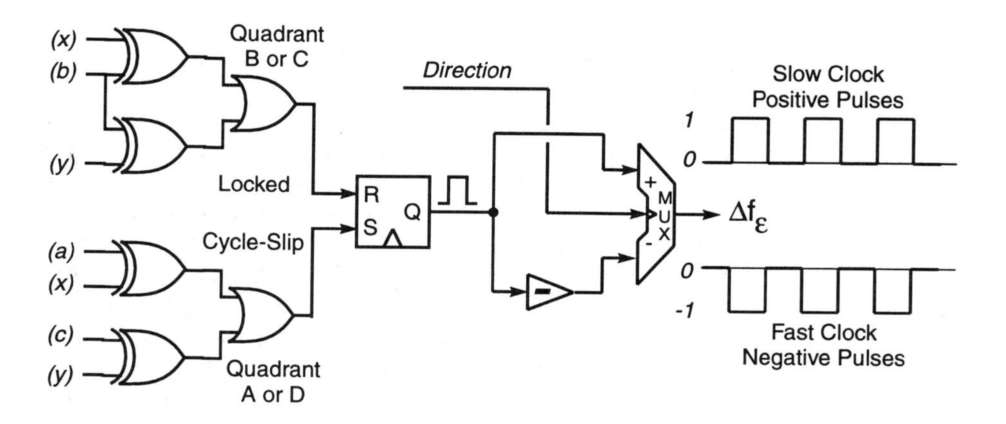

Figure 5.22 Conceptual block diagram of a frequency error detector for NRZ random data based on a four-quadrant rotational analogy.

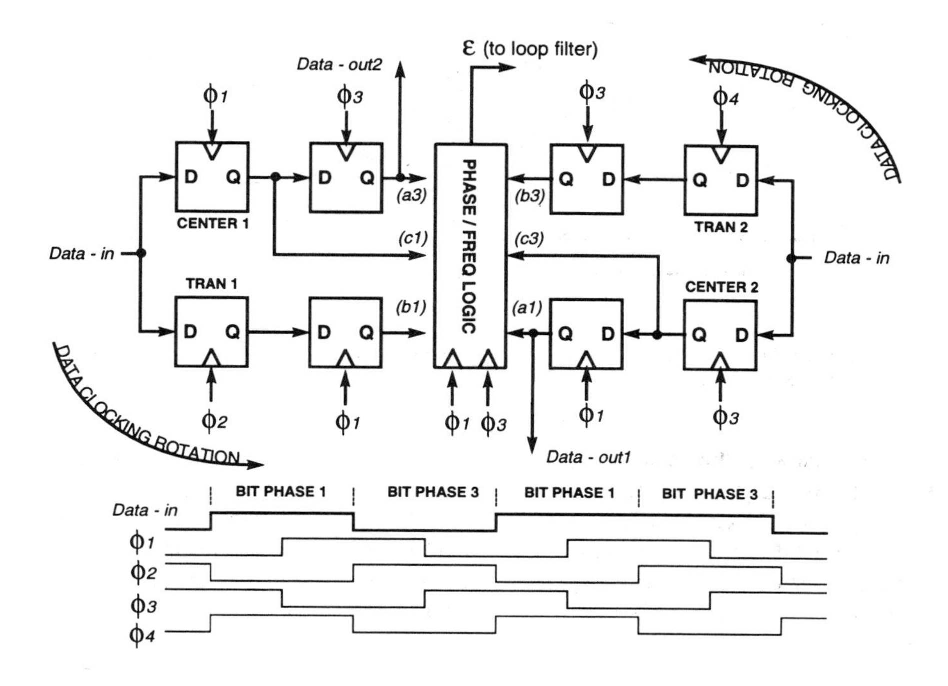

Figure 5.23 Block diagram of a 2-level interleaved implementation of Alexander's clock recovery and data retiming circuit.

either passed directly to the output, or is inverted and then passed, depending on the direction of the rotation. In this diagram we have shown that the direction signal should be set high when a B-to-A transition occurs, indicating that the clock is slow, and the transition is rotating backward. When the clock is fast, the signal is set low on a C-to-D transition. We will not show a complete schematic here. Once the rotational concept is understood, the implementation is straight forward. In the following section we will discuss a further modification to the Alexander circuit, showing how the throughput can be increased by using bit-interleaving.

5.2.2 Interleaved Alexander-Late Circuit for Higher Throughput

It is clear that Alexander's circuit is easily pipelined. If the throughput of the receiver is limited by the delay of the decision circuit, then two decision circuits can be interleaved to double the maximum bit-rate. A block diagram of an interleaved circuit is shown in Fig. 5.23. The clocking of this circuit assumes that the flop-flops have no delays.

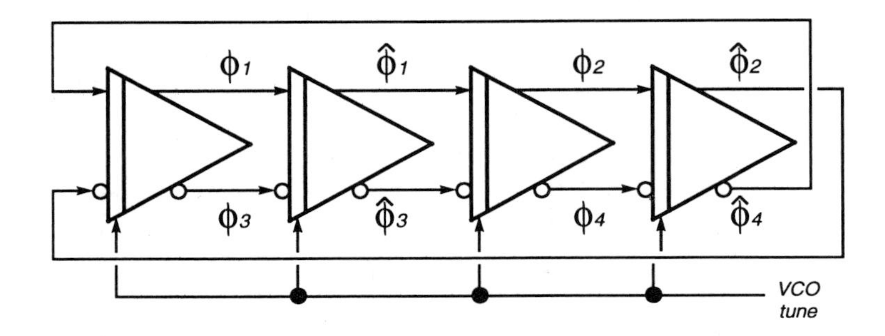

Figure 5.24 4-Stage ring VCO used to generate multi-phase clocking for bit-interleaved clock recovery scheme.

One can visualize the sampling of the data as rotating counter-clockwise around the four outer flip-flops. On clock cycle ϕ_1, (a_1), (b_1), and (c_1) are clocked into the logic PFD logic block; on phase ϕ_3, (a_3), (b_3), and (c_3) are transferred. The PFD logic makes decisions as to whether the clock was early or late, and multiplexes the result of phase 1, and phase 3, onto a signal ϵ which is sent to the loop filter, and in turn, to the VCO. Obviously at high-speed, this circuit will require embellishments to account for flip-flop delays.

Interleaved Circuit using a 4-Phase Clock to Account for Flip-Flop Delays

An efficient means of deskewing signals, before resampling is to make use of multi-phase clock signals to compensate for the decision circuit delay. A simple multi-phase VCO is shown in Fig. 5.24. This VCO is a 4-stage ring oscillator, built with differential tunable delay elements. As well as the four phases $[\phi_1, \phi_2, \phi_3, \phi_4]$, we also have an additional four signals $[\hat{\phi_1}, \hat{\phi_2}, \hat{\phi_3}, \hat{\phi_4}]$, for a total of eight possible clocking phases. Therefore, time offsets in increments of $T/4$ can be obtained by skipping the appropriate number of clock phases. If we consider an example where the decision circuit delay is close to the bit-period T, then we can obtain the clocking scheme for a high-speed circuit as shown in Fig. 5.25. Since ϕ_1 and ϕ_2 are separated by $T/2$, then a delay of T in the clocking corresponds to skipping to phase ϕ_3. Therefore, the clock phases in brackets in Fig. 5.25 have been obtained by adding a delay of T to the clocks shown in Fig. 5.23. However, we still need to add a delay of approximately T, for skewing of the signals (c_1) and (c_3). This can be accomplished by using four identical delay cells as were used in the VCO of Fig. 5.24. Since the VCO delay cells have a delay of $T/4$, feeding the control signal forward to the deskewing elements produces a delay of exactly T. However, since this delay does not have to be exact,

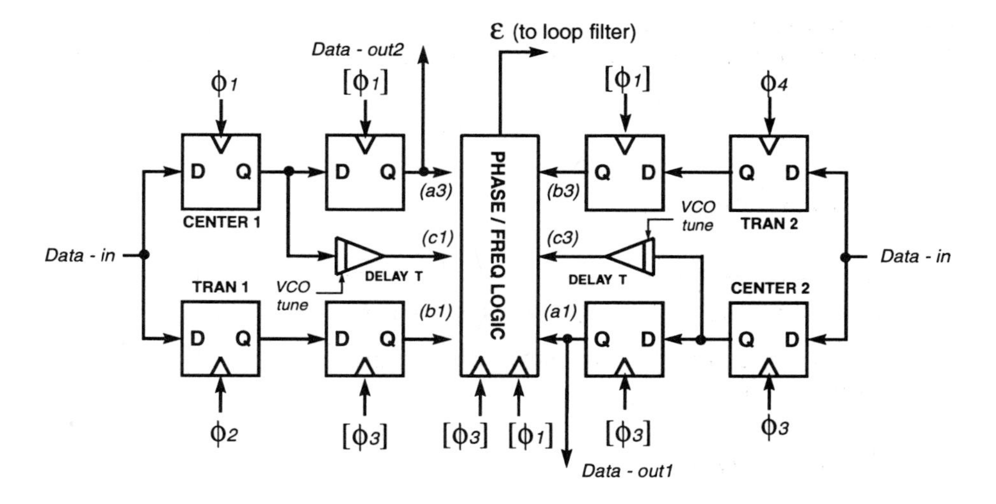

Figure 5.25 Practical circuit showing the clocking scheme required when the flip-flop delay is close to the bit-interval T.

we could use a VCO delay cell running at a quarter of the current, and at four-times the impedance level to produce a delay approximately four-times as large as the delay of a single VCO cell, thereby saving in both area, and power dissipation.

We could also implement a rotational frequency detection scheme easily with this 4-stage ring VCO. The samples offset by $T/4$ can be taken using the clock phases $[\hat{\phi}_1, \hat{\phi}_2, \hat{\phi}_3, \hat{\phi}_4]$. This interleaved circuit with frequency detection is ideally suited for circuits implemented in fine-line CMOS, where the phase/frequency detection logic can be realized efficiently, and data rates in excess of 2-Gb/s can be realized.

Maximum Speed of Interleaving

The interleaved Alexander circuit is an example of a practical self-adjusting clock recovery and data retiming circuit that is capable of operating at very high data rates. The 4-stage ring VCO has a frequency of one-half the bit-rate, and a period of $8t_{\text{VCO}}$. The bit-period T is given by

$$T = 4t_{\text{VCO}} \tag{5.15}$$

A delay cell can be realized such that the delay is between $1/f_{\max}$ and $2/f_{\max}$. Therefore, the maximum bit-rate that can be achieved is approximately in the range

$$\frac{f_{\max}}{8} < B_T < \frac{f_{\max}}{4} \tag{5.16}$$

Advantages of using this circuit are as follows.

- If all 4 outer flip-flops are matched, the clock will be automatically adjusted to the proper phase, independent of the delay and setup times of the flip-flops.
- Frequency detection can be accomplished with some additional digital logic, or by using a start-up sequence.

There are some disadvantages however.

- The phase error is obtained by making a hard decision, and, therefore, the circuit is limited to high SNR applications.
- Since the transition sample is quantized, it will randomly alternate between a high and low value. This oscillation will increase jitter in the recovered clock.

Although we mention these disadvantages based on qualitative reasoning, no analysis was performed by these authors to determine precisely at what SNR the circuit falls apart, or how much excess phase-jitter results in the clock as compared to a MAP estimator.

Limitations of Flip-Flops as Decision Circuits

Bit-interleaving can be extended to higher orders, but generally, one would not use more than two levels. The reason is that additional amplifiers and filters are also needed on the same IC chip — fabricated with the same transistors as the decision circuit. It is a safe bet, that if one can not make a decision circuit fast enough to respond within two bit periods, then the chances are that one also won't be able to build an amplifier fast enough to handle the raw data.

An additional problem occurs when a flip-flop is used as a decision circuit for high-speed applications. A decision circuit must perform three separate tasks:

1. **sampling;** the signal must be observed at a specific instant in time,

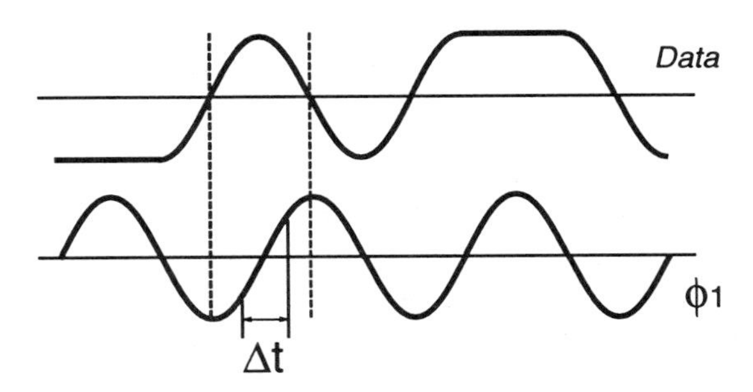

Figure 5.26 Short section of NRZ data and recovered clock.

2. **comparison;** the sampled signal must be compared to a reference and quantized (one-bit quantization in this case),

3. **regeneration;** the signal must be latched using positive feedback to regenerate the signal so that it will not change state until the next active clock phase.

In a *ideal* flip-flop each of the three functions are separated. However, when the clock rise-time is slow, these function will interfere with each other. In the limiting case the data will be changing at the same speed that the clock is rising; this can produce ambiguous results.

Example of Flip-Flop Clocking Problems at High-Speed We will illustrate the type of problems that can occur with an example. If we consider a case where the decision circuit delay is on the order of the bit-interval, then the data will be moving very fast relative to the response time of the flip-flop. A short section of a data signal and a recovered clock is shown in Fig. 5.26. A typical flip-flop works by using the clock signal to switch between a track-mode, and a regeneration-mode. In the regeneration mode, positive feedback is used to clamp the signal in one of two directions, depending on the polarity of the signal at the end of the track-mode. However, we see in a high-speed system, that the clock-edge will have a transition Δt that is a significant portion of the bit-interval. During this transition, the track-mode and regeneration-mode signals of the flip-flop will be fighting against each other. This is a serious problem when the data makes a transition. In the case illustrated, when the clock makes a positive transition, we switch from tracking to regeneration. As the clock begins a positive transition, the track-mode signal starts to become attenuated in favor of the regenerated signal. This can have the effect of reducing the SNR. For example, if the track signal is supposed to be positive, but it has not crossed the axis when the clock transition starts, then the positive feedback of the regeneration will act to keep the signal negative as the track-mode tries to push it positive. Therefore, the end result will not be as positive as it should be.

During the second half of this cycle, the regeneration will be pushing the output positive, but if the next bit is negative, the track-mode circuit plays the role of spoiler, and can prevent the regeneration from reaching a maximum positive value, especially when the data-sample was originally small in magnitude. We therefore are faced with two fundamental problems when using flip-flops as decision circuits.

- The maximum SNR of the data sample will be reduced by the interactions between the track-mode and regeneration-mode circuit.
- It is difficult to tell where the clock-edge should be placed to achieve a maximum SNR sample. Since, interactions within the flip-flop itself affects the SNR of the sample, there is no reason to believe that the optimal clocking phase will be in the center of the data eye.

Therefore, we see a definite problem with Alexander's method when using flip-flops as the decision circuit. The feedback mechanism is set up to find the point of sampling that is most random (the crossover point), and sample the data with a clock that is offset by $T/2$ from this point. Although this is a good place to sample, we are not sure if it is optimum, because interaction within the flip-flop can effect the sampling process.

From the above discussion, we see that the problem with flip-flops as high-speed decision circuits is the interaction between the track- and regeneration-modes. If we can decouple these two modes, then we can obtain a higher SNR sample, and get a better estimate of the optimal sampling phase. This decoupling can be accomplished by preceding the regeneration stage with a sample-and-hold circuit. Using a sample-and-hold will keep the signal fixed before it is regenerated by the flip-flop, thus doubling the setup time. This will allow the full benefits of interleaving to be realized. In the following sections we will present two clock-recovery, and data-retiming circuits that make use of high-speed sample-and-holds.

5.3 EARLY-LATE CIRCUIT USING A MATCHED FILTER

We saw in chapter 4 that an early-late gate clock recovery circuit can be realized using a matched-filter, together with sample-and-holds, as was illustrated in Fig. 4.57. A slightly different version of this circuit is shown here in Fig. 5.27, where we take the

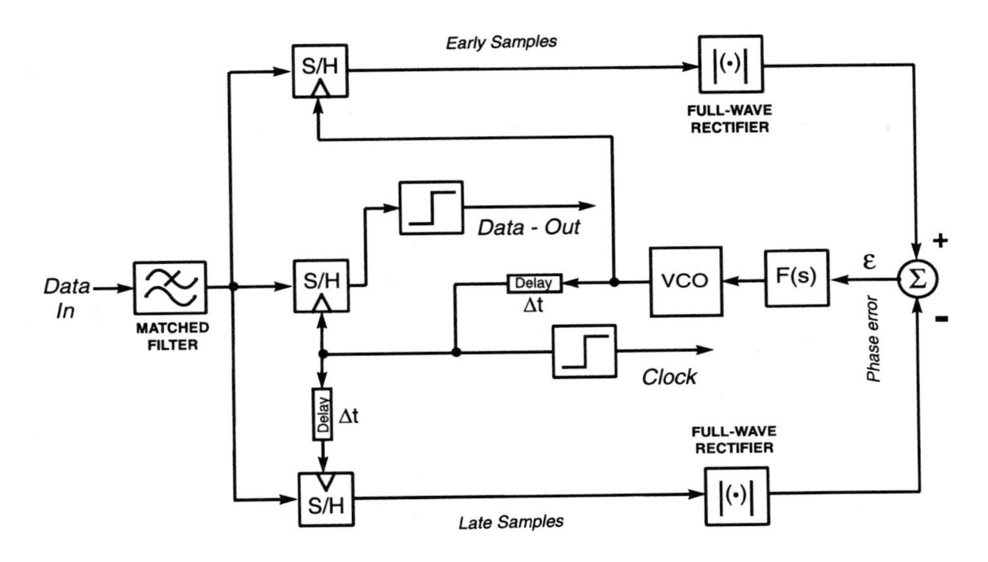

Figure 5.27 An early-late gate clock recovery and data retiming circuit using a matched filter and sample and holds to replace the integrate and dump function.

absolute value of the early and late samples to remove phase-error-polarity ambiguity caused by the random data. We recall from the discussion of early-late gates in chapter 4, that the offset Δt should be closed to $T/4$ for maximum phase-detector gain, and the resulting error signal ϵ will contain a significant ripple component when the loop is in lock. This ripple occurs because the early and late samples are skewed in time relative to each other.

It is desirable to separate the early and late samples by $T/2$ to obtain a phase-detector characteristic that is monotonic over the complete bit-interval. This was the case for the circuit of Fig. 4.57, which is repeated here in Fig. 5.28 for convenience. We can see that this circuit is very similar to Alexander's circuit with some notable exceptions.

- The flip-flop decision circuits used in Alexander's realization is replaced by a sample-and-hold followed by a limiter/regenerator. This decouples the sampling from the regeneration and improves the SNR of the data sample.

- Since the error signal in not quantized, it is a continuous function of the phase-offset. This reduces the ripple at the phase-detector output and therefore reduces the clock jitter.

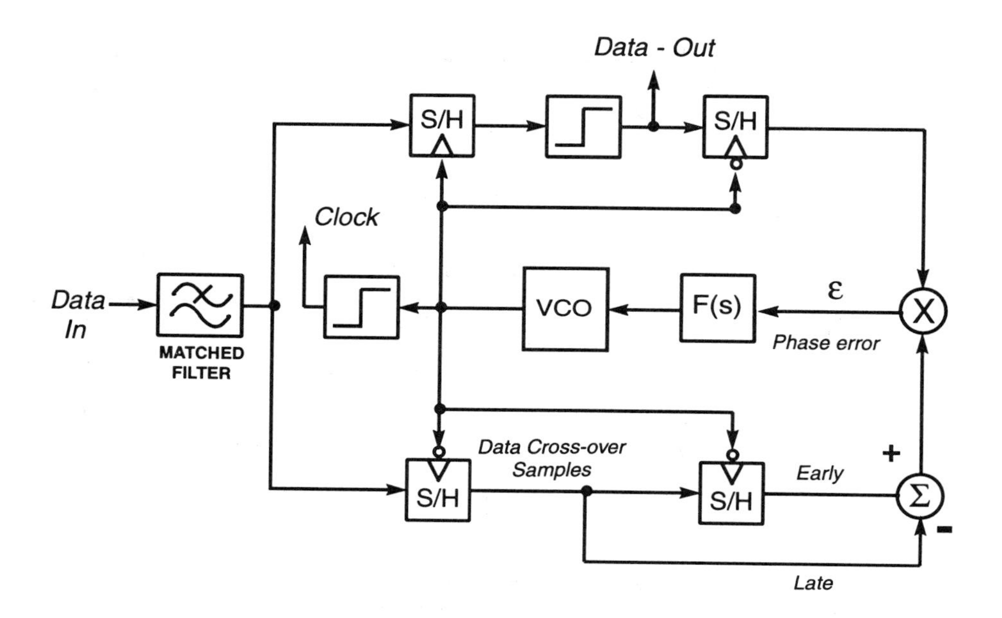

Figure 5.28 Early-late gate with matched filter and sample and holds.

- In the sampled early-late circuit, we make use of one on-time sample, and two cross-over samples (early) and (late), to produce a phase-error signal. Whereas in Alexander's circuit, two on-time samples (a) and (c), and one cross-over sample (b) was used.

In a practical implementation of the early-late circuit of Fig. 5.28, we would use bit-interleaving to increase the throughput. The VCO clock will then be running at half of the bit-rate, and the sample-and-hold functions can be realized by multiplexed track-and-hold circuits. It is left to the reader to figure out the proper clock scheme for such a bit-interleaving technique. We will turn instead to a superior circuit, that is very similar to that of Fig. 5.28. However, this new circuit adds one additional level of resampling, and by doing so, makes the phase-error signal independent of the data transition-density.

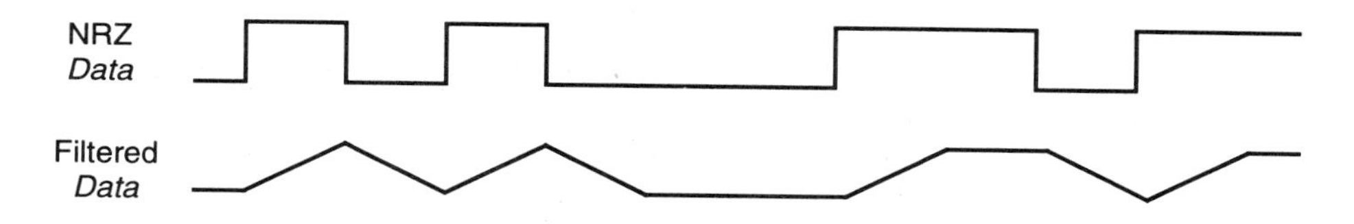

Figure 5.29 NRZ data before and after being processed by a matched filter.

5.4 HIGH-SPEED DATA TRANSITION TRACKING LOOP

At the very outset of our discussion of clock recovery for digital NRZ data, we emphasized that timing information is only present when the data makes a transition. It is therefore quite logical to ignore any output of a phase-detector during intervals when no data transition occurs. This implies that we use some type of gating mechanism to shut off the phase detector in the absence of transitions. One example of this was seen in the circuit of Cordell *et al.* [4], where a tri-state gate was used to null the phase-error when no transitions were present. Devito *et al.* [15] and Lee and Bulzacchelli [16, 17] used a technique to hold the previous phase-error value between data transitions, thereby making the phase-detector output independent of the transition-density. The circuit presented in this section is a slight modification of an early-late gate, utlizing an additional sampling stage, such that the phase-detector output is transferred to the loop-filter only when a data transition has occurred. This circuit was used in the Mars Mariner in 1969, and was patented by Anderson, Hurd, and Lindsey in 1971 [18].[3] Lindsey and Simon [19, pp. 442–457] refer to this circuit as a *data transition tracking loop* (DTTL). They also present considerable performance analysis of the DTTL.

The DTTL can be understood most easily by considering the time domain operation. Fig. 5.29 shows typical rectangular NRZ data before, and after, being passed through a matched filter. The matched filter produces a linear transition, which extends over the bit interval. Since we are only interested in what happens during transitions, we can restrict our attention to the positive data pulse shown in Fig. 5.30, from which we can determine the circuit behavior during a data transition. Recall that the ultimate goal is to position the clock-phase so that it samples the data at the point where the filtered data signal achieves a maximum SNR. Therefore, we would like to find the gradient of the on-time samples and force this value to zero. Finding the gradient can be greatly simplified by realizing that the pulse shape is essentially sinusoidal. In the case of high-speed systems, several parasitic poles near the data-rate provide additional filtering of the data. The result is that the data transitions follow nearly a sinusoidal path; for a data pattern of alternating ones and zeros, the data signal is a sinewave at half the bit-rate. For the special case of sinusoidal transitions, the gradient of the

[3] U.S. and Canadian patents are pending for high-speed realizations of DTTLs presented in this section.

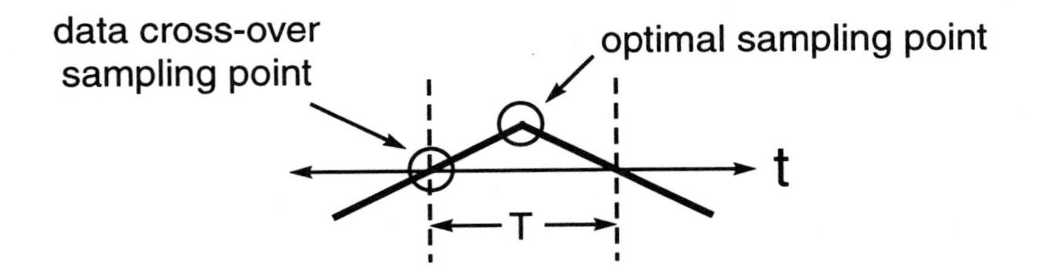

Figure 5.30 A positive data pulse showing the optimal sampling point.

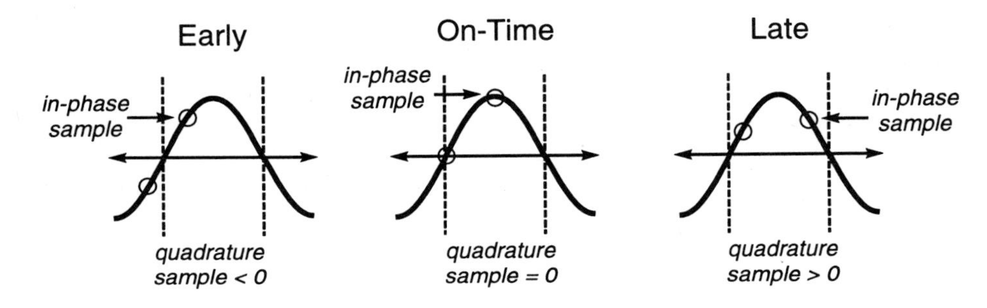

Figure 5.31 Illustration of a positive data pulse. The gradient of the on-time sample is equal to the value of the cross-over sample.

on-time samples can be found by shifting the sampling phase half of a bit-period, or 90°. This is illustrated in Fig. 5.31. In this example,

- when the clock is early the quadrature sample is negative,
- when the clock is on-time, the quadrature sample is zero,
- when the clock is late, the quadrature sample is positive.

Considering the opposite condition when the data pulse is negative, as shown in Fig. 5.32, the polarity of the data cross-over samples are reversed. We can correct for the polarity reversal by multiplying the data-cross-over samples (or quadrature samples)[4] by the retimed data. With these observations, and the fact that the the cross-over samples are transfered to the loop filter only when a data transition occurs,

[4] If the data pulse is sinusoidal, then the data-cross-over samples will be 90 degrees out of phase with the on-time samples. Because of this 90 degree phase shift we will often call these cross-over samples "quadrature samples."

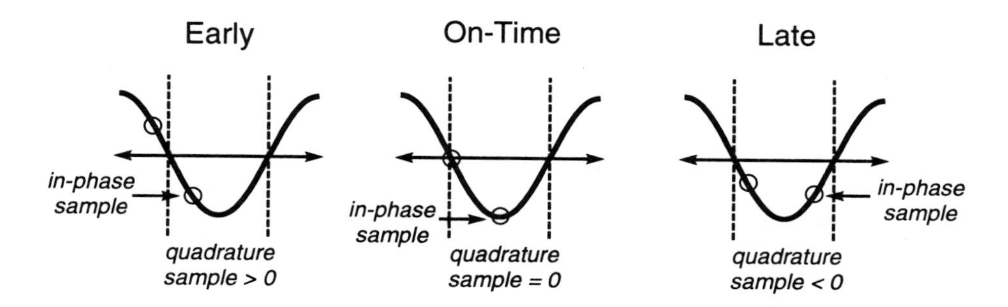

Figure 5.32 Illustration of a negative data pulse. The error signal is equal to the negative value of the cross-over sample.

the following simple list of rules for producing the desired phase-error signal can be obtained.

- If the data makes a low-to-high transition, pass the quadrature sample to the phase-detector output.
- If the data makes a high-to-low transition, pass the negative of the quadrature sample to the phase-detector output.
- If the data makes no transition, hold the previous phase-error value.

Block-Level Description of DTTL

A block diagram of a circuit that implements these rules is shown in Fig. 5.33. The clock is operating at the data rate and will experience both a positive, and a negative transition in one bit-period. The in-phase sample-and-hold (top left) samples on the positive transition of the clock, and the quadrature sample-and-hold (bottom left) samples on the negative transition of the clock. We are considering the case where the data-sampling is performed on a positive transition of the clock, in which case the quadrature samples are taken when the data crosses over the axis. The cross-over samples will contain the phase error information, but the information is valid only when a data transition occurs. The polarity of this signal also switches when the data changes value. Therefore we must post-process these sample by the sample-and-hold at the right. This circuit must sample on both a positive and negative transition in the data, and thereby ignores quadrature samples taken when no data transition occurred. Finally, the resampled phase information has the polarity ambiguity removed by multiplying the signal by the retimed data. The resulting signal ϵ is an estimate of the phase-error.

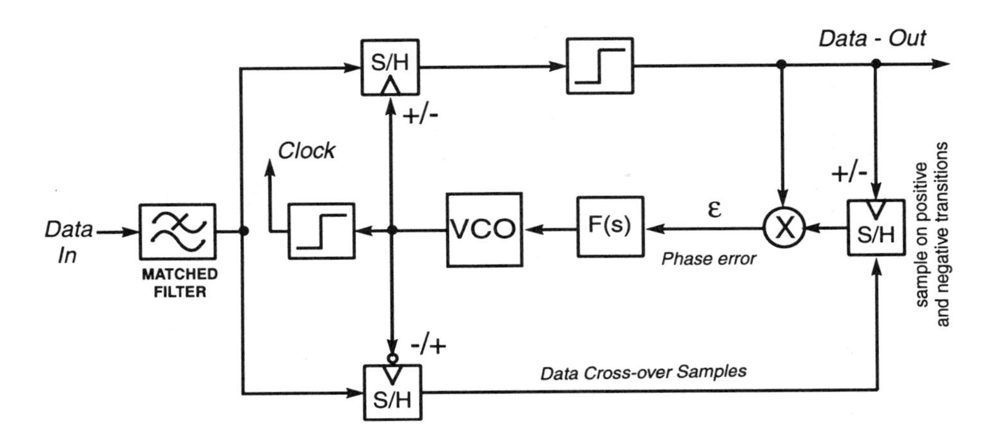

Figure 5.33 Block diagram of a data transition tracking loop (DTTL).

It is filtered by $F(s)$ and used to drive the VCO to the proper phase. In a practical implementation the multiplications should be done before the sampling, but in this idealized model it makes no difference. The block diagram is meant only to illustrate the concept.

Monotonic Phase Error Characteristic

A monotonic phase-error characteristic is desirable for several reasons: frequency acquisition will be improved, the locking range will be extended, and the phase-jitter transfer function will become more linear. To see that the resulting phase-error characteristic is monotonic over the bit interval we can first look at Fig. 5.29. Notice that the ideal NRZ data signal consists of square pulses, whereas the lowpass filtered signal transitions, from low-to-high, and vice versa, follow a linear path that extends over the entire bit-interval. Parasitic poles will smooth the sharp edges of the signal in Fig. 5.29 such that the transitions will be sinusoidal as was shown in Figs. 5.31 and 5.32.

Shape of Phase Characteristic Determined by Shape of Data Pulse The quadrature sample leads the in-phase sample by half a bit-period. The receivers job is to sample at the peaks of the data signal, and make a decision as to whether these samples are high or low. For optimal behavior we need to find these peaks. The DTTL operates on the principle that it is difficult to find signal peaks — but easy to find zero crossings. The peak of the in-phase sample is found indirectly by finding the zero-crossing of the quadrature sample. If the peak lies exactly between zero crossings, then we can sample in the middle of the zero crossing and hit the peak of the in-phase sample. In

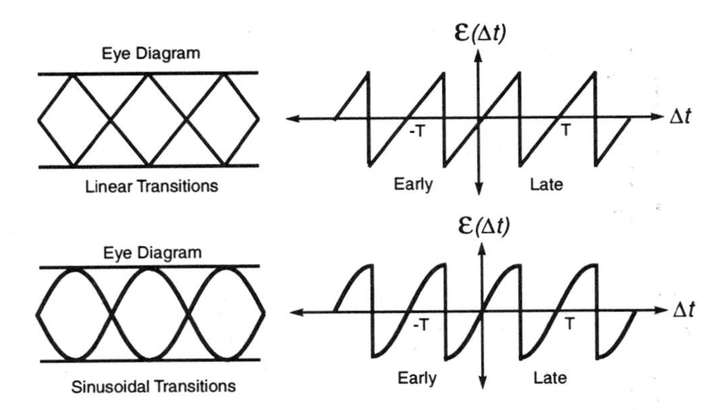

Figure 5.34 The phase-detector characteristic of a DTTL follows the shape of the data transition until it abruptly changes sign at time offsets each to $|T/2|$.

equilibrium the in-phase sample is at a peak, and the quadrature sample will be at zero. If the clock phase changes, the quadrature sample will differ from zero. Since these quadrature samples will be taken from the data itself, the shape of the data transition will give rise to the phase error function produced by the phase detector. Because the data signal is passed through a limiter before multiplying the quadrature samples, the polarity of the signal changes abruptly at a cycle-slip boundary, giving rise to a sawtooth-type phase-detector characteristic as shown in Fig. 5.34. It can be seen that the quadrature samples follow the shape of the data-transition, as a function of the phase-offset, where linear transition give rise to a sawtooth wave and sinusoidal transitions result in a switched-sinusoidal characteristic. It can also be observed that the phase-detector has the desirable property of being monotonic over the complete bit-interval $[-T/2, T/2]$.

In a practical realization, a bit-interleaving scheme as shown in Fig. 5.35 could be used. The VCO center frequency is half the data-rate, and the VCO has a quadrature output for sampling at the data cross-over points. The multiplexed track-and-hold circuits perform the function of sample-and-holding the data on both the positive and negative clock transitions. A repeatable structure, of two track-and-holds followed by a multiplexer, can be identified. Although the resampling circuit has a reversal of polarity for a negative data transition, in a fully-differential circuit this is easily realized by switching the polarity of the differential signals; thus the layout is identical to the front-end circuit, except for a cross-over in the wiring. A buffer has been added in the data cross-over arm, or quadrature arm, of the DTTL, which is used to compensate for the delay of the limiter in the in-phase arm. The matching of the delay times between buffer and limiter does not have to be accurate. What is essential, however,

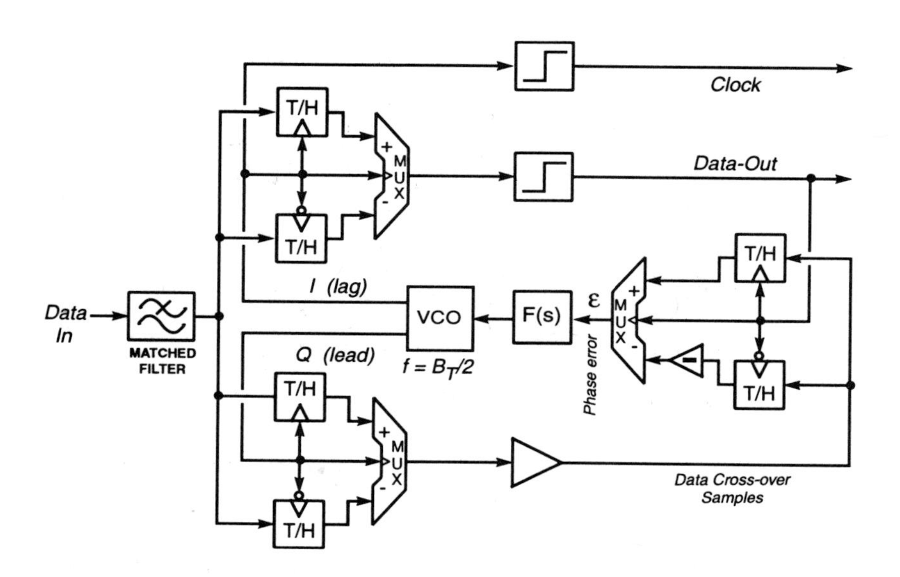

Figure 5.35 A bit-interleaved DTTL.

is that these delays are close enough in magnitude to make sure that the resampling performed by the track-and-holds on the right side of the circuit are operating on the correct quadrature samples and not a sample that is shifted in time relative to the in-phase arm. This phase detector has several desirable properties as listed below.

High-speed The circuit can function at very high speeds — limited by the speed of the track-and-hold circuits and multiplexer; both of which can operate near the limitations of the transistors.

Self-adjusting The phase-detector and decision circuit use identical building blocks. Thus the sampling phase is inherently self-adjusting, because the parasitic delays of the sampling circuits, in the in-phase and quadrature arms, will track each other.

Optimal sampling The circuit samples at $T/2$ seconds offset from the data cross-overs, which for sinusoidal pulse-shapes, or any other pulse that achieves a maximum SNR at the center of the bit-interval, is an optimal sampling point in terms of provided a MAP estimate of the data arrival time and obtaining the maximum SNR at the sampling instant.

High sensitivity Since the signal is sampled and held before a decision is made, the regeneration does not interfere with the data acquisition. Thus the SNR is not adversely affected. When flip-flops are used to sample and regenerate the data, as is the case for other circuits, the effect of the regeneration on the data sampling makes it unclear as to where the optimal sampling point occurs, and it may not be in the center of the bit-interval; this sampling uncertainty further reduces the receiver sensitivity in other approaches which use flip-flops as decision circuits.

Monotonic phase-error The phase-detector function is monotonic over the bit interval $[-T/2, T/2]$. This improves acquisition and tracking, and linearizes the jitter transfer function.

No double-frequency ripple Resampling the phase-error signal at data-transitions significantly reduces ripple in the error signal at twice the clock frequency, which reduces phase-jitter in the recovered clock.

Independent of data pattern Since the phase error is only transferred when a transition in the data occurs, the Phase-detector output is independent of the data transition-density to a first-order. This substantially reduces pattern-dependent jitter in the recovered clock.

Symmetry: Primary Advantage of DTTL for High-Speed Operation We have discussed in chapter 3 how systematic offsets in timing recovery are the dominant factors in performance degradation at high-speeds. Therefore, the primary advantage of this sampled DTTL is its symmetry, which makes the circuit insensitive to systematic errors. The residual phase-error in the recovered clock will be a result of random mismatches in the circuits and the layout, which can be maintained to a high-degree of accuracy. This circuit looks like a sampled Costas Loop, and can be viewed as a modification of a gradient-based MAP estimator as described in chapter 4. For sinusoidal data transitions, the quadrature samples give the gradient. Therefore, the DTTL provides a MAP clock-phase estimate in steady-state operation.

5.4.1 Frequency Detection in a DTTL

The DTTL has several desirable properties. However, as it stands, it can not lock to a data-signal that differs substantially from the VCO frequency. In fact, the natural acquisition of the loop can only pull-in frequency errors of the same order as the closed loop bandwidth. Since a narrow bandwidth is needed to reduce the phase-jitter (high-Q), this pull-in range will be quite narrow. At a data-rate of 10-Gb/s, the VCO frequency is 5-GHz. For an effective Q of 1000, the maximum frequency deviation is on the order of 10-MHz, or 0.2% of the VCO frequency. It is undesirable to design a VCO with a center frequency stable to within 0.2% for this application, therefore, we

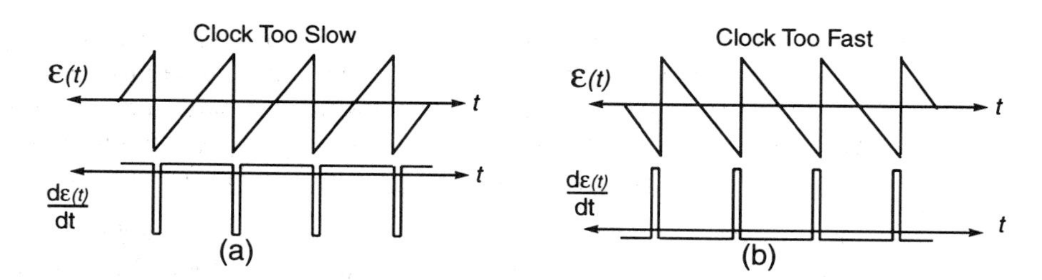

Figure 5.36 Illustration of the error signal ϵ with a sawtooth-type characteristic as a function of time and their derivatives for: (a) a slow clock, (b) a fast clock.

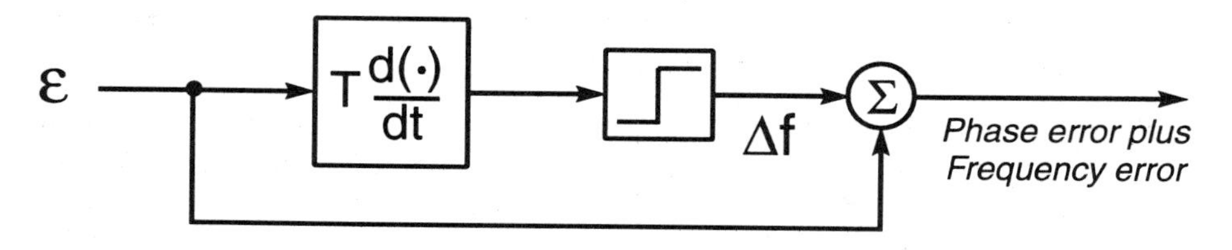

Figure 5.37 Circuit for extracting a frequency error signal from a sawtooth-type phase-error characteristic.

require the addition of a frequency detector (FD) to insure that frequency acquisition will occur upon start-up.

Several options exist for adding a frequency detector to the DTTL. We could use any of the circuits discussed thus far, as stand-alone frequency detectors, and simply add the resulting frequency error to the phase-error of the DTTL. However, since the DTTL has available in-phase data samples and quadrature cross-over samples, we have a structure that looks very much like a quadricorrelator already. All that is needed is to utilize this information wisely to provide an indication of the frequency error with a minimal addition of hardware.

We can derive a simple frequency detector by realizing that the phase-error signal ϵ is a sawtooth-type function of the phase-error. Two different conditions exist for a slow clock and a fast clock as illustrated in Figs. 5.36(a) and (b). The derivative of the error function is in the proper direction most of the time, however the dc value of the derivative is zero. A simple approach to deriving a frequency error that gives only the sign of the frequency error is to limit the derivative, as shown in Fig. 5.37. The dc value at the output of the limiter will be positive for a slow clock and negative for a fast clock. When the derivatives are passed through a hard-limiter, the result is shown in Fig. 5.38

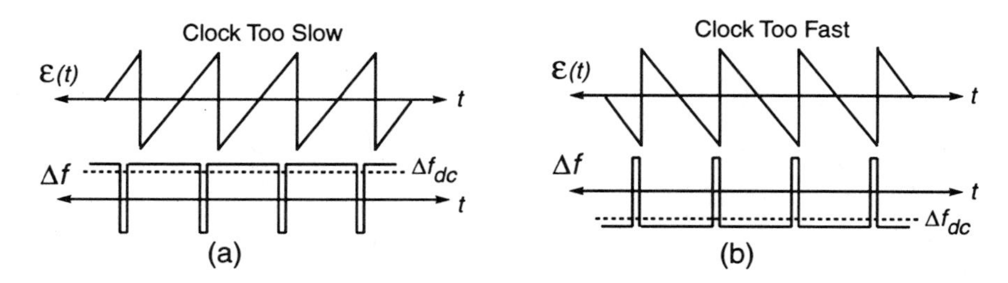

Figure 5.38 Frequency error polarity extraction from a sawtooth type phase-error function for: (a) a slow clock, (b) a fast clock.

Unique Properties of DTTL for Sawtooth Frequency Detector

The DTTL is ideally suited for implementing the sawtooth FD because the error signal ϵ is resampled and contains virtually no ripple. Therefore, only a broadband lowpass filter is needed to smooth glitches before producing the desired sawtooth function. We have seen other circuits that produce a sawtooth phase-error characteristic, such as an early-late circuit of Fig. 5.28. However, these circuits have severe ripple in the absence of data transitions, and a narrowband lowpass filter is required to reduce this ripple before a sawtooth function is obtained. The sawtooth FD approach could be applied to the early-late circuit as well. However, the FD will not be able to recognize a frequency error outside the bandwidth of the ripple-reduction filter and therefore will have a built-in range limitation. Since the DTTL does not require a narrowband ripple-reduction filter, it can recognize frequency errors of at least 10–20%, providing a significant range over which the VCO center frequency can vary and still be pulled-in by the DTTL.

MATLAB Simulation Results System-level simulations of the DTTL were performed to verify functionality. The resulting error signal ϵ is shown in Fig. 5.39 for the case of a 5% frequency error. The input signal is random NRZ data with sinusoidal transitions. It can be seen that the phase-detector characteristic is not a smooth switched sinusoid, which is an artifact of the random nature of the data; when several bits arrive without a transition, the phase error will make a larger jump after a transition finally occurs. We require filtering with a bandwidth on the order of 25% of the bit-rate to smooth this phase-detector output before further processing.

The error signal ϵ was used to derive the direction of the frequency error. A finite difference was taken, and the result was passed through a hard-limiter to obtain the signal Δf_ϵ. This signal consists of pulses with a dc value in the opposite direction of the frequency error. The resulting frequency error signal Δf_ϵ for the same condition of a 5% frequency error is shown in Figs. 5.40 (a) and (b), for a slow, and a fast clock

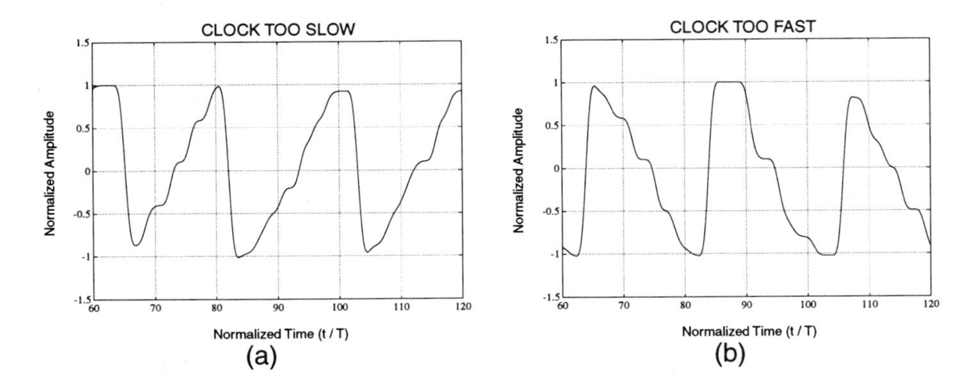

Figure 5.39 Phase-error signal resulting from a MATLAB simulation of a DTTL for frequency errors of (a) -5%, (b)+5%.

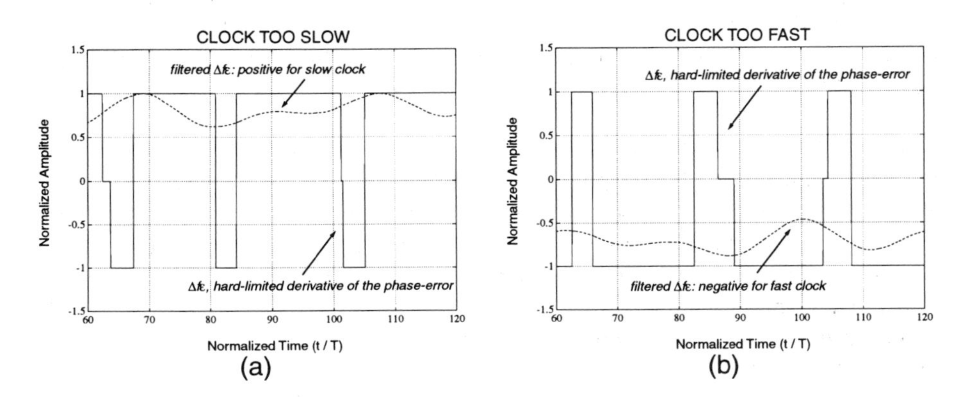

Figure 5.40 Frequency-error signal resulting from a MATLAB simulation of a DTTL for frequency errors of (a) -5%, (b)+5%.

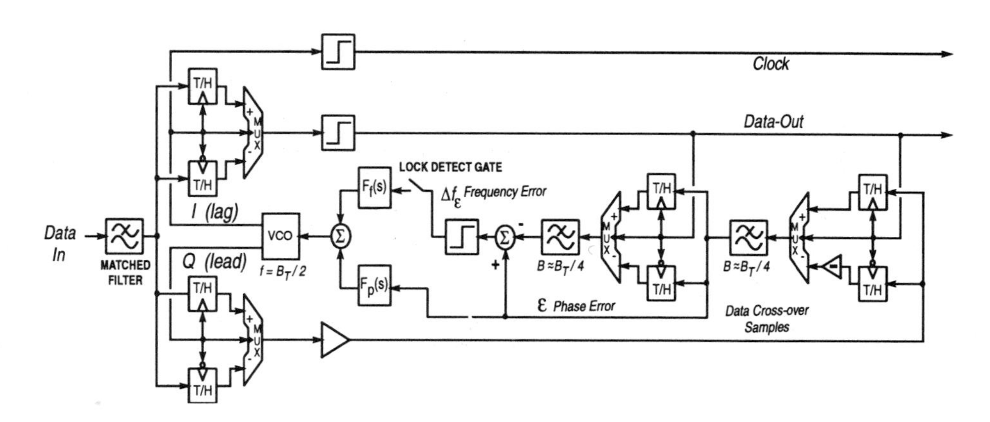

Figure 5.41 Block diagram of a DTTL with frequency detection

respectively. It can be seen that the filtered version of Δf_ϵ is positive for a slow clock, and negative for a fast clock. Various other simulation results of the DTTL will be reserved for chapter 10 where the actual transistor-level implementation is discussed.

Variations of the Sawtooth Frequency Detector

Once we make the primary observation that the slope of a sawtooth phase characteristic can give us the direction of the frequency error, then we can add various gating techniques and other embellishments to the FD to improve performance. The basic structure of a practical DTTL with frequency detection is shown in Fig. 5.41. Considerations of closed-loop stability and jitter-peaking will dictate the gain and transfer function of the filter $F_f(s)$ in the frequency-error path. When the loop is in lock, the average frequency error will go to zero. However, it will vary randomly, and can disturb the dynamics of the PFLL. To reduce the effect of the frequency error signal on the loop when in phase-lock, a lock-detector can be used to force the frequency error to zero after the phase acquisition is complete. One simple technique is to use a *dead-zone* near the point of zero phase-error. This is illustrated in Fig. 5.42, where the frequency-error is only enabled when the phase-error exceeds a given threshold. With this technique, the operation of frequency acquisition can be separated from phase tracking. During frequency acquisition, the phase-error signal will nominally equal zero, and will have no effect on the loop. Once frequency acquisition has been established, the phase-error signal takes over, and the frequency-error feedback path is broken.

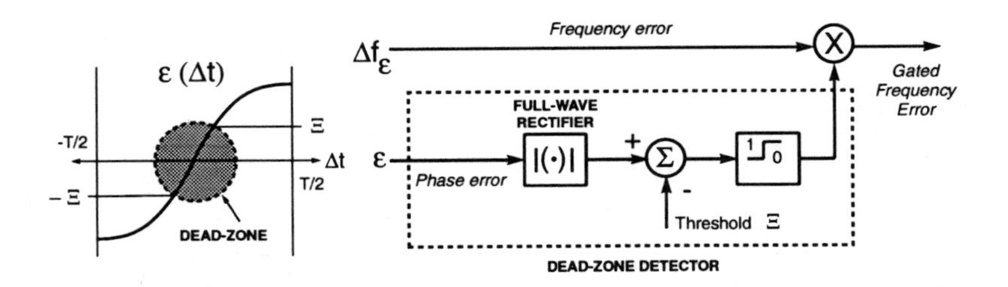

Figure 5.42 Illustration showing how the frequency error signal can be forced to zero when the phase error is within a boundary surrounding zero.

5.5 SUMMARY

In this chapter we considered some of the practical aspects of clock recovery circuits used in high-speed applications. To insure that a clock can be recovered from the data, either a frequency reference, or a frequency acquisition aid is required. A frequency reference would take the form of a stable resonator, such as a quartz crystal, whereas an acquisition aid will produce an error signal, which can drive the VCO to the correct frequency. Various frequency detection schemes were described in this chapter. A rotating wheel analogy was used extensively for conceptualizing the frequency detection operation.

A few clock recovery and data-retiming circuit were presented, all of which are self-adjusting, which is a result of the decision circuit being utilized in a balanced configurations for extracting phase-error information. At this point, the reader may be interested in some quantitative performance comparisons of several of the clock recovery techniques discussed thus far. Parameters of particular interest are as follows.

- Bit-error-rate as a function of the SNR.
- Clock phase-jitter as a function of the noise-bandwidth and the SNR.
- Frequency acquisition, and tracking ranges.
- Sensitivity to offsets in the clock phase.

Certainly these parameters are important. However, in a high-speed system, parasitics of the actual circuit will ultimately determine performance. From a practical point of view we might be more interested in comparing the following specifications of alternative architectures:

- Maximum operating speed.
- The minimum input signal level. Since regenerators have a finite gain, the input must be greater than a given level to result in a full-level output signal.
- SNR penalty compared to an ideal receiver with impulse sampling.
- Robustness against transistor model variations and temperature changes.

These quantities are all inexorably linked to the transistor-level circuit and layout. Therefore, it is difficult to evaluate system performance until, at least, the primary functions have been designed at the transistor-level, so that circuit-simulations using SPICE can be performed. However, even obtaining circuit-level simulation results are difficult. A few of these difficulties are described below.

High-Q Clock extractors are necessarily *high-Q* circuits. Therefore, thousands of clock cycles need to be simulated to examine the low frequency behavior of the circuit after lock has been established. Even longer simulations are required to extract information about cycle-slipping and frequency acquisition. For example Pottbäcker *et al.* [9, 10] reported on an 8-GHz frequency-detector with an acquisition time on the order of 1 ms. Therefore, the system needs 8-million clock cycles to acquire frequency; a horrendously long transient simulation would be required to observe this phenomenon.

Random Data To make matters worse, the data is random, and several simulations are required to determine average circuit behavior.

Random Noise We want to test the circuits performance in noise, which is also random, implying that we need to look at noise and data statistically and simultaneously — adding another dimension to an already large problem.

Cross-Coupling Coupling is *not* implicitly modeled, so it is very easy to ignore effects in simulation that could in reality be detrimental.

Review of Part I

Since complete circuit characterization from simulations are impractical, and indeed, even impossible, the strategy used by these authors is to first design prototype- and test-circuits, and use the measured results to determine the affect of various parasitics on system performance. With this knowledge of circuit parasitics the fundamental aspects of receiver design were reevaluated in terms of their applicability to high-speed applications. Basic theories needed for evaluation of circuit performance were presented in chapters 2 and 3. Various receiver block diagrams were presented in chapter 4 and practical high-speed versions of these circuits were described in this chapter. It is hoped that by providing a background in the theory, and reviewing several previously reported circuits, Part I of this bookbook has helped to unify circuit design with systems theory, and provide a basis for improved circuit designs of the future.

Introduction to Part II

To gain a full understanding of integrated fiber-optic receivers there is no substitute for doing transistor-level circuit designs. Multiple problems will arise in the design process, the solution of which generally enhances the designers understanding of both the circuit and the overall system. Therefore Part II of this book, which describes the detailed circuit design of various building blocks of a fiber-optic receiver, is necessary to provide integration and expand comprehension of the information presented in Part I.

Since Part II of this book concerns detailed circuit design issues, it is difficult to speak in general terms. Therefore we will restrict our attention to a specific technology and to a specific architecture. The technology we will use is AlGaAs/GaAs HBTs (Heterojunction Bipolar Transistors), and the architecture that we intend to implement is the interleaved DTTL described in the previous section. We present Part II as an application of the ideas presented in Part I, but we are not implying that the technology and architecture chosen are preferable in all cases. We have spent a great deal of effort to explain the underlying concepts in Part I, but the real challenge of producing integrated fiber-optic receivers is in the circuit design. We will now turn to this topic where we will concern ourselves with problems of a more physical nature.

REFERENCES

[1] Hans Ransijn and Paul O'Connor. A PLL-based 2.5-Gb/s GaAs clock and data regenerator IC. *IEEE J. Solid-State Circuits*, 26(10):1345–1353, October 1991.

[2] Kazuo Hagimoto, Yuuzou Miyagawa, Yutaka Miyamoto, Masanobu Ohhata, Tatsuhito Suzuki, and Hiroyuki Kikuchi. Over 10 Gb/s regenerators using monolithic IC's for lightwave communication systems. *IEEE J. Select. Areas Commun.*, SAC-9(5):673–682, June 1991.

[3] Peter Wennekers, Ulrich Novotny, Axel Huelsmann, Gugrun Kaufel, Klaus Koehler, Brian Raynor, and Joachim Schneider. 10-Gb/s bit-synchronizer circuit with automatic timing alignment by clock phase shifting using quantum-well AlGaAs/GaAs/AlGaAs technology. *IEEE J. Solid-State Circuits*, 27(10):1347–1352, October 1992.

[4] Robert R. Cordell, J. B. Forney, Charles N. Dunn, and William G. Garrett. A 50 MHz phase- and frequency-locked loop. *IEEE J. Solid-State Circuits*, SC-14(6):1003–1010, December 1979.

[5] Donald Richman. Color-carrier reference phase synchronization accuracy in NTSC color television. *Proc. IRE*, 42:106–133, January 1954.

[6] J. A. Bellisio. A new phase-locked timing recovery method for digital regenerators. In *IEEE Int. Conf. Commun.*, pages 10–17–10–20, Philadelphia, Pennsylvania, June 1976.

[7] F. Gardner. *Phaselock Techniques*. Wiley, New York, second edition, 1979.

[8] David G. Messerschmitt. Frequency detectors for PLL acquisition in timing and carrier recovery. *IEEE Trans. on Commun.*, COM-27(9):1288–1295, September 1979.

[9] Ansgar Pottbäcker, Ulrich Langmann, and Hans-Ulrich Schreiber. A 8 Gb/s Si bipolar phase and frequency detector IC for clock extraction. In *ISSCC Dig. Tech. Papers*, pages 162–163, San Francisco, California, February 1992.

[10] Ansgar Pottbäcker, Ulrich Langmann, and Hans-Ulrich Schreiber. A Si bipolar phase and frequecny detector IC for clock extraction up to 8 Gb/s. *IEEE J. Solid-State Circuits*, 27(12):1747–1751, December 1992.

[11] Ansgar Pottbäcker and Ulrich Langmann. An 8 GHz silicon bipolar clock-recovery and data-regenerator IC. In *ISSCC Dig. Tech. Papers*, pages 116–117, San Francisco, California, February 1994.

[12] J. D. H. Alexander. Clock recovery from random binary signals. *Electron. Lett.*, 11(22):541–542, October 1975.

[13] Richard C. Walker, Cheryl L. Stout, Jieh-Tsorng Wu, Benny Lai, Chu-Sun Yen, Tom Hornak, and Patrick T. Petruno. A monolithic 622Mb/s clock extraction data retiming circuit. *IEEE J. Solid-State Circuits*, 27(12):1805–1811, December 1992.

[14] Richard C. Walker, Thomas Hornak, Chu-Sun Yen, Joey Doernberg, and Kent H. Springer. A 1.5 Gb/s link interface chipset for computer data transmission. *IEEE J. Select. Areas Commun.*, SAC-9(5):698–703, June 1991.

[15] Lawrence DeVito, John Newton, Rosamaria Croughwell, John Bulzacchelli, and Fred Benkley. A 52 MHz and 155 MHz clock-recovery PLL. In *ISSCC Dig. Tech. Papers*, pages 142–143, San Francisco, California, February 1991.

[16] Thomas H. Lee and John F. Bulzacchelli. A 155 MHz clock recovery delay-and-phase-locked loop. In *ISSCC Dig. Tech. Papers*, pages 160–161, San Francisco, California, February 1992.

[17] Thomas H. Lee and John F. Bulzacchelli. A 155-MHz clock recovery delay- and phase-locked loop. *IEEE J. Solid-State Circuits*, 27(12):1736–1746, December 1992.

[18] T. O. Anderson, W. J. Hurd, and W. C. Lindsey. U.S. pat. no. 3,626,298; Transition Tracking Bit Synchronization System, December 1971.

[19] William C. Lindsey and Marvin K. Simon. *Telecommunication Systems Engineering*. Dover Publications Inc., New York, 1991. Dover edition first published in 1991 is an unabridged, unaltered republication of the work first published by Prentice-Hall, Inc., Englewood Cliffs, N.J., 1973 in its "Prentice-Hall Information and System Science Series.".

PART II

CIRCUIT DESIGN

Turn on your radio and let me hear the song
Switch on your electric light
Then we can get down to what is really wrong
— Van Morrison

6

HETEROJUNCTION BIPOLAR TRANSISTORS: A BRIEF INTRODUCTION

One of the first tasks to be performed in the design of ICs for fiber-optic receivers is choosing an appropriate IC technology. Primary factors to consider are speed, performance, reliability, and cost. The ultimate choice will depend on the specific application and the anticipated volume of production. For high-volume data communication applications, such as LANs (Local Area Networks) the best choice will likely be bulk CMOS, because of its low cost and adequate performance in a high SNR environment. Conversely, in low SNR applications, such as telecommunications, or in high-speed ATM (Asynchronous Transfer Mode) switching, where extra speed and performance justify a large increase in circuit costs, HEMT (High Electron Mobility Transistors) or HBTs might be used. Since HEMTs and HBTs are fabricated from III–V compounds, it is possible to integrate light sources and detectors on the same substrate with the circuitry; this can be advantageous for low-noise operation and can be used to control I/O impedance levels, thereby reducing interconnect problems.

Overview of Available IC Technologies

In the following section we will list some of the available IC technologies and discuss their advantages and disadvantages for use in high-speed serial communication links.

BJT Silicon bipolar junction transistors are versatile devices; they offer high-speed, high reliability, and relatively low cost. They could be used in both telecommunication and data communication applications at data rates from 1–10 Gb/s [1].

CMOS CMOS is well known for its low cost, high reliability, and high packing density. The speed is almost as good as BJTs; as gate lengths shrink, the speed will continue to increase. CMOS is ideal for datacom applications such as FDDI and ATM receivers. CMOS has been used for gigabit-per-second data links [2], and is currently applicable to data rates up to 2.5 Gb/s, with this number increasing as the minimum gate length drops.

BiCMOS It is often advantageous to combine the speed and high transconductance of BJTs, with the high input impedance and high packing density of CMOS [3]. BiCMOS has been utilized in a 6-GHz, 60-mW PLL, which could be used in a complete fiber-optic receiver [4, 5]. BiCMOS is more expensive than either CMOS or BJT, but is also more versatile; applications include both telecom and datacom systems operating at data rates of 1–10 Gb/s.

SOI CMOS Silicon-on-Insulator (SOI) is an emerging technology with a long history [6]. In the early eighties CMOS SOS (Silicon-on-Sapphire) was used for radiation-hardened military applications, but was too expensive for the consumer market. Recently, high quality transistors have been fabricated using a thin-film of silicon on top of an insulating oxide layer. SOI has the advantages that parasitic capacitances to the substrate are drastically reduced, if not eliminated, cross-coupling is reduced substantially, and latch-up is no longer a consideration, allowing devices to be packed extremely close to one another. The devices are also easily scaled for deep-submicron ULSI applications. Past results have been impressive, producing ring-oscillator gate delays of 13 ps! Presently the technology is not widespread and is still expensive, (this is due primarily to wafer costs; the actual processing of SOI is simpler than bulk CMOS because of the elimination of wells, well contacts, and field implants) but increased volume of production is expected to drive the costs down and make this a common technology in the future. SOI could be used in both telecom and datacom and could operate at data rates as high as 20 Gb/s.

GaAs FET GaAs field-effect transistors have been used extensively in MMICs (Monolithic Microwave Integrated Circuits) and have proven reliability. GaAs FET processing is more expensive than silicon-based technologies, but the higher speed might be attractive for some telecom applications. GaAs FETs have been used in multi-gigabit-per-second systems [7] and are applicable for data rates in the range of 1–20 Gb/s.

HEMT High Electron Mobility Transistors (HEMTs) have been used in millimeter-wave, low-noise applications, and could be used in high-speed fiber-optic receivers [8]. One disadvantage is the high-cost, but this could be offset by the fact that light sources and detectors can be integrated together on the same substrate. As long as one must use III–V compounds for electro-optic devices, it might actually be more economical to integrate, at least, some of the receiver circuitry, such as low-noise amplifiers, with the electro-optics, thereby reducing noise and allowing for controlled impedance interconnections. HEMTs would typically find application in systems operating at 10–20 Gb/s and beyond.

Technology	Data Rate	Cost	Applications
BJT	1–10 Gb/s	Low	telecom, datacom
CMOS	0–2.5 Gb/s	Low	datacom
BiCMOS	0.5–10 Gb/s	Medium	telecom, datacom
SOI CMOS	0.5–20 Gb/s	Medium	telecom, datacom
GaAs FET	1–20 Gb/s	Medium	telecom
HEMT	5– <20 Gb/s	High	telecom
HBT	5– <20 Gb/s	High	telecom

Table 6.1 Summary of applicability of various technologies to use in integrated fiber-optic receivers.

HBT Heterojunction bipolar transistors are fabricated with the same materials as HEMTs, but are bipolar transistors as opposed to FETs. HBTs are suitable for high-speed logic; V_{be} for various transistors can be matched to within a few millivolts, which allows realization of low-voltage digital circuits with acceptable noise-margins. Since HBTs can be made of III–V alloys, they can also be integrated on the same substrate with electro-optic devices (we will see an example of this in chapter 7, Fig. 7.36). HBTs are also expensive and are therefore used primarily in special telecom applications at speeds in the range of 5–20 Gb/s or higher [9, 10, 11].

Some of the key aspects of the technologies just described for use in high-speed fiber-optic receivers are summarized in table 6.1. For the remainder of this book we will be discussing circuits designed and fabricated using HBTs. The following sections introduce this technology and give SPICE models that will be used in simulations.

6.1 OVERVIEW OF HBTs

In order to achieve the high speeds required in a 10-Gb/s fiber-optic receiver, it is proposed to realize the circuitry using heterojunction bipolar transistors (HBTs). This exciting new technology has several advantages for high-speed operation, which were noted as early as 1948 by Shockley in his original patent of bipolar transistors [12], but which have only recently shown promise due to advances in molecular beam epitaxy (MBE) technology. Kroemer [13] developed the theory and analysis of HBTs that has resulted in modern devices. HBTs with f_ts in the 100 GHz range have already been reported [14, 15], and it is anticipated that devices may reach the 200 GHz range in the future [16]. Further process enhancements are expected to make HBTs a dominant technology for both high-speed analog and digital circuits. HBTs are bipolar devices with charge transport being controlled by the bulk properties of the semiconductor. This can be advantageous as opposed to field-effect transistor (FETs), with charge transport occurring at the surface of the semiconductor material, where high defect concentrations can degrade performance. In comparison to FETs, HBTs have high transconductance, can be matched closely, have low flicker noise, high output impedance, and no hysteresis effects. These properties make them attractive for analog design as well as low-voltage-swing current-mode digital circuits. In addition, HBTs exhibit high linearity, and can operate at large current densities, making them attractive for use in power amplifiers.

In this chapter we will present a brief introduction to HBTs, and give some first-order models. For the reader interested in a more detailed discussion of HBTs, the book by Ali and Gupta is an excellent place to start [17], offering chapters on both theory, and applications. Kroemer's overview of HBTs presented in 1982 is also recommended [18].

6.2 ADVANTAGES OF HBTs FOR HIGH-SPEED OPERATION

The key feature of an HBT resulting in increased speed is the formation of a heterojunction at the base-emitter interface such that the bandgap energy on the emitter side of the junction is larger than the energy gap on the base side. A band diagram of a graded base Npn (N denotes a wider band-gap than n and p) HBT is shown in Fig. 6.1, [18, p. 15], [17, p. 256]. This energy difference ΔE_g blocks reverse charge-carrier injection from the base to emitter, resulting in near unity emitter injection efficiency, independent of the doping levels. The freedom to optimize doping levels for wideband performance, without suffering a degradation in current gain (β), gives HBTs an approximate 2:1 speed advantage over comparable homojunction bipolar junction transistors (BJTs). The base doping of an HBT can be increased to lower the base resistance and increase $f_{\max}$; simultaneously, the emitter doping can be reduced, lowering the base-emitter junction capacitance C_{je}. Further improvements in speed result from using GaAs or InP as the semiconductor material. The high electron-mobility of these materials reduces the base-transit time τ_f, and the semi-insulating substrate reduces the collector-substrate capacitance C_{cs}. In addition, the high base-doping level gives rise to a large early-voltage which improves linearity.

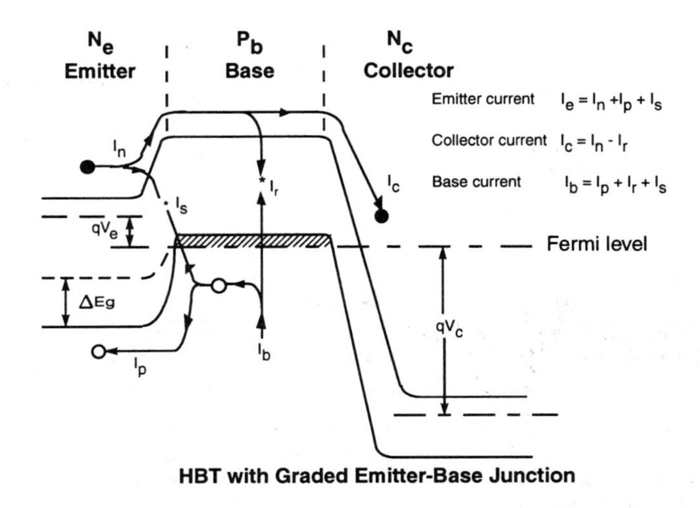

Figure 6.1 Energy band diagram of a linearly graded Npn HBT.

6.3 AlGaAs/GaAs HBTs: TYPICAL PARAMETERS

The most mature HBT technology to date is AlGaAs/GaAs, with $f_{\max}$ in the 30-50 GHz range consistently reported by TRW [19, 20], Rockwell [14], and NTT [21]. Rockwell has demonstrated prescalars operating at input frequencies up to 26.9 GHz, while TRW has fabricated a voltage-controlled oscillator (VCO) operating at 37.7 GHz. TRW has also demonstrated LSI capabilities in recent A/D converters. This level of integration has now made it feasible to integrate a fiber-optic receiver onto a single chip.

Processing of the HBT circuits presented in Part II of this book, was donated by TRW who also provided financial support for this effort.[1] A cross-section of TRWs AlGaAs/GaAs HBT process is shown in Fig. 6.2. This process also has Schottky diodes available, as well as thin-film 100 Ω/square NiCr resistors. Typical device parameters for (3μm x 10μm) and (1μm x 10μm) emitter-area devices are given in table 6.2. The small-signal SPICE model used in simulation is shown in Fig. 6.3, with typical parameters for a (3μm x 10μm) emitter device shown. The forward biased base-emitter junction voltage V_{be} is approximately 1.3–1.4 Volts, which is twice that of a silicon BJT. Since TRWs AlGaAs/GaAs HBT process was in the developmental stages when this research began, and small emitter-area devices were not reliable, the

[1] This research was supported by the University of California MICRO Program and TRW, Inc. under Grants 90-102 and 91-102.

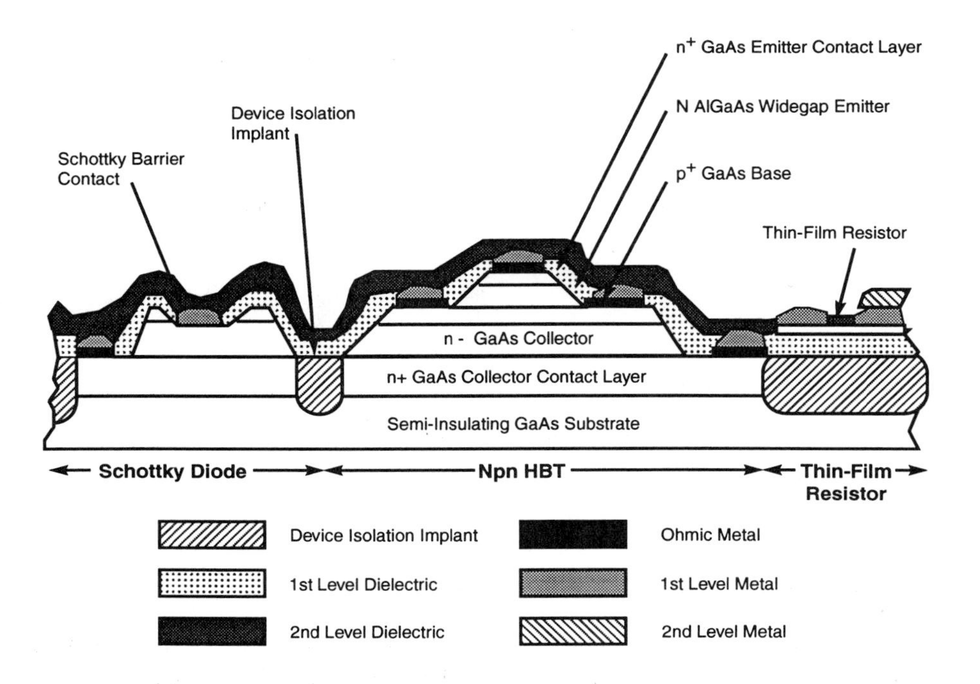

Figure 6.2 Cross-sectional view of TRW's AlGaAs/GaAs HBT process.

Emitter Area	$3\mu m \times 10\mu m$	$1\mu m \times 10\mu m$
dc Current	2 mA	2 mA
Current Density	67 (μA/μm^2)	200 (μA/μm^2)
V_{be}	1.2–1.4 V	1.2–1.4 V
C_{je}	45 fF	15 fF
r_b	50 Ω	25 Ω
r_e	10 Ω	10 Ω
Beta (β)	25–200	25–200
f_t	22 GHz	40 GHz
$f$max	30 GHz	50 GHz

Table 6.2 Typical AlGaAs/GaAs HBT device parameters.

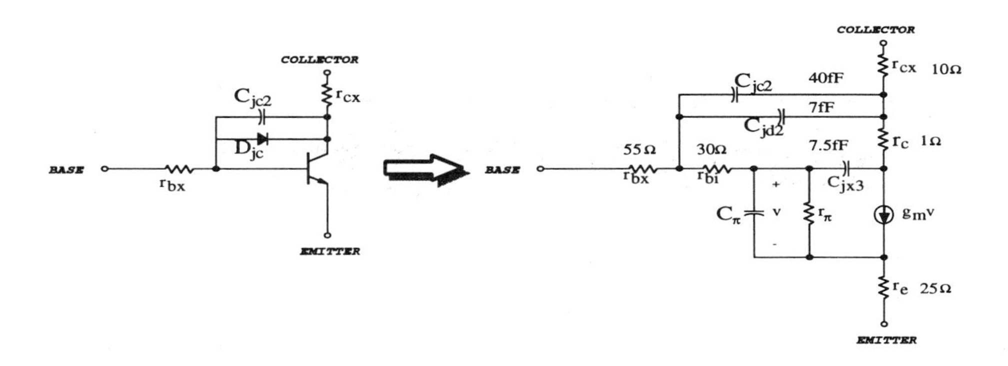

Figure 6.3 HBT SPICE subcircuit and small-signal model parameter values for a (3μm x 10μm) emitter device

conservative (3μm x 10μm) minimum device dimension was chosen to maximize yield and ensure functionality in first-generation circuits. An approximate increase in speed by a factor of two can be obtained in second-generation circuits simply by substituting (1μm x 10μm) devices for the existing transistors in the prototype circuits.

6.4 InP-BASED HBTs: TYPICAL PARAMETERS

The electron mobility in an InP-based HBT is even higher than in GaAs. Therefore, InP-based HBTs can achieve higher speed than their GaAs counterparts. Furthermore, InP has additional advantages for use in medium-scale integrated fiber-optic receivers. The bandgap energy of InP is well suited for use as a long wavelength photodetector (1.3μm and 1.55μm). This is ideal for single-mode glass optical fibers that have low losses at these wavelengths. Also InP has a thermal conductivity which is not as good as silicon, but is better than GaAs. Therefore, problems with power dissipation are somewhat relaxed in InP as opposed to GaAs. V_{be} of an InP HBT also lies between GaAs HBTs and silicon BJTs at about 1 Volt for a collector current of 4 mA.

A cross-sectional view of TRW's Npn InP-based HBT is shown in Fig. 6.4. This device can achieve a unity power gain frequency of $f_{\max} = 100$ GHz. Typical device parameters are given in table 6.3, and the corresponding small-signal model is shown in Fig. 6.5, for a bias current of 4-mA. We can calculate f_t and $f_{\max}$ for the InP-based HBT of Fig. 6.5. The expression for the unity current gain frequency is given by

$$f_t \simeq \frac{g_m}{2\pi(C_\mu + C_\pi)}, \tag{6.1}$$

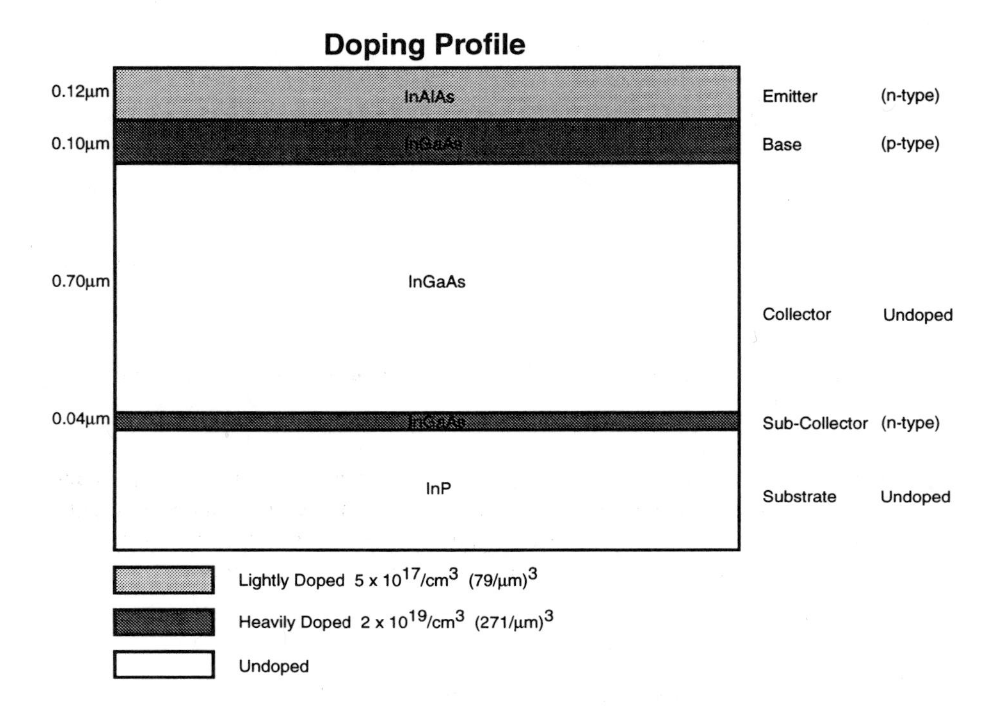

Figure 6.4 Cross-sectional view and doping profile for and Npn InP-based HBT using InAlAs as the wideband emitter.

$f$max	100 GHz
f_t	65 GHz
Breakdown Voltage V_{ce}	$V_{ce} \simeq 3$ V for I_c 1–4mA
C_π	300fF @ $I_c = 4$mA
C_μ	75fF
C_{cs}	10fF
r_b	10Ω
r_e	2Ω
r_c	3Ω
L_b	15pH
L_c	30pH

Table 6.3 Typical small-signal parameters for a $(1\mu\text{m} \times 10\mu\text{m})$ emitter InP-based HBT.

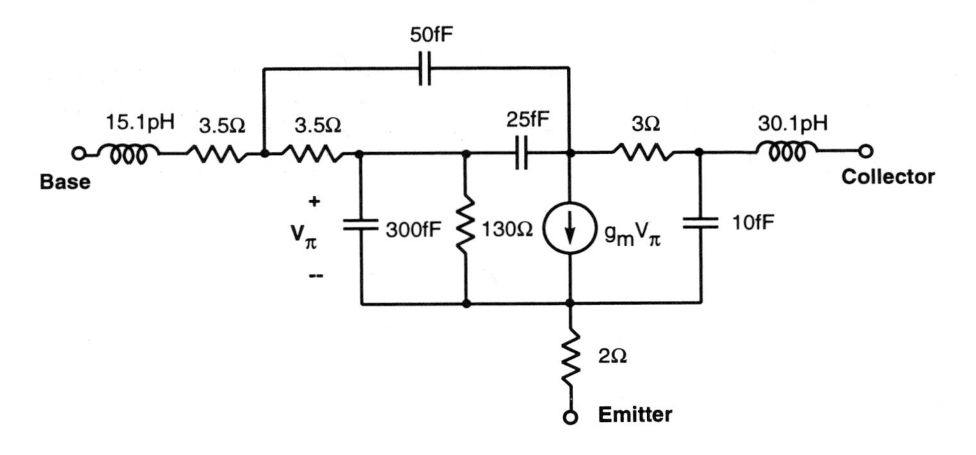

Figure 6.5 Small-signal model of an InP-based HBT with parameters shown for a bias current of 4-mA.

from which we find that $f_t = 65$ GHz ($I_c = 4$-mA, $g_m = 1/6.5\Omega$). The following expression is commonly used to approximate $f_{\max}$ [22, p.117];

$$f_{\max} \simeq \frac{1}{2}\sqrt{\frac{f_t}{2\pi C_\mu r_b}}. \tag{6.2}$$

For the values given in table 6.3, we find that $f_{\max} \simeq 70$ GHz for a 4-mA bias current.

6.5 SPICE MODELS FOR CIRCUIT SIMULATION

We conclude this brief introduction to HBTs by presenting nominal models for circuit simulation using SPICE. The reader should keep in mind, that these models are approximate, and only give a first-order indication of actual device behavior. However, the models are useful for predicting the relative importance of various parasitics on circuit performance.

6.5.1 AlGaAs/GaAs HBT SPICE Models

The maximum current gain (β), due to reverse carrier injection into the base, is controlled by the parameter B_F, which is set to 10^6. Therefore, the actual current gain in the model is dominated by recombination, and is determined by the slope-parameters N_F and N_E. The following expressions are used in the model to determine the dc currents.

$$I_c = I_S \exp\left(\frac{1}{N_F}\frac{V_{be}}{V_t}\right) \tag{6.3a}$$

$$I_b = I_{SE} \exp\left(\frac{1}{N_E}\frac{V_{be}}{V_t}\right). \tag{6.3b}$$

Therefore, the current gain is given by

$$\beta = \frac{I_c}{I_b} = \frac{I_S}{I_{SE}}\left(\frac{I_c}{I_S}\right)^{(1-N_F/N_E)}. \tag{6.3c}$$

We can now express the ratio (N_F/N_E) in terms of the current parameters as

$$\frac{N_F}{N_E} = \frac{\ln\left[I_c/(\beta I_{SE})\right]}{\ln\left[I_c/I_S\right]}. \tag{6.3d}$$

This expression can be used to find N_E when I_c, I_S, I_{SE}, N_F and β are known.

Table 6.4 gives examples of two SPICE model files for a (3μm x 10μm) emitter. The first model is typical, and the second model gives best-case parameter values. For higher current applications, dual and quad emitter devices can be used. SPICE model files for these devices are given in table 6.5.

GaAs HBTs	
Nominal Case	**Best Case**
MODEL FOR HF3X10SEB	MODEL FOR HF3X10SEB
.SUBCKT HBTSE 1 2 3	.SUBCKT HBTSE 1 2 3
Q1 5 4 3 MHBT1	Q1 5 4 3 MHBT1
RBX 2 4 55	RBX 2 4 20
RC 1 5 10	RC 1 5 3
CJX 4 5 40fF	CJX 4 5 20fF
DCJX 4 5 MDCX	DCJX 4 5 MDCX
.MODEL MHBT1 NPN	.MODEL MHBT1 NPN
+ RE=25 RC=1 RB=30	+ RE=5 RC=1 RB=15
+ CJE=45fF CJC=7.5fF XCJC=1.0	+ CJE=45fF CJC=7.5fF XCJC=1.0
+ VJE=1.25 VJC=1.0	+ VJE=1.25 VJC=1.0
+ MJE=0.5 MJC=0.5	+ MJE=0.5 MJC=0.5
+ BF=1MEG BR=0.001	+ BF=1MEG BR=0.001
+ IS=13E-24 ISE=40E-21	+ IS=13E-14 ISE=40E-21
+ NF=1.1333 NE=1.5167	+ NF=1.1333 NE=1.5167
+ TF=6ps	+ TF=3ps
.MODEL MDCX D	.MODEL MDCX D
+ IS=1E-20 CJO=7fF	+ IS=1E-20 CJO=4fF
.ENDS HBTSE	.ENDS HBTSE

Table 6.4 SPICE models for a (3μm $\times$ 10μm) GaAs HBT: nominal and best case.

GaAs HBTs	
Dual Emitter	Quad Emitter
MODEL FOR HF3X10DEB .SUBCKT HBTDE 1 2 3 Q1 5 4 3 MHBT1 RBX 2 4 27.5 RC 1 5 12 CJX 4 5 64fF DCJX 4 5 MDCX .MODEL MHBT1 NPN + RE=12.5 RC=1 RB=15 + CJE=90fF CJC=15fF XCJC=1.0 + VJE=1.25 VJC=1.0 + MJE=0.5 MJC=0.5 + BF=1MEG BR=0.001 + IS=26E-24 ISE=80E-21 + NF=1.1333 NE=1.5167 + TF=6ps .MODEL MDCX D + IS=1E-20 CJO=11.2fF .ENDS HBTDE	MODEL FOR HF3X10QEB .SUBCKT HBTQE 1 2 3 Q1 5 4 3 MHBT1 RBX 2 4 13.8 RC 1 5 10 CJX 4 5 112fF DCJX 4 5 MDCX .MODEL MHBT1 NPN + RE=6.3 RC=1 RB=7.5 + CJE=180fF CJC=30fF XCJC=1.0 + VJE=1.25 VJC=1.0 + MJE=0.5 MJC=0.5 + BF=1MEG BR=0.001 + IS=52E-24 ISE=160E-21 + NF=1.1333 NE=1.5167 + TF=6ps .MODEL MDCX D + IS=1E-20 CJO=19.6fF .ENDS HBTQE

Table 6.5 Typical SPICE models for (3μm $\times$ 10μm) dual and quad emitter GaAs HBTs.

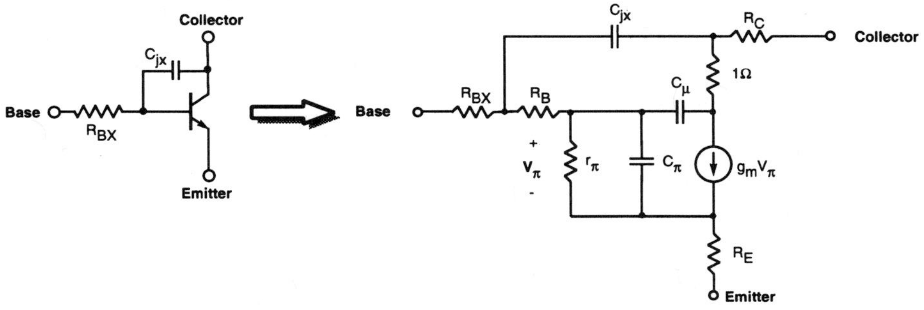

Figure 6.6 Illustration of subcircuit for InP-based HBT used in SPICE simulations.

6.5.2 InP-Based HBT SPICE Models

The subcircuit for a (1μm $\times$ 10μm) emitter InP-based HBT is illustrated in Fig. 6.6. The SPICE model file and model parameters are given in table 6.6.

6.6 SUMMARY

Fiber-optic communication can be used for various applications over a wide range of data rates. The IC technology most appropriate for a given application will depend on the speed, SNR, cost and volume of production, and several other factors relating to the overall system. In this chapter, we a gave a very brief overview of important IC technologies and indicated where these would be most effectively utilized. The circuits designed in the remainder of this book use HBTs, but the circuit design techniques are applicable to any IC technology. HBTs were chosen for their high-speed advantage. Since this work was a research project, economic considerations, such as production costs, did not enter into this decision; concerns of product development for high-volume, lower speed applications, would have exerted stronger economic pressure to utilize BJTs or CMOS. To aid the reader in understanding the circuits presented in the following chapters, an introduction to HBTs was given and simple circuit models for SPICE simulation were provided.

InP-Based HBT: $(1\mu m \times 10\mu m)$ Single-Emitter	
Model File	Parameters
MODEL FOR INP-105 .SUBCKT HBTSE 1 2 3 Q1 5 4 3 MHBT1 RBX 2 4 PRBX RC 1 5 PRC CJX 4 5 PCJX .MODEL MHBT1 NPN + RE=PRE RC=1 RB=PRB + CJE=PCJE CJC=PCJC XCJC=1.0 + VJE=1.25 VJC=1.0 + MJE=PMJE MJC=0.5 + BF=1MEG BR=0.001 + IS=PIS ISE=PISE + NF=PNF NE=PNE + TF=PTF FC=0.69 .ENDS HBTSE	PRBX = 20.0 PRB = 15.3 PRE = 15.0 PRC = 3.6 PTF = 1.80ps PCJX = 11.0fF PCJC = 14.3fF PCJE = 163.0fF PMJE = 0.500 PIS = 1.897E-11 PISE = 2.256E-10 PNF = 1.0310 PNE = 1.5460

Table 6.6 Typical SPICE model for a (1μm $\times$ 10μm) emitter InP-based HBT.

REFERENCES

[1] Hans-Martin Rein. Silicon bipolar integrated circuits for multigigabit-per-second lightwave communications. *J. Lightwave Technol.*, LT-8(9):1371–1378, September 1990.

[2] Richard C. Walker, Thomas Hornak, Chu-Sun Yen, Joey Doernberg, and Kent H. Springer. A 1.5 Gb/s link interface chipset for computer data transmission. *IEEE J. Select. Areas Commun.*, SAC-9(5):698–703, June 1991.

[3] Scott D. Willingham, Kenneth W. Martin, and A. Ganesan. A BiCMOS low-distortion 8-MHz low-pass filter. *IEEE J. Solid-State Circuits*, 28(12):1234–1245, December 1993.

[4] Behzad Razavi and James Sung. A 6GHz 60mW BiCMOS phase-locked loop with 2V supply. In *ISSCC Dig. Tech. Papers*, pages 114–115, San Francisco, California, February 1994.

[5] Behzad Razavi and JanMye James Sung. A 6GHz 60mW BiCMOS phase-locked loop with 2V supply. *IEEE J. Solid-State Circuits*, 29(12), February 1994.

[6] Jean-Pierre Colinge. *Silicon-on-Insulator Technology: Materials to VLSI.* Kluwer Academic Publishers, Boston, 1991.

[7] Hans Ransijn and Paul O'Connor. A PLL-based 2.5-Gb/s GaAs clock and data regenerator IC. *IEEE J. Solid-State Circuits*, 26(10):1345–1353, October 1991.

[8] Zhi-Gong Wang, Manfred Berroth, Jörg Seibel, Peter Hofmann, Axel Hülsmann, Klaus Köhler, Brian Raynor, and Joachim Schneider. A 19GHz monolithic integrated clock recovery using PLL and 0.3μm gate-length quantum-well HEMTs. In *ISSCC Dig. Tech. Papers*, pages 118–119, San Francisco, California, February 1994.

[9] Mehran Bagheri, Keh-Chung Wang, Mau-Chung F. Chang, Randy B. Nubling, Peter M. Asbeck, and Andy Chen. 11.6-GHz 1:4 regenerating demultiplexer with bit-rotation control and 6.1-GHz auto-latching phase-aligner IC's using AlGaAs/GaAs HBT technology. *IEEE J. Solid-State Circuits*, 27(12):1787–1793, December 1992.

[10] Klaus Runge et al. AlGaAs/GaAs HBT IC's for high-speed lightwave transmission systems. *IEEE J. Solid-State Circuits*, 27(10):1332–1341, October 1992.

[11] Junko Akagi, Yasuhiko Kuriyama, Kouhei Morizuka, Masayuki Asaka, Kunio Tsuda, Masao Obara, Hideaki Yamakawa, and Hiroyuki Ibe. AlGaAs/GaAs HBT receiver ICs for a 10 Gbps optical communication system. In *IEEE GaAs IC Symposium*, pages 45–48, New Orleans, Louisiana, October 1990.

[12] W. Shockley. U.S. pat. no. 2,569,347, 1951.

[13] Herbert Kroemer. Theory of a wide-gap emitter for transistors. *Proc. IRE*, 45(11):1535–1537, November 1957.

[14] P. Asbeck, M. Chang, J. Higgins, N. Sheng, G. Sullivan, and K. Wang. GaAlAs/GaAs heterojunction bipolar transistors: Issues and prospects of application. *IEEE Trans. Electron Devices*, 36(10):2032–2042, October 1989.

[15] R. Nottenburg, Y. Chen, M. Panish, D. Humphrey, and R. Hamm. Hot-electron InGaAs/InP heterojunction bipolar transistors with ft of 110GHz. *IEEE Electron Device Lett.*, 10(1):30–32, January 1989.

[16] Y. Chen, R. Nottenburg, M. Panish, R. Hamm, and D. Humphrey. Subpicosecond InP/InGaAs heterostructure bipolar transistors. *IEEE Electron Device Lett.*, 10(6):267–269, June 1989.

[17] Fazal Ali and Aditya Gupta, editors. *HEMTs and HBTs: Devices, Fabrication, and Circuits*. Artech House, Boston, 1991.

[18] Herbert Kroemer. Heterojunction bipolar transistors and integrated circuits. *Proc. IEEE*, 70(1):13–25, January 1982.

[19] M. E. Kim, A. K. Oki, G. M. Gorman, , D. K. Umemoto, and J. B. Camou. GaAs heterojunction bipolar transistor device and IC technology for high-performance analog and microwave applications. *IEEE Trans. Microwave Theory Tech.*, 37(9):1286–1303, September 1989.

[20] M. E. Kim et al. GaAs heterojunction bipolar transistor device and IC technology for high-performance analog and microwave applications. *IEEE Trans. Microwave Theory Tech.*, 37(9):1286–1303, September 1989.

[21] H. Ichino, N. Ishihara, Y. Yamauchi, O. Nakajima, K. Nagata, and A. Iwata. A 10-Gbit/s decision circuit using AlGaAs/GaAs HBT technology. In *ISSCC Dig. Tech. Papers*, pages 188–189, February 1990.

[22] David J. Roulston. *Bipolar Semiconductor Devices*. McGraw-Hill, New York, 1990.

7

LOW-NOISE PREAMPLIFIER: THEORETICAL ANALYSIS AND DESIGN OF A 12-GHz InP HBT PROTOTYPE CIRCUIT

The goal of this chapter is to determine the effect of noise in electronic circuits on overall system performance and to establish design criteria so that these noise contributions can be minimized. The data-bearing optical signal emanating from a fiber is converted into a small electric current via the photodetector. The purpose of a preamplifier, as shown in Fig. 7.1, is to boost this current to a high enough level such that noise added by following stages will be negligible in comparison to the signal. The added noise is most troublesome when the signal level is the smallest, and this occurs at the preamplifier input. Therefore the total system noise, and thus the receiver sensitivity, will be determined by the preamplifier's noise performance. In the following sections we will discuss physical sources of noise in electronic circuits and analyze their effect on system performance for a fiber-optic preamp. We will also discuss the circuit design, and simulated results, of a low-noise preamplifier designed using InP-based HBTs. This circuit contains an integrated positive-intrinsic-negative (PIN) photodetector for detection of light at a wavelength of 1.3-μm. We will show in the following analysis that the reduction of parasitic capacitance afforded by integrating the photodetector and preamplifier on the same chip, as in this circuit, improves the noise performance of the receiver.

7.1 SOURCES OF NOISE

We saw in chapter 3 that the performance of a fiber-optic receiver is limited by random fluctuations in the received signal. The transmitted light signal consists of discrete energy packets or photons emitted randomly from a source. This quantum nature of light imposes a limit, such that approximately 20 photons are required on average for each *one* symbol to achieve an error probability of 10^{-9}. This is known as the *Quantum*

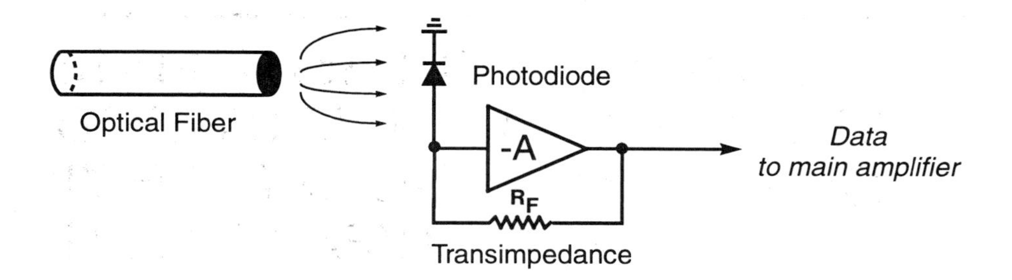

Figure 7.1 Block diagram of a transresistance fiber-optic preamplifier.

Limit in an optical receiver, and for a 10-Gb/s system corresponds to an optical power of -48.1 dBm for a wavelength of 1.30-μm, as was shown in table 3.3. In a practical receiver, however, the optical power must be increased by several orders of magnitude above the quantum limit in order to overcome the noises inherent in the receiver's electronics. The charge transport process in electronic circuits is analogous to the transmission of photons, and its random fluctuations obey similar statistics. We will now discuss the sources of random noise in electronic circuits, and give expressions for their power spectral densities (PSDs) as a function of frequency.

7.1.1 Shot Noise

Electrical current consists of discrete charges moving randomly within a conductor. The average current in a given time interval is determined by the average number of charges crossing a surface. The instantaneous current, however, will vary around this average value. These random variations are known as *Shot Noise*, and they represent a fundamental limitation on the accuracy of any electronic circuit. Shot noise was first described by Schottky and given its name (*Schroteffekt*) in 1918 in a classic paper written in Germany at the end of World War I [1].

Shot noise is present whenever charges cross a barrier, such as a depletion region in a pn junction or free-space in a vacuum tube. Recombination processes also exhibit shot noise, for example, the base current of a bipolar transistor. The statistical variations of electron emissions, or recombinations, can be derived assuming that these processes are governed by a Possion process [2, ch. 7], [3, ch. 16]. For a large range of frequencies, the noise PSD[1] is essentially constant with a value given by the *Schottky Equation*;

$$\hat{S}(f) = qI \qquad (A^2/\mathrm{Hz}), \tag{7.1}$$

[1] The term "power spectral density" (PSD) will be used even though the units will be in (A^2/Hz) or (V^2/Hz). However, to be proper we should actually say that this is the power dissipated in a 1Ω resistor.

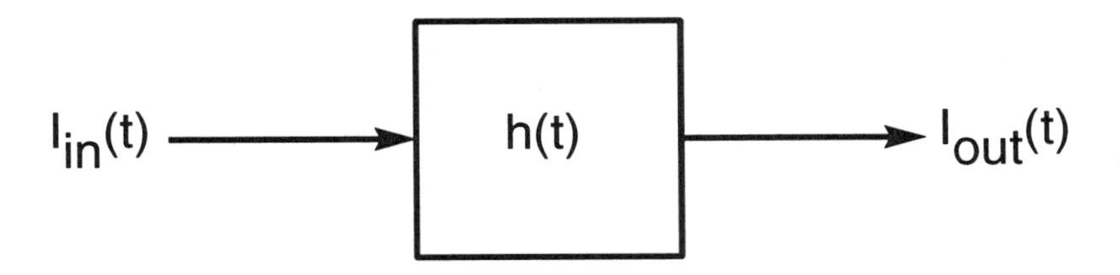

Figure 7.2 Linear network.

where q is the charge on an electron, and I is the average current. This noise power goes to zero for frequencies much greater than $1/\tau_t$, where τ_t is the transit time of a charge across the boundary. Normally, the transit frequency is quite a bit higher than the maximum usable frequency of the device $f_{\max}$. Therefore, assuming the noise PSD to be constant, or *white*, for all frequencies is an accurate approximation. The noise power in a bandwidth of B Hertz, is given by

$$\langle i_n^2 \rangle = qI(2B) \qquad (A^2). \tag{7.2}$$

Using the white noise approximation the spectrum is no longer bandlimited, and the noise will have infinite power. However, we always view the effect of the shot-noise in the context of a bandlimited system so that the noise power will be finite.

In previous chapters we have dealt with the two-sided noise PSD, defined as the Fourier transform of the autocorrelation function. However, for circuit design, we traditionally use the one-sided PSD, defined as twice the two-sided PSD. Therefore the one-sided PSD for shot noise is simply

$$\boxed{S(f) = 2qI \qquad (A^2/\mathrm{Hz}).} \tag{7.3}$$

This shot-noise spectrum is illustrated in Fig. 7.3, where the rms noise current at the output of an ideal bandpass filter with a bandwidth ($B = 1/T_B$) is given by

$$i_{\text{rms}} = \sqrt{2qIB} = \sqrt{I(2q/T_B)}. \tag{7.4}$$

So we see that the rms noise current is the geometric mean of the dc current and the current produced by two charges in a time T_B.

Example Shot-Noise Calculations

We will now perform a simple calculation that will help provide additional insight into the shot-noise mechanism. For a current of value I, the average number of charges $\overline{N}$

To remind us that this is actually not "power," we will use the notation S(f) denoting "spectral density," as opposed to P(f) for "power spectral density."

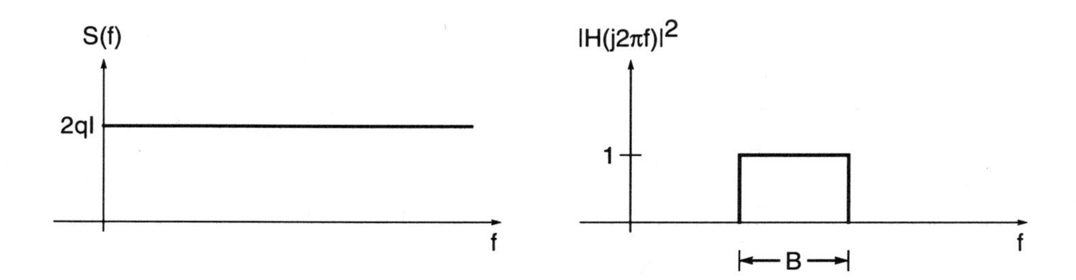

Figure 7.3 One-sided spectral density of shot-noise and an ideal bandpass filter with a bandwidth of B.

crossing a surface in time T_B is

$$\overline{N} = \frac{I}{q/T_B} \tag{7.5}$$

We can represent the instantaneous current in time T_B as $i(t, \cdot)$, which is a random process equal to

$$i(t, \cdot) = I + i_n(t, \cdot), \tag{7.6}$$

where $i_n(t, \cdot)$ is a zero-mean noise current. We will make use of the autocorrelation function $r_s(\tau)$, which can be calculated as the inverse Fourier transform of the PSD. For shot-noise, this is equal to an impulse function of weight qI;

$$r_i(\tau) = qI\delta(\tau). \tag{7.7}$$

The autocorrelation function is the expected value of two points on the same noise sample function separated by a time τ. For a zero-mean wide-sense stationary process, the autocorrelation function is defined as

$$r_i(\tau) \triangleq E[i_n(t, \cdot)i_n(t + \tau, \cdot)]. \tag{7.8}$$

When the result of this expression is an impulse function, we interpret this to mean that the noise waveform is so *wiggly* that any two samples, no matter how closely spaced in time, are uncorrelated. Furthermore, the correlation of two samples taken at the same time is infinite, or in other words, the noise process has infinite power, as it must because the PSD is non-zero for all frequencies. In reality, the noise will be correlated for small time separations on the order of τ_t (the transit time across a shot-noise boundary) and the autocorrelation function is a large spike that is finite over a small interval. However, approximating this spike with an impulse function of equal area is quite accurate, and it is often convenient for analysis as we will soon see.

To obtain a meaningful result in our noise analysis, we must always associate the white noise with some bandlimiting process. A simple bandlimiting operation is to take a moving average of the noise over a time T_B. If we do this, we obtain a new random process defined by

$$\hat{i}(t,\cdot) \triangleq I + \hat{i}_n(t,\cdot) \tag{7.9}$$

where $\hat{i}_n(t,\cdot)$ is the average of $i_n(t,\cdot)$ over a time T_B given by

$$\hat{i}_n(t,\cdot) = \frac{1}{T_B}\int_t^{t+T_B} i_n(t_1,\cdot)dt_1. \tag{7.10}$$

We would now like to find the ratio of the rms noise current to the signal current. We can calculate the rms noise as follows,

$$\langle i_n^2\rangle = E\left[\hat{i}_n(t,\cdot)\right]^2 = E\left[\frac{1}{T_B}\int_0^{T_B} i_n(t_1,\cdot)dt_1\right]^2 \tag{7.11a}$$

$$= \frac{1}{T_B^2}\int_{t_1=0}^{T_B}\int_{t_2=0}^{T_B} E\left[i_n(t_1,\cdot)i_n(t_2,\cdot)\right]dt_1dt_2 \tag{7.11b}$$

$$= \frac{1}{T_B^2}\int_{t_1=0}^{T_B}\int_{\tau=-t_1}^{T_B-t_1} E\left[i_n(t_1,\cdot)i_n(t_1+\tau,\cdot)\right]dt_1d\tau \tag{7.11c}$$

$$= \frac{qI}{T_B^2}\int_{t_1=0}^{T_B}\int_{\tau=-t_1}^{T_B-t_1} \delta(\tau)d\tau dt_1 \tag{7.11d}$$

$$= \frac{qI}{T_B^2}\int_{t_1=0}^{T_B} dt_1 \tag{7.11e}$$

$$= \frac{qI}{T_B}. \tag{7.11f}$$

From (7.5) we find that the dc current can be expressed in terms of the average number of charges crossing a boundary in time T_B as

$$I = \overline{N}\left(\frac{q}{T_B}\right). \tag{7.12}$$

Substituting this for the noise variance into (7.11) we obtain

$$\langle i_n^2\rangle = \overline{N}\left(\frac{q}{T_B}\right)^2. \tag{7.13}$$

The rms value is just the square root of $\langle i_n^2\rangle$;

$$i_{\mathrm{rms}} = \sqrt{\overline{N}}\frac{q}{T_B}. \tag{7.14}$$

The ratio of the rms noise current to the signal current from (7.12) and (7.14) is therefore

$$\boxed{\frac{i_{\text{rms}}}{I} = \frac{1}{\sqrt{N}}.} \tag{7.15}$$

The ratio of the noise to the signal current is just the inverse of the square root of the expected number of charges crossing a boundary in time T_B. This has appeared before; we saw in chapter 2 that the standard deviation of the average of N independent identically distributed random variables was also reduced by $\sqrt{N}$. Since random noise fluctuations add incoherently, their average approaches zero as the square-root of the number of observations, whereas the signal adds coherently and approaches a steady-state, non-zero average value. The noise can be reduced to an arbitrarily low level by increasing the expected number of charges passing the boundary. This can be accomplished either by increasing the nominal current I, or by lengthening the averaging interval T_B. Therefore increasing the SNR requires either a higher bias current or a lower bandwidth.

Averaging functions other than the simple moving average can be easily analyzed in the frequency domain. The general expression for the noise variance is found by the one-sided integral

$$\boxed{\langle i_n^2 \rangle = 2qI \int_0^\infty |H(j2\pi f)|^2 df,} \tag{7.16}$$

where $H(j2\pi f)$ is the Fourier transform of the impulse response $h(t)$ of a linear noise filter. For a simple one-pole filter with a frequency response given by

$$|H(j2\pi f)| = \frac{1}{1 + j(f/f_{3dB})}, \tag{7.17}$$

the noise-current variance is

$$\langle i_n^2 \rangle = 2qIf_{3dB} \int_0^\infty \frac{1}{1+x^2} dx \tag{7.18}$$

$$\langle i_n^2 \rangle = 2qIf_{3dB} \int_0^{\pi/2} d\theta \tag{7.19}$$

$$\langle i_n^2 \rangle = 2qI\left[(\pi/2) f_{3dB}\right], \tag{7.20}$$

and the effective noise bandwidth of the filter is $(\pi/2)f_{3dB}$, or 1.57 times the 3dB frequency.

7.1.2 Thermal Noise

Another source of white noise in electronic circuits results from the conversion of momentum, due to the kinetic energy of charged particles, into a potential difference that can be seen across the terminals of a resistor. This energy conversion involves Boltzmann's constant ($k = 1.38054 \times 10^{-23}$ J/°K), and is proportional to the absolute temperature. The resulting electrical fluctuations are referred to as thermal noise. This effect was also discussed by Schottky in 1918 [1]. In 1928, Johnson was able to observe this noise in the laboratory [4]. Johnson then discussed his results with Nyquist, who derived the expressions describing thermal noise based on thermodynamic considerations [5]. These results follow directly from analysis of black-body radiation [6, pp. 98–102].

Oliver summarizes the results of Nyquist and shows by the principles of statistical mechanics and quantum mechanics that the two-sided power-spectral-density delivered, to a noiseless resistor at absolute zero, from a noisy resistor of equal value at a temperature T, is given by [7, p. 134, eq. (16)]

$$S_T(f) = \frac{hf}{e^{hf/kT} - 1} \qquad \text{(Watts/Hz)}, \tag{7.21}$$

where h is Planck's constant ($h = 6.624 \times 10^{-34}$ Js). For $hf << kT$ (7.21) is approximately equal to kT, which is independent of frequency f. To get a feeling for the relative magnitude of the values kT and hf, which might have more physical significance to a circuit designer, we can normalize by the charge on an electron. The familiar term $kT/q = V_T$ is the thermal voltage, which is equal to 25.9-mV at 300°K; we find that hf/q equals 4.14-μV at 1-GHz. Writing (7.21) in normalized form gives

$$S_p(f) = kT\left[\frac{(hf/q)/(kT/q)}{e^{(hf/q)/(kT/q)} - 1}\right] = kT\left[\frac{\epsilon}{e^{\epsilon} - 1}\right], \tag{7.22}$$

where ϵ is the ratio $(hf/q)/(kT/q)$. At a frequency of 100-GHz, $hf/q = 0.414$-mV, $\epsilon = 0.016$, and the noise power spectral density is 0.992(kT) at 300°K. Therefore an error of only 1% is incurred if we assume the spectral density is flat (white noise) all the way to 100-GHz. With this white-noise assumption, the power delivered to the cold resistor R_0 from the resistor R is equal to $kT\Delta f$ for all frequencies, where Δf is the frequency interval. This situation is illustrated in Fig. 7.4, where the noise is represented by a series voltage source. The load resistor R_0 is equal in value to the source resistor R. For this case, the output voltage V_0 is half of the noise source voltage, and the power delivered in a bandwidth of B Hertz is

$$P_0 = kTB = \left(\frac{v_n}{2}\right)^2 \frac{1}{R}. \tag{7.23}$$

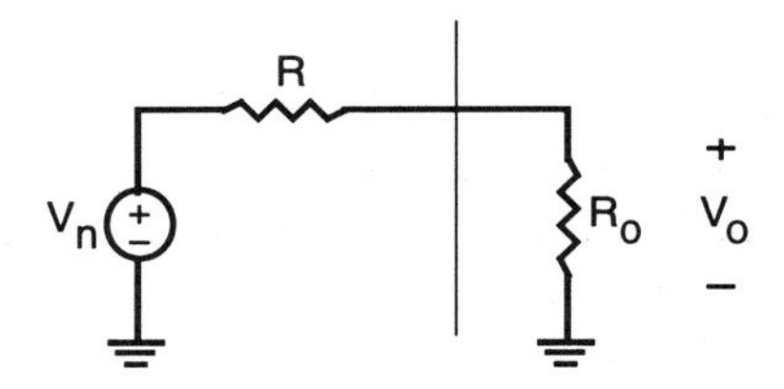

Figure 7.4 Illustration of the thermal noise of a resistor R at a temperature T delivering power to a load at temperature zero.

Therefore the expected squared noise voltage is found to be

$$v_n^2 = 4kTRB. \tag{7.24}$$

Since the thermal noise has a constant spectral density to very high frequencies, we can find from (7.24) that the value of the PSD is given by

$$\boxed{S_v(f) = 4kTR \qquad (\mathrm{V}^2/\mathrm{Hz}).} \tag{7.25}$$

We could also have represented the noise as a current source in parallel with the resistor. The value of the noise current is then

$$i_n = \frac{v_n}{R} = \frac{\sqrt{4kTRB}}{R} \tag{7.26}$$

$$i_n^2 = \frac{4kTB}{R}. \tag{7.27}$$

The equivalent thermal noise current spectral density is therefore given by

$$\boxed{S_i(f) = \frac{4kT}{R} \qquad (\mathrm{A}^2/\mathrm{Hz}).} \tag{7.28}$$

Example Thermal Noise Calculation To illustrate a simple thermal noise calculation we can consider a parallel combination of a resistor and capacitor as shown in Fig. 7.5 The transfer function is that of a one-pole lowpass filter with a noise bandwidth as found in (7.18) to be $(\pi/2)f_{3dB}$. Therefore the expected voltage squared is

$$\langle v_n^2 \rangle = 4kTR\left[\frac{\pi}{2}\left(\frac{1}{2\pi RC}\right)\right] \tag{7.29}$$

$$\langle v_n^2 \rangle = \frac{kT}{C}. \tag{7.30}$$

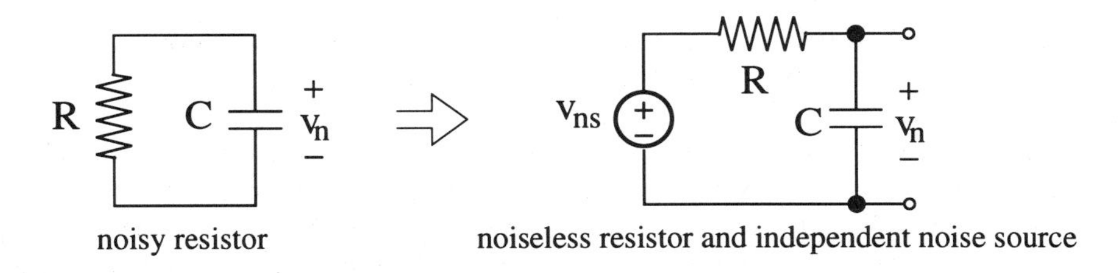

Figure 7.5 A capacitor in parallel with a noisy resistor shown modelled as a voltage noise source and a noiseless resistor.

This result is independent of the resistance value, because as R is increased the noise magnitude increases, but the bandwidth decreases, keeping the noise power constant. Since all physical capacitors will have a low shunt conductance, this noise will be associated with any capacitor and is known as kT/C (kT over C) noise. If we again normalize by q we can get a better feel for the order of magnitude of this noise. The noise variance can be written as

$$\langle v_n^2 \rangle = \frac{kT}{C} = \left[\frac{kT}{q}\right]\frac{q}{C}. \tag{7.31}$$

We can define a voltage $V_C \stackrel{\Delta}{=} q/C$ as the voltage required to place a single charge of value q on the capacitor C. The rms noise voltage across the capacitor is therefore given by the geometric mean of V_C and the thermal voltage V_T.

$$v_{\text{rms}} = \sqrt{V_T V_C}. \tag{7.32}$$

For a specific example, at 300°K the thermal voltage is 25.9-mV, or putting this in terms of decibels with a millivolt reference V_T=28.27-dBmV. For a 1-pF capacitor, V_C=0.16-μV or -75.92-dBmV. The rms noise in dBmV is just the average of the voltages V_C and V_T in dBmV; $v_{\text{rms}} = 1/2(28.27 - 75.92) = -23.82$-dBmV or 64.4-$\mu$V. The noise is reduced by 10-dB/decade as C increases.

7.1.3 Frequency Dependant Noise Sources

Shot noise and thermal noise are present in all circuits. These are both white noise sources with a constant spectral density extending to frequencies well beyond the useful operating frequency of the devices used. Shot noise is fundamental to the charge transport process and occurs because charges are carried in discrete packets. Shot-noise is therefore the ultimate limiting factor on the accuracy of any electronic circuit. Thermal noise, however, can be reduced to arbitrarily low levels by reducing the ambient temperature, but this is nearly always too expensive for practical solutions and is reserved for only the most critical applications.[2]

In addition to the unavoidable white noise sources, there are also noises due to imperfections in the materials that trap and release charges at random intervals. This type of noise has a frequency dependent spectrum with most of its energy concentrated at low frequencies. One such frequency-dependent noise is called (1/f) noise because the shape of its PSD varies inversely with frequency. This is also known as flicker noise because of the effect it has on visual displays.

Flicker noise is dependent upon the choice of materials, the processing purity, and the concentration of defects. In a FET device, current is carried near the semiconductor's surface, and the flicker noise is considerably higher than it is in a bipolar device because the concentration of defects at the surface is higher than in the bulk semiconductor. Flicker noise is typically characterized by the *corner frequency*. This is the frequency where the flicker noise is equal to the white noise in the device. For HBTs this corner frequency is in the kilohertz range. Since we are concerned with broadband receivers with effective bandwidths of 10-GHz, the contribution of the flicker noise at low frequencies as compared to the broadband white noise is negligible. Therefore, in the analysis to follow, we will ignore flicker noise and other low frequency noise sources such as burst-noise or popcorn-noise. Ignoring low-frequency noise is justifiable in amplifiers for the reasons just discussed. However, we must keep in mind that in other circuits with which we will be dealing, such as oscillators and mixers, low-frequency noise is converted to high-frequency noise and can therefore not be ignored.

7.2 RELATIONSHIP BETWEEN NOISE AND RECEIVER SENSITIVITY

Our task is to design a fiber-optic receiver that has the highest sensitivity possible. In other words we want the receiver to be able to function at very low optical input power levels. The higher the receiver's sensitivity, the farther away receivers can be physically spaced and still maintain an acceptable level of performance (bit-error-rate $< 10^{-9}$). We saw in chapter 3 that the energy in a photon of wavelength λ is given by

$$e_{\text{ph}} = \frac{hc}{\lambda}, \tag{7.33}$$

[2] Progress in superconductors has lead to the development of lower cost and lighter weight closed-cycle refrigerators, which might find their way into a wider range of applications in the future — low-noise amplification is certainly one such application.

where h is Plank's constant and c is the speed of light. The numerical value of the photon energy is

$$e_{\text{ph}} = \frac{198.6 \times 10^{-12}(\text{nJ} \cdot \mu\text{m})}{\lambda}. \tag{7.34}$$

The optical power is obtained by dividing the energy by the time interval,

$$p_{\text{ph}} = \frac{e_{\text{ph}}}{T_B}; \tag{7.35}$$

and for a bit-period of 100-ps, the optical power is

$$p_{\text{ph}} = \frac{1.986(\text{nW} \cdot \mu\text{m})}{\lambda}. \tag{7.36}$$

We also saw in chapter 3 that the signal to noise ratio (SNR) of the sample statistics required to meet the performance objective was

$$\text{SNR} \geq \frac{P_{\text{sig}}}{P_{\text{noise}}} = 6^2. \tag{7.37}$$

The current produced by the photodetector in the bit-interval T_B can be determined from the average number of photons-per-bit incident on the detector $\overline{n}$. If we define the quantum efficiency η as the ratio of the number of photons that generate electron-hole pairs to the number of photons arriving at the receiver, then the output current of the photodiode for a *one* symbol is expressed as

$$I_{\text{sig}} = \frac{\eta \overline{n}_1 q}{T_B}, \tag{7.38}$$

where $\overline{n}_1 = 2\overline{n}$. The average signal current is half of this value because the light is a unipolar source. The output current of a photodetector is illustrated in Fig. 7.6; the average current is just

$$I_{\text{av}} = \frac{\eta \overline{n} q}{T_B}. \tag{7.39}$$

This average current establishes the reference from which the signal is compared to the noise. To achieve the desired performance, the difference between the signal and its average level must be greater than Q_{SNR} multiplied by the rms current;

$$\frac{\eta \overline{n} q}{T_B} \geq (Q_{\text{SNR}}) i_{\text{rms}}, \tag{7.40}$$

where Q_{SNR} depends upon the noise filter used and the performance criteria. For a bit-error-rate (BER) of 10^{-9}, Q_{SNR} is is approximately 6, which is just the square root of the SNR.

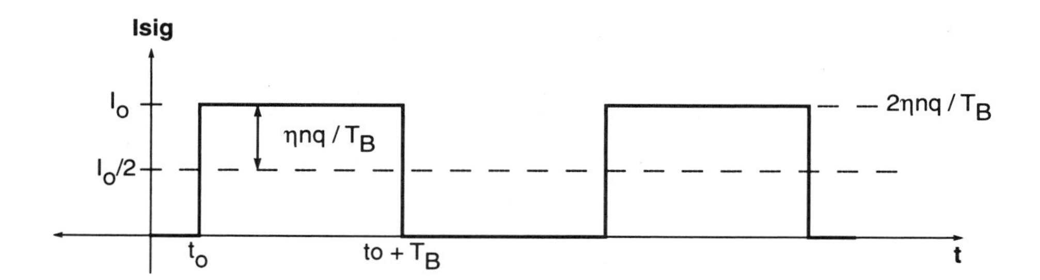

Figure 7.6 Current output from a photodetector.

The average optical power for an equal number of ones and zeros is given by

$$P_{\text{av}} = \frac{P_{\text{on}}}{2} + \frac{P_{\text{off}}}{2}, \tag{7.41}$$

which is simply half of the *on* power; therefore

$$P_{\text{av}} = \left(\frac{\overline{n}_1}{2}\right)\frac{hc}{\lambda T_B} = \frac{\overline{n}hc}{\lambda T_B}. \tag{7.42}$$

Substituting the average power for $\overline{n}$ into (7.40) we obtain

$$\eta P_{\text{av}}\left(\frac{q\lambda}{hc}\right) \geq (Q_{\text{SNR}})i_{\text{rms}}. \tag{7.43}$$

(7.43) gives us an expression for the minimum average optical power in terms of the equivalent input rms noise current, the SNR (Q_{SNR}), the quantum efficiency of the photodetector, and physical constants. Substituting numerical values for the physical constants we obtain

$$\boxed{P_{\text{av}} = 1.242\left(\frac{\mu\text{W}\cdot\mu\text{m}}{\mu\text{A}}\right)\left[\frac{(Q_{\text{SNR}})i_{\text{rms}}}{\eta\lambda}\right].} \tag{7.44}$$

For the case of Q_{SNR}=6, $\lambda = 1.3\mu\text{m}$, $\eta = 0.5$, and $i_{\text{rms}} = 1\mu\text{A}$, the average optical power is 11.46 μW or -19.4 dBm. We can represent this as a power penalty over the quantum limit, which from table 3.3 is seen to be -48.1 dBm. Therefore, the increased power above the quantum limit needed for this receiver is 28.7 dB. We will now analyze the noise performance of a transresistance preamplifier to find the rms current noise, which is a function of the bit-rate and circuit parameters. Then from (7.44) we will be able to determine the minimum optical power needed to meet our performance objective.

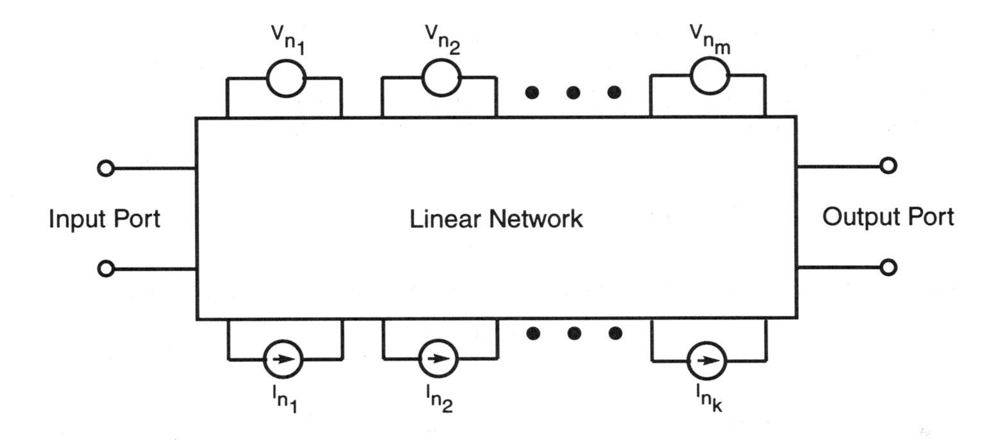

Figure 7.7 Arbitrary linear network to illustrate noise analysis.

7.3 CALCULATIONS OF NOISE IN LINEAR CIRCUITS

In the analysis of noise in circuits we will assume that the noise is a wide-sense stationary random process. This turns out to be a valid assumption in most cases. Any changes in the noise statistics are usually related to temperature or aging, which vary much more slowly than the signal, so the noise can be considered stationary within the observation interval of interest. We will also assume that all sources of noise are mutually uncorrelated. This too is a valid assumption and is due to the huge number of charges moving within a circuit. We wouldn't expect the thermal movement of changes in one resistor to be synchronized with the random shot emissions of electrons crossing a depletion region in another part of the circuit. Although these events are related by physical laws, the mechanisms governing their random fluctuations are, at least statistically speaking, independent. With these two assumptions; namely,

- the noise is wide-sense stationary,
- all noise sources are mutually uncorrelated,

we can analyze the noise behavior of circuits quite easily. Consider the linear network shown in Fig. 7.7. The network shows M noise voltages and K noise currents. From the theory of random processes, we know that if there were only one noise source (for example v_{n_1}) the noise spectral density at the output would be

$$S_{out_{v_1}}(f) = S_{v_1}(f)\left|H_{v_1}(j2\pi f)\right|^2, \tag{7.45}$$

where $S_{v_1}(f)$ is the spectral density of the noise source, and $H_{v_1}(j2\pi f)$ is the transfer function of the network from v_{n_1} to the output. Since we have assumed all noise sources are independent, we can find the spectral density at the output by summing each of the individual noise contributions. Therefore, at the output

$$S_{out}(f) = \sum_{m=1}^{M} S_{v_m}(f) \left| H_{v_m}(j2\pi f) \right|^2 + \sum_{k=1}^{K} S_{i_k}(f) \left| R_{i_k}(j2\pi f) \right|^2 . \quad (7.46)$$

The noise power at the output is found by integrating $S_{out}(f)$ over all frequencies;

$$\sigma_n^2 = \int_0^\infty S_{out}(f) df. \quad (7.47)$$

If we substitute (7.46) for $S_{out}(f)$, we realize that the total noise variance can be written as

$$\sigma_n^2 = \sum_{m=1}^{M} \sigma_{v_m}^2 + \sum_{k=1}^{K} \sigma_{i_k}^2 . \quad (7.48)$$

In other words, the output noise power is the sum of the individual power contributions from each noise source, and the rms noise is the square root of the sum of the powers. Therefore, if we have a noise source producing a power of unit value, the rms noise will also be unity. If we have two identical independent noise sources, the power will double, and the rms noise will be equal to $\sqrt{2}$.

Often circuit designers are more interested in the rms noise than the noise power, and it is common to see the noise spectral density expressed as the square root of the PSD. For example, a current noise of $(16 \times 10^{-24}\text{A}^2/\text{Hz})$ is expressed as $(4\text{pA}/\sqrt{\text{Hz}})^2$. Therefore, to find the rms noise we take the square root of the bandwidth and multiply by the square root of the PSD. For a bandwidth of 10 GHz the rms noise is

$$i_{\text{rms}} = (4\text{pA}/\sqrt{\text{Hz}})\sqrt{10\text{GHz}} = 0.4\mu\text{A}. \quad (7.49)$$

Since the output noise is a function of the gain of the circuit, it is also common to express the total output noise as an equivalent noise source at the input. This is an analytical convenience because we are not necessarily interested in the actual value of the noise, but rather the signal-to-noise ratio (SNR). A large noise at the output is not a problem when the signal is also large. Expressing the total output noise as an equivalent input noise is useful for directly comparing the noise contributions of incoming signals. As an example, we can consider the previous circuit that produced an equivalent input current noise PSD of $(4\text{pA}/\sqrt{\text{Hz}})^2$ in a bandwidth of 10 GHz, which corresponded to an rms noise of 0.4 μA. If an SNR of 6^2 is required, then the input rms signal must be 6 times larger than the input noise, or at least 2.4 μA.

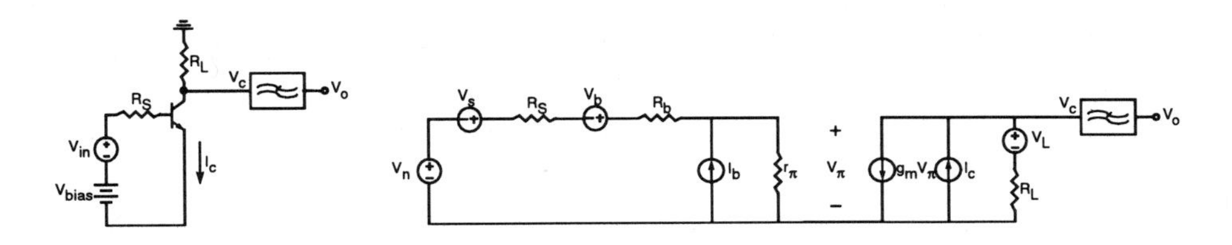

Figure 7.8 Circuit diagram and small signal model of a one transistor amplifier showing the white noise sources.

A Simple Example: One Transistor Amplifier

We will now demonstrate how noise can be calculated in a simple example, and we will get a feel for the relative magnitudes of the different types of noise. A simple transistor amplifier is shown in Fig. 7.8. The small signal equivalent circuit is shown with all of the white noise sources. We will now calculate the noise power at the collector due to each of the noise sources.

The thermal noise due to the load resistor R_L adds directly to the collector voltage. Therefore, the gain is unity, and the spectral density is given by

$$S_{c_{R_L}}(f) = 4kTR_L(1)^2 = 4kTR_L. \tag{7.50}$$

The collector voltage due to the shot-noise in the collector i_c is simply

$$v_c = R_L i_c. \tag{7.51}$$

The spectral density of the noise at the collector is then

$$S_{c_{i_c}}(f) = 2qI_c(R_L)^2, \tag{7.52}$$

where I_c is the bias current in the collector. The gain from the input voltage source is the same as the gain from the thermal noises due to the source and base resistances, and is given by

$$A = \frac{r_\pi}{r_\pi + R_s + r_b} g_m R_L,$$
$$A = \gamma_\pi g_m R_L, \tag{7.53}$$

where,

$$\gamma_\pi \triangleq \frac{r_\pi}{r_\pi + R_s + r_b}. \tag{7.54}$$

Therefore, the spectral density of the noise at the output due to the thermal noise of R_s and r_b is

$$S_{c\,R_s}(f) = 4kTR_sA^2, \quad (7.55)$$

$$S_{c\,r_b}(f) = 4kTr_bA^2. \quad (7.56)$$

Finally, the bias current shot noise produces a collector voltage of magnitude

$$v_c = i_b[(R_s + r_b) \parallel r_\pi]g_m R_L. \quad (7.57)$$

The average base current is just the collector bias current divided by the current gain β. Therefore, the spectral density at the collector is given by

$$S_{c\,i_b}(f) = 2q(I_c/\beta)\,[g_m R_L]^2\,[(R_s + r_b) \parallel r_\pi]^2. \quad (7.58)$$

The total spectral density at the collector can be written as follows

$$S_c(f) = 4kT\left[A^2(R_s + r_b) + R_L\right] + 2qI_c\left[R_L^2 + \frac{(g_m R_L)^2[(R_s + r_b) \parallel r_\pi]^2}{\beta}\right] \quad (7.59)$$

We can now express this noise as an equivalent voltage source at the input. The gain from the collector to the input is $(1/A)$. Therefore, we need to divide $S_c(f)$ by the square of the voltage gain to obtain the spectral density at the input,

$$S_{in}(f) = S_c(f)/A^2. \quad (7.60)$$

Hence,

$$S_{in}(f) = 4kT\left[R_s + r_b + \frac{R_L}{A^2}\right] + \frac{2qI_c}{A^2}\left[R_L^2 + \frac{(g_m R_L)^2[(R_s + r_b) \parallel r_\pi]^2}{\beta}\right] \quad (7.61)$$

For the case of a low impedance source $(R_s + r_b \ll r_\pi)$, γ_π is approximately unity, and the voltage gain is simply $g_m R_L$. Therefore, the input spectral density reduces to

$$S_{in}(f) = \underbrace{4kT\left[R_s + r_b + \frac{1}{g_m A}\right]}_{\text{thermal noise}} + \underbrace{2qI_c\left[\frac{1}{g_m^2} + \frac{(R_s + r_b)^2}{\beta}\right]}_{\text{shot noise}}. \quad (7.62)$$

To compare the relative contribution of the shot noise and thermal noise terms, we multiply and divide the shot noise by kT. Remembering that the thermal voltage V_T is given by

$$V_T = \frac{kT}{q} \quad (25.86\text{mV @ } 300^\circ K), \quad (7.63)$$

we can write the noise spectral density as

$$S_{in}(f) = 4kT\left[R_s + r_b + \frac{1}{g_m A}\right] + 4kT\left(\frac{I_c}{2V_T}\right)\left[\frac{1}{g_m^2} + \frac{(R_s + r_b)^2}{\beta}\right]. \quad (7.64)$$

Recalling that the transconductance of a bipolar device is given by I_c/V_T, we finally obtain the desired expression for the spectral density of the equivalent input noise voltage for the circuit of Fig 7.8, which is valid for low and medium frequencies;

$$S_{in}(f) = 4kT\left[R_s + r_b + \frac{1}{g_m A} + \frac{1}{2g_m} + \frac{g_m(R_s + r_b)^2}{2\beta}\right]. \quad (7.65)$$

For a transistor with the parameters given in table 7.1, we obtain the following numerical values for the input referred noise spectral density

$I_c =$	1mA
$B =$	100
$R_L =$	$2.6K\Omega$
$r_b =$	50Ω
$R_s =$	50Ω
$A =$	100
$g_m =$	$1/26\Omega$

Table 7.1 Parameters for one transistor amplifier.

$$\begin{aligned} S_{in}(f) = \quad & 4kT\,[50\Omega + && \text{Thermal noise due to } R_s \\ & 50\Omega + && \text{Thermal noise due to } r_b \\ & 0.26\Omega + && \text{Thermal noise due to } R_L \\ & 13\Omega + && \text{Shot noise due to } I_c \\ & 1.92\Omega] && \text{Shot noise due to } I_b. \end{aligned} \quad (7.66)$$

The noise can also be expressed as an equivalent resistance,

$$S_{in}(f) = 4kTR_{eq} = 4kT(115\Omega). \quad (7.67)$$

For this example the equivalent noise resistance is 115Ω. Looking at the numerical contributions we see that the noise due to the base current is smaller than the collector current shot noise. This could be reversed, however, if the source resistance were increased or if the current gain β were reduced. Also notice that the noise due to the resistive load is negligible. This would not be the case if an active load were used. Thus, we usually find resistive loads at the front-end of a low-noise amplifier.

Now we will evaluate the rms noise voltage to get a feel for typical magnitudes. It is useful to remember that

$$4kT(1\Omega) = (0.129\text{nV}/\sqrt{\text{Hz}})^2. \tag{7.68}$$

Therefore, for an equivalent noise resistance of 115Ω, the spectral density is given by

$$S_{in}(f) = (1.39\text{nV}/\sqrt{\text{Hz}})^2. \tag{7.69}$$

For an ideal lowpass filter of bandwidth Δf, the variance of the input-referred noise voltage is simply

$$\langle v_n^2 \rangle = S_{\text{in}}(f)\Delta f. \tag{7.70}$$

The rms noise is just its square root, and it is given by

$$v_{\text{rms}} = 1.39\text{nV}\sqrt{\Delta f}. \tag{7.71}$$

The equivalent noise voltages for various bandwidths are given in table 7.2; the rms noise increases with the bandwidth at a rate of 10-dB/decade.

v_{rms}	Δf
139 nV	10 kHz
4.40μ V	10 MHz
139μ V	10 GHz

Table 7.2 Equivalent input voltage noise for various bandwidths.

This simple example serves to illustrate the concepts used in noise analysis. However, we have ignored parasitics that will alter the contribution of the noise at various frequencies. We will find, in the analysis of a preamplifier for a fiber optic receiver to follow, that the input capacitance is a crucial parameter. The reason is that the capacitance reduces the gain of the signal but has no effect on the collector current shot noise, thus reducing the SNR — the input-referred noise spectral density will, therefore, increase with frequency.

7.4 TRANSRESISTANCE PREAMPLIFIER NOISE ANALYSIS

The noise analyses of preamplifiers for fiber-optic receivers have appeared throughout the literature. Personick was the first to provide a detailed characterization of the noise of preamplifiers with high impedance sources [8, 9, 10]. In recent years papers by Muoi [11], Brian and Lee [12], and Kasper and Campbell [13] have reviewed the state of the art in fiber-optic receiver design and outlined the results of detailed noise analysis. All give similar results to those obtained by Personick with various differences depending on the optimization criteria used. Textbooks on the subject [14, 15, 16, 17] usually have a chapter devoted to receiver design, and they also give the fundamental results.

In this chapter, our goal is to present the noise analysis of a transresistance preamplifier and use the results to optimize the performance of an InP circuit. Although the general noise analysis has appeared elsewhere, we will consider it once again and point out some subtleties that have not been mentioned in the literature. The analysis to follow is meant to be tutorial in nature, and it stresses a physical understanding of the noise mechanisms. Aside from presenting the expressions for the noise, we will show how the results were derived and also explain the trade-offs made between noise performance, bandwidth, and stability.

We will first state some results, explain the physical meaning of the terms, and present simulation results that are in agreement with the expressions given. The following table defines all of the terms used in the forthcoming analysis:

detector capacitance	C_d
stray capacitance	C_s
detector-plus-stray capacitance	$C_{ds} = C_d + C_s$
total bipolar capacitance	$C_{TB} = C_{ds} + C_\pi + C_\mu$
parasitic capacitance, bipolar	$C_\beta = C_{ds} + g_e + C_\mu$
total FET capacitance	$C_{TF} = C_{ds} + C_{gss} + C_{gs} + C_{gd}$
excess noise factor, FET	Γ
bipolar current gain	β
bipolar base resistance	r_b
data rate	f_b
maximum bipolar f_t	$f_F = 1/2\pi\tau_F$
normalized integration constant	I_2
normalized integration constant	I_3
integration constant ratio	$m = I_3/I_2[1 + 2/A_1]$
first-stage voltage gain	$A_1 = g_m R_c$
open-loop voltage gain	$A = A_1 A_2$
base resistance equivalent frequency	$f_{rb} = 1/2\pi C_{ds} r_b$

7.4.1 General Considerations and Interpretations

In this section we will present the noise analysis of a transresistance preamplifier and discuss the physical interpretation of the results. In sections to follow we will go into considerably more detail, but it is necessary that the reader grasps the fundamental problems before proceeding to more complicated issues.

In some of the early fiber-optic receivers, data-rates were low enough (100 Mb/s), compared to the bandwidth of transistors, such that optimization of noise performance concentrated on effects that were dominant at low and intermediate frequencies. Results of these optimizations led to rules-of-thumb that were used in the design of preamplifiers. However, when the data-rates approach the transistor's limitations, high-frequency effects take over, and the conditions under which the noise is minimized changes. In the first-order analysis to follow we will determine the dominant noise sources at low and intermediate frequencies. We present this primarily for historical reasons so that the reader can better understand some of the early literature on

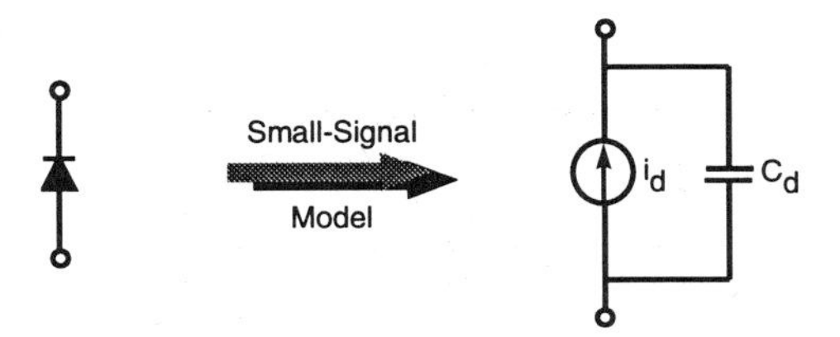

Figure 7.9 A PIN photodetector.

the subject. Later we will broaden the bandwidth and see how the noise changes as the data-rate approaches $f_{\max}$.

The photodetector is a reversed-biased diode as shown in shown in Fig. 7.9. The small-signal model consists of the photo-generated current i_d and the detector capacitance C_d. The natural feedback configuration for this input source is a transresistance amplifier (current in $\rightarrow$ voltage out). A current amplifier could work, provided that its noise is low enough; however, we will show that a common-base current-buffer actually has more noise than a common-emitter configuration, because it has no current gain in the first stage; therefore noise contributions from following stages will be large in comparison to the signal.

We will consider the generic transresistance configuration shown in Fig. 7.10. Thermal noise due to the feedback resistor is represented by a shunt current source with a spectral density of $4kT/R_F$. (The gain stage is considered noiseless for the time being, and the output impedance is assumed to be zero.) The transfer function is therefore given by

$$R_T(s) = \frac{-v_o}{i_d + i_{R_F}} = \frac{R_F/[1+\epsilon(s)]}{1 + \left[\frac{\epsilon(s)}{1+\epsilon(s)}\right] sC_{in}R_F}. \tag{7.72}$$

For low B_T we can assume that the overall amplifier speed is dominated by the time constant due to C_{in} and R_F. The gain stage is broadband in comparison, and the loss in the passband is simply $\epsilon(s) \approx \epsilon_0 = 1/A_0$, or the inverse of the dc gain. As a result, the closed-loop transresistance is

$$R_T(s) \approx \frac{R_F}{1 + s[\epsilon_0 C_{in} R_F]}, \tag{7.73}$$

which is a first-order system with a dominant pole due to the time-constant $(\epsilon_0 C_{in} R_F)$. From this expression we can find the first fundamental trade-off in the transimpedance configuration — low-noise operation vs. high-speed performance. The 3-dB frequency

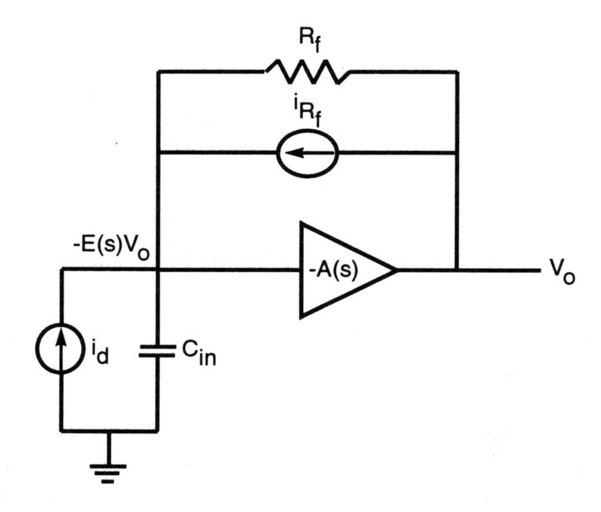

Figure 7.10 A transresistance amplifier figure.

for the amplifier is

$$f_{3dB} = \frac{1}{2\pi\epsilon_0 C_{in} R_F} = \frac{A_0}{2\pi C_{in} R_F}, \tag{7.74}$$

which is proportional to the open-loop voltage gain. The capacitance C_{in} is primarily dominated by the detector capacitance C_d and is typically quite large ($\sim$ 0.5 pF). The reason for a large C_d is due to the numerical aperture of the optics — namely, the inability to focus the light on a small area. As a result, the surface-area of the photodetector is large (a diode area of 50 μm x 50 μm is typical) which leads to a high capacitance. The bandwidth is fixed by the data rate. The gain A_0, however, is under the designer's control but will have a practical limit (40–60 dB). We will see later on, at high frequencies, that assuming a constant gain is no longer valid, and that the dc gain will actually need to be quite a bit lower ($\sim$ 10) to meet stability requirements. For the case of low data rates, R_F determines the bandwidth; therefore, to meet the speed requirement there will be an upper limit on the feedback resistor R_F given by

$$R_F < \frac{A_0}{2\pi f_{3dB} C_{in}}, \tag{7.75}$$

which is the first half of the noise-bandwidth trade-off. The second half arises from the noise current of the feedback resistor. The input referred noise spectral density due to R_F is

$$S_{i_{RF}}(f) = \frac{4kT}{R_F}. \tag{7.76}$$

As a first-order approximation, the rms noise current due to R_F referred to the input is

$$\langle i_n^2 \rangle = \frac{4kTf_{3dB}}{R_F}. \tag{7.77}$$

If a maximum value is set such that $\langle i_n^2 \rangle \leq \overline{i_m^2}$, we obtain the lower limit

$$R_F > \frac{4kTf_{3dB}}{\overline{i_m^2}}. \tag{7.78}$$

The maximum bandwidth occurs when the upper and lower limits are identical. Putting these limits together,

$$\frac{4kTf_{3dB}}{\overline{i_m^2}} < R_F < \frac{A_0}{2\pi f_{3dB} C_{in}}, \tag{7.79}$$

the maximum operating frequency for a given noise level is just

$$f_m = \sqrt{\frac{A_0}{2\pi C_{in}} \frac{\overline{i_m^2}}{4kT}}. \tag{7.80}$$

For an input capacitance of 1 pF, the maximum 3-dB bandwidth at 300° K is

$$f_m = [3.09(\text{GHz}/\mu\text{A})] i_{\text{rms}} \sqrt{A_0}. \tag{7.81}$$

Typically one is forced to make the bandwidth high at the expense of the noise. In this case, the noise resistance is chosen to be as large as possible and still meet the bandwidth requirements. Therefore, the feedback resistor is chosen to be

$$R_F = \frac{A_0}{2\pi f_{3dB} C_{in}}, \tag{7.82}$$

or for a 1 pF capacitor,

$$R_F = [159(\Omega \cdot \text{GHz})] \frac{A_0}{f_{3dB}}. \tag{7.83}$$

The resulting noise for this feedback resistance is

$$i_{\text{rms}} = f_{3dB} \sqrt{4kT \cdot \frac{2\pi C_{in}}{A_0}}, \tag{7.84}$$

and, for an input capacitance of 1 pF at 300° K, the rms noise current is

$$i_{\text{rms}} = \frac{f_{3dB}}{\sqrt{A_0}} (0.323\mu\text{A}/\text{GHz}). \tag{7.85}$$

Summary of First-Order Analysis

The main points to remember in this first-order analysis are that:

- Noise power due to R_F increases as the *square* of the bandwidth — rms noise increases linearly with the bandwidth.
- Noise and bandwidth limitations are caused by the high values of capacitance of the photodetectors.
- Increasing dc gain (which reduces the voltage swing at the input node and hence the effective capacitance C_{in}/A_0) reduces the noise by allowing a larger feedback resistor to be used.

This discussion has outlined some of the basic concepts in the design of low-noise transresistance amplifiers for low or intermediate frequencies. However, the analysis does not hold for high-speed circuits because

- The assumption that the amplifier is wide-band in comparison to the dominant pole is violated.
- The assumption that the noise is predominantly due to the feedback resistor is violated.

In the following section, the noise contribution from all of the internal noise sources within the amplifier will be derived. It will be shown that, at low frequencies, the noise is dominated by the feedback resistor's thermal noise and the shot-noise of the BJT's base current. However, we will find that other terms (collector-current shot noise and base-resistance thermal noise) will dominate at higher frequencies.

7.4.2 Detailed Analysis of a Transresistance Preamplifier

Performing a complete small-signal noise analysis can get complicated — even for very simple circuits. The analysis can be simplified by realizing that many of the noise sources in the circuit will have little or no effect on the overall noise, so these sources can be ignored from the start. In this section we will proceed with the noise analysis one step at a time, making approximations as we go along; then we will check the final analytical result against simulations as further justification for making these approximations.

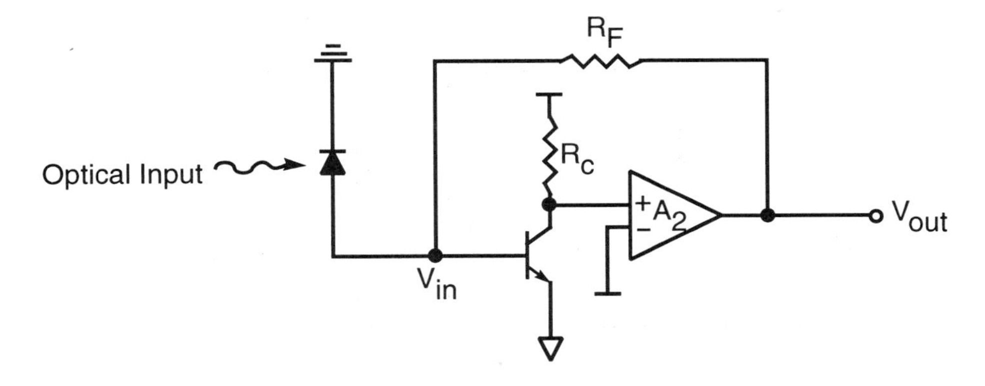

Figure 7.11 Block diagram of a transresistance preamp with a reverse-biased diode photodetector.

Transresistance Calculation

We begin by finding the most important parameter of the amplifier, the transresistance. A schematic of a typical transresistance preamplifier using a common-emitter front-end is shown in Fig. 7.11. The analysis of feedback amplifiers is greatly simplified using Bode's *Asymptotic Gain Formula* and *Blackman's Impedance Formula*, which dispenses with the *Shunt-Shunt, Series-Shunt* nonsense.[3] Rosenstark [18] gives a very clear and concise explanation of the derivation and application of these formulas. It can be shown [18, p.16] that the transresistance of a feedback amplifier can be written in the form

$$R_T = R_\infty \left[\frac{L(s)}{1 + L(s)} \right] + R_0 \left[\frac{1}{1 + L(s)} \right], \tag{7.86}$$

where L(s) is the loop gain or return-ratio,[4] R_∞ is the transconductance of the amplifier when the loop gain approaches infinity, and R_0 is the transconductance when $L(s)$ approaches zero; R_0 represents forward signal transmission through the feedback path — a term that is usually not accounted for in the simplified block-diagram formulation of feedback circuits.

The loop gain $L(s)$ can easily be calculated for the small-signal model of Fig. 7.12 by replacing the controlled source v with a voltage of value v_1 and finding the returned

[3] Attempts to explain feedback in circuits, which have bidirectional signal propagation, in terms of unidirectional block diagrams are often inadequate.

[4] There may be more than one loop in an amplifier. Furthermore the loop gain can be different for different loops when local feedback is present. However, as long as R_∞, R_0, and $L(s)$ are all calculated with respect to the same loop, the overall result R_T will be unique, and can be obtained regardless of the loop one chooses to consider.

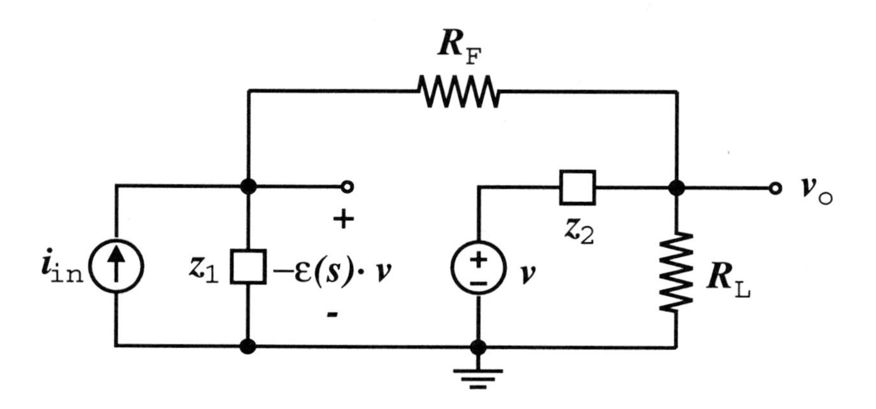

Figure 7.12 Small-signal model of a transresistance amplifier for the purpose of calculating the loop gain and overall transresistance.

voltage v, where $\epsilon(s)v$ is the voltage across the control impedance z_1 and $\epsilon(s)$ is the loss of the amplifiers $\epsilon(s) = 1/A(s)$. The return-ratio is then given by $L(s) = -v_1/v$. It can be seen by inspection that

$$-\epsilon(s)v = v_1 \frac{Z_1}{R_L \parallel Z_2 + R_F + Z_1}\left(\frac{R_L}{R_L + Z_2}\right); \quad (7.87)$$

multiplying both sides by $A(s)$, and dividing by v_1 we obtain the following expression for the loop gain:

$$L(s) = -\frac{v}{v_1} = \frac{A(s)Z_1}{R_L \parallel Z_2 + R_F + Z_1}\left(\frac{R_L}{R_L + Z_2}\right). \quad (7.88)$$

If the impedance Z_1 is large compared to the feedback resistor R_F, and if the impedance Z_2 is small compared to the load R_L, we see that the loop gain $L(s)$ reduces to the forward gain of the amplifier $A(s)$ as expected.

To find the overall transconductance we'll consider the asymptotic limits of R_T as $L(s)$ goes from infinity to zero. For infinite loop gain ($L \to \infty$), the voltage across the control impedance Z_1 is zero; therefore, no current can enter Z_1 and all the input current must pass through R_F, which has one node anchored to zero volts via Z_1. The output voltage is, therefore, $v_o = -R_F i_{\text{in}}$, and $R_\infty = -R_F$.

In the other extreme, when the loop-gain is equal to zero ($L \to 0$), the load impedance will consist of $R_L \parallel Z_2$, and v_o can be determined by a simple current division ratio.

$$R_0 = R_L \parallel Z_2 \left[\frac{Z_1}{Z_1 + R_F + R_L \parallel Z_2}\right] \quad (7.89)$$

The transresistance of the amplifier for an arbitrary value of the loop-gain is simply given by

$$\boxed{R_T = -R_F \frac{L}{1+L} + R_0 \frac{1}{1+L},} \tag{7.90}$$

where L is given by (7.88) and R_0 is given by (7.89).

Output Impedance

The output impedance can be found easily using *Blackman's Impedance Formula*:

$$Z_{\text{out}} = Z^o_{\text{out}} \frac{1+L_{sc}}{1+L_{oc}}, \tag{7.91}$$

where L_{sc} is the loop-gain L calculated when the output port is short-circuited, L_{oc} is the loop-gain when the output port is open-circuited, and Z^o_{out} is the impedance at the output port when the loop gain is equal to zero. We can easily see that short-circuiting the output port will null the loop-gain, therefore, $L_{sc} = 0$; open-circuiting the output port doesn't change the original circuit, therefore, $L_{oc} = L$. Z^o_{out} can be found easily by inspection as a parallel combination

$$Z^o_{\text{out}} = (R_L \parallel Z_2) \parallel [R_F + Z_1] \tag{7.92}$$

Hence the output impedance of the amplifier with feedback is just

$$\boxed{Z_{\text{out}} = \frac{(R_L \parallel Z_2) \parallel [R_F + Z_1]}{1+L}} \tag{7.93}$$

Typically Z^o_{out} will be dominated by Z_2 which will be the impedance looking into the emitter or the source of a transistor ($1/g_m$ at low frequencies), which will be a lower impedance than either R_L or R_F. This impedance is further reduced by the negative feedback, so for the usable bandwidth of the amplifier, the condition $R_T >> Z_{\text{out}}$ will generally hold true.

Noise due to the Feedback Resistor and Load Resistor

We will now consider the noise due to the resistors R_F and R_L. A small-signal model of the transresistance amplifier with the noise sources included is shown in Fig. 7.13 The output voltage can be found as a superposition of each current source acting individually.

$$v_o = i_{in} R_T + i_F R_T - i_F Z_{\text{out}} + i_L Z_{\text{out}} \tag{7.94}$$

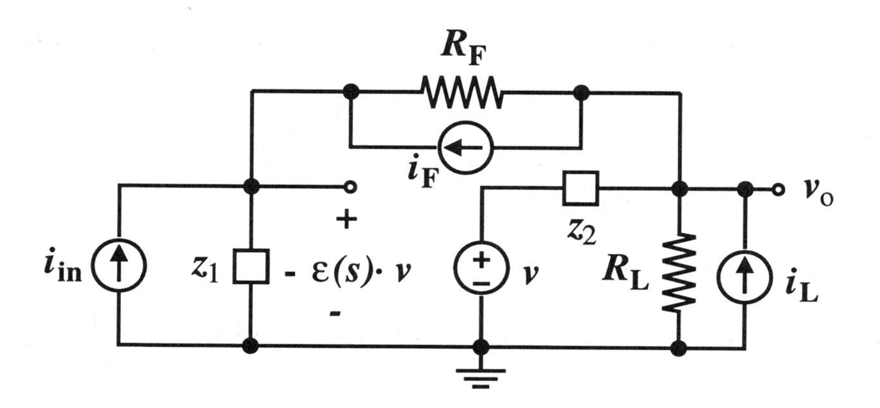

Figure 7.13 Small-signal model of a transresistance preamplifier showing the thermal-noise current sources due to R_F and R_L.

The noise sources can be represented as currents in parallel with the input source, provided that they produce the same noise level at the output as the original sources. Therefore the equivalent noise source due to i_F acting at the input has a value

$$i_{nF} = i_F \frac{R_T - Z_{\text{out}}}{R_T} = 1 - Z_{\text{out}}/R_T \tag{7.95}$$

$$i_{nF} \simeq i_F \left[1 + \frac{Z^o_{\text{out}}}{LR_F}\right] \tag{7.96}$$

The input-referred current due to thermal noise in the load resistor is just

$$i_{nL} = i_L \frac{Z_{\text{out}}}{R_T} \tag{7.97}$$

$$i_{nL} \simeq -i_L \frac{Z^o_{\text{out}}}{LR_F} \tag{7.98}$$

We can define a parameter δ to be the ratio of the output impedance to the transresistance,

$$\delta \triangleq \frac{Z_{\text{out}}}{R_T} \simeq \frac{Z^o_{\text{out}}}{LR_F}. \tag{7.99}$$

The equivalent spectral densities of the thermal noise sources at the input due to R_F and R_L are simply

$$S_{nF} = \frac{4kT}{R_F}(1+\delta)^2 \tag{7.100}$$

$$S_{nL} = \frac{4kT}{R_L}(\delta)^2 = \frac{4kT}{R_F}\left[\frac{R_F}{R_L}\delta^2\right] \tag{7.101}$$

The combined spectral density of the thermal noise due to R_F and R_L is then given by

$$S_{nFL} = \frac{4kT}{R_F}\left[1 + 2\delta + \delta^2(1 + R_F/R_L)\right]. \tag{7.102}$$

At low and intermediate frequencies $\delta \simeq 1/L(g_m R_F) << 1$, where g_m is the transconductance of the output transistor. Therefore, we usually ignore noise due to R_L, and approximate the spectral density of noise due to R_F as just $4kT/R_F$.

Noise due to the Second Gain Stage

We can consider the noise due to the second gain stage in a similar manner. Consider again Fig. 7.13 and let the noise current source i_L represent the total noise of the second stage amplifier referred to its input. If this noise has a spectral density of $S_2(f)$, then the equivalent spectral density at the input of the first stage is simply

$$S_{n2}(f) = S_2(f)\delta^2 = S_2(f)\left[\frac{Z^o_{\text{out}}}{LR_F}\right]^2 \simeq S_2(f)\left[\frac{1}{L}\frac{1}{g_m R_F}\right]^2 \tag{7.103}$$

Since the second stage amplifier will be realized using transistors and bias-current-levels similar to the first stage, the equivalent noise currents at the input of each stage will be of the same order of magnitude. Therefore, when the second stage noise is referred again to the input of the first stage it will be reduced by the factor δ^2 and will be insignificant compared to noise generated in the first stage, provided that $\delta^2 << 1$. Since the signal is amplified by the first stage, the second stage can have a larger noise than the first and this will still not degrade the overall SNR. This is true provided that $i_{rms2}/i_{rms1} << L(g_m R_F)$. This allows the noise requirements of the second stage to be relaxed, but it can not be relaxed to the extent that the second-stage noise becomes significant.

One potential problem in the above analysis is that the output impedance Z_{out} affects the magnitude of the input-referred noise current of the second stage amplifier. To remove this dependence, we can represent the noise from the second stage as a voltage. The noise current can be expressed in terms of this noise voltage as a Norton equivalent;

$$i_2 = v_2/Z_{\text{out}}. \tag{7.104}$$

Therefore, a current of value

$$i_{n2} = \frac{v_2}{Z_{\text{out}}}\left[\frac{Z_{\text{out}}}{R_T}\right] = \frac{v_2}{R_T} \tag{7.105}$$

will produce the same output as the voltage v_2, and the spectral density of the noise at the input is divided by ${R_T}^2$;

$$S_{ni2}(f) = S_{v2}/|R_T|^2 \tag{7.106}$$

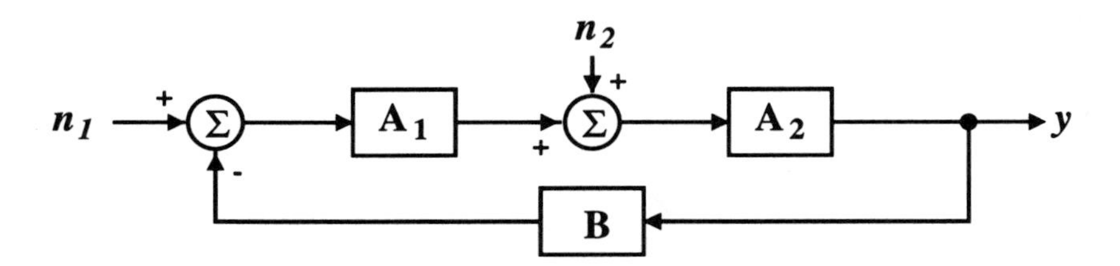

Figure 7.14 Block diagram of a two stage amplifier with negative feedback and two sources of noise.

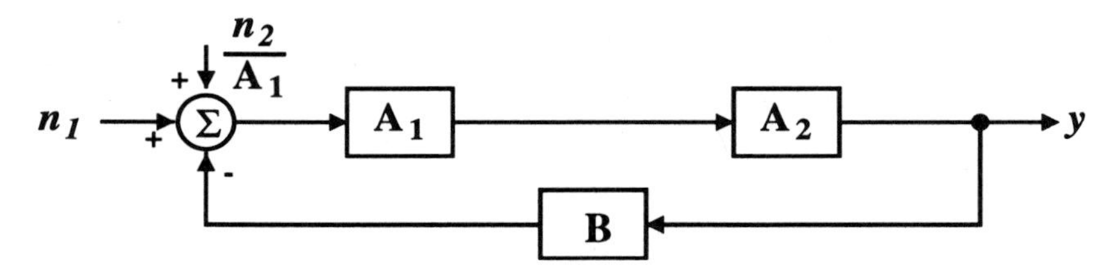

Figure 7.15 Block diagram of a two stage amplifier with negative feedback and two noise sources, where the second noise source has been referred to the input.

Therefore the equivalent noise spectral density is obtained by dividing by the squared magnitude of the gain.

In general, to refer a noise spectral density to the input of a circuit we must divide by the square of the gain from the input to the noise source. This can be illustrated simply in block-diagram form in the feedback circuit of Fig. 7.14. If the noise signal n_2 is moved to the left of the first gain stage A_1, as shown in Fig. 7.15, it will experience an extra gain of A_1. Therefore, the equivalent noise acting at the input is just n_2/A_1, and the spectral density is reduced by the square of A_1, such that.

$$S_{n2}(f) = S_2(f)/{A_1}^2. \quad (7.107)$$

Technique for Determining Equivalent Noise Sources

In a real circuit, referring noise sources to the input becomes more complicated than just dividing by the square of the gain. The signal propagation is bidirectional; furthermore, the effect of the load impedance must be taken into account. Despite their limitations, the block diagrams of Figs. 7.14 and 7.15 can be used as a guide for a more general

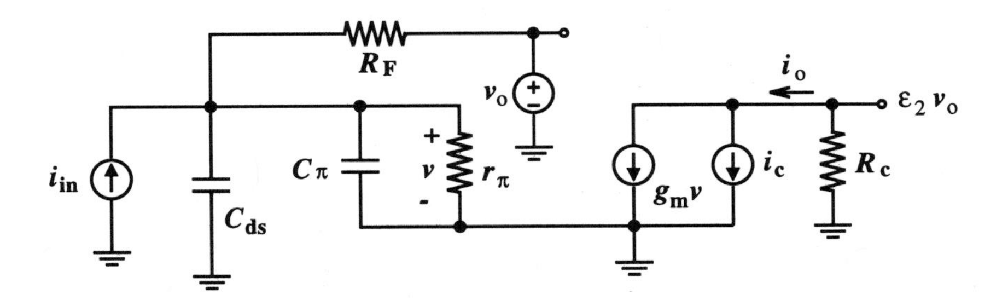

Figure 7.16 Small-signal model of a two-stage transresistance preamplifier; the output impedance is equal to zero and the second stage gain is $A_2 = 1/\epsilon_2$.

analysis technique. The gain from n_1 to the output y in Fig. 7.14 is given by

$$H_1 \triangleq \frac{y}{n_1} = \frac{A_1 A_2}{1 + A_1 A_2 B}, \tag{7.108}$$

whereas the gain from the second noise source is

$$H_2 \triangleq \frac{y}{n_2} = \frac{A_2}{1 + A_1 A_2 B}. \tag{7.109}$$

The magnitude squared of the ratio of these gains gives the factor by which the spectral density of the second noise signal will be reduced when referred to the input. This ratio is obviously $1/A_1$, however, we will write it explicitly as

$$\frac{H_2}{H_1} = \frac{A_2}{A_1 A_2} \frac{1 + A_1 A_2 B}{1 + A_1 A_2 B}. \tag{7.110}$$

Notice that H_2/H_1 is just the ratio of the numerators, or the feedforward paths, of each signal. The denominator of the two transfer function are identical. Therefore, any changes in the feedback parameter B, which alters the natural frequencies of the amplifier will not effect the equivalent noise source.

Matrix Formulation The above technique will now be illustrated for the transresistance preamplifier shown in Fig. 7.16. For this circuit we can write node equations at the base and collector to obtain

$$\begin{bmatrix} G_F + g_\pi + sC_\pi + sC_{ds} & -G_F \\ g_m & \epsilon_2 G_c \end{bmatrix} \begin{bmatrix} v \\ v_o \end{bmatrix} = \begin{bmatrix} i_{in} \\ -i_c \end{bmatrix} \tag{7.111}$$

If we were to use a first-order approach we could divide i_c by the current-gain β to refer this noise current to the base side. However, the current gain is frequency dependent and is altered by the presence of feedback. Therefore we will have to consider the individual gains from i_{in} and i_c to the output and find the ratio of these gains. Using Kramer's Rule we can find the transresistance.

$$R_T \triangleq \frac{v_o}{i_{in}} = \frac{-g_m}{\Delta}, \tag{7.112}$$

where Δ is the determinant of the admittance matrix in (7.111). We can likewise find the gain from the noise source i_c to the ouput;.

$$R_{Tc} \triangleq \frac{v_o}{i_c} = \frac{-[G_F + g_\pi + sC_{TB}]}{\Delta}, \tag{7.113}$$

where $C_{TB} = C_\pi + C_{ds}$ in this case. An equivalent noise source can be placed at the input provided that it produces the same level of noise at the output, such that the following condition is satisfied,

$$S_{nc}(f) \left| \frac{g_m}{\Delta} \bigg|_{s=j2\pi f} \right|^2 = S_c(f) \left| \frac{G_F + g_\pi + sC_{TB}}{\Delta} \bigg|_{s=j2\pi f} \right|^2 . \tag{7.114}$$

Therefore,

$$S_{nc}(f) = S_c(f) \left[\frac{(G_F + g_\pi)^2 + (2\pi f C_{TB})^2}{g_m^2} \right], \tag{7.115a}$$

and after substituting $S_c(f) = 2qI_c$ and $g_\pi = g_m/\beta$, we obtain

$$S_{nc}(f) = 2qI_c \left[\left(\frac{1}{g_m R_F} + \frac{1}{\beta} \right)^2 + \left(\frac{2\pi f C_{TB}}{g_m} \right)^2 \right] \tag{7.115b}$$

We can see that when $R_F \to \infty$ (no feedback) and $C_{TB} \to 0$ (no frequency dependence) the result reduces to

$$S_{nc}(f) = \frac{2qI_c}{\beta^2} \quad \text{(first-order)}, \tag{7.116}$$

which is the same result we would have obtained with a first-order analysis.

We can compare the result obtained in (7.115) with the shot-noise contribution of the base current, which has a spectral density of the form

$$S_{nb}(f) = 2qI_b = \frac{2qI_c}{\beta}. \tag{7.117}$$

Under normal circumstances, $\beta >> 1$ and $g_m R_F >> 1$, we will find that $S_{nb}(f) >> S_{nc}(f)$ at low frequencies. Therefore, we can neglect the constant term in (7.115) and the input-referred collector current shot-noise has a spectral density given by

$$\boxed{S_{nc}(f) \simeq 2qI_c \left(\frac{2\pi f C_{TB}}{g_m} \right)^2} \tag{7.118}$$

Therefore, the effective noise due to i_c will increase with frequency.

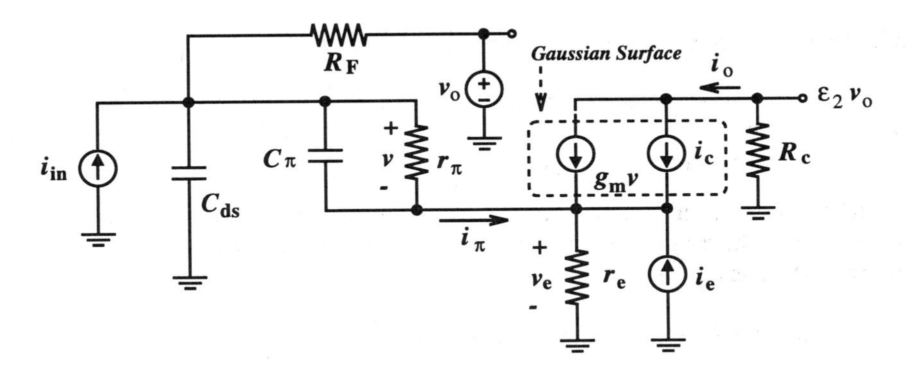

Figure 7.17 Small-signal model of a two-stage transresistance preamplifier with an emitter degeneration resistor; the output impedance is equal to zero and the second stage gain is $A_2 = 1/\epsilon_2$.

Effect of Emitter Resistance on Noise Performance

Most circuit designers are aware that emitter degeneration resistors produce local feedback and can alter the gain and natural frequencies of the amplifier considerably. We therefore need to determine whether or not the presence of an emitter resistor r_e will alter the noise performance of the amplifier. There are two primary questions that we need to consider:

1. How does the presence of an emitter resistor alter the performance of other noise sources in the circuit?
2. How does the thermal noise, generated by the emitter resistor r_e itself, affect the overall amplifier noise.

We can answer both of these questions by performing an analysis similar to that of the previous section for the circuit shown in Fig. 7.17.

Effect of Emitter Resistor on Other Noise Sources We will find in the analysis to follow that emitter resistors as large as 50Ω will have little effect on the contribution of other noise sources to the total noise, the reason being that the emitter degeneration resistor affects the gains R_T and R_{Tc} equally, so that their ratio is unchanged. To show this we can write node-equations for the circuit of Fig. 7.17 as done previously. However, it is more convenient in this case to write cut-set equations in terms of the variables v, v_o, and v_e. The equations are given below,

$$\begin{bmatrix} G_F + g_\pi + sC_\pi + sC_{ds} & -G_F & G_F + sC_{ds} \\ g_m & \epsilon_2 G_c & 0 \\ -(g_\pi + sC_\pi) & \epsilon_2 G_c & ge \end{bmatrix} \begin{bmatrix} v \\ v_o \\ v_e \end{bmatrix} = \begin{bmatrix} i_{in} \\ -i_c \\ i_e \end{bmatrix}. \quad (7.119)$$

The third row of the matrix is obtained by writing KCL equations at the emitter, where we realize that the current leaving the Gaussian surface must be equal to the current entering, which is $i_o = -\epsilon_2 G_c v_o$.

The transresistance is found, again using Kramer's rule,

$$R_T = \frac{-\begin{vmatrix} g_m & 0 \\ -(g_\pi + sC_\pi) & g_e \end{vmatrix}}{\Delta} = \frac{-g_m g_e}{\Delta}. \quad (7.120)$$

The gain from i_c to v_o is similarly given by

$$R_{Tc} = \frac{-\begin{vmatrix} G_F + g_\pi + sC_\pi + sC_{ds} & G_F + sC_{ds} \\ -(g_\pi + sC_\pi) & g_e \end{vmatrix}}{\Delta} \quad (7.121)$$

Since Δ appears in the denominator of both gains R_T and R_{Tc}, the ratio of gains is just the ratio of the feedforward paths.

$$\frac{R_{Tc}}{R_T} = \frac{g_e[G_F + g_\pi + sC_\pi + sC_{ds}] + (g_\pi + sC_\pi)(G_F + sC_{ds})}{g_m g_e} \quad (7.122)$$

This ratio can be written as $\delta + s\tau_1 + (s\tau_2)^2$, where the constant term δ is given by

$$\delta = \frac{1}{g_m R_F} + \frac{1}{\beta}\left(1 + \frac{r_e}{R_F}\right). \quad (7.123)$$

Since $r_e << R_F$ this term is virtually the same as the constant term for $r_e = 0$. The term involving s reduces to

$$s\tau_1 = \frac{sC_\pi}{g_m}\left[1 + \frac{r_e}{R_F}\right] + \frac{sC_{ds}}{g_m}\left[1 + \frac{r_e}{r_\pi}\right] \quad (7.124)$$

Since $r_e << R_F, r_\pi$, this first-order term in s is also similar to the case where $r_e = 0$.

The primary effect of r_e on the other noise sources is to add another frequency dependent term;

$$(s\tau_2)^2 = \frac{s^2 C_\pi C_{ds} r_e}{g_m}. \quad (7.125)$$

Substituting $s \to j2\pi f$ we find the magnitude squared is given by

$$\left|\frac{R_{Tc}}{R_T}\right|^2 = \left[\delta - (2\pi f\tau_2)^2\right]^2 + (2\pi f\tau_1)^2 \tag{7.126a}$$

$$= \delta^2 + (2\pi f\tau_1)^2\left[1 - 2\delta(\tau_2/\tau_1)^2\right] + (2\pi f\tau_2)^4 \tag{7.126b}$$

$$\simeq (2\pi f\tau_1)^2 + (2\pi f\tau_2)^4 \tag{7.126c}$$

Quantitative Limit on Emitter Resistance The presence of the emitter degeneration resistor adds a term to the input referred noise, which multiplies the collector current shot-noise spectral density by the frequency dependent term $(2\pi f\tau_2)^4$. However, the effect of r_e will be negligible provided $(2\pi f\tau_1)^2 > (2\pi f\tau_2)^4$ for the frequencies of interest. The frequency at which these two terms are equal is

$$f_e = \frac{\tau_1}{2\pi\tau_2^2}. \tag{7.127a}$$

Substituting for τ_1 and τ_2 we find that

$$f_e = \frac{1}{2\pi}\frac{C_\pi + C_{ds}}{g_m}\frac{g_m}{C_\pi C_{ds} r_e} = \frac{1}{2\pi C_e r_e}, \tag{7.127b}$$

where C_e is the series combination of C_π and C_{ds}. For the case of $C_{ds} = C_\pi = 0.5$-pF and $r_e = 10\Omega$, we find that f_e=64-GHz. Since the system bandwidth will be much less than 64-GHz, the term due to $(2\pi f\tau_2)^4$ can be ignored.

We could express (7.127) as an inequality in terms of r_e.

$$r_e < \frac{1}{2\pi f C_e}. \tag{7.128}$$

If r_e is less than this limit, it will have little effect on the collector current shot noise. For the case $C_e = 0.25$-pF and $f = 10$-GHz, we find r_e should be less than 64Ω. Since r_e is generally only a few ohms, we can ignore its effect on the collector current shot noise. Using the same analysis technique we can also show that the effect of r_e on all other noise sources in the circuit is negligible for typical values of parasitic emitter resistance.

Thermal Noise due to Emitter Resistance

To complete the analysis we need to determine the contribution of the input referred noise due to the thermal noise of r_e. We can find this easily for a first-order analysis by considering the equivalent circuit of Fig. 7.18. In this circuit the thermal noise current

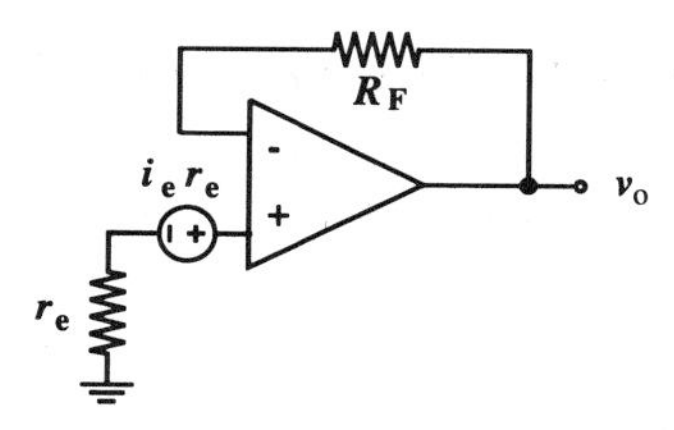

Figure 7.18 Simplified schematic of the a transresistance amplifier, where the input current is set equal to zero, and the amplifier is driven by the noise voltage at the positive terminal.

i_e can be expressed as a Thévenin equivalent voltage $r_e i_e$. If the voltage gain of the voltage amplifier is reasonably high, the circuit looks like a follower; therefore, the output voltage v_o is also $r_e i_e$. Since we know the transconductance is approximately $-R_F$, a current at the input of value i_{ne} will produce the same output voltage, provided that $-i_{ne} R_F = r_e i_e$ or

$$i_{ne} = -i_e (r_e / R_F); \tag{7.129}$$

the spectral densities are related by the square of this ratio

$$S_{ne}(f) = S_e(f) \left[\frac{r_e}{R_F} \right]^2 = \frac{4kT}{r_e} \left[\frac{r_e}{R_F} \right]^2 = \frac{4kT}{R_F} \left[\frac{r_e}{R_F} \right]. \tag{7.130}$$

The feedback resistor R_F will contribute a noise term of $4kT/R_F$. Therefore, the thermal noise, due to R_F and r_e combined, has a spectral density of

$$S_F(f) + S_{ne}(f) = \frac{4kT}{R_F} \left[1 + \frac{r_e}{R_F} \right]. \tag{7.131}$$

For $r_e << R_F$ the contribution due to r_e is negligible and this noise current has a spectral density of approximately $4kT/R_F$.

High Frequency Effects of Thermal Noise due to Emitter Resistance The previous analysis is correct for low and intermediate frequencies, however, at higher frequencies r_e can contribute a much larger noise. We can determine the amount of noise produced as a function of frequency by returning to the network equations (7.119), which describe Fig. 7.17. We find that the gain from i_e to the output v_o is

$$R_{Te} = \frac{v_o}{i_e} = \frac{-\begin{vmatrix} G_F + g_\pi + sC_\pi + sC_{ds} & G_F + sC_{ds} \\ g_m & 0 \end{vmatrix}}{\Delta} \tag{7.132}$$

The ratio of the gains is then given by

$$\frac{R_{Tc}}{R_T} = \frac{-g_m(G_F + sC_{ds})}{g_m g_e} = -\left[\frac{r_e}{R_F} + sC_{ds} r_e \right], \tag{7.133}$$

and the magnitude squared of this ratio is

$$\left|\frac{R_{Tc}}{R_T}\right|^2 = \left(\frac{r_e}{R_F}\right)^2 + (2\pi f C_{ds} r_e)^2. \tag{7.134}$$

This consists of a constant term and a frequency dependent term. The constant term is the same that was obtained with the first-order analysis and can be ignored. Therefore the input-referred noise spectral density due to thermal noise in r_e is given by

$$S_{ne}(f) = S_e(f)\left|\frac{R_{Tc}}{R_T}\right|^2 \simeq S_e(f)(2\pi f C_{ds} r_e)^2. \tag{7.135}$$

We saw previously in (7.118) that

$$S_{nc}(f) = 2qI_c\left(\frac{2\pi C_{TB}}{g_m}\right)^2; \tag{7.136}$$

we would like to express $S_{ne}(f)$ in a similar form for comparison. After some manipulation we find

$$\boxed{S_{ne}(f) = 2qI_c\,[2g_m r_e]\left(\frac{2\pi f C_{ds}}{g_m}\right)^2 = 2qI_c\left[\frac{2V_e}{V_T}\right]\left(\frac{2\pi f C_{ds}}{g_m}\right)^2}, \tag{7.137}$$

where V_e is the dc voltage drop across r_e ($V_e = r_e I_c$), and V_T is the thermal voltage $\simeq$ 26-mV. The ratio of the noise spectral densities of emitter-resistance-thermal-noise to collector-current-shot-noise is

$$x_c^e = \frac{S_{ne}(f)}{S_{nc}(f)} = 2g_m r_e (C_{ds}/C_{TB})^2. \tag{7.138}$$

If thermal noise due to r_e is to be negligible, x_c^e must be small. For a representative case, $(C_{ds}/C_{TB})^2 = 1/2$, we obtain the following simple restriction on the maximum size of r_e.

$$\boxed{r_e < \frac{1}{g_m} = \frac{V_T}{I_c}} \tag{7.139}$$

Therefore, for a bias current of 2-mA, $1/g_m = 13\Omega$ and the thermal noise due to r_e can be neglected if $r_e << 13\Omega$. In what follows we will assume that r_e is small enough such that its noise contributions can be ignored. If this is not the case, the effect of noise due to r_e can be determined from (7.126) and (7.137).

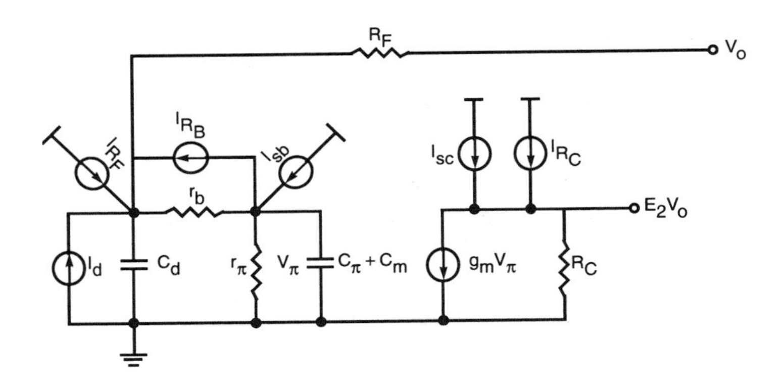

Figure 7.19 Small signal model for two-stage preamp with dominant noise sources.

7.4.3 Summary of Results for the Common-Emitter Configuration

We are now in a position to determine the complete input referred noise current from all the noise sources in the amplifier. A schematic of a typical transresistance preamplifier using a common-emitter front-end was shown in Fig. 7.11. A small signal model of this amplifier, with all the non-neglible noise sources, is shown in Fig. 7.19. The spectral density of the input noise current is found to be

$$S_{n_B}(f) = \frac{4kT}{R_F} + \frac{2qI_C}{\beta} + 4kTr_b\left(2\pi f C_{ds}\right)^2 + \left[2qI_C + \frac{4kT}{R_C}\right]\left(\frac{2\pi f C_{TB}}{g_m}\right)^2 \tag{7.140}$$

We have discussed each of these terms previously, expect the noise due to the base resistance r_b. This noise has the same functional form as the noise due to r_e. However, r_b is generally much larger than r_e and can not be ignored.

Simulated and calculated values of this spectrum are shown in Fig. 7.20(a). This plot shows that (7.140) can be used to accurately predict noise performance over the useful bandwidth of the device. The first term in (7.140) is due to thermal noise in the feedback resistor. The second term results from base-current shot noise. These two terms are dominant at low frequencies, and their respective spectral densities are plotted in Fig. 7.21.

The remaining terms of (7.140) all increase with frequency. The third term is due to thermal noise in the base resistance. This is normally smaller than the other frequency dependent noises, but it can be large if care is not taken to minimize r_b. This noise

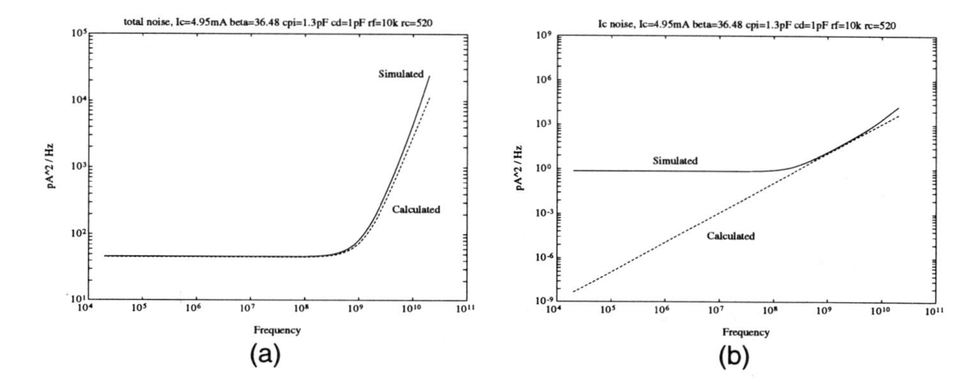

Figure 7.20 Input referred noise spectral density of a transresistance preamplifier with a common-emitter front-end: (a) total noise, (b) contribution due to shot noise.

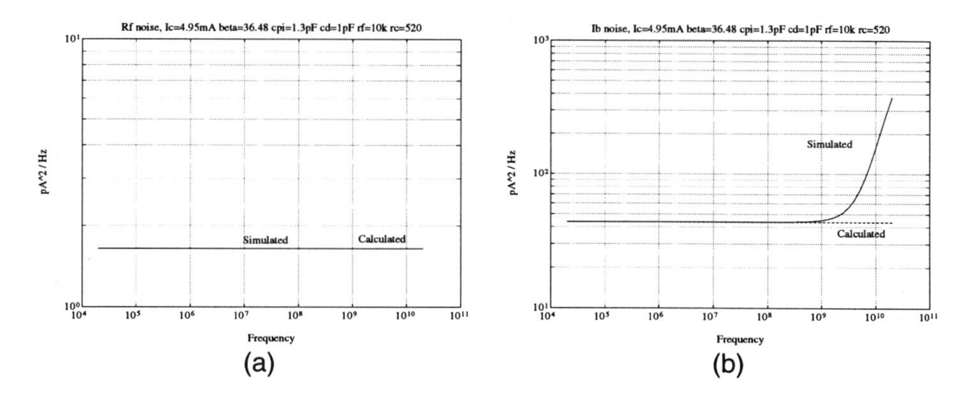

Figure 7.21 Input referred noise spectral density of a transresistance preamplifier with a common-emitter front-end due to: (a) R_F thermal noise, (b) I_b shot noise.

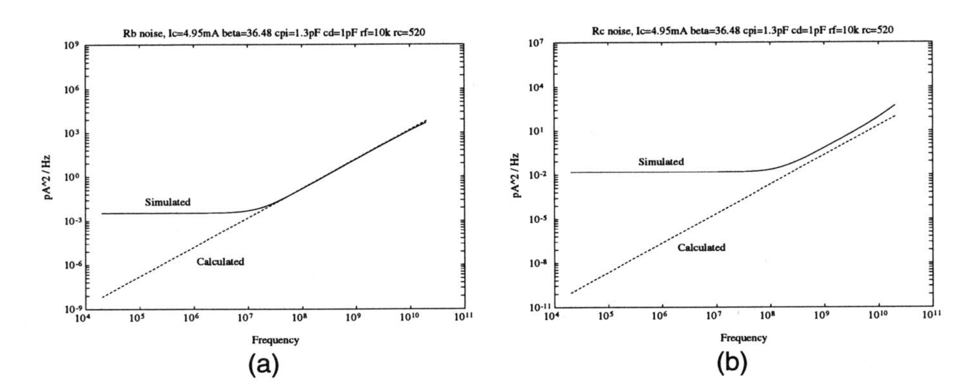

Figure 7.22 Input referred noise spectral density of a transresistance preamplifier with a common-emitter front-end due to: (a) r_b thermal noise, (b) R_c thermal noise.

is plotted in Fig. 7.22(a), and the thermal noise due to R_c is shown in Fig. 7.22(b). The fourth term results from collector-current shot noise. This is the dominant noise at high frequencies, and its spectral density was shown in Fig. 7.20(b), where it can be compared directly with the total noise. We have omitted the dc contributions of these frequency-dependent terms because they are much less than the noise terms due to R_F and i_b. However, the noise begins to increase at a break frequency determined by the dominant pole of the open-loop amplifier, because the forward gain, which keeps the input-referred noise low, starts to fall off. (This effect is similar to the reduction of the common-mode-rejection ratio or the power-supply-rejection ratio of an opamp.)

For an FET input device the result is

$$S_{n_F}(f) = \frac{4kT}{R_F} + \left[4kT\Gamma g_m + \frac{4kT}{R_C}\right]\left(\frac{2\pi f C_{TF}}{g_m}\right)^2. \tag{7.141}$$

(7.141) contains no contributions due to the base-current shot noise and thermal noise in the base resistor that we saw in (7.140). For a broadband amplifier, the frequency dependent terms become increasingly important. Therefore it is desirable to minimize the following terms:

$$\tau_{\text{Bipolar}} = C_{TB}/g_m, \tag{7.142}$$

$$\tau_{\text{FET}} = C_{TF}/g_m. \tag{7.143}$$

These are equivalent delay times which are determined by the capacitance to transconductance ratio. With all else equal, the device with the highest speed (the lowest C/g_m ratio) will exhibit the lowest noise.

Qualitative Explanation for Frequency Dependence of Noise

A photodiode generates electron-hole pairs in proportion to the number of photons incident on the device. We would like to make use of all of these electron-hole pairs to evaluate the strength of the incoming signal. However, many of these charges are lost at the outset, and are used to fill empty lattice sites in the depletion region of the photodiode. The reason these charge carriers are lost is because a voltage swing must appear at the input of the amplifier to steer the current through the feedback resistor. This voltage can only be generated by the accumulation of charge in the depletion layer. At higher frequencies more current from the diode will be lost to the depletion capacitance and the signal strength will weaken.

The noise power at the output will contain a constant term due to the shot-noise from the collector current. As the signal is reduced in magnitude at higher frequencies, the input referred noise power will therefore increase — a weaker signal with constant noise at the output is equivalent to a constant signal with increasing noise at the input. The equivalent input noise will increase with the square of the capacitance. The noise will also depend on the transconductance, which determines how large the input voltage has to swing in order to switch a fixed amount of current. If the transconductance is high, only a small voltage is needed; therefore, few charges are lost in the depletion capacitance and a larger percentage can be detected as a signal at the output. We have reasoned that the rms noise will increase with the capacitance and with frequency, but will be inversely proportional to g_m. The noise power will vary with the square of these quantities. Therefore, we expect a term in the input-referred noise spectral density of the form $(C/g_m)^2 f^2$.

7.4.4 Results for the Common-Base Configuration

Before presenting methods for optimizing the noise performance, we will compare the previous results with those obtained for a different circuit topology. An alternative circuit architecture uses a current buffer. One example of this technique uses a common-base input stage and is shown in Fig. 7.23. The input referred PSD of this structure can be found to be

$$\begin{aligned} S_{n_{CB}}(f) = \frac{4kT}{R_F} + \frac{2qI_C}{\beta}\left[1 + (2\pi f C_{ds} r_b)^2\right] 4kTr_b\left(2\pi f C_{ds}\right)^2 + \\ 2qI_C\left(\frac{2\pi f C_{TB}}{g_m}\right)^2 \frac{4kT}{R_c}\left[\frac{1}{\alpha^2} + \left(\frac{2\pi f C_{TB}}{g_m}\right)^2\right]. \end{aligned} \tag{7.144}$$

We can recognize that this is the same spectral density that was obtained for the common emitter circuit with an added contribution due to the thermal noise in the

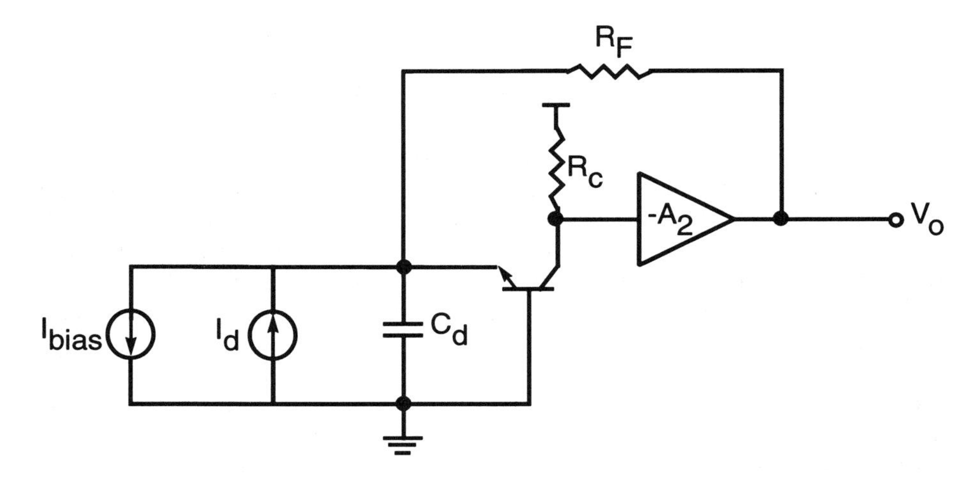

Figure 7.23 Block diagram of a transresistance preamp with a common-base front-end.

collector resistor R_c. The term α is the current efficiency parameter given by

$$\alpha \triangleq \frac{\beta}{1+\beta}, \tag{7.145}$$

and it is normally close to unity. Therefore the noise PSD can be written as

$$S_{n_{CB}}(f) = S_{n_B}(f) + \frac{4kT}{R_c}, \tag{7.146}$$

where $S_{n_B}(f)$ is the PSD of a common-emitter front-end. These spectral densities are compared in Fig. 7.24. The increased noise of the common-base stage occurs because there is no current gain from the input to the collector. Therefore, the noise from R_c adds directly to the signal instead of β times the signal. Likewise, the noise current from the second stage will also add directly to the noise of the signal; the result will be that the common-base configuration will have at least twice the noise of a common-emitter configuration.

7.4.5 Effect of a Cascode Transistor on the Noise Performance

We saw in the previous section that a large input capacitance increases the noise. Cascoding is a common technique used to reduce the input capacitance of a negative gain stage, and it is illustrated in Fig. 7.25. If the cascode transistor Q_2 were not there, the base-collector capacitance of the transistor Q_1 would be multiplied by $(1+g_{m1}R_c)$

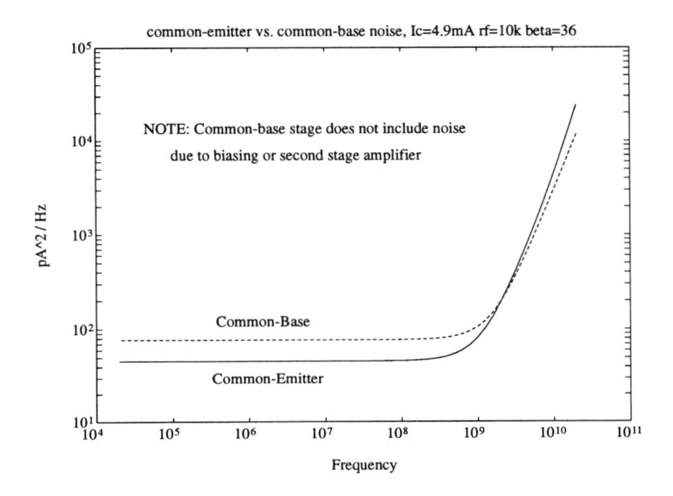

Figure 7.24 Comparison of the input current noise spectral densities for a common-emitter and common-base front-end. The higher noise of common-base is to due the thermal noise in R_c.

when referred to the input. This *Miller* capacitance will degrade the frequency response and the noise performance.

Using a cascode device keeps the collector voltage of Q_1 at a low impedance so that the Miller multiplication factor is unity. This will help the noise performance by reducing C_{in}, but we must determine whether the noise sources from the cascode device increase the noise more than the reduction in C_{in} reduces the noise.

It is quite easily seen that the cascode device Q_2 will have negligible noise compared to the input device Q_1. We can draw a Gaussian surface around Q_2 such that the sum of the ac currents entering the surface must equal the sum of the ac currents leaving the surface. When considering noise performance, we set the input signal to zero. Looking into the collector of Q_1 we see a current source with no ac current flowing. Therefore, to satisfy Kirchoff's law for our surface, the base current of Q_2 must flow through the load resistor, and the output current contribution due to Q_1 is β times the base current. Therefore, the noise current contribution from Q_2 at the load R_c is a factor of β less than the noise current due to Q_1, and the noise power is reduced by β^2.

This result has been derived without looking into the details of the current flow within the device. One might be concerned about the collector current shot-noise of Q_2. The key to the noise reduction of this term is the negative feedback of the emitter-follower

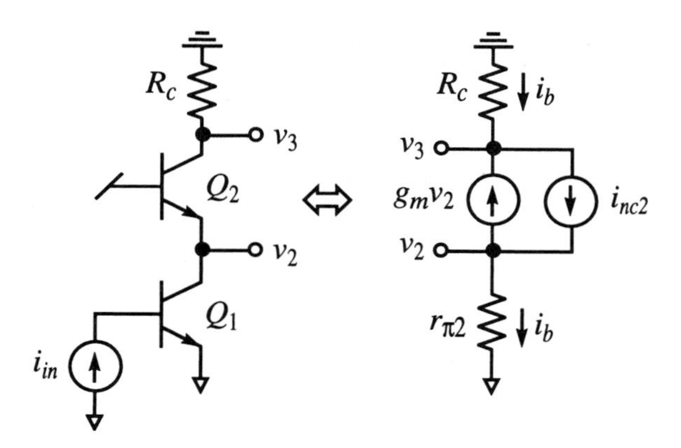

Figure 7.25 Small-signal noise analysis of cascoded device.

configuration of Q_2. Due to the high impedance at the emitter of Q_2, the noise current i_{nc2} is forced to recirculate. The noise current raises the voltage v_2 which reduces the collector current through negative feedback in such a way as to cancel out its own noise. Writing the KCL equations at v_2:

$$i_{nc2} = i_b + \beta_2 i_b \Rightarrow i_b = \frac{i_{nc2}}{1 + \beta_2}. \tag{7.147}$$

Since i_b must also be the current that is left over in the load resistor,

$$v_3 = \frac{-R_c i_{nc2}}{1 + \beta_2}. \tag{7.148}$$

It is this reduction in the added noise voltage from the cascode collector-current shot noise by a factor of $1 + \beta_2$ that enables a cascode device to be used without adding significantly to the circuit's noise ($\sim 5\%$) over an ideal current buffer. However, the reduction in the Miller capacitance afforded by using the cascode device improves the noise performance substantially. In the following analysis we will ignore the Miller capacitance since we know that we can use a cascode transistor to eliminate its effect on the input capacitance.

7.5 COMPARISON OF BIPOLAR AND FET AMPLIFIERS

In this section, we will give expressions for the optimal noise performance of a bipolar and an FET preamplifier. Although the noise analysis is straightforward, the algebra

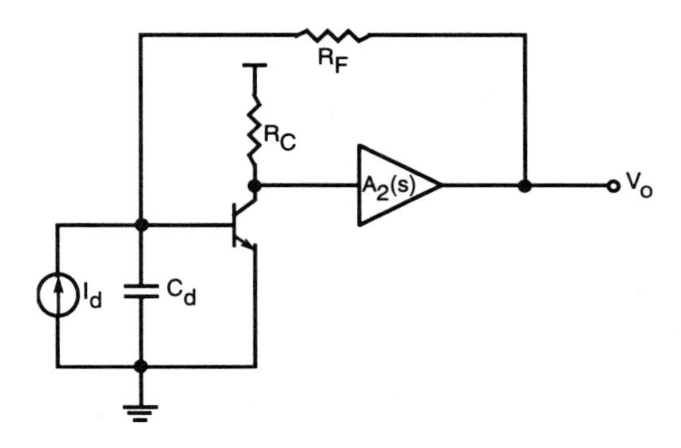

Figure 7.26 Schematic diagram of a two-stage transresistance preamplifier.

is tedious, and the main aspects of the problem can be lost in the details. Instead of going through a complete derivation, we will just state the results and offer simulations to verify the validity of the expressions given. We will be considering a preamp of the type shown in Fig. 7.26. A cascode device is usually inserted to reduce the Miller capacitance, but this is not shown in the diagram.

The input-current noise spectral density for the bipolar front-end from (7.140) can be written as follows,

$$S_{n_\beta}(f) = \frac{4kT}{R_F} + \frac{2qI_c}{\beta} + 4kTr_b(2\pi fC_{ds})^2 + \left[2qI_c + \frac{4kT}{R_c}\right]\left(\frac{2\pi fC_{TB}}{g_m}\right)^2 \quad (7.149)$$

and for an FET device the result from (7.141) is

$$S_{n_F}(f) = \frac{4kT}{R_F} + \left[4kT(\Gamma g_m) + \frac{4kT}{R_c}\right]\left(\frac{2\pi fC_{TF}}{g_m}\right)^2. \quad (7.150)$$

For an arbitrary noise filter with a transfer function of $H(j2\pi f)$, we can define integration constants normalized to the data-rate f_b as

$$I_2 = \frac{1}{f_b}\int_0^\infty |H(j2\pi f)|^2 df, \quad (7.151a)$$

$$I_3 = \frac{1}{f_b^3}\int_0^\infty f^2|H(j2\pi f)|^2 df. \quad (7.151b)$$

When the above noise spectral densities pass through a noise filter that has the normalized integration constants, I_2 and I_3, the variance of the input-current noise for a bipolar device will be

$$\langle i_{n\beta}^2 \rangle = \frac{4kT}{R_F} I_2 f_b + \frac{2qI_c}{\beta} I_2 f_b + 4kTr_b(2\pi C_{ds})^2 I_3 f_b^3 + \left[2qI_c + \frac{4kT}{R_c}\right] \left(\frac{2\pi C_{TB}}{g_m}\right)^2 I_3 f_b^3, \tag{7.152}$$

and for an FET device

$$\langle i_{nF}^2 \rangle = \frac{4kT}{R_F} I_2 f_b + \left[4kT(\Gamma g_m) + \frac{4kT}{R_c}\right] \left(\frac{2\pi C_{TF}}{g_m}\right)^2 I_3 f_b^3. \tag{7.153}$$

7.5.1 Optimization of a Bipolar Device for Low Noise

Each of the above equations can be optimized for minimum noise. Consider the bipolar device first, recall that

$$C_{TB} = C_{ds} + C_{je} + C_\mu + g_m \tau_F,$$
$$g_m = \frac{I_c}{V_T}.$$

The noise-current variance can be written as

$$\langle i_{n\beta}^2 \rangle = \frac{4kT}{R_F} I_2 f_b + 4kTr_b(2\pi C_{ds})^2 I_3 f_b^3 + \frac{4kTg_m}{2\beta} I_2 f_b + 4kT\left[\frac{1}{2} + \frac{1}{A_1}\right] I_3 f_b^3 (2\pi)^2 \frac{(C_{ds} + C_{je} + C_\mu + g_m \tau_F)^2}{g_m}, \tag{7.154}$$

where A_1 is the voltage gain of the first stage. We can define a capacitance $C_\beta \triangleq C_{ds} + C_{je} + C_\mu$, which is the input capacitance of a bipolar device due to parasitic junctions and stray capacitances. We can separate the total capacitance into the sum of the junctions, the stray capacitance, and the base charge storage capacitance $g_m \tau_F$. Rewriting 7.154 using C_β gives,

$$\langle i_{n\beta}^2 \rangle = 4kT\left\{\frac{I_2 f_b}{R_F} + \frac{(2\pi r_b C_{ds})^2 I_3 f_b^3}{r_b} + \frac{g_m}{2\beta} I_2 f_b + \left[\frac{1}{2} + \frac{1}{A_1}\right] I_3 f_b^3 (2\pi)^2 \left[\frac{C_\beta^2}{g_m} + 2C_\beta \tau_F + g_m \tau_F^2\right]\right\}. \tag{7.155}$$

We wish to optimize this expression by choosing the transconductance that results in the minimum noise. Rewriting one more time to show explicitly the optimization with

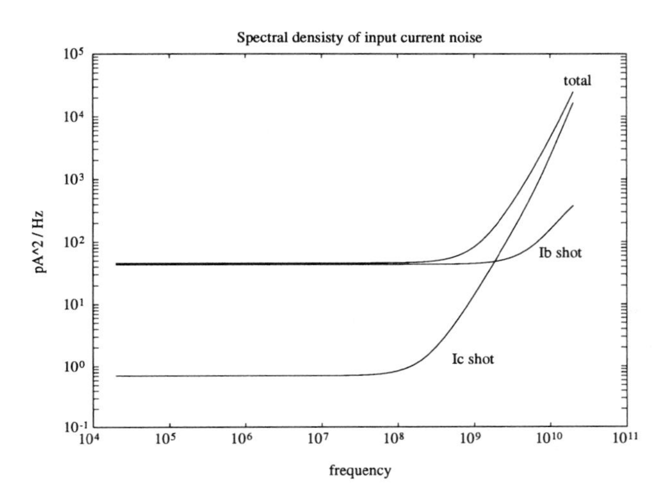

Figure 7.27 Total input referred current noise spectral density showing shot noise contributions from the base and collector currents.

respect to g_m, we get

$$\begin{aligned}\langle i_{n\beta}^2\rangle &= 4kT\left[\frac{I_2 f_b}{R_F} + r_b(2\pi C_{ds})^2 I_3 f_b^3 + \left[1+\frac{2}{A_1}\right] I_3 f_b^3 (2\pi)^2 C_\beta \tau_F\right] \quad \text{constant}\\ &+ 4kT\left[\frac{g_m}{2}\left[\frac{I_2 f_b}{\beta} + \left[1+\frac{2}{A_1}\right] I_3 f_b^3 (2\pi)^2 \tau_F^2\right]\right] \quad \text{linear term}\\ &+ 4kT\left[\frac{1}{2g_m}\left[1+\frac{2}{A_1}\right] I_3 f_b^3 (2\pi)^2 C_\beta^2\right] \quad \text{inverse term} \end{aligned} \tag{7.156}$$

This optimization is illustrated graphically in Fig. 7.27. Increasing the bias current (and thus g_m) increases the corner frequency where the collector current shot noise begins to increase. However, this also increases the low-frequency base-current shot noise. The optimization procedure adjusts the contribution of each of these terms until the total noise is minimized. At the optimal g_m, the linear term will equal the inverse term. This occurs for

$$g_{m_{opt}}\left[\frac{f_b}{\beta} + \frac{m f_b^3}{f_F^2}\right] = \frac{1}{g_{m_{opt}}} m f_b^3 (2\pi C_\beta)^2; \tag{7.157}$$

where we have defined

$$f_F \triangleq \frac{1}{2\pi\tau_F} \tag{7.158}$$

as the maximum f_t of the bipolar transistor, and

$$m \triangleq \frac{I_3}{I_2}\left[1+\frac{2}{A_1}\right]. \tag{7.159}$$

From (7.157) we obtain the transconductance that minimizes the noise as

$$g_{m_{opt}} = 2\pi C_\beta f_b \left[\frac{1}{\sqrt{1/\beta m + f_b^2/f_F^2}}\right], \tag{7.160}$$

which is proportional to the data rate and C_β. The bias current required to achieve this transconductance is just

$$I_{C_{opt}} = g_{m_{opt}} V_T. \tag{7.161}$$

At low data-rates, or low β,

$$g_{m_{opt}} = 2\pi C_\beta f_b \sqrt{\beta m}, \tag{7.162}$$

whereas for high data rates $g_{m_{opt}}$ increases until it reaches the limit

$$g_{m_{opt}} = 2\pi C_\beta f_F. \tag{7.163}$$

We can gain insight into this optimization process if we consider the relative magnitudes of the base charge-storage capacitance compared to the junction and stray capacitance C_β. At low data rates, the optimum base-charge storage capacitance is

$$C_{qb_{opt}} = g_{m_{opt}}\tau_F = C_\beta \frac{f_b\sqrt{\beta_m}}{f_F}, \quad \text{(at low data rates)}. \tag{7.164}$$

This base-charge storage capacitance increases with the data-rate until it is equal to the parasitic junction capacitance C_β giving

$$C_{qb_{opt}} = g_{m_{opt}}\tau_F = C_\beta, \quad \text{(at high data rates)}. \tag{7.165}$$

The optimum noise variance can be found by substituting the optimal g_m back into the noise expression. Realizing that at the optimum the linear term is equivalent to the inverse term, the noise is just the constant term plus double the inverse term.

$$\begin{aligned}\langle i^2_{n\beta_{opt}}\rangle &= 4kT\left[\frac{I_2 f_b}{R_F}\right] \\ &+ 4kT\left[2\pi C_{ds} I_3 \frac{f_b^3}{f_{rb}}\right] \\ &+ 4kT\left[2\pi C_\beta\left[1+\frac{2}{A_1}\right] I_3\left[f_b^2\sqrt{\frac{1}{\beta_m}+\frac{f_b^2}{f_F^2}}+\frac{f_b^3}{f_F}\right]\right]\end{aligned} \tag{7.166}$$

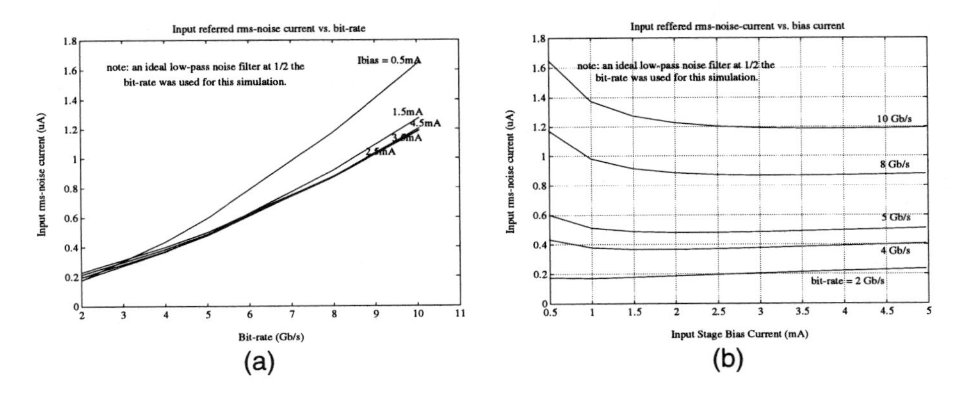

Figure 7.28 Optimum noise for a bipolar device: (a) noise vs. bit-rate, (b) noise vs. bias current.

At low data rates, 7.166 gives

$$\langle i^2_{n\beta_{opt}} \rangle \approx 4kT \left[\frac{I_2 f_b}{R_F} \right] + 4kT \left[\frac{2\pi C_\beta \left[1 + \frac{2}{A_1}\right] I_3 f_b^2}{\sqrt{\beta_m}} \right]. \tag{7.167}$$

At high data rates, 7.166 gives

$$\langle i^2_{n\beta_{opt}} \rangle \approx 4kT \left[2\pi I_3 f_b^3 \left[\frac{C_{ds}}{f_{rb}} + \frac{2C_\beta \left[1 + 2/A_1\right]}{f_F} \right] \right]. \tag{7.168}$$

At low data rates β controls the noise because the base current term is important. At higher data rates τ_F (or the base-charge-storage capacitance) controls the noise. In each case we increase I_c as much as possible until limits controlled by either β or τ_F are reached.

The optimum noise as a function of the bit rate is shown in Fig. 7.28(a) for various bias currents. Fig. 7.28(b) plots the optimum noise for various bit rates as a function of bias current. It can be seen that the bias current has a *shallow* optimum. Moving the bias current slightly away from the optimal value doesn't degrade the noise appreciably. This shallow optimum comes about because as the bias current is increased, so also is the base charge storage capacitance $g_m \tau_F$. We increase the bias current so as to maximize the transconductance-capacitance ratio (C_{TB}/g_m). At low current this ratio is dominated by junctions and strays since C_π is small. As the current is increased further, the ratio becomes dominated by the term (C_π/g_m), which approaches τ_F in

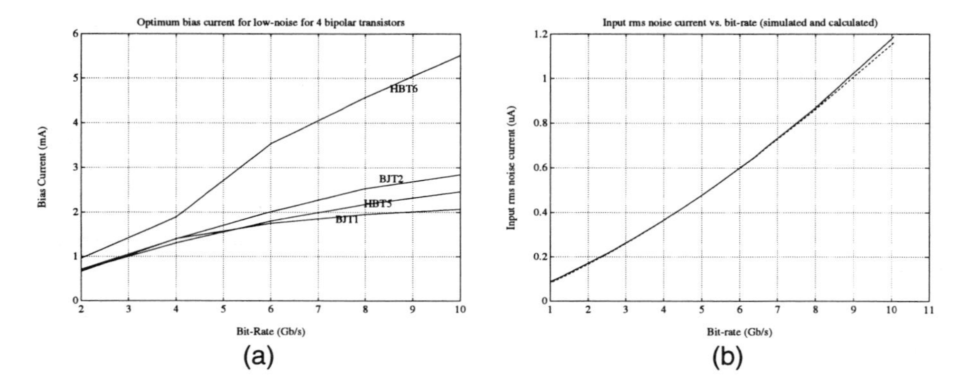

Figure 7.29 (a) Optimum bias current vs. bit-rate for various transistors. (b) Calculated and simulated rms current noise of HBT5 vs. bit-rate.

the limit. Therefore, transistors with the smallest forward transit times (highest f_t) will exhibit better noise performance, and they will be able to run at a higher current before the term $(g_m \tau_F)$ becomes comparable to the parasitic input capacitance. The optimum bias current for various devices is shown in Fig. 7.29(a) as a function of the bit rate. Faster transistors will have lower noise, and will operate at a higher optimal bias current. The rms current noise for the device labeled HBT5 in Fig. 7.29(a) is plotted in Fig. 7.29(b). Both the simulated and calculated results are given. The calculated result (dashed lines) is virtually coincident with the simulated value, showing that the results given thus far are, at least, in agreement with SPICE simulations.

7.5.2 Optimization of an FET Device for Low Noise

The noise of the FET has no minimum like that of the bipolar device. In the FET, the bias current can be increased until power dissipation limits are reached. However, there is an optimum size of device to be used. The best choice of device size is to pick the width W such that the total parasitic gate capacitance is proportional to the the detector-plus-stray capacitance;

$$C_{gss} + C_{gs} + C_{gd} = \alpha C_{ds}, \tag{7.169}$$

where $.2 \leq \alpha \leq 1$. For minimum noise we will see that $\alpha = 1$, but other considerations (such as power dissipation and bandwidth optimization) may dictate the choice of α [19]. With this definition, the total input capacitance can be expressed as $C_{TF} = C_{ds}(1+\alpha)$. Recalling that the f_t of an FET device is

$$f_t = \frac{g_m}{2\pi(C_{gss} + C_{gs} + C_{gd})} = \frac{g_m}{2\pi\alpha C_{ds}}, \tag{7.170}$$

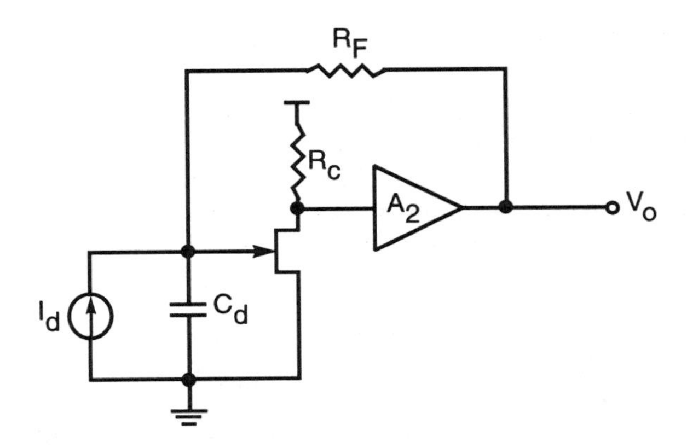

Figure 7.30 Transresistance amplifier with an FET front-end.

the FET noise can be written as

$$\langle i_{nF}^2 \rangle = \frac{4kT}{R_F} I_2 f_b + 4kT \left[\Gamma + \frac{1}{A_1} \right] \frac{(2\pi C_{ds}(1+\alpha))^2}{g_m} I_3 f_b^3, \quad (7.171)$$

or in terms of f_t this is

$$\langle i_{nF}^2 \rangle = \frac{4kT}{R_F} I_2 f_b + 4kT \left[2\Gamma + \frac{2}{A_1} \right] 2\pi C_{ds} I_3 \frac{f_b^3}{f_t} \left(1 + \frac{1}{2} \left(\alpha + \frac{1}{\alpha} \right) \right) (7.172)$$

The 3 dB bandwidth of the preamplifier will be proportional to the bit rate γfb. To minimize noise in R_F we make it as large as possible so that it still meets the bandwidth requirements;

$$\frac{2\pi C_{TF} R_F}{A} = \frac{1}{\gamma fb},$$

$$R_F = \frac{A}{2\pi C_{TF} \gamma fb}. \quad (7.173)$$

Defining a new constant $I_\gamma \triangleq I_2 \gamma$, we can write the input referred current noise variance as

$$\langle i_{nF}^2 \rangle = 4kT \left[\frac{2\pi C_{TF}}{A} I_\gamma f_b^2 + \left[2\Gamma + \frac{2}{A_1} \right] 2\pi C_{ds} I_3 \frac{f_b^3}{f_t} \left(1 + \frac{1}{2} \left(\alpha + \frac{1}{\alpha} \right) \right) \right] (7.174)$$

We recognize in (7.174) that the device scaling parameter α appears in a linear- and inverse-term. The optimal value occurs when these two terms are equal at $\alpha = 1$.

Determination of the Dominant Noise Term

We would now like to determine which terms of (7.174) are dominant. The first is due to thermal noise in the feedback resistor and is proportional the square of the data rate. The last two terms are due to channel thermal noise and thermal noise in the load resistor, respectively. These two terms are proportional to the cube of the data rate and will eventually overpower the contributions from R_F. The capacitance parameters are given by

$$\begin{aligned} C_{TF} &= C_{ds} + C_{gss} + C_{gs} + (1 + A_1)C_{gd}, \\ C_{TF} &= C_{ds}(1 + \alpha) + A_1 C_{gd}, \end{aligned} \tag{7.175}$$

where $(1 + A_1)C_{gd}$ is the Miller capacitance. We can define a capacitance $C_m \triangleq A_1 C_{gd}$ that will go to C_{gd} if a cascode transistor is used. The cross-over data rate, where the gate-induced voltage due to thermal noise in the channel becomes dominant, will occur when

$$\frac{C_{TF}}{A} I_\gamma < \left[2\Gamma + \frac{2}{A_1}\right] C_{ds} I_3 \frac{f_b}{f_t}(1 + 1/2(\alpha + 1/\alpha)),$$

$$\left[\frac{C_{ds}(1+\alpha)}{A} + \frac{C_m}{A}\right] I_\gamma < \left[2\Gamma + \frac{2}{A_1}\right] I_3 \frac{f_b}{f_t}(1 + 1/2(\alpha + 1/\alpha))C_{ds}. \tag{7.176}$$

Therefore the thermal noise due to R_F starts to become negligible for normalized frequencies greater than

$$\frac{f_b}{f_t} > \frac{\left[\frac{1+\alpha}{A} + \frac{C_m/C_{ds}}{A}\right]}{\left[2\Gamma + \frac{2}{A_1}\right](1 + 1/2(\alpha + 1/\alpha))} \times \frac{I_\gamma}{I_3}. \tag{7.177}$$

Putting in some numbers for a noise-less cascode stage with $\alpha = 1$ and $C_m = 0$, the channel thermal noise dominates for

$$\begin{aligned} f_b &> \frac{f_t}{2A\Gamma}\frac{I_\gamma}{I_3}, \quad I_\gamma/I_3 \approx 10, \\ f_b &> \frac{5f_t}{A\Gamma}. \end{aligned} \tag{7.178}$$

As a rough example for $\Gamma = 2$ and $A = 40dB$, the channel thermal noise dominates when

$$f_b > \frac{f_t}{40}. \tag{7.179}$$

This example shows that as the data rate gets close to the speed limitation of the transistor f_t, the frequency-dependent noise will become dominant. The actual cross-over frequency will depend on the gain A and on the noise filtering parameters I_γ/I_3.

7.5.3 Qualitative Expression Comparing Bipolar and FET Devices

The noise contributions of a bipolar front-end can now be compared to the noise of an FET front-end. Consider the high frequency noise contributions. The noise of a bipolar preamplifier at the optimal bias current is given by

$$\langle i_{n_{\beta_1}}^2 \rangle \approx 4kT \left[2\pi C_\beta \left[1 + \frac{2}{A_1} \right] I_3 f_b^3 \left[\frac{1}{f_b} \sqrt{1/\beta_m + f_b^2/f_F^2} + \frac{1}{f_F} \right] \right], \quad (7.180)$$

assuming r_b is made small enough to be insignificant. (This can be done for $r_b \approx 10\Omega$ for moderately sized $[40 \times 12] \mu m^2$ HBT devices.) And the noise of an FET is

$$\langle i_{nF_1}^2 \rangle \approx 4kT \left[2\pi C_{ds} \left[2\Gamma + \frac{2}{A_1} \right] I_3 f_b^3 \left[\frac{(1 + 1/2(\alpha + 1/\alpha))}{f_t} \right] \right], \quad (7.181)$$

We can define the ratio of noise powers as

$$N_1 \triangleq \frac{\langle i_{nF_1}^2 \rangle}{\langle i_{n_{\beta_1}}^2 \rangle}. \quad (7.182)$$

For $A_1 \geq 10$, $\alpha = 1$, and ignoring $2/A_1$, we get

$$N_1 = \frac{\frac{C_{ds}(2\Gamma)2}{f_t}}{\frac{C_\beta \left[1 + \sqrt{f_F^2/\beta m f_b^2 + 1} \right]}{f_F}},$$

$$N_1 = \frac{f_F}{f_t} \frac{C_{ds}}{C_\beta} 4\Gamma \left[\frac{1}{1 + \sqrt{f_F^2/\beta m f_b^2 + 1}} \right],$$

$$N_1 = \frac{f_F}{f_t} \left[\frac{1}{1 + \frac{C_{je} + C_\mu}{C_{ds}}} \right] 4\Gamma \left[\frac{f_b}{f_b + \sqrt{f_b^2 + f_F^2/\beta m}} \right]. \quad (7.183)$$

We recall that in this ratio f_t is the unity-current-gain frequency for the FET and f_F is the maximum unity-current-gain frequency of the bipolar device given by $f_F \triangleq 1/(2\pi\tau_F)$. For small data rates this favors FET devices;

$$N_1 = \frac{f_b \sqrt{\beta m}}{f_t} \left[\frac{1}{1 + \frac{C_{je} + C_\mu}{C_{ds}}} \right] 4\Gamma. \quad (7.184)$$

For large data rates (this favors the fastest devices with bias towards BJTs)

$$\lim_{f_b \to \infty} N_1 = \frac{f_F}{f_t} \left[\frac{1}{1 + \frac{C_{je} + C_\mu}{C_{ds}}} \right] 2\Gamma. \quad (7.185)$$

Defining another noise ratio, N_2, that includes the noise due to the base resistance, we get

$$\langle i^2_{n\beta_2} \rangle = 4kT \left[2\pi C_{ds} I_3 \frac{f_b^3}{f_{rb}} \right]$$
$$+ 4kT \left[2\pi C_\beta \left[1 + 2/A_1\right] I_3 \left[f_b^2 \sqrt{1/\beta m + f_b^2/f_F^2} + f_b^3/f_F \right] \right] \quad (7.186)$$

$$\langle i^2_{F_1} \rangle = 4kT \left[2\pi C_{ds} I_3 \frac{f_b^3}{f_t} \left[2\Gamma + 2/A_1\right] (1 + 1/2(\alpha + 1/\alpha)) \right]. \quad (7.187)$$

As an example, for $\alpha = 1$ and A_1 large

$$\frac{1}{N_2} = \frac{\langle i^2_{n\beta_2} \rangle}{\langle i^2_{F_1} \rangle} = \frac{f_t}{f_{rb} 4\Gamma} + \frac{1}{N_1}, \quad (7.188)$$

$$\frac{1}{N_2} = \frac{f_t}{4\Gamma} \left[\frac{1}{f_{rb}} + \frac{1}{f_F} \left(1 + \frac{C_{je} + C_\mu}{C_{ds}} \right) \left(1 + \sqrt{f_F^2/\beta m f_b^2 + 1} \right) \right]. \quad (7.189)$$

Minimizing this gives

$$\frac{1}{N_{2_{\min}}} = \lim_{\substack{f_b \to \infty \\ \beta \to \infty}} \frac{1}{N_2} = \frac{f_t}{2\Gamma f_F} \left[\frac{f_F}{2 f_{rb}} + \left(1 + \frac{C_{je} + C_\mu}{C_{ds}} \right) \right]. \quad (7.190)$$

$$\frac{1}{N_{2_{\min}}} = \frac{f_t}{2\Gamma f_F} \left[\frac{C_{ds} r_b}{2\tau_F} + \left(1 + \frac{C_{je} + C_\mu}{C_{ds}} \right) \right], \quad (7.191)$$

For $f_t = g_{m_F}/(2\pi C_{ds})$, we get

$$\frac{1}{N_{2_{\min}}} = \frac{g_{m_F}}{4\Gamma} \left[r_b + \frac{2\tau_F}{C_{ds}} \left(1 + \frac{C_{je} + C_\mu}{C_{ds}} \right) \right]. \quad (7.192)$$

The conclusion is that if $1/N_{2_{\min}}$ is less than one by a significant amount, then the bipolar device can have lower noise than the FET if β and/or the data-rate is high enough. It should be clear that for an HBT with low r_b and high f_F (low transit time τ_F) $1/N_{2_{min}}$ will be less than one.

7.6 InP PREAMPLIFIER

We can now make use of the results of the preceding analysis to optimize the performance of a transresistance preamplifier. We have shown that to minimize the noise at

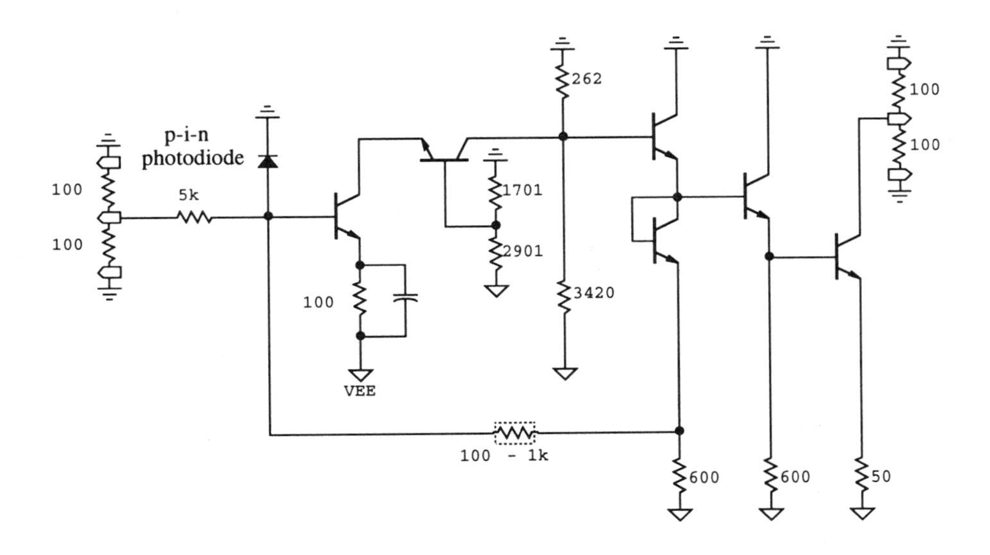

Figure 7.31 A low-noise InP transimpedance preamplifier.

high frequencies we need to maximize the transconductance to capacitance (g_m/C) ratio. This ratio can be improved by using a high-speed device and by minimizing the parasitic capacitances of the photodetector and the stray interconnect capacitance. We have therefore chosen to implement an electro-optical InP integrated low-noise transimpedance preamplifier, as shown in Fig. 7.31. InP HBTs have extremely fast transit times ($\tau_F \simeq 1$ps). Also a PIN photodetector matched to the low-loss wavelength of single-mode glass optical fibers ($\lambda \simeq 1.3\mu$m) can be integrated on the same chip, thus substantially reducing interconnect capacitance.

7.6.1 Circuit Design

A minimal design was used for this prototype amplifier to maximize yield. The amplifier uses a single cascode stage with a dc gain of approximately 40. The high impedance node is buffered by an emitter follower and level shifted before it is connected to the feedback resistor. The bias voltages, and thus the bias currents, are dependent on the base-emitter junction voltage ($\sim$ 1V). The bias current of the input stage is approximately 4mA.

In a broadband amplifier the natural frequencies of the circuit can be close together, which can cause stability problems. A root locus is a useful tool to track the locations

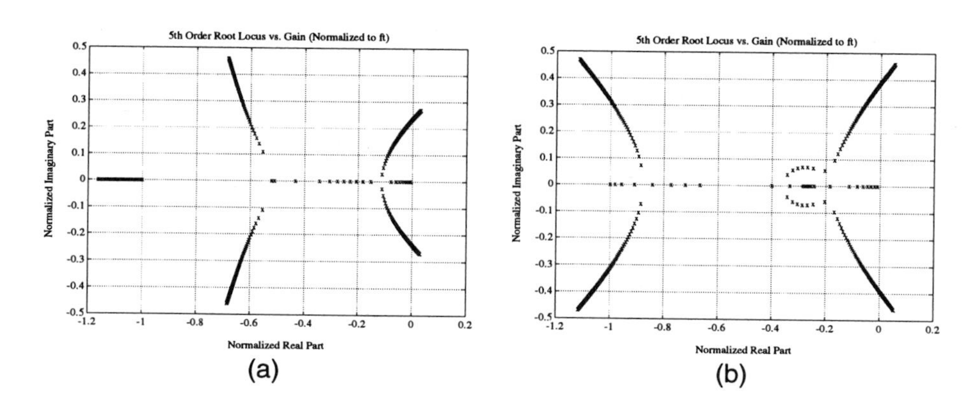

Figure 7.32 Root Locus of a transresistance preamplifier as a function of the loop gain for a 5th order system.

of the natural frequencies of the system as a function of the loop gain. Two root loci are shown in Fig. 7.32. We can see from Fig. 7.32 that the dominant poles bend toward the right half plane and can cause the circuit to become unstable. By calibrating the root locus plot, we can determine the value of the loop gain required to achieve the desired pulse response.

Aside from the PIN diode that will serve as an optical detector, a provision was made for testing with an electrical input. A 50Ω termination is provided at the input that consists of two 100Ω resistors in parallel. A series resistor of $5k\Omega$ was used to simulate a the photodetector current when an electrical input is used. The output buffer is a common emitter stage with a 50Ω termination at the collector.

7.6.2 Simulated Results

The transconductance and the output noise spectral density are shown in Fig. 7.33. This output noise is the input noise filtered by the transfer function of the amplifier. The output noise spectrum has a peak near the data rate. We used a similar shaped colored noise spectrum in chapter 3, where we considered the effect of various filtering functions on receiver performance in the presence of colored noise. Now we have provided the justification for using such a spectral density in the analysis.

We are also interested in the pulse response of the circuit. The preamplifier may have very low noise, but if it rings, it can result in a narrow *eye* that degrades the BER. The simulation results for a pseudo-random data input are shown in Fig. 7.34. We can

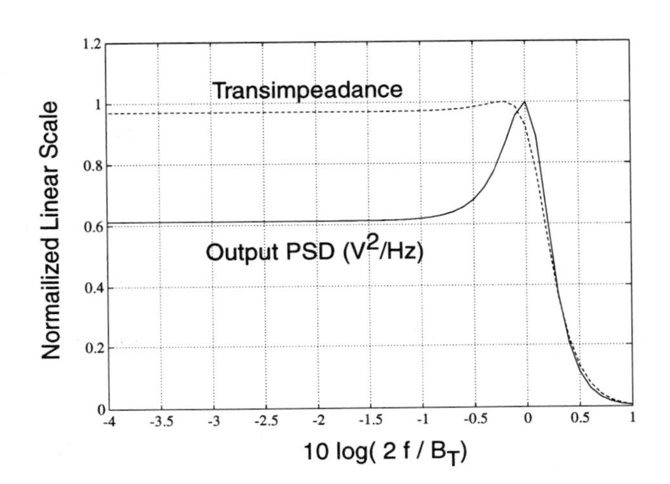

Figure 7.33 Simulated transresistance and output noise spectrum vs. frequency.

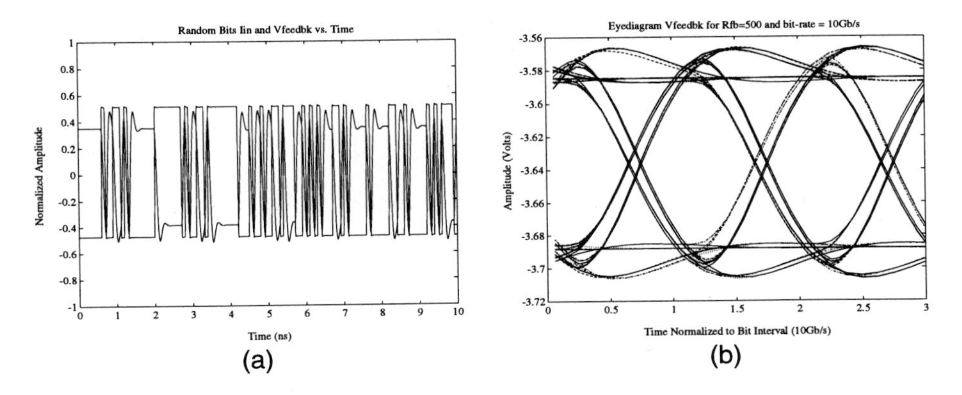

Figure 7.34 Simulation result of InP preamp for a data rate of 10 Gb/s with a feedback resistor of 500Ω: (a) time response, (b) eye diagram.

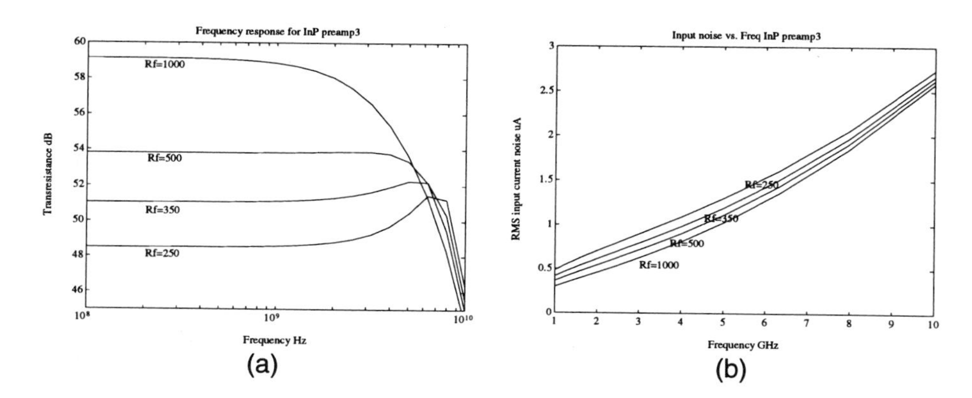

Figure 7.35 Simulation results of InP preamplifier for various feedback resistances: (a) frequency response (b) rms noise current.

see from the eye diagram in Fig. 7.34(b) that the bandwidth is too narrow and causes intersymbol interference that reduces the eye opening.

The bandwidth and equivalent noises are shown in Fig. 7.35 for various values of the feedback resistor. In this design we can achieve a 3 dB bandwidth of 10 GHz with a 500Ω feedback resistor and a detector capacitance of $C_{ds} = 0.5$pF. The equivalent rms noise at the input for a 10 Gb/s system is approximately 1.25μA. We can therefore find the minimum optical power needed to achieve a given BER. From (7.44) the minimum power is

$$P_{\text{av}} = 1.242 \left(\frac{\mu\text{W} \cdot \mu\text{m}}{\mu\text{A}} \right) \left[\frac{(Q_{\text{SNR}}) i_{\text{rms}}}{\eta\lambda} \right]. \tag{7.193}$$

For $Q_{\text{SNR}} = 6$, $\lambda = 1.3\mu$m, the minimum power needed is

$$P_{\text{av}} = \frac{7.165}{\eta} \quad (\mu\text{W}), \tag{7.194}$$

where η is the quantum efficiency of the photodetector. For $\eta = .717$, $P_{\text{av}} = 10\mu$W or -20 dBm.

7.6.3 Measured Results

This preamplifier was processed at TRW. A microphotograph of the circuit is shown in Fig. 7.36. Because the transistor parameters vary substantially in this developmental InP-based HBT process, a tunable feedback resistor was used to ensure that the desired frequency response can be obtained. The feedback resistor can be changed by selectively breaking air-bridge metal lines that shunt segments of the resistor.

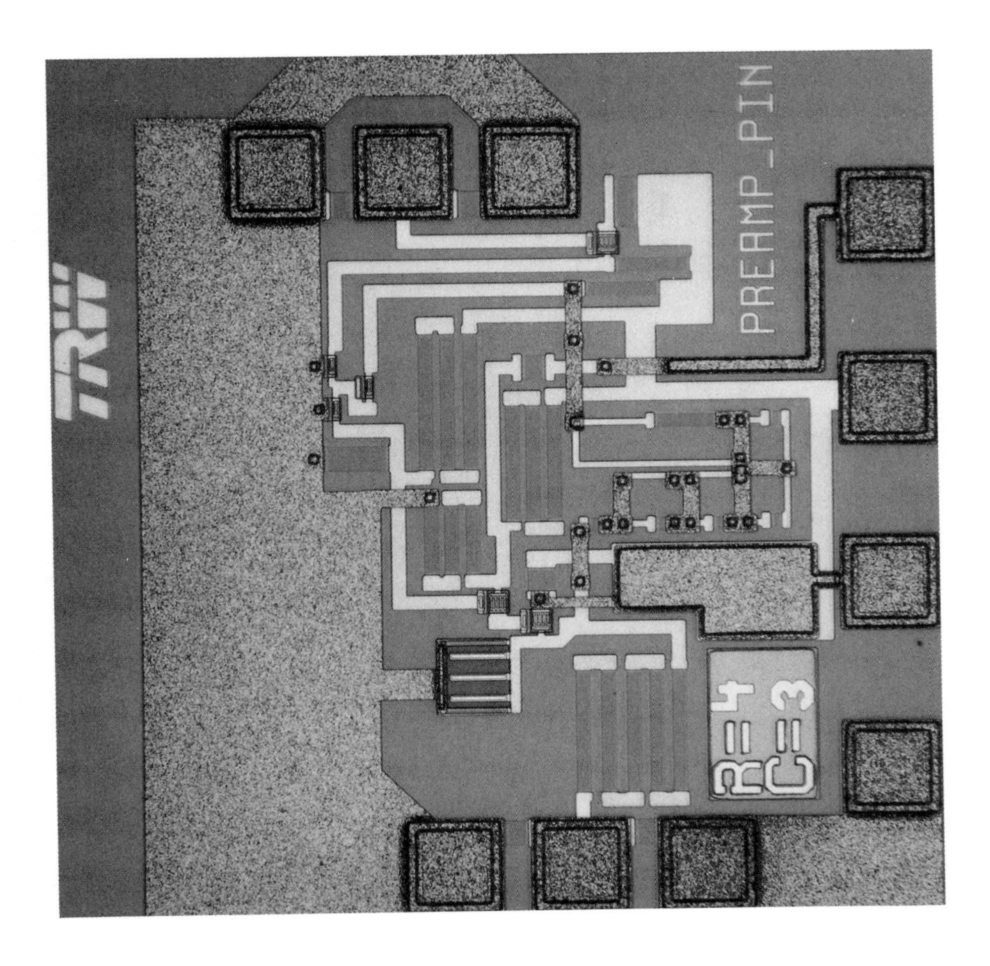

Figure 7.36 Microphotograph of InP integrated PIN photodetector and transresistance preamplifier.

Measured device parameters show an f_t of 70-GHz and an $f_{\max}$ of 100-GHz. This performance surpassed the models used in simulation and would result in a preamplifier bandwidth of 12–15 GHz. The preamplifier described is currently under test and measured results will be reported elsewhere.

REFERENCES

[1] Walther Schottky. Über spontane Stromschwankungen in verschiedenen Elektrizitätsleitern. *Ann. Phys.*, 57:541–567, 1918.

[2] Wilbur B. Davenport, Jr. and William L. Root. *An Introduction to the Theory of Random Signals and Noise*. IEEE Press, New York, 1987. IEEE PRESS edition of a book published by McGraw Hill Book Company in 1958 under the same title.

[3] Athanasios Papoulis. *Probability, Random Variables, and Stochastic Processes*. McGraw Hill, New York, 1965.

[4] John Bertrand Johnson. Thermal agitation of electricity in conductors. *Phys. Rev.*, 32:97–109, 1928.

[5] H. Nyquist. Thermal agitation of electric charge in conductors. *Phys. Rev.*, 32:110–113, 1928.

[6] Charles Kittel and Herbert Kroemer. *Thermal Physics*. W. H. Freeman and Company, New York, second edition, 1980.

[7] B. M. Oliver. Thermal and quantum noise. In Madhu S. Gupta, editor, *Electrical Noise: Fundamentals & Sources*, chapter 3, pages 129–148. IEEE Press, New York, 1977.

[8] Stewart D. Personick. Receiver design for digital fiber optic communication systems, part I and II. *Bell Syst. Tech. J.*, 52(6):843–886, July 1973.

[9] R. G. Smith and S. D. Personick. Receiver design for optical fiber communication systems. In Henry Kressel, editor, *Semiconductor Devices for Optical Communication*, chapter 4, pages 86–160. Springer-Verlag, New York, 1980.

[10] Stewart D. Personick. *Fiber Optics Technology and Applications*. Plenum Press, New York, 1985.

[11] Tran Van Muoi. Receiver design for high-speed optical-fiber systems. *J. Lightwave Technol.*, LT-2(3):243–267, June 1984.

[12] Mike Brian and Tien-Pei Lee. Optical receivers for lightwave communication systems. *J. Lightwave Technol.*, LT-3(6):1281–1300, December 1985.

[13] Bryon L. Kasper and Joe C. Campbell. Multigigabit-per-second avalanche photodiode lightwave receivers. *J. Lightwave Technol.*, LT-5(10):1351–1364, October 1987.

[14] Gerd Keiser. *Optical Fiber Communications*. McGraw-Hill, Inc., New York, second edition, 1991.

[15] J. Gowar. *Optical Communication Systems*, chapter 14–15. Prentice-Hall, London, 1984.

[16] M. J. N. Sibley. *Optical Communications*. McGraw-Hill, New York, 1990.

[17] Clemens Baack, editor. *Optical Wideband Transmission Systems*. CRC Press, Boca Raton, Florida, 1986.

[18] Sol Rosenstark. *Feedback Amplifier Principles*. Macmillan Publishing Co., New York, 1986.

[19] Asad A. Abidi. On the optimum of gigahertz FET transimpedance amplifiers. *IEEE J. Solid-State Circuits*, SC-22(6):1207–1209, December 1987.

8

VOLTAGE CONTROLLED OSCILLATORS AND HIGH-SPEED TESTING

HBT processes using III–V semiconductors are still in their infancy as compared to silicon technologies; in the early stages of this research (1989) it was not uncommon for model parameters of GaAs HBTs to vary by 100% or more from run-to-run. It was, therefore, essential to design test chips to aid in process evaluation and modeling. In this chapter we will briefly describe two voltage controlled oscillator (VCO) circuits used for this purpose. The first is a four-stage ring oscillator, and the second is an emitter-coupled multivibrator.

Testing of circuits in the gigahertz range can be quite troublesome. Although microwave designers are familiar with high-speed testing techniques, this information is not well known among designers of traditional analog circuits. Therefore, in the later part of this chapter we will present a brief introduction to high-speed testing techniques. References will be provided where additional information about this important topic can be found.

8.1 FOUR-STAGE RING VCO

A useful circuit for evaluating device performance is a variable frequency ring oscillator. A four-stage, fully-differential VCO is shown in Fig. 8.1(a). By measuring the frequency of oscillation we can determine the delay times of the internal circuits as a function of the bias current. We can also use this information to fine-tune SPICE models to accurately predict switching speeds. The delay cell, shown in Fig. 8.1(b), is a differential pair with a resistive load; reversed biased base-emitter junctions are used as variable load capacitances for adjustment of the delay time, which can additionally be altered by the bias current I_B. As oscillation requires an odd number of inversions

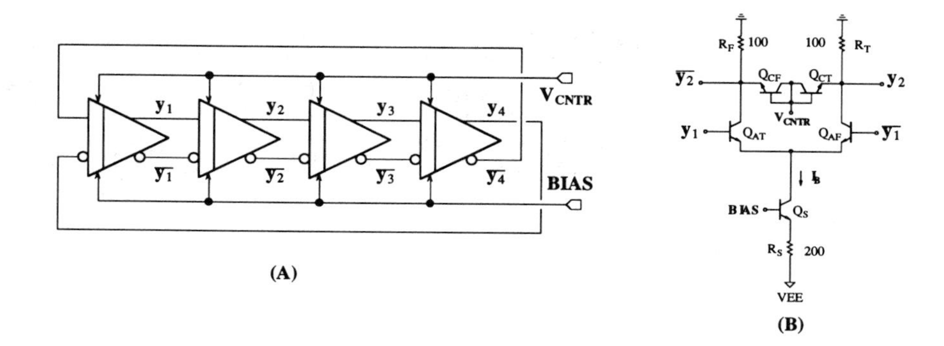

Figure 8.1 (a) four-stage ring VCO. (b) differential delay cell

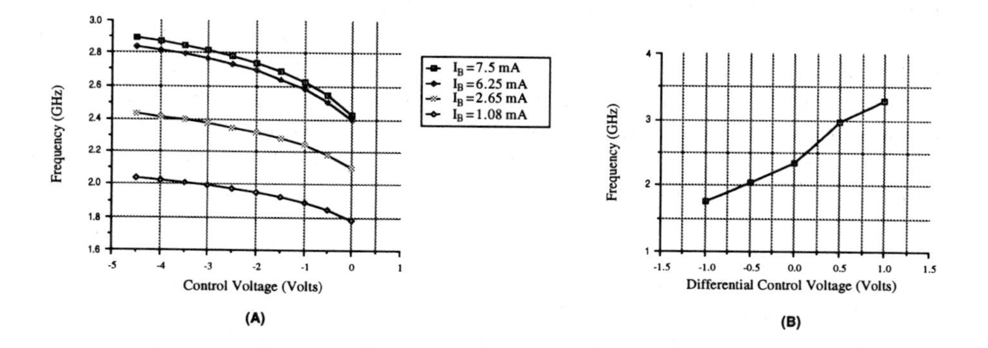

Figure 8.2 Frequency vs. control for: (a) four-stage ring VCO at various bias currents, (b) emitter-coupled multivibrator.

in the signal path, the positive and negative inputs to one of the differential delay cells are interchanged. The use of an even number of delay elements is advantageous because quadrature shifted versions of any output signal are available. By using fully-differential delay cells, the effect of power-supply noise on the oscillating frequency is minimized, thereby reducing the phase jitter.

The frequency of oscillation was measured as a function of the control voltage for various values of I_B. The results are plotted in Fig. 8.2(a); the center frequency can be varied from 2.5 GHz $\pm 16\%$ by adjusting the bias current. Furthermore, V_{CNTR} can be used to vary the frequency by an additional $\pm 5\%$.

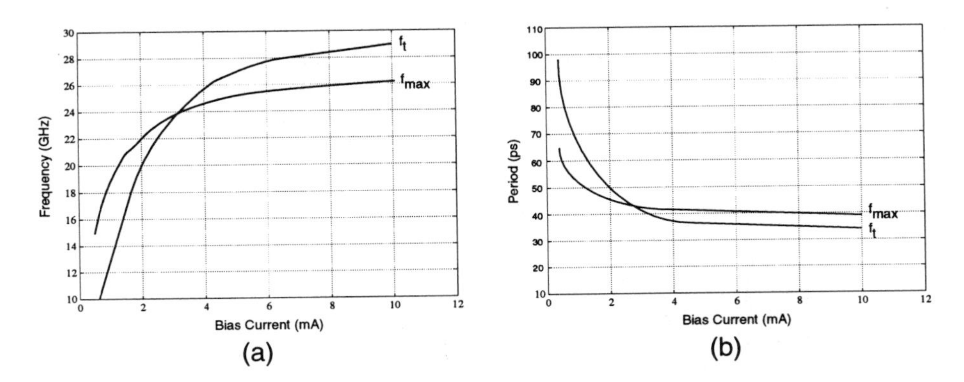

Figure 8.3 (a) $f_{\max}$ and f_t vs. bias current. (b) $1/f_{\max}$ and $1/f_t$ vs. bias current.

The circuit was fabricated in an AlGaAs/GaAs HBT process using (3μm x 10μm) minimum emitter area devices. Nominal SPICE models for this process were gleaned using several characterization methods (dc current-voltage characteristics, S-parameter measurements, and analytical techniques based on doping profiles) to obtain nominal parameter values. Starting from these nominal models, the base-resistance and base-collector capacitance were determined by optimization. Model parameters were adjusted with the aid of the optimizer of HSPICE so as to fit simulations to measured data. The resulting models were given in table 6.4.

The period of oscillation for an n-stage oscillator is given by

$$T = 2nt_d, \tag{8.1}$$

where t_d is the delay of each cell. The transistor parameters $f_{\max}$ and f_t for the process are plotted in Fig. 8.3. We can deduce that the delay time t_d obtained from Fig. 8.2(a) is approximately equal to $1/f_{\max}$. Hence, a first-order estimate of the oscillating frequency of the VCO is given by

$$f_o \simeq \frac{f_{\max}}{2n}. \tag{8.2}$$

For this prototype VCO the maximum oscillating frequency of the transistors, $f_{\max}$, is about 22 GHz, from which we estimate f_o to be 2.75 GHz. For TRW's InP process with (1μm x 10μm) transistors and an $f_{\max}$ of 100 GHz, the 4-stage VCO could achieve a frequency of oscillation of approximately 12.5 GHz based on this simple estimation. Later, in section 8.4, we will derive a more accurate expression for estimating delay times.

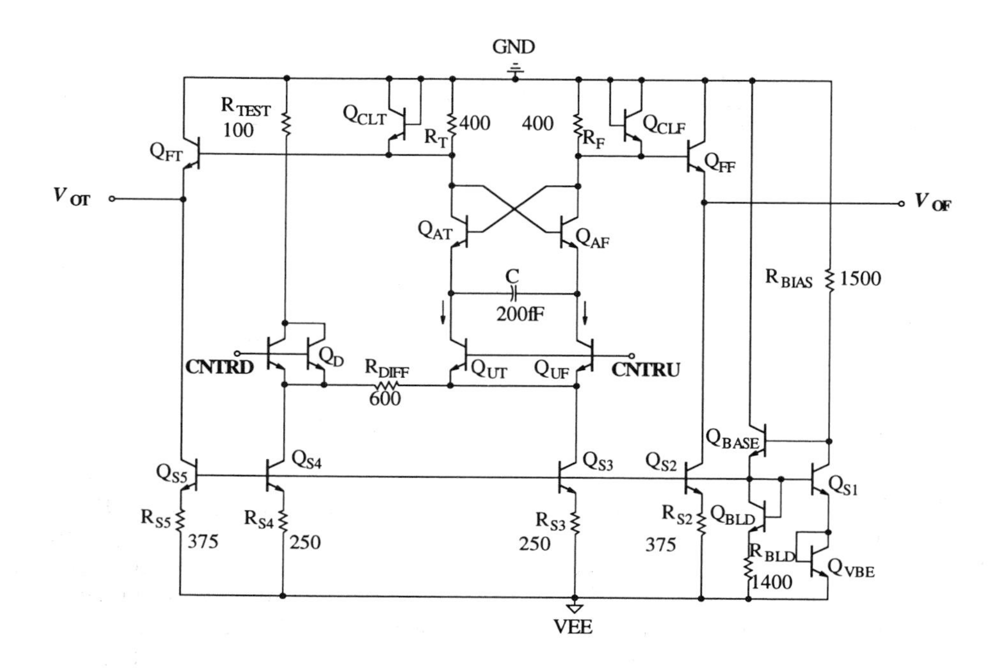

Figure 8.4 An emitter-coupled multivibrator VCO

8.2 EMITTER-COUPLED MULTIVIBRATOR VCO

The second VCO designed was an emitter-coupled multivibrator [1]. The schematic for this circuit is shown in Fig. 8.4. The cross-coupled transistors provide positive feedback for a high loop gain, and the clamping diodes, Q_{CLT} and Q_{CLF}, act to limit the amplitude of the oscillation. Frequency tuning is accomplished by varying the oscillating current, I_{OSC}, which alters the charging time of the emitter capacitor (200 fF). This current is controlled by a differential voltage across the inputs CNTRU and CNTRD. Measured results of this circuit, fabricated in the same HBT process as before, are given in Fig. 8.2(b). The tuning range of this circuit is 2.5 GHz $\pm 32\%$. It can be seen that the frequency, as a function of the control voltage, deviates from a linear response; this can result from the distortion in the voltage-to-current conversion in the degenerated differential pair, and it could also be due to the fact that the common-mode voltage is not fixed, which would lead to different values of parasitic capacitances, as a function of the control voltage, on the emitters of Q_{AT} and Q_{AF}. Moreover, for a relaxation oscillator, where the regeneration time is negligible, the frequency of oscillation will ideally be linearly proportional to I_{OSC}; however, when

	Ring VCO	Multivibrator VCO
Maximum Frequency	2.9 GHz	3.3 GHz
Tuning Range	2.35±23%	2.5 GHz±32%
Linearity	2.5%	5.4%
Power Supply Voltage	-8V	-8V
Power Dissipation	34–240 mW	178 mW
Active Area	300μm×400μm	300μm×400μm
Second Harmonic	-31 dBc	-11.9 dBc
Third Harmonic	-30 dBc	-14.0 dBc
Power 100 kHz Offset	-57 dBc	-53.5 dBc

Table 8.1 Measured VCO results.

the regeneration and relaxation times are comparable, as is the case in a high-speed oscillator, the relationship between frequency and current is less obvious, and linearity should not be expected.

8.3 COMPARISON OF RING AND EMITTER-COUPLED VCO

Measured results of the two oscillators are summarized in table 8.1. In order to obtain measurements in a 50Ω environment, the output buffer of Fig. 8.5 was used. The buffer circuit was operated at a bias current of approximately 10 mA, which provided a maximum differential output voltage of 250 mV. The two circuits have similar characteristics: the emitter-coupled multivibrator VCO having a larger maximum frequency and tuning range than the ring VCO, whereas the ring oscillator has slightly less phase-jitter (less power at 100-KHz offset from the center frequency). For extraction of the clock signal from digital data of a known rate, a VCO center frequency is a constant factor times the data-rate, and the tuning range must be large enough so as to center the VCO within the required frequency for clock recovery over worst case processing and temperature variations. The tuning range of the above VCO circuits (> ±20%) is adequate for this application. Microphotographs of the four-stage ring oscillator, and the emitter- coupled multivibrator are shown in Figs. 8.6 and 8.7, respectively.

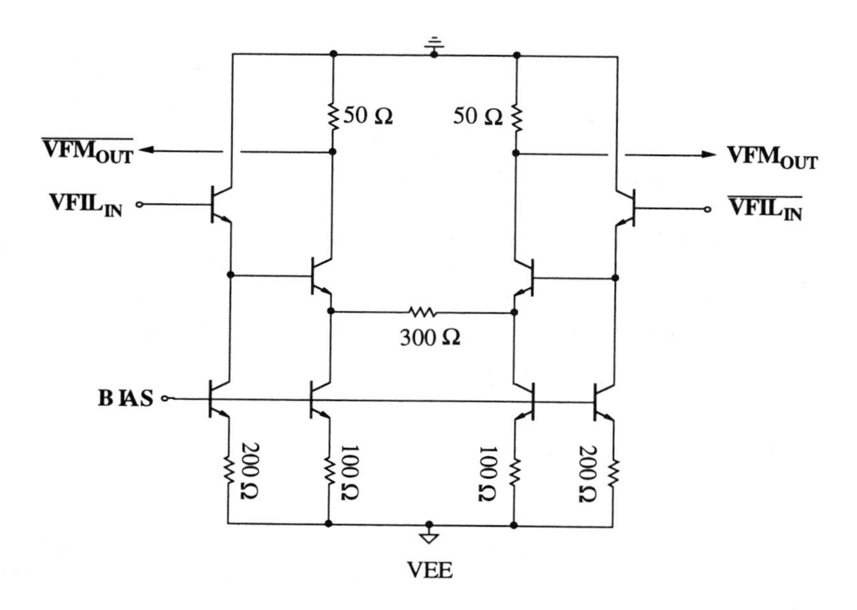

Figure 8.5 50Ω output buffer.

8.4 TIMING ESTIMATION

In the previous section we gave a very simple estimate for the frequency of oscillation that is useful for a first-order estimate of circuit performance. However, we would like to know which parameters are most important in limiting the frequency of operation of the VCOs. By deriving an expression that accurately predicts circuit behavior, we can use it both as an aid for extracting model parameters, and as a means for optimizing circuit performance. In this section we will present one method of estimating the timing of a ring oscillator that is applicable when the gain of each delay-stage is small ($\simeq 2$).

The basic delay cell of a current-mode ring oscillator is shown in Fig. 8.8. The dc transfer characteristic for the differential pair, ignoring base current, is given by the familiar hyperbolic tangent function [2].

$$V_{\text{out}} = IR\left[\tanh\left(\frac{V_{\text{in}}}{2V_T}\right)\right], \tag{8.3}$$

which is plotted in Fig. 8.9. We can see that the linear range of input signals in approximately $[-4V_T, 4V_T]$. At a temperature of 300°K this corresponds to a voltage range of about $[-100\text{mV}, 100\text{mV}]$.

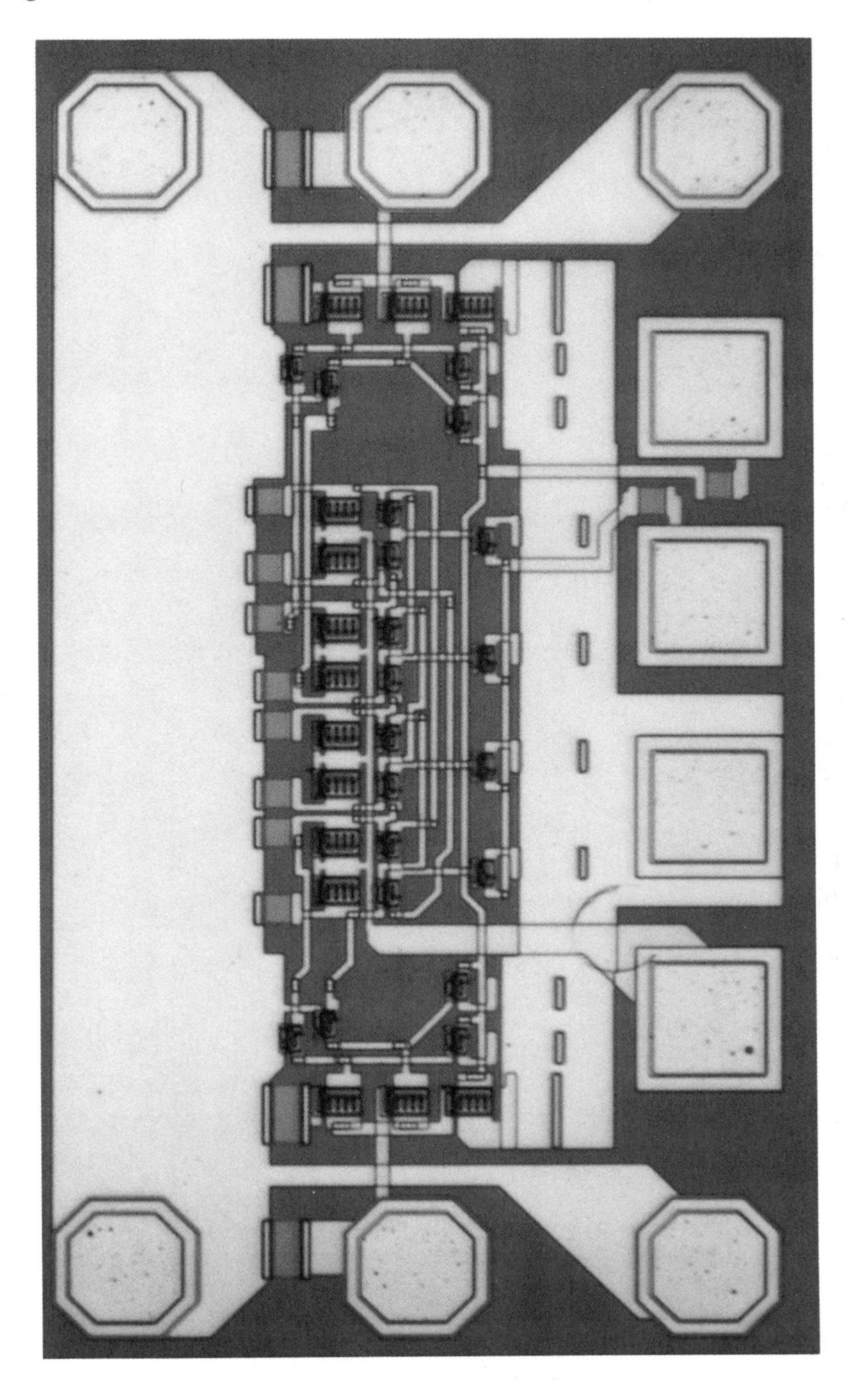

Figure 8.6 Microphotograph of an HBT four-stage ring VCO.

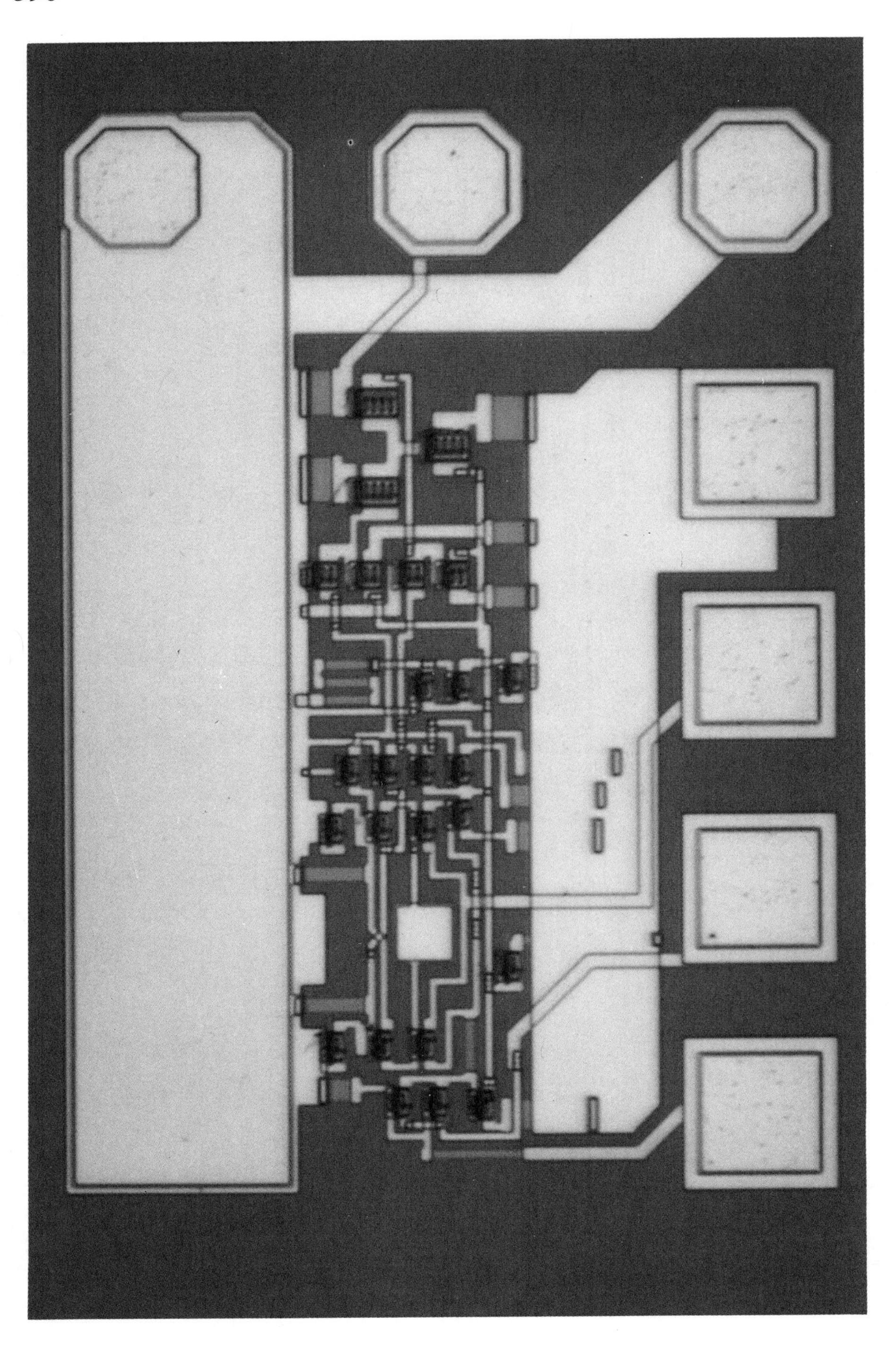

Figure 8.7 Microphotograph of an HBT emitter-coupled multivibrator VCO.

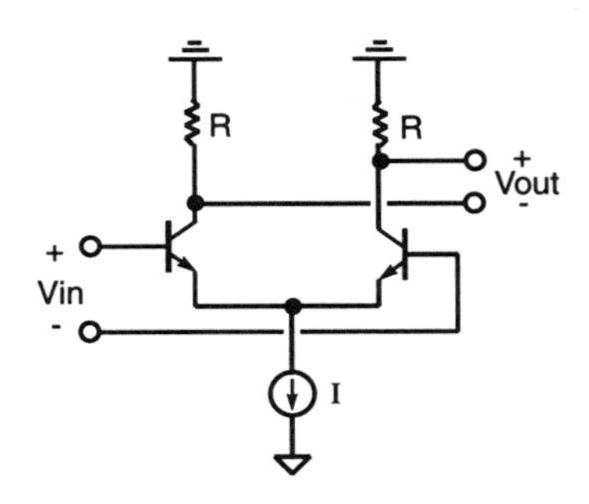

Figure 8.8 Differential delay cell.

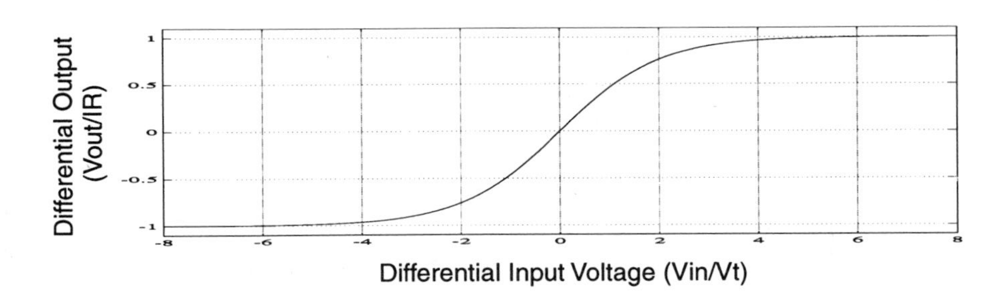

Figure 8.9 Hyperbolic tangent transfer curve of a bipolar differential pair amplifier.

The maximum gain of the circuit, A_0, occurs at the balanced point when both the differential input and output voltages are equal to zero. The value of the gain at this point is $IR/2V_T$. We can write the transfer characteristic in terms of A_0;

$$V_{\text{out}} = \left[2V_T A_0 \tanh\left(\frac{V_{\text{in}}}{2V_T}\right)\right]. \tag{8.4}$$

The clamping voltage is $2V_T A_0$, and for a gain of two this circuit clamps at $|V_{\text{out}}| = 4V_T$. Therefore the output voltage of the delay-cell is in compliance with its linear input range. When used in a ring oscillator, the output of one cell will drive the input of the next, and no delay-cell will be *overdriven* such that the voltage excursions are far beyond that necessary to switch the current. As a result all switching transistors will remain in the linear operating region nearly all of the time, and linear small-signal analysis can be used to predict circuit performance.

Period of oscillation

Now that we have justified using small-signal analysis, we can make use of familiar circuit analysis techniques to estimate the frequency of operation. We will use a lumped time-constant approach. Although lumping the effect of each natural frequency into one effective pole is unsatisfactory in predicting phase-lag or the pulse response in a feedback system, it is a useful approximation for estimating circuit delay times and usually gives reasonable results.

Lumping all poles into a single time constant, we can model the delay cell as a first-order system with a transfer function given by

$$H(s) = \frac{A_0}{1 + s\tau}. \tag{8.5}$$

In steady-state operation at the frequency of oscillation (f_o), each delay cell will contribute (π/n) radians of phase lag. This gives us a simple means of finding f_o in terms of τ. The phase of a delay cell at the frequency f_o is given by

$$-\angle H(j2\pi f_o) = \tan^{-1}(2\pi f_o \tau) = \frac{\pi}{n}. \tag{8.6}$$

Therefore the oscillation frequency is that value which produces the proper phase lag, and is given by

$$\boxed{f_o = \frac{\tan(\pi/n)}{2\pi\tau}.} \tag{8.7}$$

The equivalent delay time per stage is then

$$\boxed{t_d = \frac{\tau(\pi/n)}{\tan(\pi/n)}.} \tag{8.8}$$

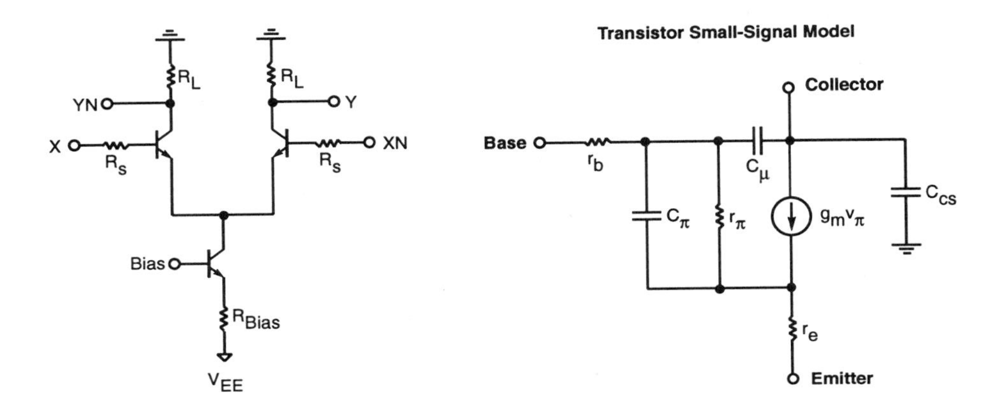

Figure 8.10 Differential pair amplifier together with the small-signal transistor model used in hand calculations.

For the special case of n=4, we obtain the following results,

$$f_o = \frac{1}{2\pi\tau} \tag{8.9}$$

$$t_d = \tau(\pi/4). \tag{8.10}$$

Explicit Expressions for the Lumped Time Constant

Now we need to find an expression for the lumped time constant in terms of device parameters. Derivation of the transfer function of the differential pair is straightforward. However, we will skip the derivation and simply give the results. The circuit we will analyze is shown in Fig. 8.10 with the small signal model used in hand calculations. The differential transfer function has the following form;

$$A(s) = A_0 \frac{1 - s\tau_{n1} - (s\tau_{n2})^2}{1 + s\tau_{d1} + (s\tau_{d2})^2 + (s\tau_{d3})^3}. \tag{8.11}$$

The midband gain is given by

$$A_0 = g_m R_L \left[\frac{\gamma_\pi}{1 + \gamma_\pi r_e (g_m + g_\pi)} \right], \tag{8.12}$$

where the base voltage reduction factor γ_π has been defined as

$$\gamma_\pi \triangleq \frac{r_\pi}{R_s + r_b + r_\pi}. \tag{8.13}$$

We will now give expressions for all of the time-constants in (8.11). The first-order time-constant in the denominator is the sum of all the capacitances in the circuit multiplied by the equivalent resistance seen across their terminals. This is our lumped time-constant, and we will separate it into four terms such that

$$\tau_{d1} = \tau_\pi + \tau_{m\mu} + \tau_{o\mu} + \tau_{cs} \tag{8.14}$$

where;

$$\begin{array}{ll} \tau_\pi & \text{is due to } C_\pi, \\ \tau_{m\mu} & \text{is the Miller effect at the input,} \\ \tau_{o\mu} & \text{is the Miller effect at the output,} \\ \tau_{cs} & \text{is due to } C_{cs}. \end{array} \tag{8.15}$$

These individual delay contributions are given by the following expressions

$$\tau_\pi = C_\pi \left[\frac{(R_s + r_b) \parallel r_\pi + \gamma_\pi r_e}{1 + \gamma_\pi r_e (g_m + g_\pi)} \right] \tag{8.16a}$$

$$\tau_{m\mu} = C_\mu \left[(R_s + r_b) \parallel r_\pi \left[\frac{1 + (g_m + g_\pi)(R_L + r_e)}{1 + \gamma_\pi r_e (g_m + g_\pi)} \right] \right] \tag{8.16b}$$

$$\tau_{o\mu} = C_\mu \left[\frac{\gamma_\pi [1 + r_e (g_m + g_\pi)]}{1 + \gamma_\pi r_e (g_m + g_\pi)} \right] R_L \tag{8.16c}$$

$$\tau_{cs} = C_{cs} R_L. \tag{8.16d}$$

The lumped time-constant t_{d1} is the only parameter we need in our first-order model. However, we will give expressions for all of the other time constant values for completeness.

The second-order time-constant in the denominator can be shown to be

$$\begin{aligned} \tau_{d2}^2 = \; & R_L (R_s + r_b) \left[\frac{\gamma_\pi}{1 + \gamma_\pi r_e (g_m + g_\pi)} \right] \times \\ & \left[(C_\pi C_\mu + C_\pi C_{cs}) \left[1 + \frac{r_e}{R_s + r_b} \right] + C_\mu C_{cs} \left[1 + r_e (g_m + g_\pi) \right] \right], \end{aligned} \tag{8.17}$$

and the third-order time-constant is given by

$$\tau_{d3}^3 = R_L (R_s + r_b) r_e \left[C_\pi C_\mu C_{cs} \right] \left[\frac{\gamma_\pi}{1 + \gamma_\pi r_e (g_m + g_\pi)} \right]. \tag{8.18}$$

The numerator time-constant expressions are found to be

$$\tau_{n1} = \frac{C_\mu}{g_m} \left[1 + r_e (g_m + g_\pi) \right] \tag{8.19}$$

$$\tau_{n2}^2 = \frac{C_\mu}{g_m} C_\pi r_e. \tag{8.20}$$

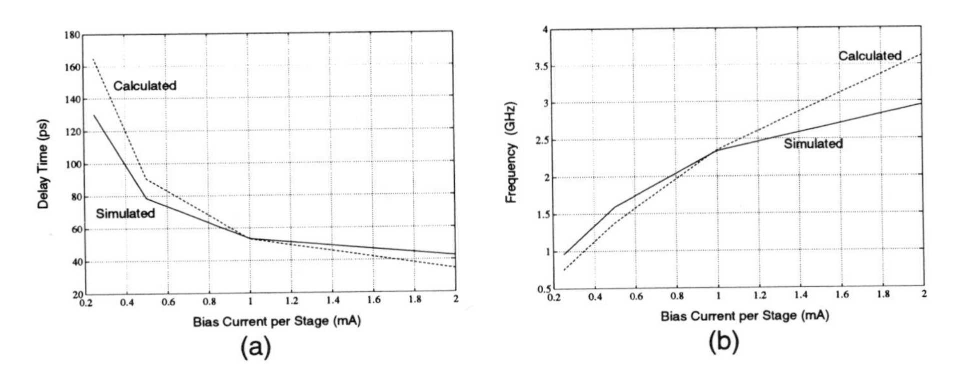

Figure 8.11 Simulations version lumped time-constant approximation for a 4-stage ring oscillator vs. bias current: (a) delay time of a single cell, (b) oscillation frequency.

Comparison with Simulations

This timing estimate was compared to simulations, and the results are plotted in Fig. 8.11. It can be seen that the estimate is accurate to within about 10%. More important than the accuracy of the estimate is the prediction of the relative contributions to the delay-time of each of the various parasitics. Furthermore, the estimate predicts the variations of the oscillation frequency under various bias conditions. Given the expression for the lumped time-constant, we can determine which parasitic is most important and adjust bias conditions and device geometries accordingly to achieve optimal performance.

8.5 HIGH-SPEED TESTING

For circuits operating at frequencies in the gigahertz range, testing procedures and packaging techniques are just as important, if not more, than the actual circuit design. Small parasitics become important at these high frequencies and can dominant circuit performance. For example, a typical bond-wire inductance is about 0.6 nH. The magnitude of the impedance due to the bond-wire at 1 GHz is 3.8 Ω. However, at 10 GHz this increases to 38 Ω, and is similar in magnitude to circuit impedances. Furthermore, a bonding pad capacitance of 0.5 pF is typical. This gives a shunt impedance of 320 Ω at 1 GHz, reducing to 32 Ω at 10 GHz. Therefore, at high speeds we require techniques to minimize parasitics. We will also try to use the parasitics to our advantage, such as incorporating bond-wires as part of an inductive peaking scheme.

In this section we will present a brief introduction to high-speed testing procedures. In the gigahertz range all of the testing and packaging is done in a controlled impedance environment, using transmission lines of one type or another. For instance we will use coaxial cables for interconnecting test equipment and circuit boards. Within the circuit board, both microstrip and coplanar transmission lines will be used. To minimize reflections, transmission lines should be terminated. These termination should be placed as close to the active circuitry as possible to minimize lead inductances. There are several references on transmission lines that the reader can consult. Elliott's recent book [3] on guided waves in microwave circuit provides an excellent treatment of the subject. The books by Ott [4] and Bakoglu [5] also give valuable information on noise reduction, parasitics, and packaging techniques.

8.5.1 Wafer Probing

The testing procedure usually begins by probing the wafer. This often gives good results because there are no parasitics associated with packaging at this point. Microwave probes can be used up to 40 GHz. These probes use a coplanar transmission line with a controlled impedance (usually 50 Ω) all the way to the tip. The probe usually has three connections in a ground-signal-ground arrangement. Referring to the microphotographs of the two VCO circuit in Figs. 8.6 and 8.7, the reader can see high-frequency output pads arranged in a ground-signal-ground configuration for wafer probing. A 50 Ω terminating resistor can also be seen between the center conductor and ground. A better termination method was used in the preamplifier circuit of Fig. 7.36, where the termination resistor is composed of two 100 Ω resistors in parallel. This configuration allows the current density to remain symmetric at the end of the transmission line, and improves the response at high-frequencies.

A high-speed wafer probing setup is illustrated in Fig. 8.12. Microwave probes are mounted on a probe station. The coplanar transmission line at the tip of the probe is converted to a coaxial line, and an SMA connector is used to interface the signal to test equipment. The top view of a microwave probe from Cascade Microtech is shown in Fig. 8.13. The body of the probe is approximately one inch long. Because of the large physical size of the probe it is difficult to probe more than 4 high-speed signals at the same time.

8.5.2 Surface-Mount Packages for Testing

To be used in the real world, circuits must be packaged. High-speed packages are generally much more expensive than their low frequency counterparts because of the

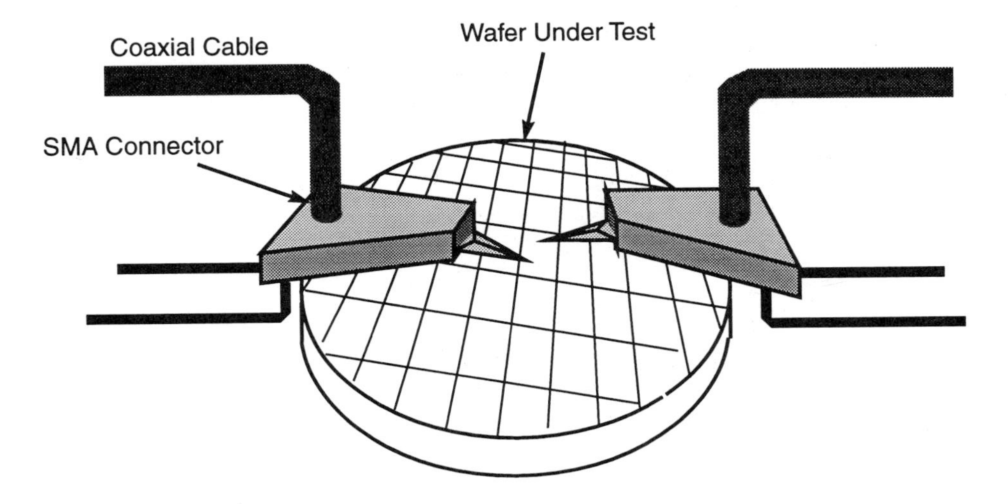

Figure 8.12 Illustration of wafer probing using controlled impedance microwave probes.

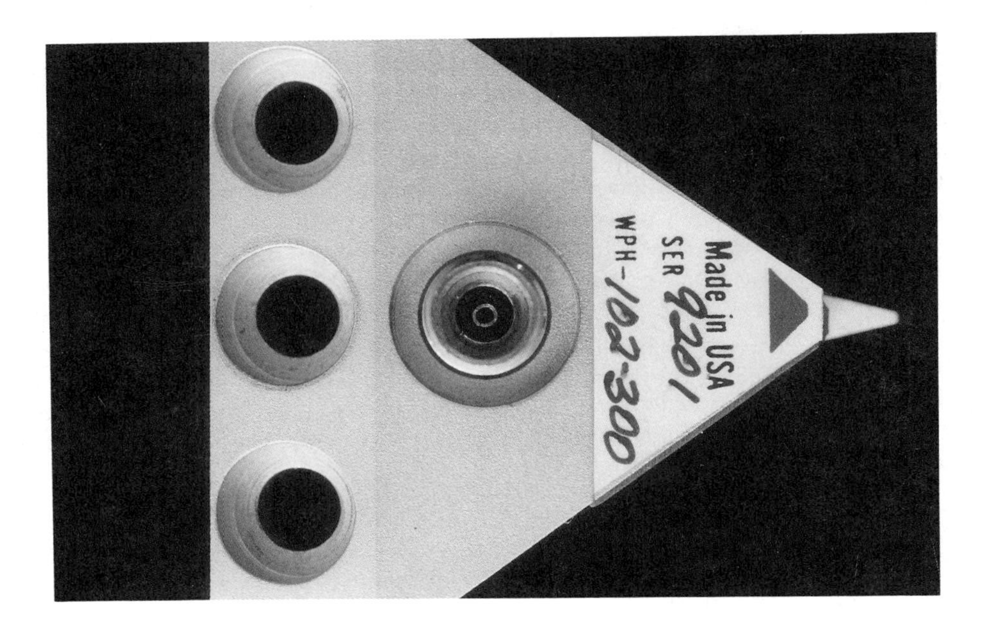

Figure 8.13 Top view of a microwave probe from Cascade Microtech.

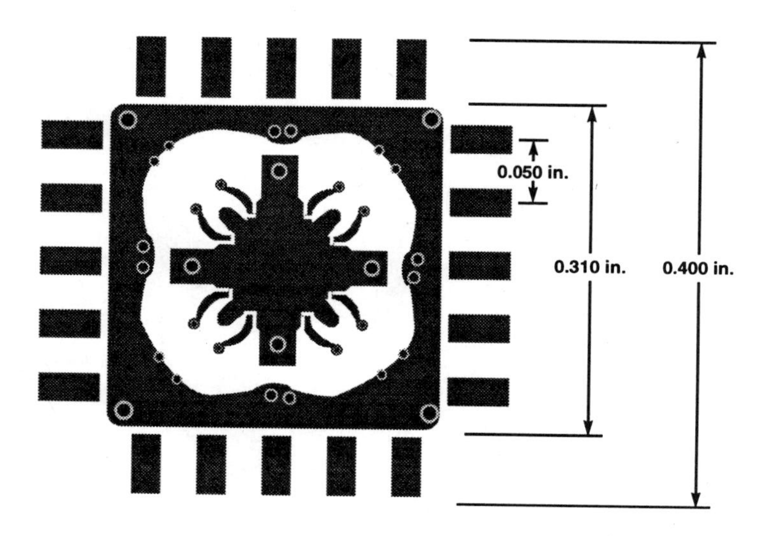

Figure 8.14 TEKPAC^{TM} surface mount package for dc-18 GHz available from Tektronix.

quality of the materials needed, and the accuracy of the dimensions required to maintain controlled impedance transmission lines. In this research we have made extensive use of the TEKPAC^{TM}: a surface mount test package available from Tektronix Inc. This package has been described in the literature [6, 7]. An illustration of a TEKPAC^{TM} is shown in Fig. 8.14. The package has 8 signal lines: two per side. Each of the signal pins are flanked by ground pins. The large metal area in the middle is a ground plane. The signal lines are routed along the back of the package and are connected to the top side through via holes.

After dicing, a test chip can be mounted on the surface of the TEKPAC^{TM}. This author used silver epoxy for mounting. The epoxy was applied, and the chip was affixed, allowing the epoxy to cure approximately 2 hours in an oven at 150°C. Wirebonds are then made from the IC pads to the ground-plane and signal lines of the TEKPAC^{TM}. An illustration of a TEKPAC^{TM} with an IC mounted on the surface is shown in Fig. 8.15. This shows 2 wires bonded to the ground plane, and 4 signal lines bonded to 4 of the 8 available signal lines.

Evaluation Kit for Surface-Mounted ICs

In order to interface signals to and from the chip, an evaluation kit is also available from Tektronix Inc. The evaluation kit is designed so that the TEKPAC^{TM} can be

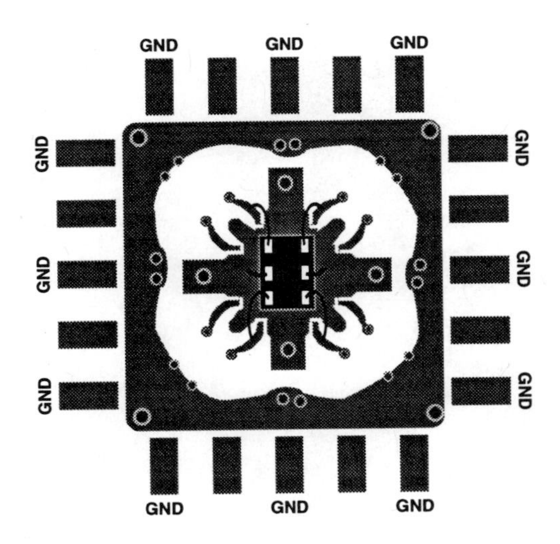

Figure 8.15 Bounding diagram of an IC surface mounted with silver epoxy on a TEKPAC^{TM} platform.

placed in the center as shown in Fig. 8.16. Connections are made from the signal lines of the TEKPAC^{TM} to the microstrip transmissions lines of the evaluation kit by pressure. The microstrip lines are distributed to 8 separate SMA connectors on the sides the evaluation kit as shown in Fig. 8.17.

To apply pressure to make the electrical connection, a plastic ring is placed on the pins of the TEKPAC^{TM}. The evaluation kit has a lid with a piece of rubber affixed inside the lid, which makes contact with the plastic ring as the lid is screwed into place, applying the pressure needed to make electrical contact to the signal pins. The evaluation kit with the lid in place is shown in Fig. 8.18. When the lid is closed, the evaluation kit provides a grounded shield for the circuit under test against optical and electro-magnetic interference. Signals can now be routed easily to test equipment and to couplers using coaxial cables via the SMA connectors.

8.5.3 Microstrip Transmission line Hybrid Circuit

The TEKPAC^{TM} is a very useful package. Several chips can be mounted and bonded. All of the test equipment can be connected and calibrated. Then different chips can be tested simply by *popping* a new TEKPAC^{TM} into the evaluation kit without having to disconnect any wires, or do any soldering. However, when more than 8 test signals are

Figure 8.16 Photo of evaluation kit with VCO chip mounted on a TEKPAC™.

Figure 8.17 Evaluation kit for TEKPAC™; approximately 1.75 in per side.

Figure 8.18 Evaluation kit for TEKPAC^{TM} with lid.

required, a custom hybrid test circuit may be needed. If one has access to the proper facilities it is often a simple matter to make your own hybrid test board.

A microstrip line can be etched on a DURIOD^{TM} board. Rubylith masks can be cut with the aid of simple CAD tools, and the pattern can be transferred, either directly, or by photographic reduction, to the test board. A skilled technician can layout and etch a test board within 30 minutes. Various connectors and test fixtures can be purchased from microwave component vendors. 180° hybrid couplers and can be used to take the sum or difference of two signals in the GHz range; this is useful for performing differential to single-ended conversions. Microwave passive components, such as chip-resistors and capacitors, are also available from various vendors. Small chip resistors can be mounted directly on the surface of the board to provide terminations, thus minimizing inductive leads. Chip-capacitors can be placed as close to the component as possible for power supply and bias line decoupling.

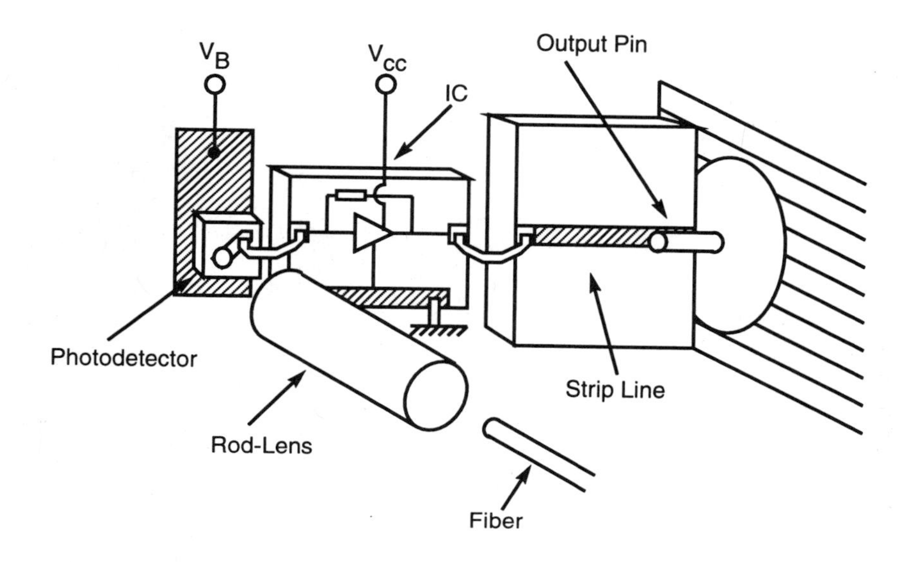

Figure 8.19 Illustration of a possible packaging technique for a multigigabit per second fiber-optic preamplifier module.

8.5.4 Packaging

We will now take a brief look at some packaging techniques that have been described in the literature to improve high speed performance. Fujita *et al.* [8] described a 5-Gb/s fiber-optic receiver module, and provided a useful illustration of a high-speed packaging technique. An avalanche photodiode (APD) was packaged together with a low-noise transresistance preamplifier. A diagram of the receiver module is shown in Fig. 8.19 illustrating how one might go about packaging such a circuit.

A 10-GHz mixer for coherent optical systems was recently reported by Fujita *et al.* [9, 10]. In the packaging of the circuit, an improved bonding technique, illustrated in Fig. 8.20, was utilized. The straightforward bonding diagram for a chip bonded to a microstrip transmission line is shown in Fig. 8.20(a). The bond-wire to the signal line is excessively long and has an inductance of approximately 0.6 nH. This long inductor can be broken in half and bonded to the top plate of a capacitor with its bottom plate connected to ground. This is illustrated in Fig. 8.20(b). The equivalent circuits for these connections are shown in Fig. 8.21. By choosing the center capacitor properly so that its characteristic impedance together with the 0.3 nH inductors is similar to the characteristic impedance of the transmission line (240 fF for a 50Ω line), a substantial improvement in performance can be obtained.

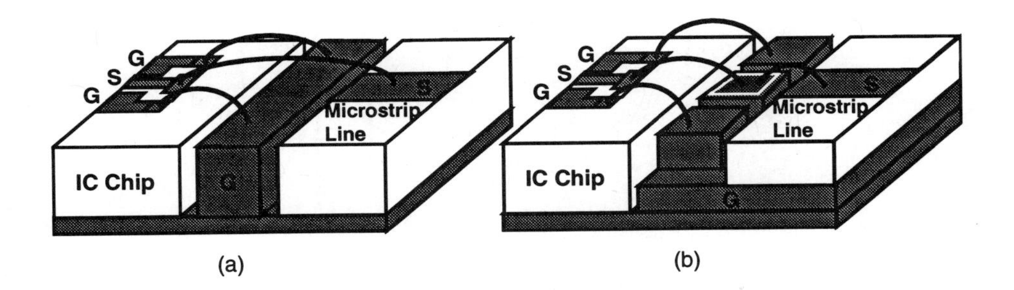

Figure 8.20 High-speed packaging technique.

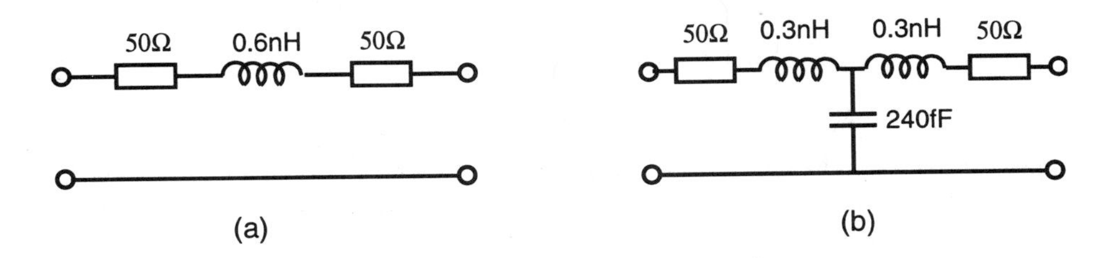

Figure 8.21 Circuit model for high-speed packaging technique.

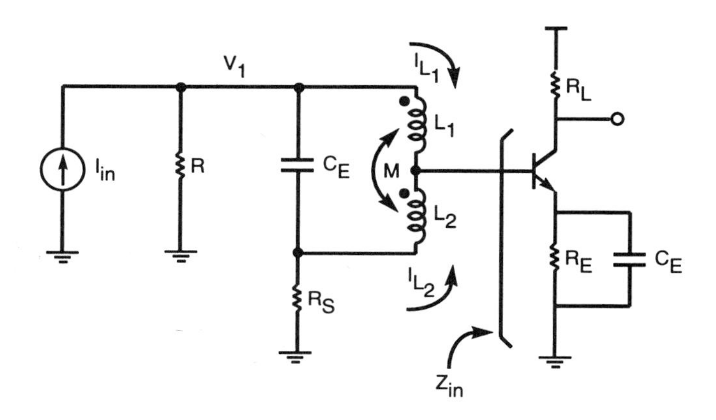

Figure 8.22 T-coil bonding configuration for constant input impedance.

Inductive Peaking and the T-Coil

In high-speed packaging we try to minimize parasitic capacitance and inductance. However, once we realize that we must live with a certain amount of inductance and capacitance, other techniques have been developed that use the reactance of these parasitics in an intelligent way so that circuit performance is not degraded, and in some instances is even enhanced. Distributed amplifiers are an example of this technique, where the input capacitance of the amplifier is utilized to construct a quasi delay-line. Inductive peaking [11, pp. 334–344] is another technique whereby bond-wire inductances are used to *peak* the frequency response, thus broad-banding the circuit.

A very clever technique for providing a constant input impedance over all frequencies is known as a T-coil. This circuit is illustrated in Fig. 8.22. The T-coil uses coupled inductors and a shunt capacitor. The energy is distributed among the inductors, shunt capacitor, and input capacitance of the transistor in such a way that the impedance seen by the transmission line is constant and real for all frequencies. A T-coil can be approximated be using two bond-wires. The mutual coupling between the bond-wires is rather weak, but it is strong enough to implement the broadband matching needed to maintain a relatively constant impedance. For additional information on high-speed packaging the reader is referred to the paper by Ellenberger [12].

8.5.5 Differential Design

Until now we have not explicitly stated that differential design is preferable at high speeds. This should be fairly self evident. However, we will make a few comments to point out some of the more prominent benefits. Within the chip, parasitics that become dominant at high speeds, and seriously degrade the performance of a single-ended circuit occur on both signal lines of a differential circuit, and are reduced by the common-mode rejection. Also, in a differential circuit, the switching of currents only occurs within the chip where distances, and thus inductance are small. Therefore, high-speed ac current don't need to be delivered through long, high-inductance paths. Differential I/O lines can be routed in close proximity to each other, and the total current passing through a surface intersecting both signal lines will be constant. Ferrite beads can be placed outside of the differential signal lines, choking common-mode signals, while allowing differential signals to pass. All external bias and supply lines carry dc current, where the sum of the differential currents are always constant. This minimizes fluctuation on these lines and simplifies power supply and bias decoupling.

8.5.6 Testing Procedures

Network Analyzer An excellent source of information on testing procedures can be obtained from manufacturers of high-speed test equipment. A network analyzer is a versatile measurement system that can perform several types of complex measurements quickly. Hewlett Packard's HP 8720A is one such system, and the user's guide provides useful information about testing procedures [13].

Spectrum Analyzer A gigahertz spectrum analyzer is invaluable in any high-speed measurement laboratory. One particular unit is the HP 8562A from Hewlett Packard [14]. Several useful measurement techniques can be found in the user's guide and in application notes.

Sampling Scope A sampling oscilloscope is useful for looking at periodic waveforms and performing time domain measurements. Time domain reflectometry (TDR) is complementary to frequency domain techniques, and is useful for narrowband network characterization. The 11801A from Tektronix is a sampling scope with a TDR sampling head. Applications notes and a user's guide are available from distributors [15].

Application Notes Suppliers of high-speed circuits are also a useful sources of information on high-speed testing and packaging. A seminar on high-speed design was given by Analog Devices in 1989, and the notes are published in a volume that is available through the company [16]. Manufactures of accessories such as bias-tees, couplers, cables, connectors, and microwave active and passive components usually have well written application notes describing the proper usage of these devices. Application notes, such as those given in [17, 18, 19] are also valuable sources of information on microwave design and testing techniques.

REFERENCES

[1] Alan B. Grebene. *Bipolar and MOS Analog Integrated Circuit Design*. Wiley, New York, 1984.

[2] Paul R. Gray and Robert G. Meyer. *Analysis and Design of Analog Integrated Circuits*. Wiley, New York, 1977.

[3] Robert S. Elliott. *An Introduction to Guided Waves and Microwave Circuits*. Prentice Hall, Englewood Cliffs, NJ, 1993.

[4] Henry W. Ott. *Noise Reduction Techniques in Electronic Systems*. Wiley, New York, second edition, 1988.

[5] H. B. Bakoglu. *Circuits, interconnections, and Packaging for VLSI*. Addison-Wesley, Reading, Massachusetts, 1990.

[6] Geoffrey Herrick and Keith E. Jones. Surface-mount pack houses GaAs MMICs. *Microwave and RF*, 25, June 1986.

[7] Keith E. Jones, Gary S. Barta, and Geoffrey C. Herrick. A 1 to 10 GHz tapered distributed amplifier in a hermetic surface mount package. In *IEEE GaAs IC Symposium*, pages 137–140, Monterey, California, November 1985.

[8] S. Fujita, T. Suzaki, A. Matsuoka, S. Miyazaki, T. Torikai, T. Nakata, and M. Shikada. High sensitivity 5 Gbit/s optical receiver module using Si IC and GaInAs APD. *Electron. Lett.*, 26(3):175–176, February 1990.

[9] Shuichi Fujita, Yuhki Imai, Yasuro Yamane, and Hiroshi Fushimi. DC-10GHz mixer and amplifier GaAs ICs for coherent optical heterodyne receiver. In *IEEE ISSCC Dig. Tech. Papers*, pages 122–123, San Francisco, California, February 1991.

[10] Shuichi Fujita, Yuhki Imai, Yasuro Yamane, and Hiroshi Fushimi. DC to 10-GHz mixer and amplifier GaAs IC's for coherent optical heterodyne receiver. *IEEE J. Solid-State Circuits*, 26(12):1847–1852, December 1991.

[11] Dennis L. Feucht. *Handbook of Analog Circuit Design*. Academic Press, Inc., Harcourt Brace Jovanovich, Publishers, San Diego, 1990.

[12] J. Ellenberger. Packaging faster silicon circuits. *Microwave and RF*, 27:121–124, August 1988.

[13] *HP 8720A Microwave Network Analyzer: Users Guide*. Hewlett Packard.

[14] *HP 8562A Spectrum Analyzer: Users Guide*. Hewlett Packard.

[15] *Tektronix 11801A Sampling Oscilloscope: Users Guide*. Tektronix.

[16] *High-Speed Design Seminar*, 1989. Notes compiled by The High-Speed Data Converter Group, Analog Devices, Greensburo, North Carolina. Published by Analog Devices, Inc., Available through ADI distributors.

[17] S-parameter design, April 1972. Hewlett Packard, Application Note 154.

[18] Richard W. Anderson. S-parameter techniques for faster, more accurate network design, 1967. Hewlett Packard, Application Note 95-1.

[19] Neal C. Silence. The Smith Chart and its usage in RF design. *RF Design*, pages 85–88, April 1992.

9

6-GHz PHASE-LOCK LOOP USING AlGaAs/GaAs HBTs

In this chapter, a fully integrated 6 GHz phase-locked-loop (PLL), fabricated using AlGaAs/GaAs heterojunction bipolar transistors (HBTs), is described [1]. The PLL is an important test circuit that verifies functionality of key circuit building-blocks of a multigigabit-per-second clock recovery circuit for fiber optic communication systems. The PLL consists of a frequency quadrupling ring voltage controlled oscillator (VCO), a balanced phase-detector, and a lag-lead loop filter. The closed-loop bandwidth is approximately 150 MHz. The tracking range was measured to be greater than 750 MHz at zero steady-state phase-error. The non-aided acquisition range is approximately 300 MHz, or twice the closed loop bandwidth. The minimum emitter-area of the AlGaAs/GaAs HBTs was 3μm x 10μm, and the devices exhibited a unity current-gain frequency of f_t = 22 GHz, and a unity power-gain frequency of f_{max} = 30 GHz for a bias current of 2 mA. The speed of the PLL can be doubled by using 1μm x 10μm emitters in next generation circuits. The chip occupies a die area of 2mm x 3mm and dissipates 800mW with a supply voltage of -8V. Each of the circuits composing the PLL will be described in the following sections.

9.1 FREQUENCY QUADRUPLING RING VCO

A frequency quadrupling ring VCO was designed and fabricated separately from the PLL [2, 3]. This VCO, illustrated in Fig. 9.1, has two quadrature outputs at twice the ring frequency, and one output at four times the ring frequency. The core of this VCO is a four-stage ring oscillator. When an even number (n) of matched delay elements is used, each pair of taps separated by $n/2$ stages will be 90 degrees out of phase. For example, y_1 and y_3 are quadrature pairs, as are y_2 and y_4. When each of these pairs are mixed, the resulting signals, I and Q, are at twice the ring frequency, and are

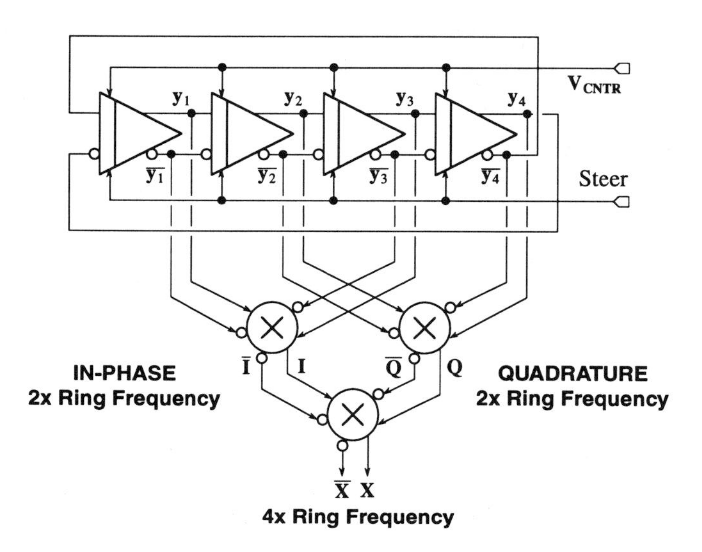

Figure 9.1 VCO with I and Q, in-phase and quadrature, double frequency outputs, and a quadrupled frequency output, X.

themselves in quadrature. Another level of frequency doubling can also be performed by mixing I and Q to obtain a signal, X, at four times the ring frequency.

The mixing arrangement can be implemented in a variety of ways. If the signals y are binary-valued, the multipliers are equivalent to exclusive-OR gates, and the cascade of multipliers can be represented as $(y_1 \oplus y_3) \oplus (y_2 \oplus y_4)$, which is logically equivalent to $y_1 \oplus y_2 \oplus y_3 \oplus y_4$. Therefore, for binary-valued signals, a modulo-two-sum of each tap will generate a signal at 4-times the ring oscillator frequency. Razavi and Sung [4, 5] used this approach in a 6-GHz BiCMOS PLL, which dissipated only 60-mW. The modulo-two-sum was accomplished using a novel technique, which is applicable to ring oscillators with an odd number of stages. For a three-stage oscillator there are 6 possible states for the taps (y_1, y_2, y_3); if these taps could be chosen arbitrarily there would be 8 possible states, however (-1,1,-1) and (1,-1,1) can not occur in a ring-oscillator structure. Therefore, the modulo sum $y_1 \oplus y_2 \oplus y_3$ is equivalent to the algebraic sum $y_1 + (-y_2) + y_3$ for the six valid states of the ring oscillator. By representing the taps y as currents, and summing them at a common node, a frequency tripler can be realized.

The delay cell of the VCO core is shown in Fig. 9.2. This circuit uses a differential current steering input (STEER) for coarse adjustment of the VCO frequency, and

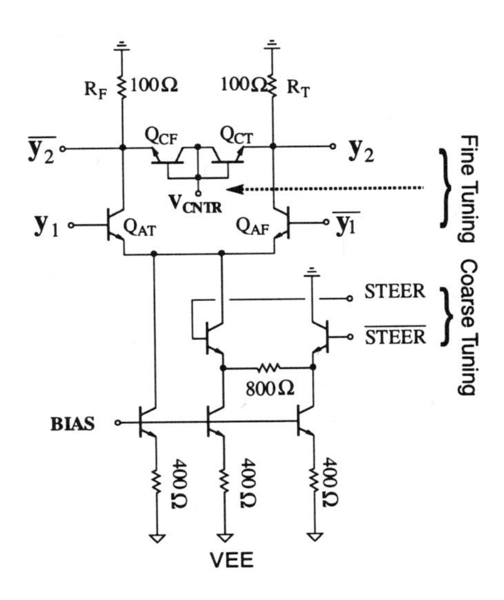

Figure 9.2 Ring oscillator delay cell with differential current steering inputs for coarse tuning and a reversed bias diode for fine tuning

a control voltage (V_{CNTR}) of reverse-biased base-emitter junction capacitances for frequency fine tuning. Balanced differential design helps to minimize jitter due to common-mode noise and especially due to power supply coupling, which is a major source of jitter in high-frequency oscillators. Simulation results reveal that a single delay cell, terminated with a source resistor of value r_b and a load resistor of $C_\pi / C_\mu g_m$, achieves a delay time of approximately $1/f_{\text{max}}$, where f_{max} is the unity-power gain frequency of the transistor given approximately by

$$f_{\text{max}} \simeq \frac{1}{2}\sqrt{\frac{f_t}{2\pi C_\mu r_b}}. \tag{9.1}$$

To ensure oscillations, a gain greater than unity is required, and the load resistor must be increased accordingly. This increases the delay time of the ring oscillator cell, as do the emitter-follower buffer stages inserted before the frequency doubling mixers, resulting in an actual delay time of the loaded ring oscillator cell of between $1.5/f_{\text{max}}$ and $2/f_{\text{max}}$, depending on bias conditions. Therefore, the ring frequency f_1 is such that

$$\frac{f_{\text{max}}}{16} < f_1 < \frac{f_{\text{max}}}{12}, \tag{9.2a}$$

Maximum Frequency	6.8 GHz
Power Dissipation	300–400 mW
Tuning Range (Steer)	6.25 GHz $\pm$ 400 MHz
Gain (Steer)	2π(440 MHz) / mA
Tuning Range (Vcntr)	$\pm$ 200 MHz
Gain (Vcntr)	2π(100 MHz) / Volt
Temperature Coefficient	1 MHz / $^\circ$ C (uncompensated)
Phase-Jitter	< 1 degree (rms)
Spectral Content	-100 dBc/Hz @ 100 kHz offset

Table 9.1 Measured results of the VCO.

and the 4x signal achieves a maximum frequency in the range

$$\frac{f\text{max}}{4} < f_4 < \frac{f\text{max}}{3}. \tag{9.2b}$$

Measured results of the VCO are summarized in Table 9.1. The maximum obtainable frequency is 6.8 GHz. The tuning range is plotted in Fig. 9.3(a) as a function of the bias current per delay cell, and in Fig. 9.3(b) as a function of the reversed biased diode voltage. The VCO can be tuned by approximately 1 GHz by altering the bias current, and by 500 MHz by modulating the load capacitance diode. A microphotograph of the VCO is shown in Fig. 9.4.

9.2 FULLY-BALANCED MIXER

Frequency doubling and phase detection are performed by a fully symmetric circuit with the property of equal delay paths for each input signal [6]. Half of this circuit is a Gilbert multiplier, or equivalently, a current-mode exclusive-NOR gate as shown in Fig. 9.5. When a single Gilbert multiplier is used as the complete mixer, differences in signal propagation delays between the top-level and bottom-level input differential pairs results in an effective phase-shift between the two signals being multiplied. This causes a steady-state phase-error when the multiplier is used as a phase-detector in a PLL, reducing both the tracking and acquisition ranges. This phase lag also gives rise to a dc offset voltage at the output of a frequency doubler when quadrature signals

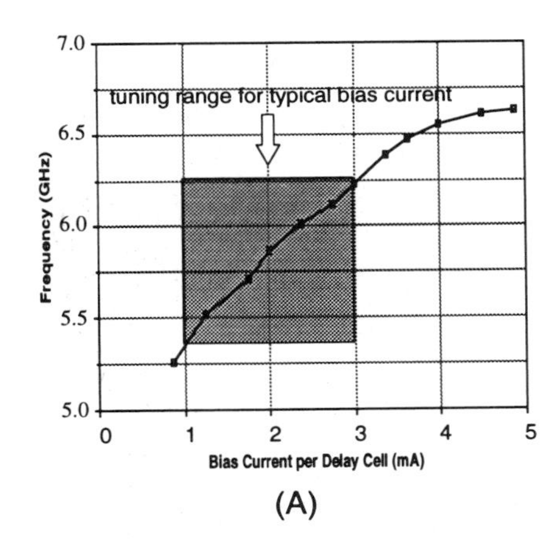

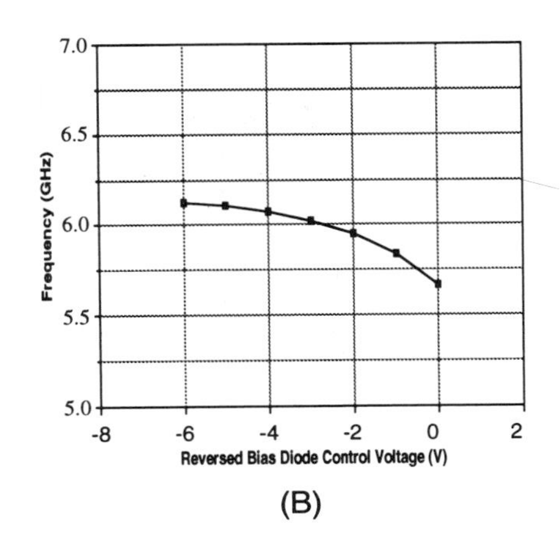

Figure 9.3 VCO measured results: (A) frequency vs. bias current, (B) frequency vs. control voltage.

are multiplied. For this particular HBT process, the delay-time difference between a signal applied to the top differential pair and a signal applied to the bottom, is on the order of 15 ps. This corresponds to a phase-lag of 32 degrees at 6 GHz, which is unacceptable.

By modeling the Gilbert multiplier of Fig. 9.5 as an ideal multiplier with an input phase difference, the circuit of Fig. 9.6 illustrates how two such mixers can be used in antiparallel to cancel the phase offset. Each mixer is identical, but their inputs are interchanged. Therefore the resulting phase-errors produced by the two mixers will be equal in magnitude, but opposite in sign. Summing the result of each mixer, the phase-error can be eliminated to the degree of matching accuracy of the two mixers. The fully symmetric circuit of Fig. 9.7 implements this phase-error compensation by summing the output current of the two Gilbert multipliers at the load resistor. Razavi and Sung [4, 5] use a similar technique, but they add a clever modification to allow the use of low-voltage power supplies, thereby reducing the power dissipation substantially.

9.3 LOOP FILTER

The loop filter sets the PLL's closed-loop bandwidth as well as its dynamic response. Considerations in designing a loop filter are stability, frequency acquisition range, and

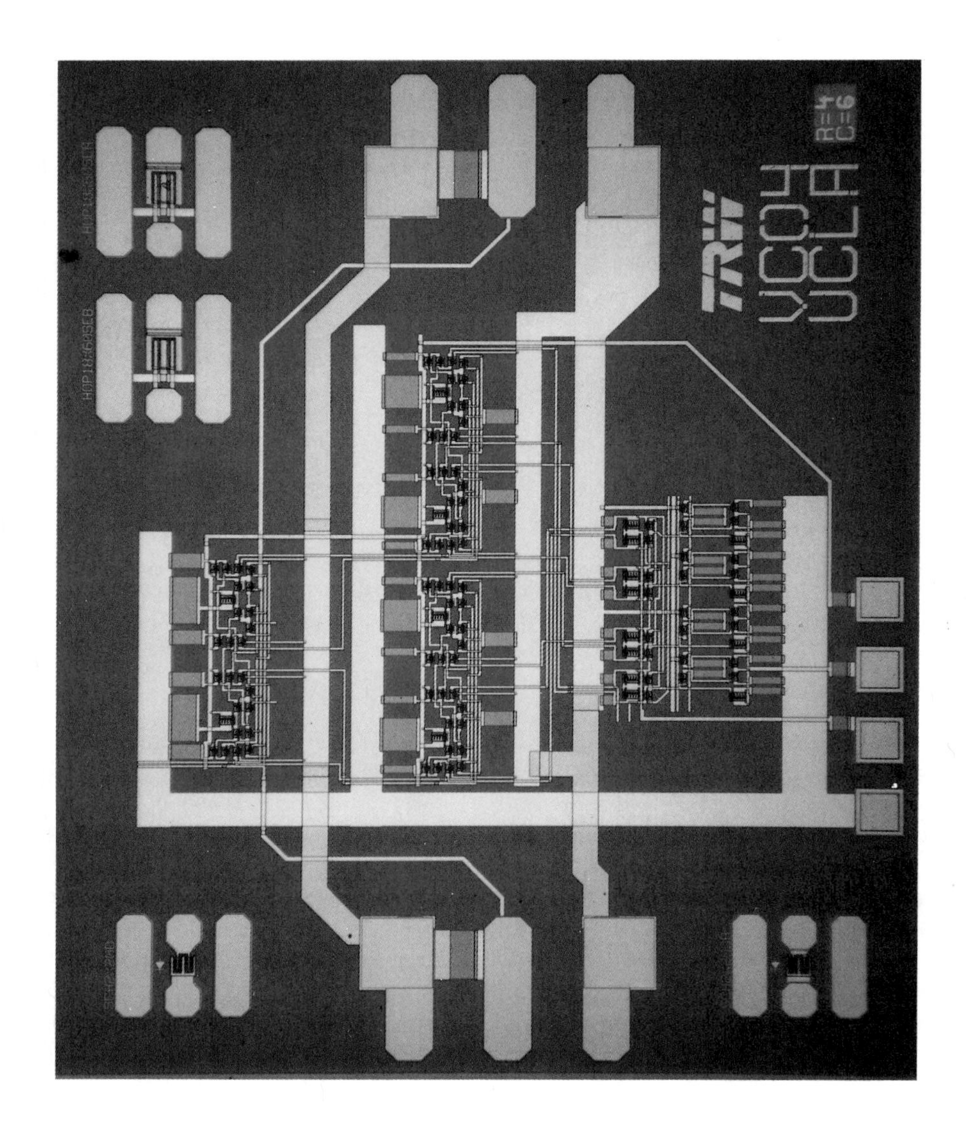

Figure 9.4 Microphotograph of Frequency Quadrupling VCO

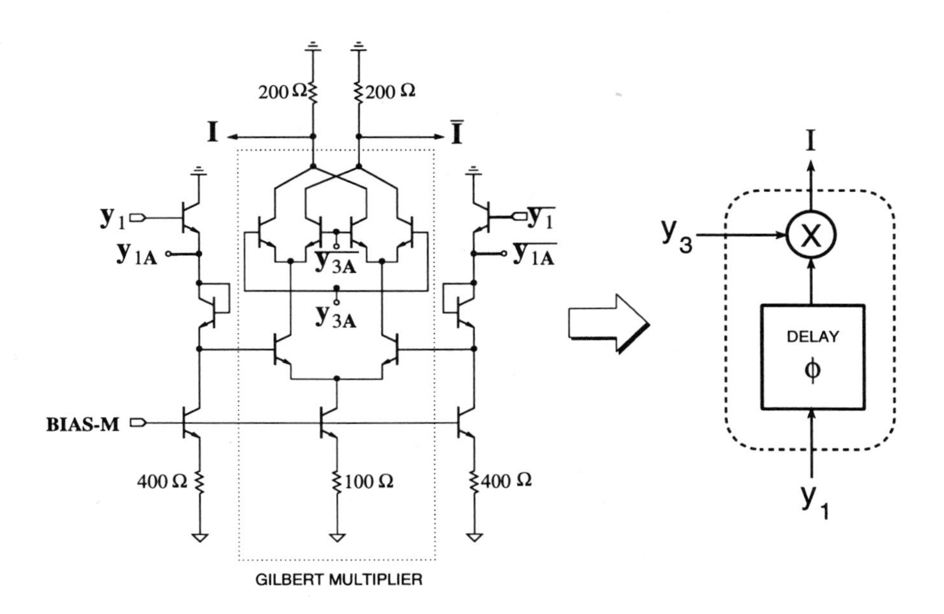

Figure 9.5 Gilbert multiplier

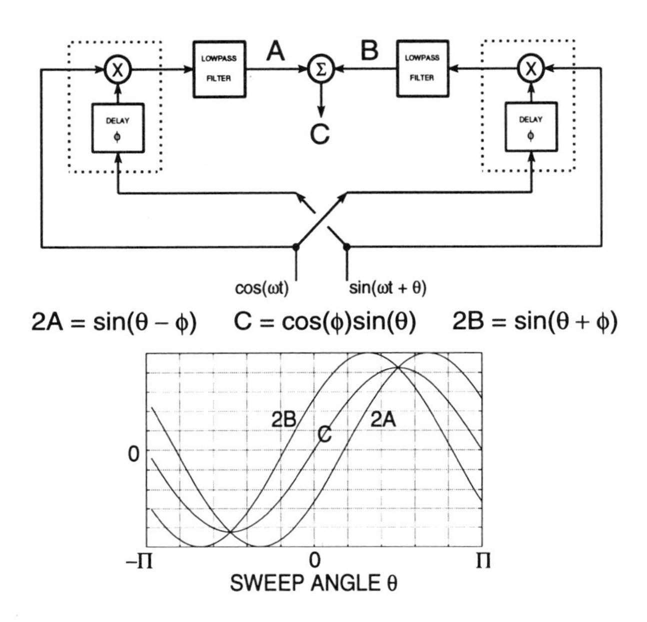

Figure 9.6 Technique for compensating phase-lag using two matched Gilbert multipliers.

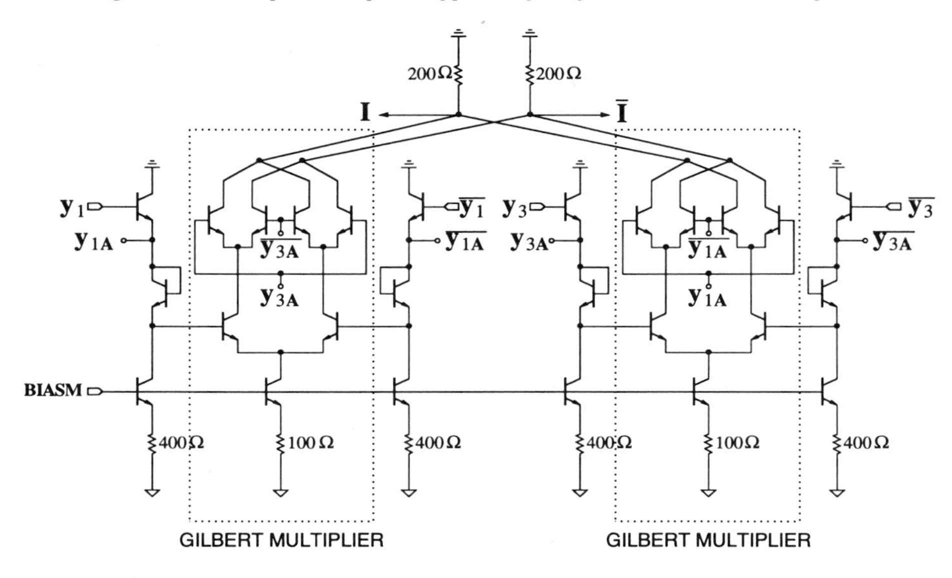

Figure 9.7 Fully-balanced mixer using two Gilbert multipliers in parallel.

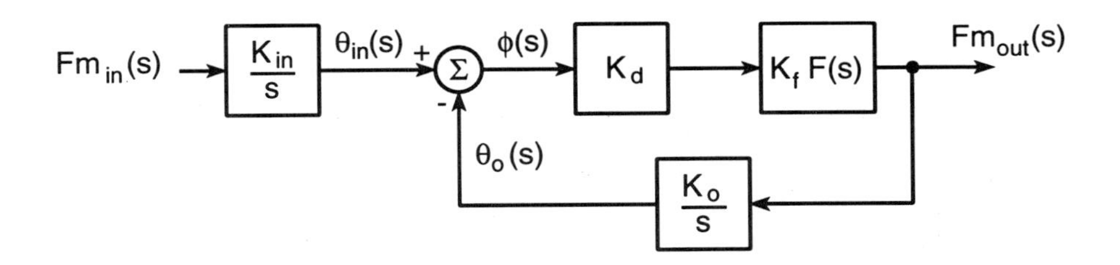

Figure 9.8 Linearized Small-Phase-Error Model of PLL

phase-jitter suppression. The familiar linearized small-phase-error model of a PLL is shown in Fig. 9.8, where $F(s) = \frac{F_N(s)}{F_D(s)}$ is the transfer function of a loop filter with unity dc-gain. K_d, K_f, and K_o are the gains of the phase-detector, loop-filter, and VCO respectively. A frequency-modulation input signal is also shown with a gain of K_{in}. The closed loop transfer function of the PLL for a general loop filter is given by

$$H_\theta(s) = \frac{\theta_o(s)}{\theta_{in}(s)} = \frac{\dfrac{K_d K_f K_o F(s)}{s}}{1 + \dfrac{K_d K_o K_f F(s)}{s}} = \frac{K_d K_f K_o F_N(s)}{s F_D(s) + K_d K_f K_o F_N(s)}. \tag{9.3}$$

Defining a gain $\Omega_k \triangleq K_d K_f K_o$ (rad/s), then for a lag-lead loop filter of the form

$$F(s) = \frac{1 + s\tau_z}{1 + s\tau_p}, \tag{9.4}$$

the resulting closed loop transfer function is 2nd order, and is given by

$$H_\theta(s) = \frac{\Omega_k(1 + s\tau_z)}{s^2\tau_p + s(1 + \Omega_k\tau_z) + \Omega_k}. \tag{9.5}$$

It is useful to express the loop parameters in terms of the undamped natural frequency, $\omega_n = 2\pi f_n$ and the damping ratio ζ.

$$H_\theta(s) = \frac{1 + \dfrac{s}{\omega_n}\left(2\zeta - \dfrac{\omega_n}{\Omega_k}\right)}{1 + \dfrac{s}{\omega_n}2\zeta + \left(\dfrac{s}{\omega_n}\right)^2} \tag{9.6}$$

where

$$\omega_n^2 = \frac{\Omega_k}{\tau_p}$$

$$\zeta = \frac{1}{2}\left[\frac{\omega_n}{\Omega_k} + \omega_n\tau_z\right].$$

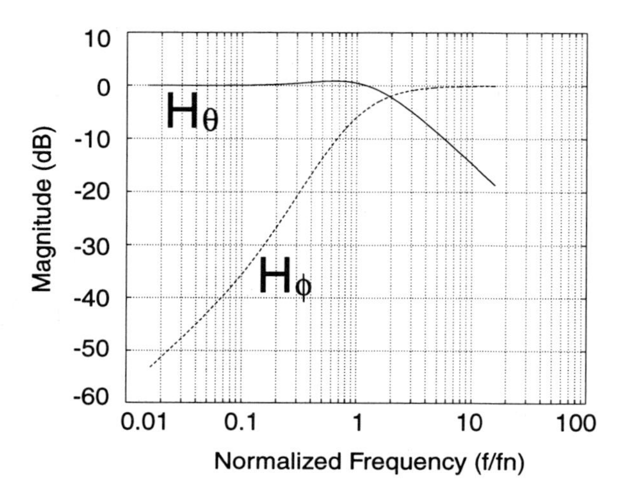

Figure 9.9 Magnitude response of PLL closed-loop transfer functions H_θ, and H_ϕ.

The transfer function for frequency modulated signals is identical to the phase-modulation transfer function except for a constant term:

$$H_{F_m}(s) = \frac{Fm_{\text{out}}(s)}{Fm_{\text{in}}(s)} = \frac{K_{in}}{K_o} H_\theta(s). \tag{9.7}$$

Another important transfer function relates the phase-error, $\phi(s)$ to the input phase.

$$H_\phi(s) = \frac{\phi(s)}{\theta_{in}(s)} = \frac{\frac{s}{\omega_n}\left(\frac{\omega_n}{\Omega_k}\right) + \left(\frac{s}{\omega_n}\right)^2}{1 + \frac{s}{\omega_n}(2\zeta) + \left(\frac{s}{\omega_n}\right)^2} \tag{9.8}$$

The magnitudes of $H_\theta(s)$ and $H_\phi(s)$ are plotted in Fig. 9.9 as a function of the normalized frequency variable for the case of $\zeta = 1$ and $\Omega_k >> \omega_n$. The loop filter has a limited bandwidth so that the PLL attenuates modulations of the carrier frequency above the undamped natural frequency of the loop f_n. The two transfer functions H_θ and H_ϕ have interesting interpretations as regards to phase-jitter filtering. If we assume that the input to the PLL contains phase-jitter, but the VCO of the PLL is jitter-free,

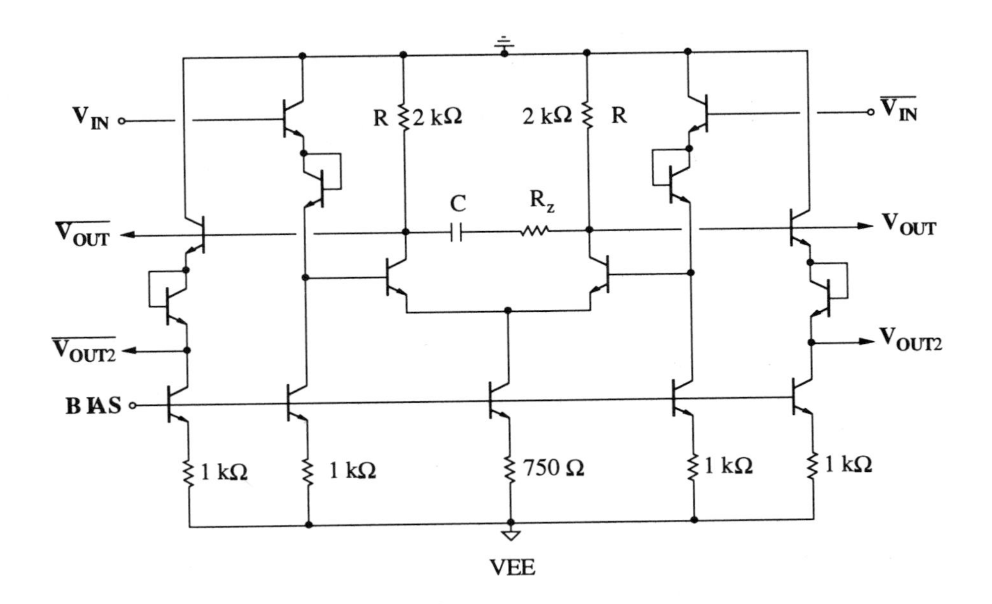

Figure 9.10 Differential Lag-Lead Loop Filter

then the VCO output will be modulated by the input phase-jitter. However, the original jitter will be filtered by the lowpass function H_θ. Therefore, to reduce the jitter of the PLL VCO one should reduce the PLLs closed-loop bandwidth. Conversely, if the input signal is assumed to be jitter-free, and the PLL VCO has significant free-running phase-jitter, then the negative feedback of the loop will act to modulate the VCO in such a way as to cancel its own phase-jitter. The PLL will be able to track and suppress self-jitter within the loop bandwidth. The resulting closed-loop VCO jitter will then be the original jitter filtered by the highpass function of H_ϕ. In this case jitter is reduced by increasing the loop bandwidth.

The circuit of Fig. 9.10 approximates a lag-lead characteristic. The small-signal transfer function for this filter, ignoring higher-order poles due to parasitics, is given approximately by

$$\frac{\Delta V_{OUT}}{\Delta V_{IN}} = g_m R \left[\frac{1 + sCR_z}{1 + sC\,(R_z + 2R)}\right] = g_m R \left[\frac{1 + s\tau_z}{1 + s\tau_p}\right] \tag{9.9}$$

where $$\tau_z = R_z C = \frac{2\zeta}{\omega_n} - \frac{1}{\Omega_k},$$

Loop Parameters			Component Values		
K_d	=	69 (mV/rad)	τ_p	=	5.5 (ns)
K_f	=	25	τ_z	=	1.5 (ns)
K_O	=	$2\pi 800$ (Mrad/s/V)	C	=	1.0 (pF)
Ω_k	=	8685 (Mrad/s)	R	=	2.0 (kΩ)
			R_z	=	1.5 (kΩ)

Table 9.2 Loop parameters and component values.

$$\tau_p = 2RC + \tau_z = \frac{\Omega_k}{{\omega_n}^2}.$$

Since the loop filter is integrated with the PLL, the maximum capacitor value is limited by area constraints to about 1pF. The loop parameters and corresponding filter component values are given in Table 9.2 for the design goals of f_n = 200 MHz and ζ = 1. The parasitic poles of the loop filter provide additional lowpass filtering of the 12 GHz double frequency ripple from the output of the phase detector, reducing ripple-induced phase jitter. However, these higher-order poles also add excess phase-lag which reduces the loop phase margin, and possibly cause ringing in the transient response. Simulations predict an overshoot in the step response of 5%, corresponding to an equivalent damping factor of ζ = 0.7, which is approximately a 2-pole Butterworth response.

9.4 OUTPUT BUFFER AND BIAS CIRCUITS

The output buffer is shown in Fig. 9.11. It consists of a pair of emitter-follower buffers, followed by a degenerated differential pair with 50 Ω on-chip load resistors. The nominal bias current is approximately 11 mA, which results in a maximum differential output voltage swing of 550 mV. Since the maximum anticipated differential input signal to the buffer is 2 V, a 300 Ω emitter degeneration resistor is used to accommodate a differential input signal of up to 3 V.

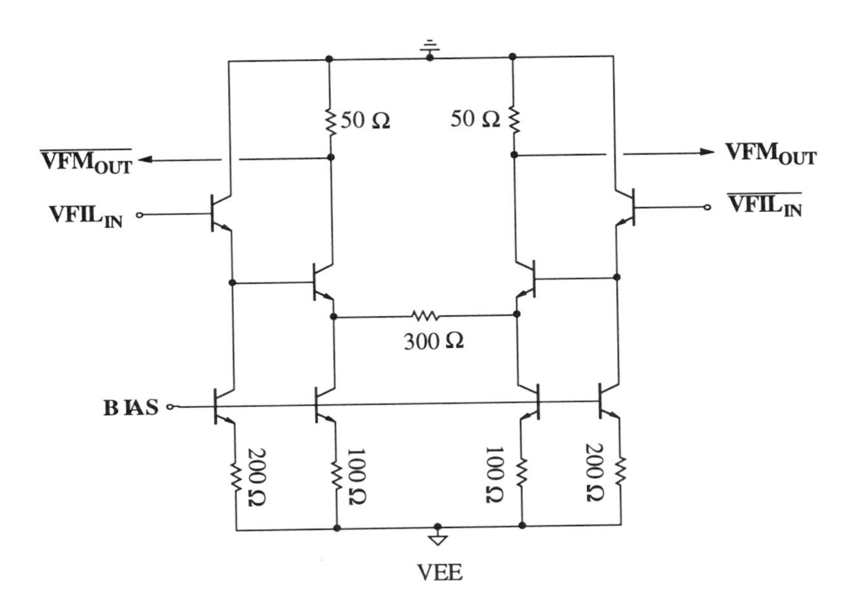

Figure 9.11 Emitter Degenerated Output Buffer with 50 Ω On-Chip Load Resistors

Bias Circuits

Three identical bias circuits are used, one of which is illustrated by Fig. 9.12. Separate circuits bias the mixers, the VCO-core delay cells, and the output stage. These bias circuits provide a nominal bias voltage of V_{be} + 550 mV when V_{FORCE} is open circuited, but can be altered from V_{be} + 400 mV to V_{be} + 2.5 V if V_{FORCE} varies from VEE to GND.

9.5 RESULTS

A block diagram of the PLL circuit is shown in Fig. 9.13. A microphotograph of the complete PLL is shown in Fig. 9.14. To facilitate testing, an identical VCO was fabricated to provide an on-chip signal source. Testing of the chip was accomplished by frequency modulating the input VCO (STEER) signal, and monitoring the buffered control voltage, (FM_{OUT}), of the PLL VCO. These measurements were repeated for different values of V_{CNTR}, which adds stress to the loop by creating an initial frequency offset. The tracking range was measured by starting with the PLL in lock, and slowly changing the FM input voltage until a loss of lock occurred. The acquisition

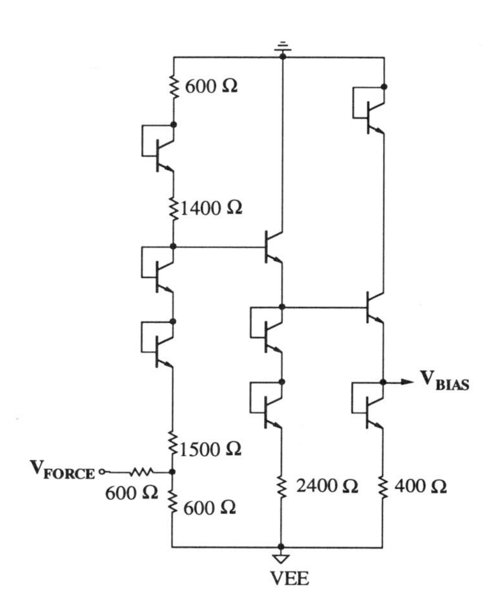

Figure 9.12 Bias circuit

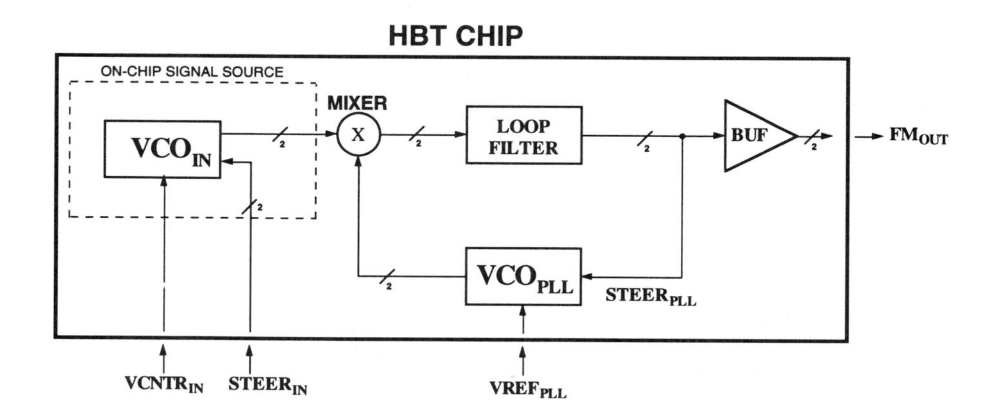

Figure 9.13 Block diagram of 6 GHz HBT phase-locked-loop.

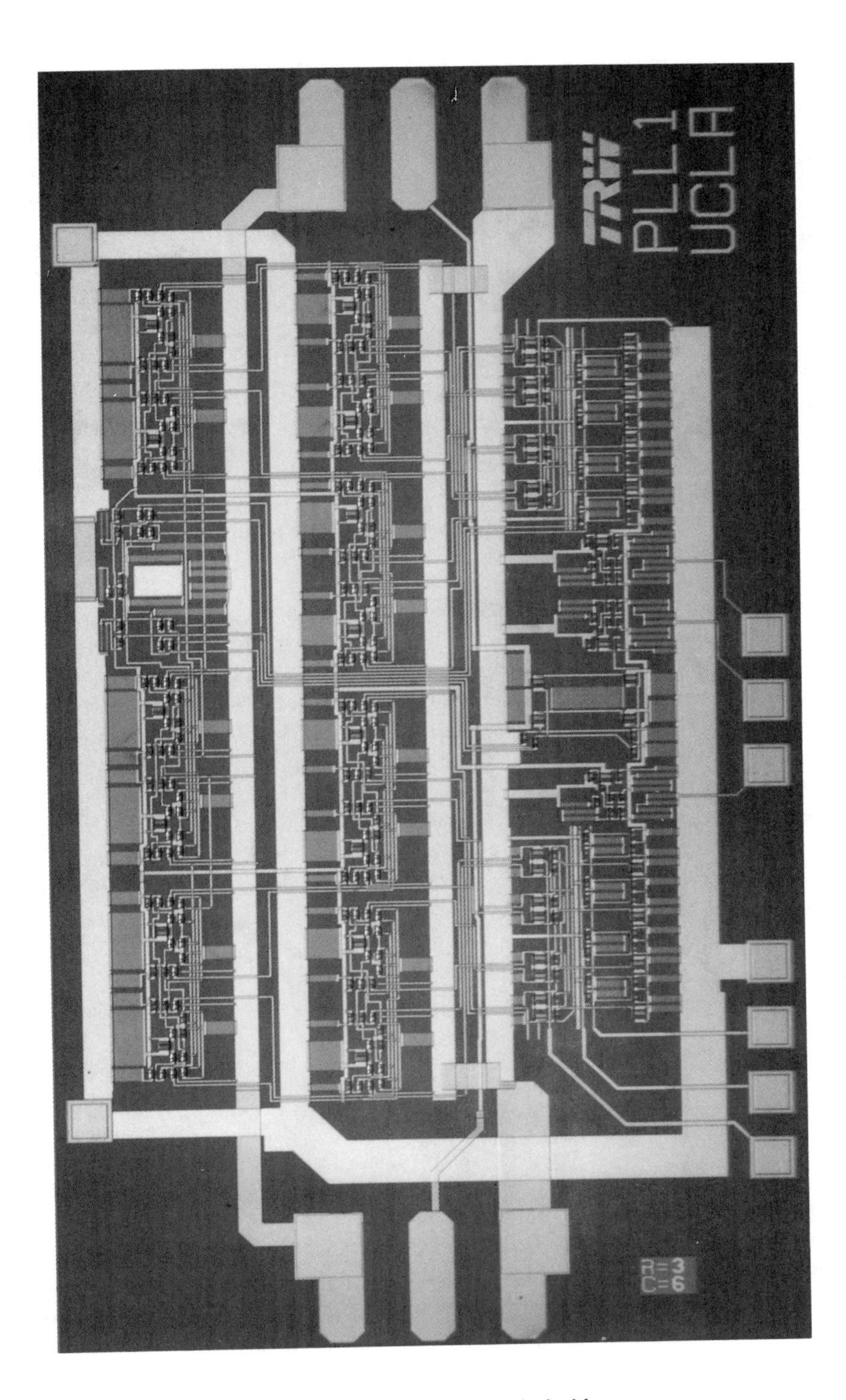

Figure 9.14 Microphotograph of 6 GHz HBT phase-locked-loop.

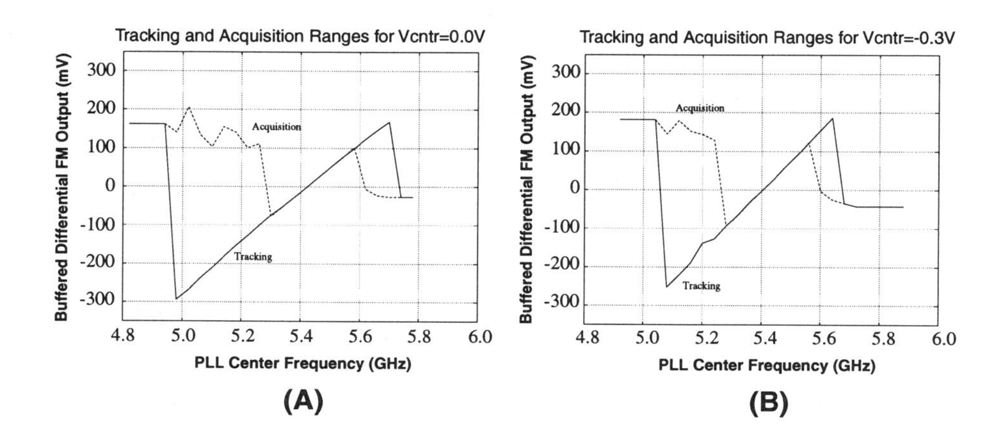

Figure 9.15 DC tracking and acquisition ranges for V_{CNTR} = 0.0 V and -0.3 V.

range was measured by starting with the loop out of lock and varying the FM input until lock was established. The tracking and acquisition ranges are plotted in Fig. 9.15(a) for V_{CNTR} = 0.0 Volts. The tracking range for this condition is 750 MHz, and the acquisition range is approximately 300 MHz. Fig. 9.15(b) shows a plot of the same ranges for V_{CNTR} = -0.3 V, which adds a frequency offset and therefore, a steady-state phase-error to the loop. In this case the tracking range is reduced to about 550 MHz, while the acquisition range is slightly less than 300 MHz. Fig. 9.16(a) shows a measured FM output waveform of the loop dynamically losing and regaining lock in response to modulation of V_{CNTR} by a 2.4 V peak-to-peak sinewave at 1 KHz. Gardner gives expressions for the maximum frequency deviation from the VCO center, Δf_p, that can be "pulled-in" by the self-acquisition of the loop [7]. Expressed in terms of circuit parameters,

$$|\Delta f_p| \simeq \frac{\Omega_k}{2\pi}\sqrt{2F(0)F(\infty)} = \frac{\Omega_k}{2\pi}\sqrt{2\tau_z/\tau_p}, \tag{9.10}$$

and in terms of loop parameters

$$|\Delta f_p| \simeq 2f_n\sqrt{\frac{\zeta\Omega_k}{\omega_n} - 1/2}. \tag{9.11}$$

For the loop parameters given in Table 9.2, $|\Delta f_p| \simeq 1.02$ GHz. Although, (9.10) takes into account the sinusoidal phase-detector characteristic, it assumes that Ω_k is constant over the entire acquisition range. In this particular circuit, Ω_k results from a cascade of two differential pairs (the loop-filter and the current steering VCO), and therefore has the functional form of a double-nested hyperbolic tangent, which reduces the gain

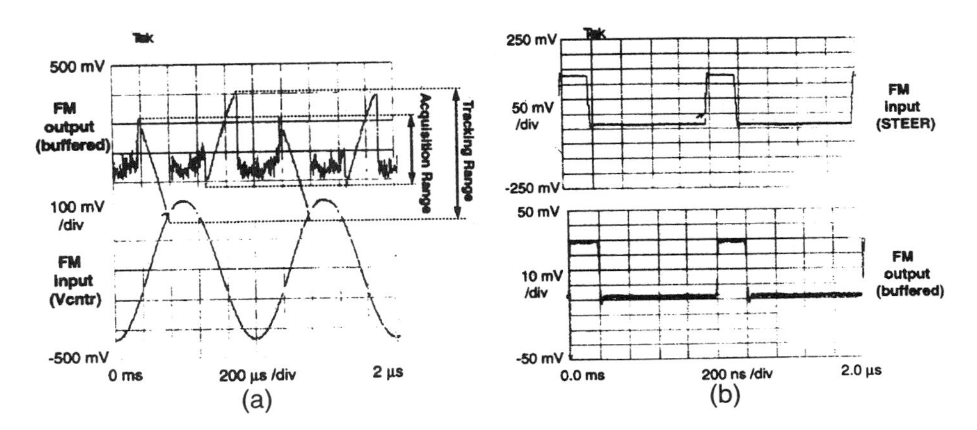

Figure 9.16 (a) Measured FM output showing PLL dynamically losing and reacquiring lock in response to a 2.4 Vpp 1 KHz sinewave modulation of V_{cntr}. (b) Frequency step input (STEER) and buffered PLL VCO input signal (FM).

substantially at the extremes of the tuning range. For an interval of 90% of the tuning range, the average gain, $\overline{\Omega_k}$ is a factor of 4 less than Ω_k at the center frequency of the VCO. Replacing Ω_k in (9.10) with $\overline{\Omega_k}$ gives an acquisition range of $\pm$ 250 MHz, which is still significantly greater than the measured acquisition range ($|\Delta f_a| \simeq$ 150 MHz). This discrepancy is due to offsets and noise in the actual circuit. In the presence of a large frequency error the dc value from the phase detector error signal is quite small, and must be accumulated in the loop filter over several cycles, building up a voltage that tunes the VCO. Such a small error signal is defeated by offsets and noise, and no tuning signal accumulates; as a result the PLL can not acquire.

The time required to "pull-in" a frequency of Δf is given by

$$T_p(\Delta f) \simeq \frac{1}{2\pi f_n}\frac{1}{2\zeta}\left(\frac{\Delta f}{f_n}\right)^2, \tag{9.12}$$

which shows that the acquisition time is proportional to the square of the initial frequency offset. For a frequency error equal to the theoretical limit of the acquisition range, $\Delta f = \Delta f_p$, and after substituting for f_n and ζ,

$$T_p(\Delta f_p) \simeq 2\tau_p\left[\frac{\Omega_k \tau_z}{1+\Omega_k \tau_z}\right]. \tag{9.13}$$

For the usual case of $\Omega_k \tau_z >> 1$,

$$T_p(\Delta f) \simeq 2\tau_p\left(\frac{\Delta f}{\Delta f_p}\right)^2. \tag{9.14}$$

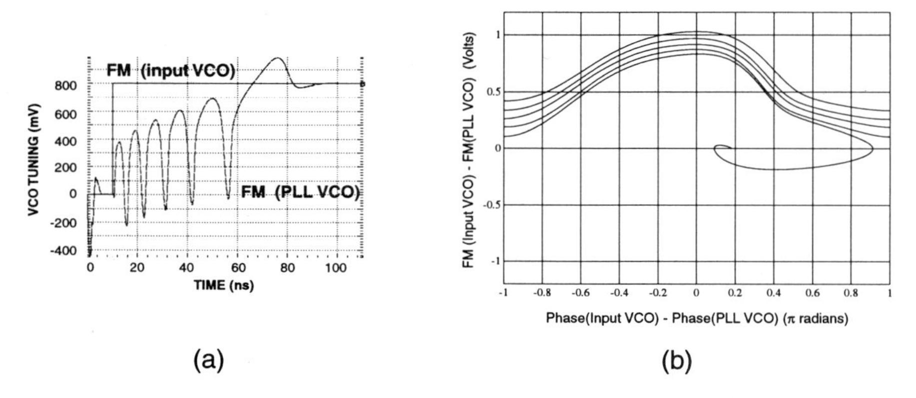

Figure 9.17 Cycle-slipping behavior during frequency acquisition of the PLL simulated using SPICE: (a) time domain behavior, (b) phase-plane portrait.

This expression shows that the acquisition time depends only on the initial frequency error and the time-constant of the dominant-pole of the loop filter. For $\Delta f = 150$ MHz, $T_p = 0.25$ ns. However, (9.14) is not valid for small frequency errors lying within the locking range of the PLL, or for frequencies close to the edge of the acquisition range as Fig. 9.17(a) illustrates. This plot shows the simulation results of frequency acquisition for Δf slightly less than Δf_p. For this case the acquisition time is 60 ns, which is more than a factor of 10 greater than that predicted by (9.14). The phase-plane portrait for this simulation is shown in Fig. 9.17(b), where it can be seen that the loop settles to a steady-state phase offset of 32 degrees which is an artifact of the finite dc gain Ω_k.

$$\theta_{\text{steady-state}} = \frac{2\pi\Delta f}{\Omega_k} \tag{9.15}$$

The linear tracking behavior and noise bandwidth can be determined by using small-signal modulations around the locking point. The measured closed-loop bandwidth varied from 100 MHz to 200 MHz, depending on the steady-state phase error, with ζ ranging from 0.5–1.0. The change in closed-loop bandwidth is due to the compression nonlinearities mentioned previously. Loop gain is reduced in the presence of a steady-state phase-error by the sinusoidal phase-detector, the differential loop filter, and the current steering VCO control. In addition, there is some amplitude modulation of the VCO with frequency which also reduces the loop gain. Fig. 9.16(b) shows the PLLs pulse response for a 175 mV, 200 ns pulse to the positive current-steering FM input.

Transistor Count	300
Die Area	2mm × 3mm
Supply Voltage	-8 V
Power Dissipation	800 mW
Maximum Center Frequency	6.8 GHz
Closed Loop Bandwidth	100–200 MHz
Effective Selectivity Q_{eq}	17–34
Tracking Range	700 MHz
Acquisition Range	300 MHz
Acquisition Time (Δf = 150 MHz)	0.25 ns*

Table 9.3 Summary of measured PLL results, (* Simulated).

9.6 SUMMARY

The measured results of the PLL are summarized in Table 9.3. A fully-integrated PLL has been fabricated using AlGaAs/GaAs HBTs. The chip contains over 300 transistors. A doubling of the speed of this PLL can be obtained in second generation circuits by substituting 1μm x 10μm devices for the 3μm x 10μm minimum emitter-area transistors used. This PLL is a fundamental building block for multigigabit-per-second clock recovery circuits for use in fiber-optic communication systems.

REFERENCES

[1] Aaron W. Buchwald, Kenneth W. Martin, Aaron K. Oki, and Kevin W. Kobayashi. A 6GHz integrated phase-locked loop using AlGaAs/GaAs heterojunction bipolar transistors. *IEEE J. Solid-State Circuits*, 27(12):1752–1762, December 1992.

[2] Aaron W. Buchwald and Kenneth W. Martin. A high-speed voltage-controlled oscillator with quadrature outputs. *Electron. Lett.*, 27(4):309–310, February 1991.

[3] Kenneth W. Martin and Aaron W. Buchwald. Differential-logic ring oscillator with quadrature outputs, U.S. pat. no. 5,180,994, January 1993.

[4] Behzad Razavi and James Sung. A 6GHz 60mW BiCMOS phase-locked loop with 2V supply. In *ISSCC Dig. Tech. Papers*, pages 114–115, San Francisco, California, February 1994.

[5] Behzad Razavi and JanMye James Sung. A 6GHz 60mW BiCMOS phase-locked loop with 2V supply. *IEEE J. Solid-State Circuits*, 29(12), February 1994.

[6] L. Schmidt and Hans-Martin Rein. New high-speed bipolar XOR gate with absolutely symmetrical circuit configuration. *Electron. Lett.*, 26:430–431, 1990.

[7] F. Gardner. *Phaselock Techniques*, chapter 5. Wiley, New York, second edition, 1979.

10

CLOCK RECOVERY AND DATA RETIMING IC: CIRCUIT DESIGN AND SIMULATION RESULTS

In this final chapter we return to system-level issues and blend them with circuit design constraints to produce a clock-recovery and data retiming IC. We will not present a detailed circuit, but rather outline the design procedure and give preliminary simulation results, both at the system- and transistor-level. As was discussed at the end of chapter 5, characterization of clock recovery circuits by simulation is difficult for two primary reasons. First, the input signal consists of random data plus noise; therefore, typical performance measures are based on statistical techniques, which require several data samples. Second, the clock recovery circuit is narrow-band compared to the data-rate, requiring thousands, or even millions, of bit-periods to be observed before the clock phase is altered. Nonetheless, simulation can predict the maximum speed of operation and is useful in optimizing the circuits dynamic response. Several aspects of a clock recovery system have been simulated, and some of the results will be presented in this chapter. We will first present system-level simulations, which are used to evaluate various architectures under ideal conditions. Then we will show how these architectures can be implemented as ICs and give preliminary circuit simulation results.[1]

10.1 SYSTEM-LEVEL SIMULATIONS

The critical aspect of a high-speed clock recovery loop is its insensitivity to parasitics. This was discussed in chapters 4 and 5. The performance of a given architecture

[1] Readers should be advised that the simulations presented in this chapter ignore electro-magnetic coupling of adjacent circuits and other couplings through power supplies and bias lines. Such coupling can cause several adverse effects, such as injection locking of the PLL. Although simulations are useful for fine-tuning and evaluating circuit performance, many parasitic effects are difficult to model and are masked by simulation. Therefore, in high-speed analog design there is no substitute for building and testing actual ICs.

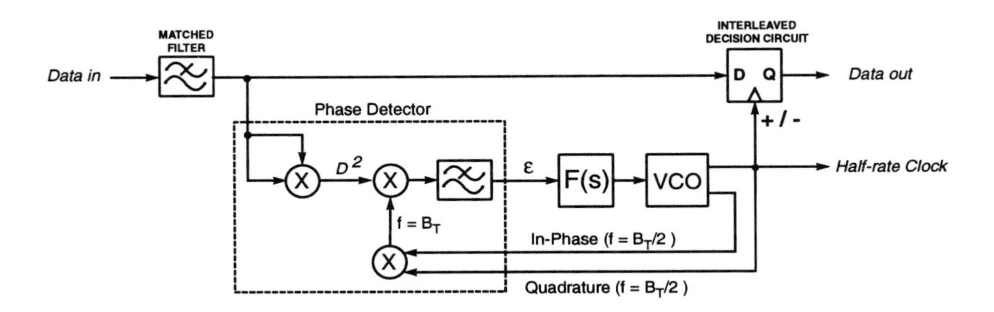

Figure 10.1 Block diagram of a clock recovery PLL using an LPF and squaring for edge detection of the random data.The recovered clock is at half the data rate.

will be highly dependent upon the circuit realization. However, we can first consider an idealized case where we ignore the parasitics. We can then perform system-level simulations to observe overall loop behavior without getting bogged down in all of the second-order effects. This section will present some system-level simulations that were performed assuming idealized circuit blocks. These simulations were executed using MATLAB.

10.1.1 Squaring Loop

We will first consider a simple architecture as shown in Fig. 10.1. This circuit low-pass filters the random NRZ data and then squares it to produce pulses for each data transition. The phase difference between the data transitions and the recovered clock is detected with a multiplier and a low-pass filter. The loop filter utilizes a lag-lead structure; the resulting second order system was designed to have a damping ratio of $\zeta = 1$. The closed-loop bandwidth was purposely chosen to be much greater than what would actually be used. This was done so that the dynamic behavior of the loop can be observed without having to run the simulations for an extended period of time.

Results for a Periodic Input with No Noise

Simulations with maximum data (a square-wave at a frequency of $B_T/2$) and no additive noise are shown in Fig. 10.2. The phase-error, frequency error, and clock waveforms are given in Figs. 10.2 (a), (b), and (c), respectively. The phase-plane portrait, which shows the phase-error plotted as a function of the frequency error, is shown in Fig. 10.2(d). This simulation primarily illustrates linear behavior. The

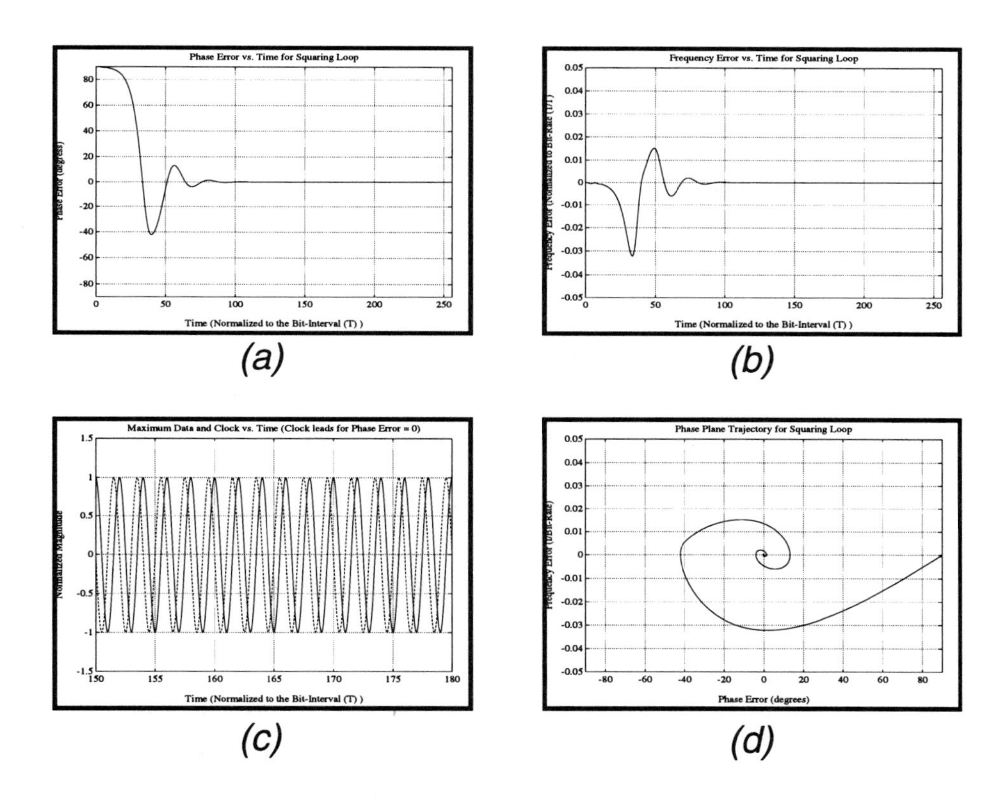

Figure 10.2 Simulation of a squaring clock recovery circuit for maximum data with no noise and with no frequency error: (a) phase-error, (b) frequency error, (c) clock waveform, (d) phase-plane portrait.

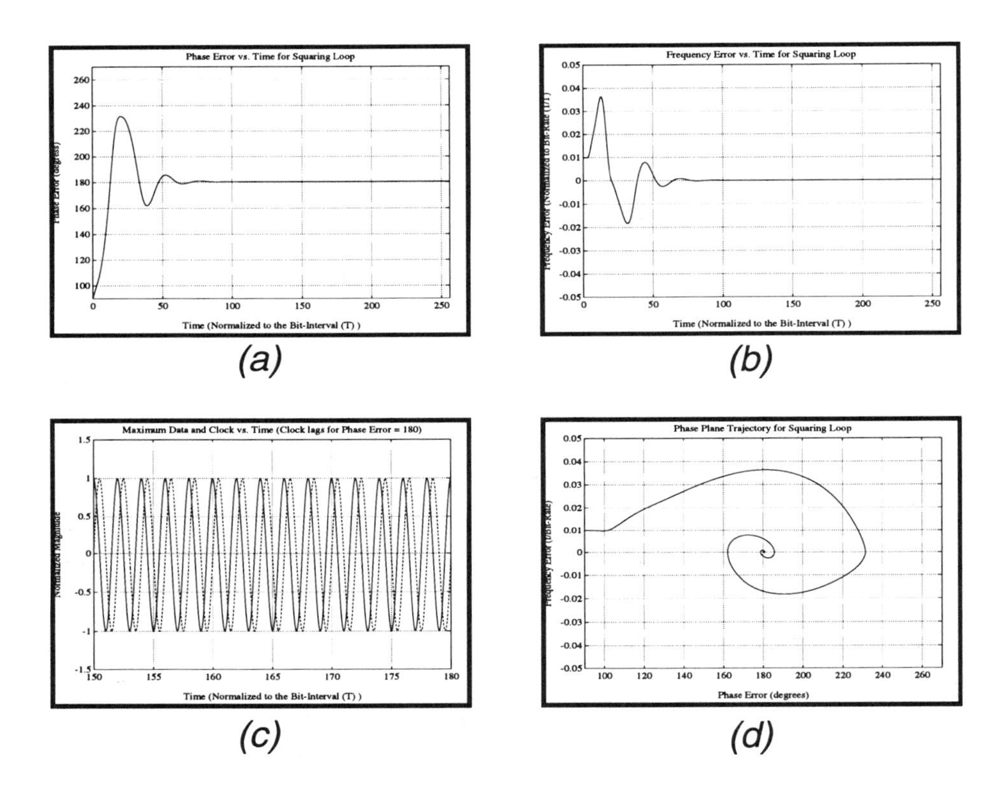

Figure 10.3 Simulation of a squaring clock recovery circuit for maximum data with no noise and with no frequency error: (a) phase-error, (b) frequency error, (c) clock waveform, (d) phase-plane portrait.

loop acquires phase lock without a cycle-slip, and the phase-plane trajectory reaches a steady-state at the origin. Similar results are shown in Fig. 10.3. The difference between this simulation and the former is that the initial phase was shifted by 180°.

When the data-rate and the center frequency of the VCO are not identical, a steady-state phase error will result. This stresses the loop, and it, in turn, reduces the acquisition and tracking ranges. The phase-error also results in a timing error in the recovered clock. This reduces the SNR of the sampling point and degrades performance. Fig. 10.4 shows the result of a simulation with a 1% frequency error. The VCO tuning voltage must differ from zero to match the data-rate, and the resulting phase error is just the tuning voltage divided by the product of the phase detector and loop filter gains.

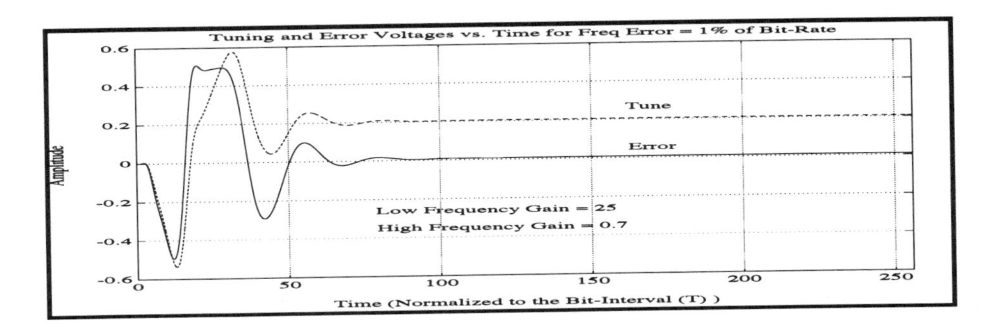

Figure 10.4 Simulation of a squaring loop with maximum data, no noise, and an initial 1% frequency error.

Simulations with Random Data and Noise

Now that we have observed the general dynamic behavior of the loop, we can randomize the input signal and add noise to see how this affects the loop behavior. Simulations of the squaring clock recovery circuit for random data and an SNR of 10^2 are shown in Fig. 10.5. The phase error is shown in Fig. 10.5(a). It shows the same general behavior as in the simulations shown in Figs. 10.2 and 10.3; however, the phase is modulated due to the random data and the additive noise. A histogram of the phase noise, after lock has been achieved, is shown in Fig. 10.5(b). The time waveforms of the random data and the recovered clock area are shown in Fig. 10.5(c), and the phase-plane portrait, which shows the steady-state phase noise clearly as a *blob* near the origin, is given in Fig. 10.5(d).

Simulations of the clock recovery circuit are shown in Figs. 10.6(a) and (b), where the random data and the resulting clock signal are plotted for two different SNRs, and the eye-diagrams for these simulations are shown in Figs. 10.6(c) and (d), respectively. It can be seen from the eye-diagrams that the recovered clock has a nominal transition in the middle of the bit interval, as it should. The random phase-jitter in the clock is seen to be larger for the higher SNR, as expected. A characteristic of bi-phase signaling is that the clock is equally likely to lock to a positive or a negative transition.

10.1.2 Simulations of a Digital Transition Tracking Loop (DTTL)

We presented the digital transition tracking loop (DTTL) in chapter 5, where we stated several of its desirable properties for high-speed clock recovery. In this section we will demonstrate that the idealized circuit provides the desired functionality. Later, we will

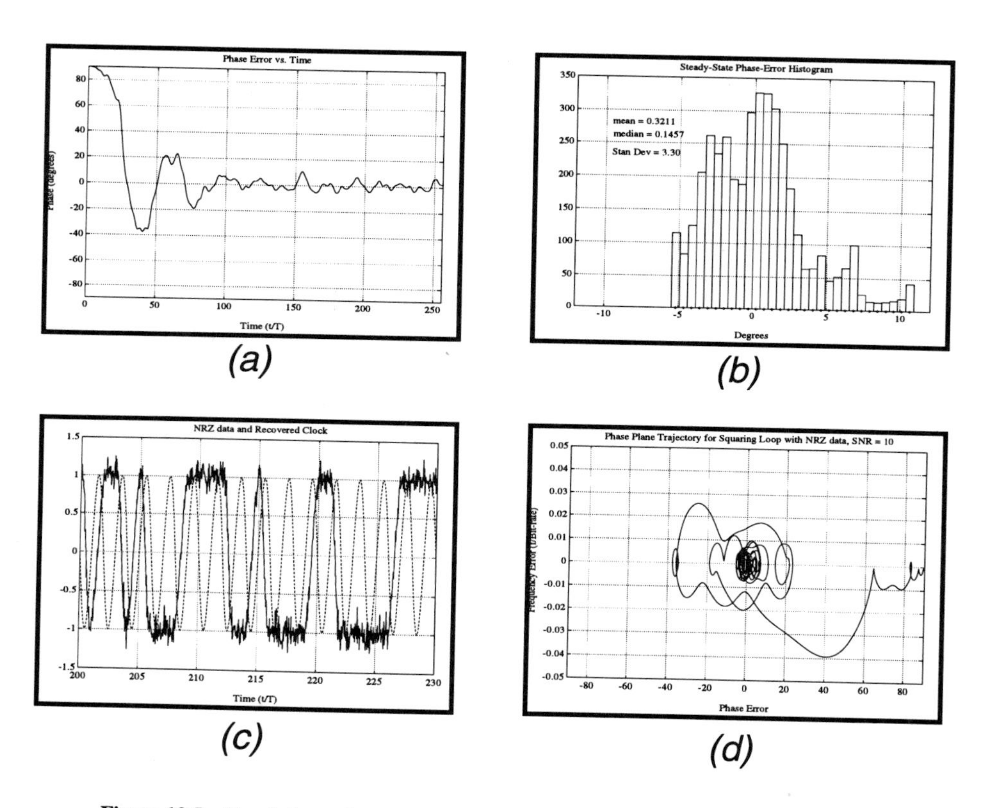

Figure 10.5 Simulations of a squaring clock recovery loop for an SNR of 10^2.

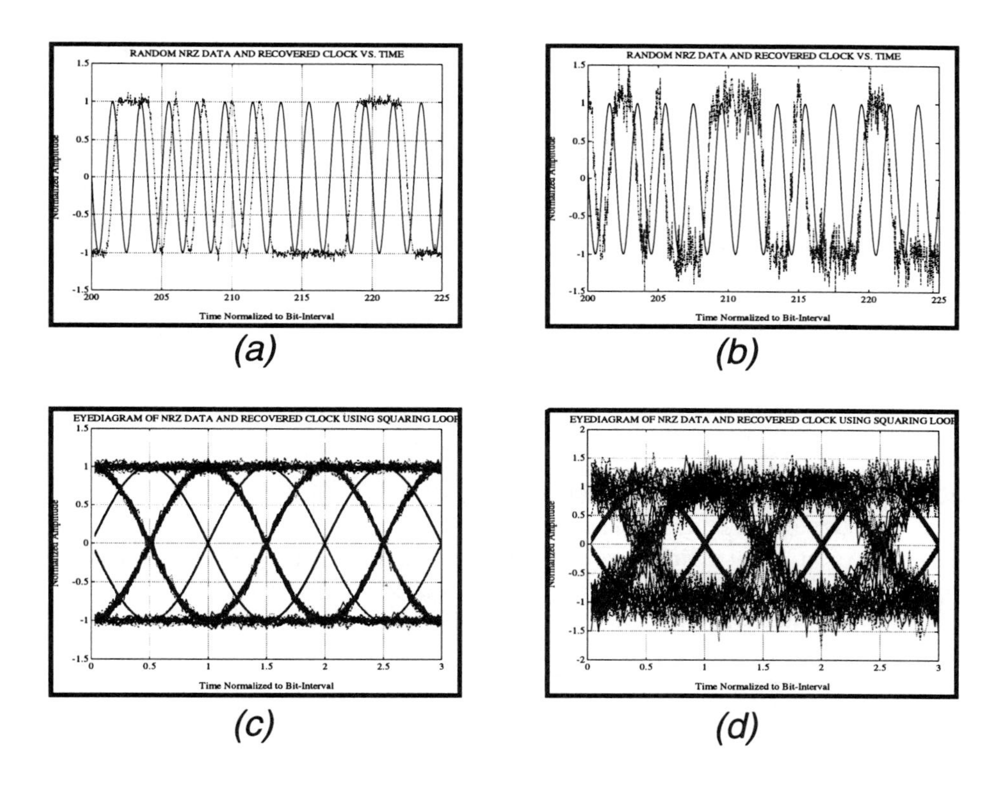

Figure 10.6 Simulations of squaring loop for two different SNRs: (a) and (b) are the resulting time waveforms, (c) and (d) are the eye-diagrams.

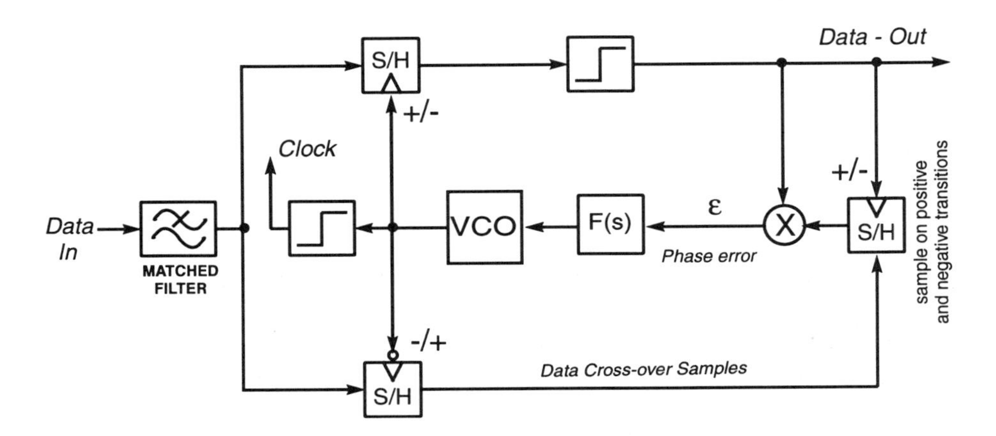

Figure 10.7 Block diagram of a digital transition tracking loop.

simulate the DTTL using actual transistors and compare the results to the idealized model.

Phase Detector Characteristic

A block diagram of a DTTL is shown in Fig. 10.7. As a first order of business we need to determine whether the phase detector characteristic has the sawtooth function that we expect. To obtain the phase error characteristic, the DTTL was simulated open-loop. The VCO frequency was set to be either slightly less or slightly greater than the data-rate. Therefore the phase error will increase, or decrease with time, and we can plot the output of the phase detector ϵ as a function of time to obtain the phase detector characteristic, as shown in Fig. 10.8 for the case of a 5% frequency error. Notice that the characteristic $\epsilon(t)$ is indeed a sawtooth function, as was predicted in chapter 5.

Interleaving and Frequency Detection

In reality, the DTTL would be interleaved as shown in Fig. 10.9. We can add frequency detection to the circuit, as was shown in chapter 5. The frequency detection operates by passing the derivative of the phase-error function through a limiter and a lowpass filter. This operation is illustrated in Fig. 10.10, where the derivative is approximated by a finite difference of ϵ taken at one bit-period intervals. The resulting error signal for a $\pm 5\%$ frequency error is shown in Fig. 10.11, where it can be seen that the frequency error signal is positive for a slow clock and negative for a fast clock.

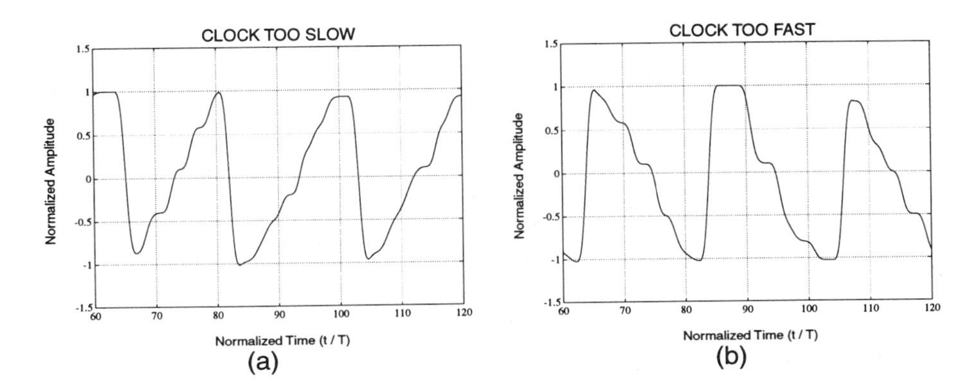

Figure 10.8 Phase-error signal resulting from a MATLAB simulation of a DTTL for frequency errors of (a) -5%, (b)+5%.

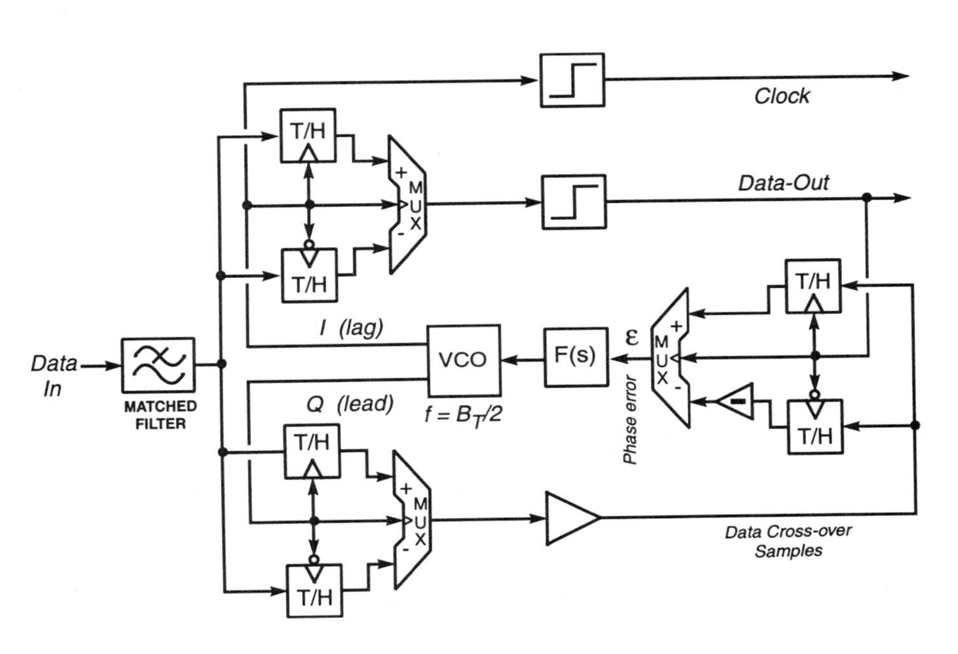

Figure 10.9 Block diagram of a bit-interleaved DTTL.

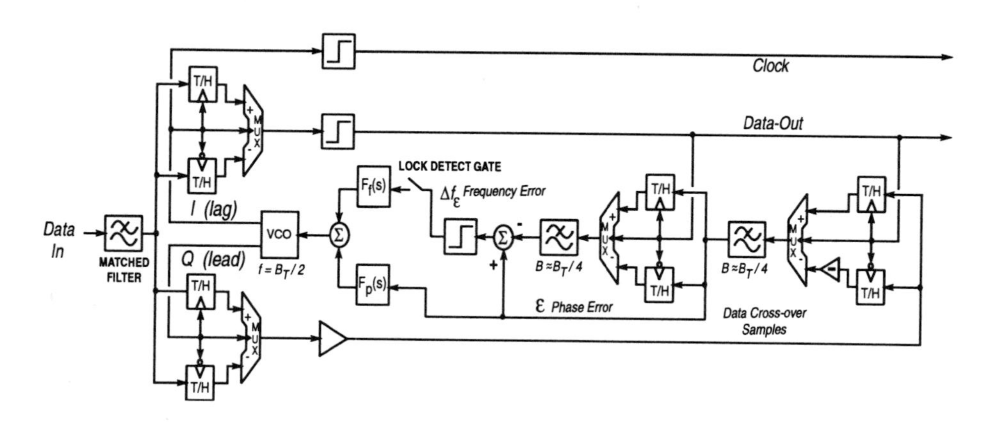

Figure 10.10 Block diagram of a bit-interleaved DTTL with frequency detection.

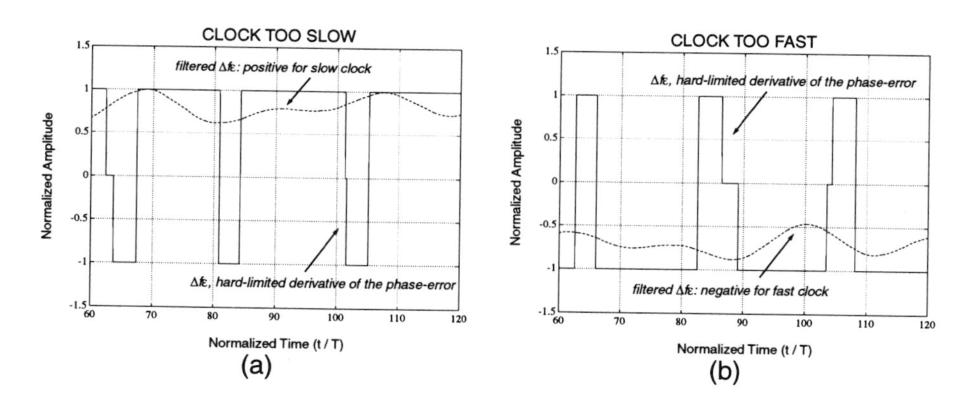

Figure 10.11 Frequency-error signal resulting from a MATLAB simulation of a DTTL for frequency errors of (a) -5%, (b)+5%.

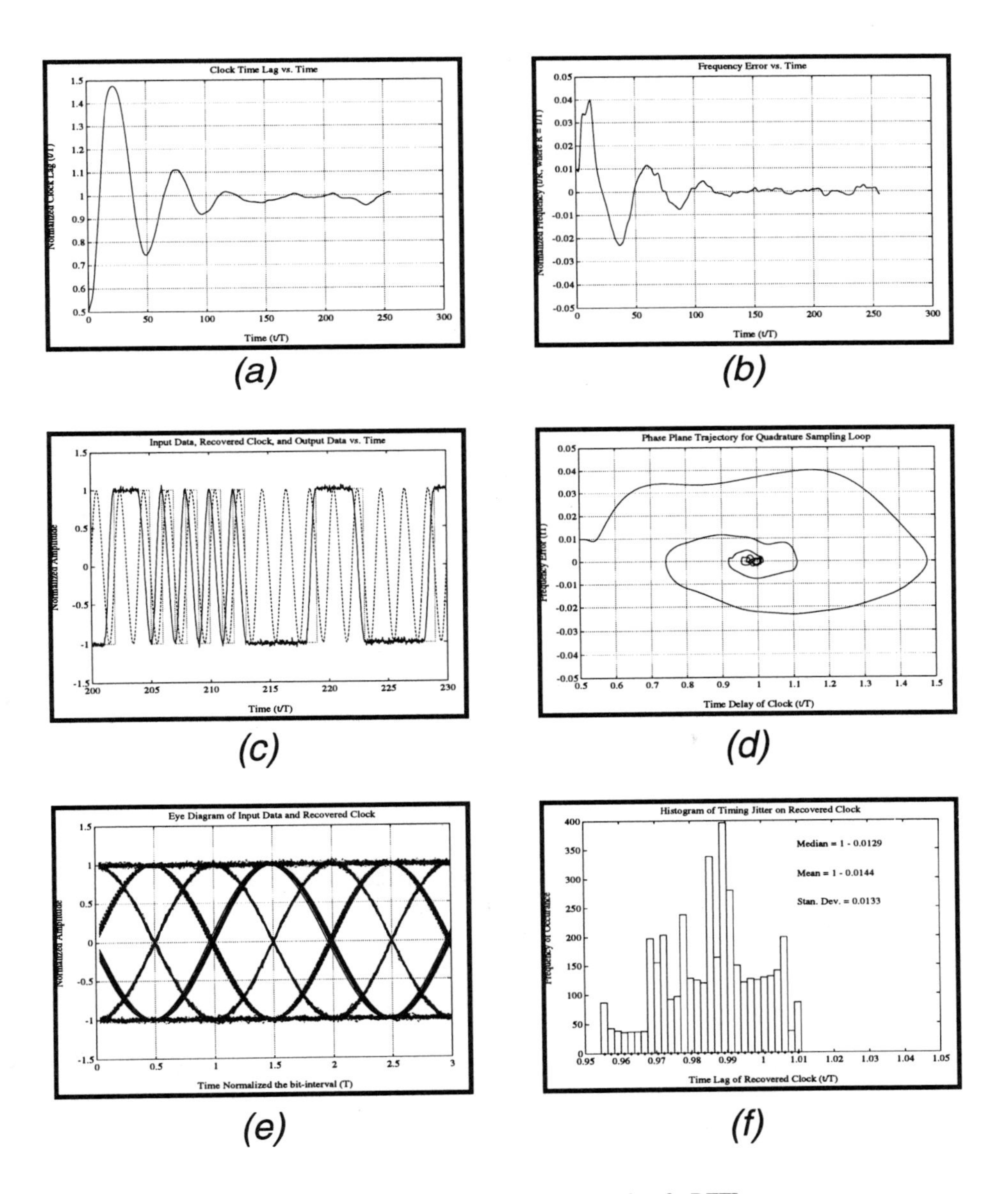

Figure 10.12 Plots of simulations' results of a DTTL.

System level simulation results of the ideal DTTL are shown in Fig. 10.12 for an SNR of 50^2. The fundamental behavior of the DTTL for ideal components is similar to the operation of the squaring loop. The resulting eye diagram of Fig. 10.12(e) illustrates that the recovered clock has zero-crossings at the middle of the bit-interval. The histogram of the phase error on Fig. 10.12(f) shows that the rms phase error is approximately 1% for an SNR of 50^2. The actual phase jitter in the system will depend upon the closed-loop bandwidth of the DTTL.

10.2 CIRCUIT-LEVEL SIMULATIONS

Now that we have verified the functionality of the architecture of the DTTL, we can perform circuit level simulations to determine whether it has the expected low sensitivity to parasitic delays. To obtain preliminary results, we will use standard circuit building blocks. Later we can modify the circuitry to optimize performance. However, these first-order results give a good indication of the maximum operating speed of the circuit. Circuit simulations were performed using the models for an AlGaAs/GaAs HBT process with emitter areas of $(3\mu\text{m} \times 10\mu\text{m})$. At a typical bias current of 2 mA, f_{max} and f_t are approximately 25 GHz. These models were given in chapter 6.

The track-and-hold circuit used in this first-order simulation is shown in Fig. 10.13 [1, 2]. Other diode-bridge sampling circuits could be used and slew-rate enhancement can be added to reduce settling-time [3]. A SPICE simulation of the track-and-hold circuit at a sampling rate of 4 GS/s (8 Gb/s in an interleaved circuit) is given in Fig. 10.14.

To multiplex the interleaved in-phase and quadrature samples back to a serial signal, we used the simple current-mode switch shown in Fig. 10.15. Alternating samples are passed to the output resistors by steering the bias current through the appropriate differential pair, under the control of the clock signal.

A latch is used in the final stage of the decision circuit to boost the gain and provide regeneration. This allows us to improve the speed by using a smaller hold capacitor and a lower gain in the sampling circuit. The design of a current-mode latch is well known, and we used the straightforward approach in this simulation. A schematic of this latch is shown in Fig. 10.16.

Results of SPICE simulations of the DTTL designed using the circuits just described are shown in Figs. 10.17 and 10.18 for data rates of 2 Gb/s and 5 Gb/s, respectively.

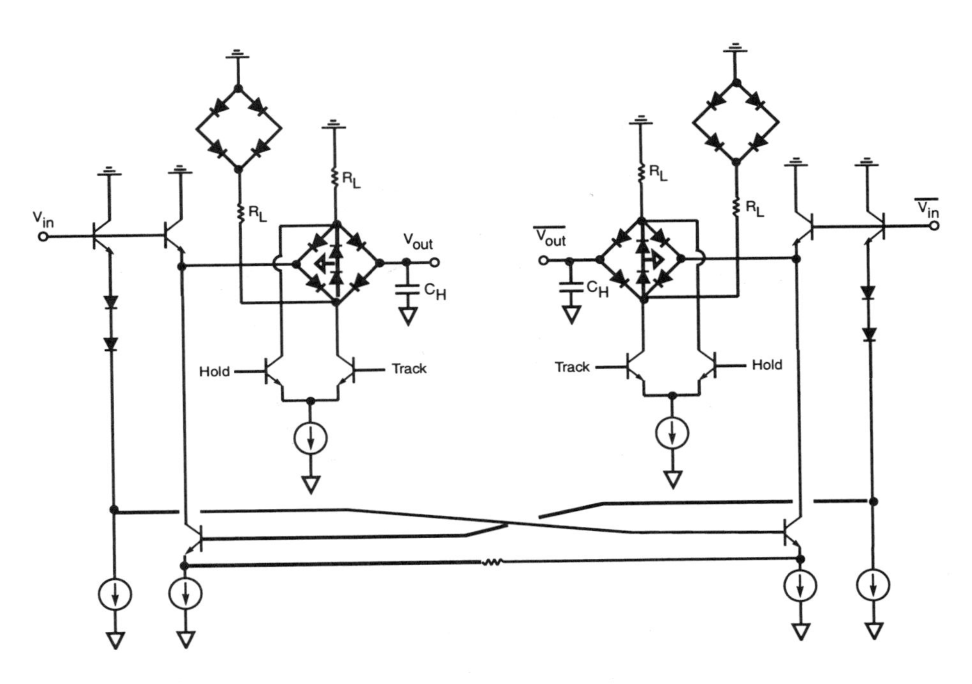

Figure 10.13 Diode-bridge sample and hold circuit.

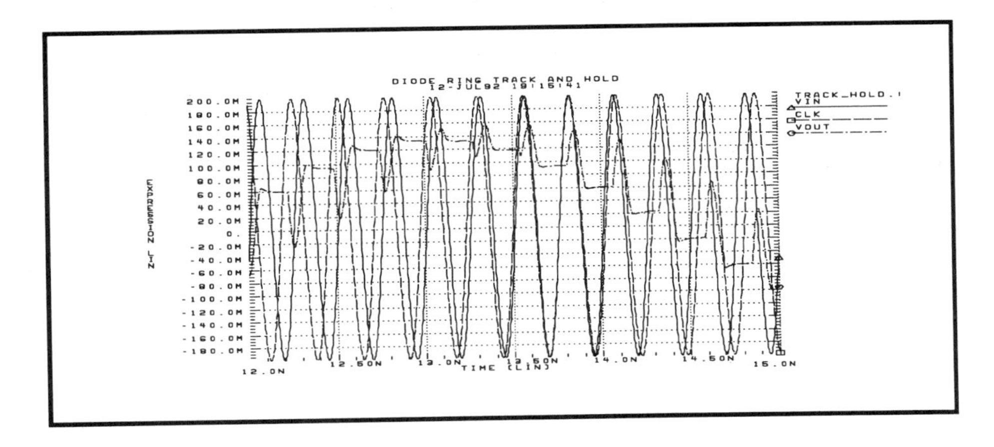

Figure 10.14 SPICE simulation results of a diode-bridge track-and-hold circuit.

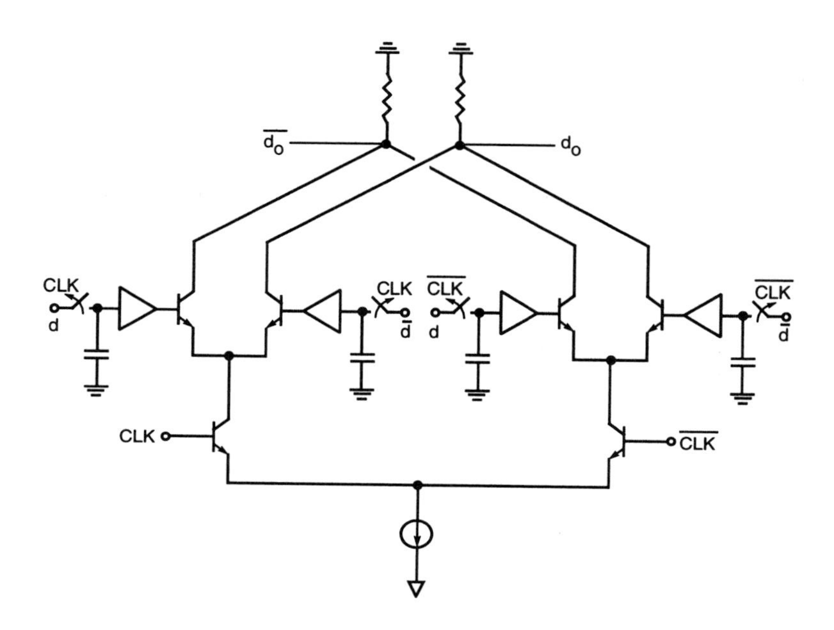

Figure 10.15 Schematic diagram of a 2:1 multiplexor.

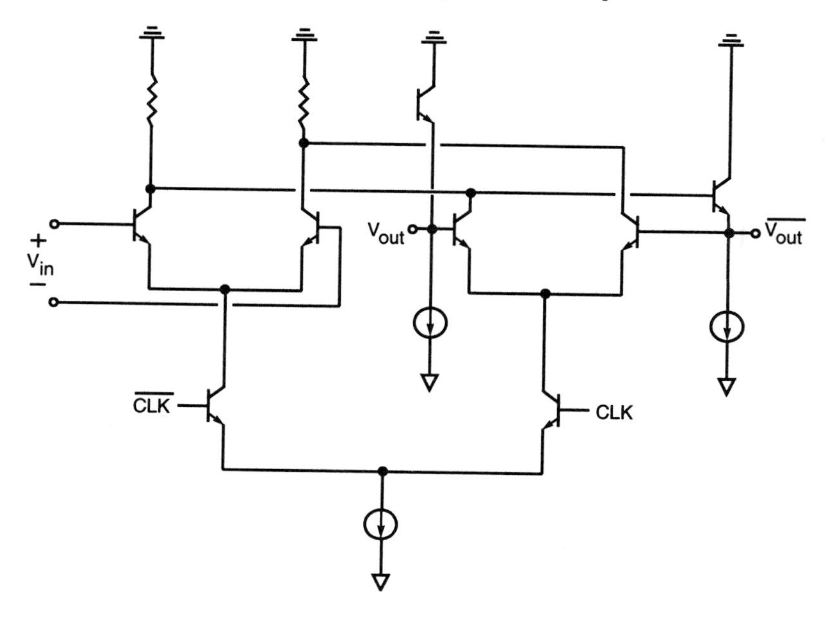

Figure 10.16 Latch for increased gain and data regeneration.

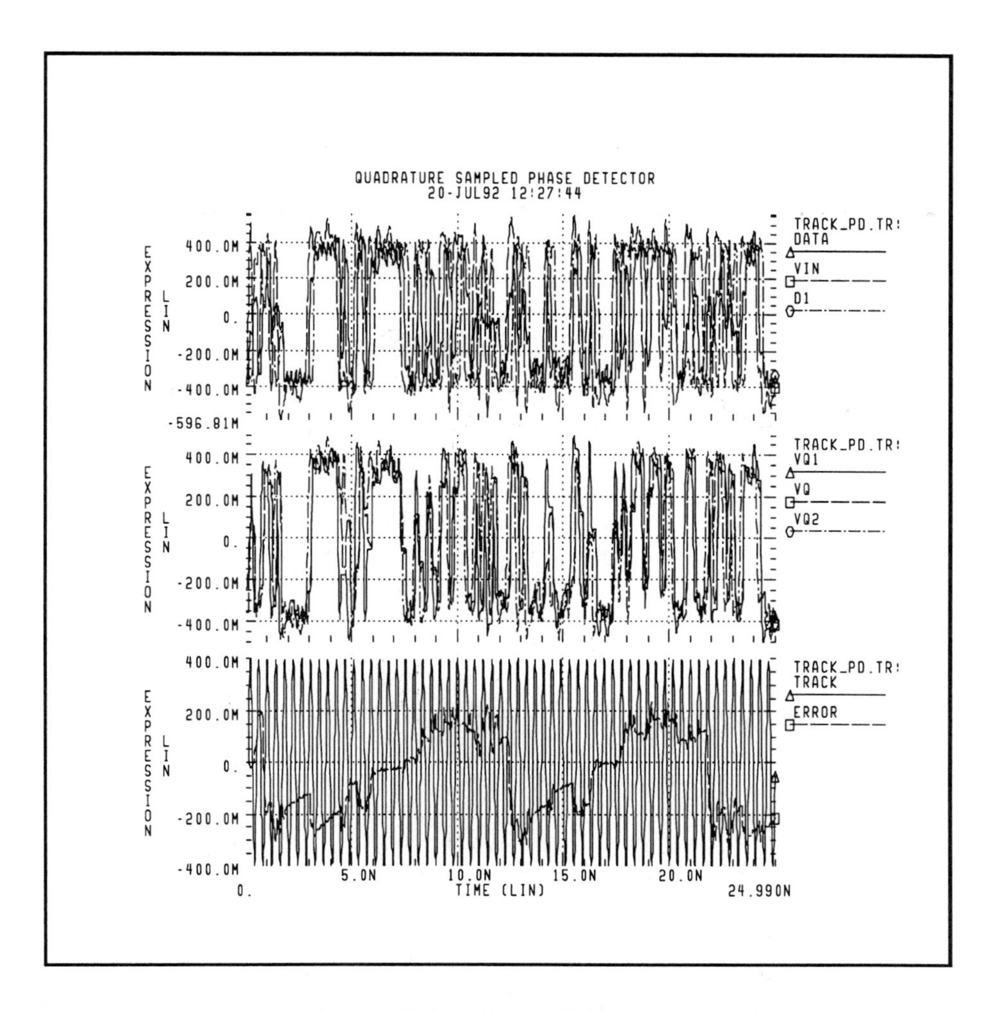

Figure 10.17 SPICE Simulation of a DTTL operating at 2 Gb/s.

It is difficult to separate the waveforms in these plots. The top plot in each of these figures is the input and output data. The middle plot shows the in-phase and quadrature samples of the DTTL, and the bottom plot shows the phase error between the input data and the clock signal. For these simulations, the loop was broken to determine whether the proper phase-error function can be obtained. We can distinguish a somewhat noisy sawtooth phase-error function in the bottom plots of Figs. 10.17 and 10.18. For these simulations, conservative models were used ($f_{\max} \sim 25$GHz). The simulations show functionality at a data rate of 5 Gb/s or $(f_{\max}/5)$. Therefore, if an advanced InP-based

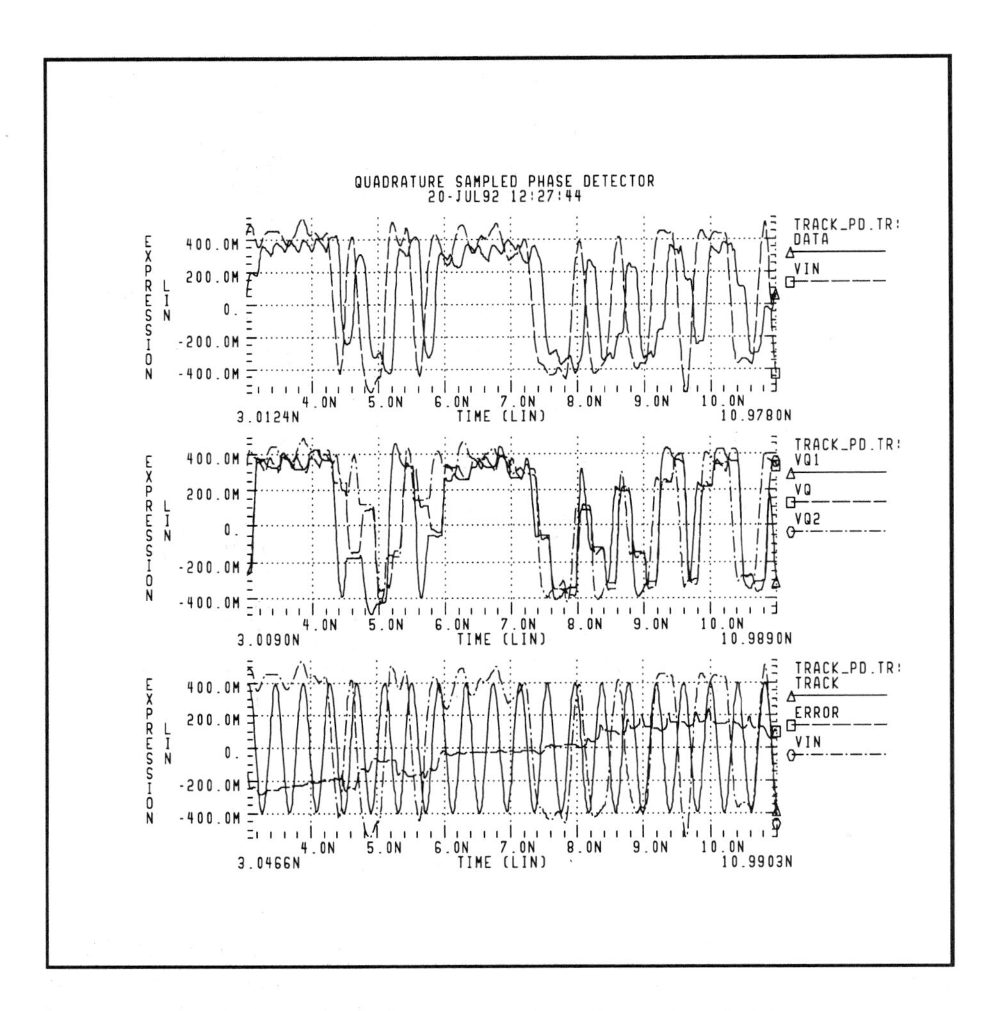

Figure 10.18 SPICE Simulation of a DTTL operating at 5 Gb/s.

HBT process were used with $f_{\max} = 100\text{GHz}$, data-rates up to 20 Gb/s could be accommodated by this circuit.

10.3 FURTHER RESEARCH

To conclude, we will briefly outline some open problems related to the realization of an integrated 10 Gb/s fiber-optic receiver.

Photodetector

In this book we have not discussed the photodetector thoroughly. It has been assumed that either a PIN photodiode or an avalanche photodiode (APD) will be used. Work needs to be done to optimize the quantum efficiency η of the photodetector and to optimize the dimensions in order to obtain the highest sensitivity receiver. Aside from η, it was shown in chapter 7 that one of the key parameters in the overall receiver sensitivity was the parasitic capacitance of the photodetector. Improving lens systems, so that a more optical energy can be focused onto a smaller area, can have a significant impact on system performance.

Preamplifier

A detailed noise analysis for the preamplifier was presented. However, the SNR of the test statistic was not optimized. This involves taking into account both the noise performance, and the effect of the amplifier's pulse response on the data eye. For example, an amplifier can exhibit peaking in its pulse response. This will typically increase the noise bandwidth, however it can also increase the signal magnitude at the sampling instant, such that the SNR of the sample is increased. Further work is needed to determine guidelines concerning the optimization of noise performance within the context of a receiver for random NRZ data. This works also needs to take into account the postamplifier, as discussed below.

Postamplifier

The Postamplifier performs noise filtering and signal conditioning. Optimal performance of the receiver must take into account the noise filtering of this stage. Aside from noise filtering, the postamplifier must have an automatic gain control feedback loop so as to always output a constant signal level to the clock recovery circuit. This stage will determine the dynamic range of the overall receiver. Other important functions of the postamp are to provide dc restoration of the signal and to convert the single-ended signal to a balanced differential signal for subsequent processing.

Optimization of Building Blocks

The primary focus of this work was to design first-generation prototype circuits. We have not focused much attention on optimizing the speed once the architecture is chosen. After a functional receiver is demonstrated, further work will be needed to add embellishments to the circuits to improve speed and to reduce the sensitivity to temperature and power supply variations.

Effective SNR Improvement Using a Sample-and-Hold

It was stated in chapter 5 that using a sample-and-hold circuit before the decision circuit would improve the effective SNR as compared to using a decision circuit, which consists of only a regenerative latch with no holding function. However, no quantitative results were given. To determine the SNR improvement, test circuits will have to be built to directly compare the two schemes.

Comparison of Competing Clock Recovery Schemes

Several methods for clock recovery in broadband systems were presented in chapter 5. It would be interesting to compare the performance of various approaches in a real system. This would require the fabrication and evaluation of several different clock recovery circuits. It is possible that some parasitics that were overlooked by these authors may make one circuit better than the DTTL. However, at this point the DTTL with frequency detection appears to be the best approach for recovering a clock from high-speed random NRZ data.

Evaluation of Actual Circuits

At the time of this writing, several communication links, operating at gigabit-per-second rates are being realized as integrated circuits, with many more expected in the near future. Several questions as to the preferred IC technology and the preferred architectures for such circuits have yet to be resolved. Eventually a large volume of these ICs will be designed and deployed for such applications as,

- fiber-optics, Optical disks, ATM switches,

- magnetic disk-drive electronics,
- wireless communication and personal communication,
- high-speed data communication over metallic media such as coaxial cables and twisted pairs,
- others.

Various architectures and circuit building blocks will certainly emerge and become widespread, whereas others will disappear. However, with such a wide variety of applications, no *one* approach will be used in all cases. Designers will have to understand trade-offs in cost, speed, performance, power dissipation, etc. to best utilize available resources for a specific application. It is our hope that designers of future high-speed communication circuits, when faced with these trade-offs, will find the information in this book useful.

REFERENCES

[1] William T. Colleran, H. T. Phan, and Asad A. Abidi. A 10-bit, 100 MS/s pipelined A/D converter. In *IEEE ISSCC Dig. Tech. Papers*, San Francisco, California, February 1993.

[2] William T. Colleran and Asad A. Abidi. A 10-b, 75-MHz two-stage pipelined bipolar A/D converter. *IEEE J. Solid-State Circuits*, 28(12):1187–1199, December 1993.

[3] M. H. Wakayama, H. Tanimoto, T. Tasai, and Y. Yoshida. A 1.2-μm BiCMOS sample-and-hold circuit with a constant-impedance, slew-enhanced sampling gate. *IEEE J. Solid-State Circuits*, 27(12):1697–1708, December 1992.

INDEX

Please put me back in water.
I am Paddle-to-the-Sea.
— HOLLING CLANCY HOLLING

Please put me back in water.
I am Paddle-to-the-Sea.
— Holling Clancy Holling